Final LIMTECH Colloquium and International Symposium on Liquid Metal Technologies 2017

IOP Conference Series: Materials Science and Engineering
Volume 228

Dresden, Germany
19 - 20 September 2017

Editors:

Gunter Gerbeth
Robeert Stieglitz

ISBN: 978-1-5108-4648-7
ISSN: 1757-8981

Printed from e-media with permission by:

Curran Associates, Inc.
57 Morehouse Lane
Red Hook, NY 12571

Some format issues inherent in the e-media version may also appear in this print version.

Printed by Curran Associates, Inc. (2017)

For permission requests, please contact the Institute of Physics
at the address below.

Institute of Physics
Dirac House, Temple Back
Bristol BS1 6BE UK

Phone: 44 1 17 929 7481
Fax: 44 1 17 920 0979

techtracking@iop.org

Additional copies of this publication are available from:

Curran Associates, Inc.
57 Morehouse Lane
Red Hook, NY 12571 USA
Phone: 845-758-0400
Fax: 845-758-2633
Email: curran@proceedings.com
Web: www.proceedings.com

Table of contents

Volume 228

Final LIMTECH Colloquium and International Symposium on Liquid Metal Technologies

19–20 September 2017, Dresden, Germany

Published online: 3 August 2017

Preface

Generic Problems and Phenomena

Energy Conversion and Storage

Metals and Semiconductor Processing

Measurement Techniques

IOP Conf. Series: Materials Science and Engineering **228** (2017) 011001 doi:10.1088/1757-899X/228/1/011001

HELMHOLTZ
ASSOCIATION

LIMTECH Alliance

Preface: Final LIMTECH Colloquium and International Symposium on Liquid Metal Technologies (LIMTECH) Sept. 19-20, Dresden Germany

This volume of IOP Conference Series – Materials Science and Engineering represents a selection of contributions presented at the Final LIMTECH Colloquium and International Symposium on Liquid Metal Technologies (LIMTECH) held at Helmholtz-Zentrum Dresden – Rossendorf (Dresden, Germany) on Sept. 19-20, 2017.

Inspired by both the growing societal interest and awareness in energy and resource efficiency and the enhanced measurement and computational capabilities being available nowadays, liquid metal research has recently regained increased attendance. Helmholtz-Zentrum Dresden - Rossendorf (HZDR) and the Karlsruhe Institute of Technology (KIT), both being members of the German Helmholtz Association of Research Centres, therefore took the initiative to synthesize a multi-directional program in order to

- bridge the Helmholtz activities within the Helmholtz energy R&D programs,
- foster collaboration with the knowledge driven university research,
- utilize jointly the infrastructures of Helmholtz centres and universities,
- stimulate collaboration with industrial partners and international research groups,
- enhance talent development in a competitive ambience.

To match the individual interest of the contributing partners and to provide a sustainable perspective for the contributing parties and associations the Helmholtz Alliance LIMTECH bundled the R&D activities on ***Li***quid ***M***etal ***Tech***nologies (LIMTECH) in a three pillars concept addressing fundamental research aspects, energy and storage applications as well as materials processing activities. These pillars were complemented by a ***Y***oung ***I***nvestigators ***G***roup (YIG) focusing on measurement techniques for liquid metal flows, a central element of all experiments in the field.

LIMTECH was funded during October 2012 – September 2017 by the German Helmholtz Association. The funding comprised the Helmholtz centres HZDR, KIT, German Aerospace Center (DLR) and Forschungszentrum Jülich (FZJ) and the following German university partners: Technical University Dresden, Technical University Ilmenau, Leibniz University Hannover, TU Bergakademie Freiberg, University of Potsdam – Leibniz Institute for Astrophysics Potsdam, Georg-August University Göttingen, and RWTH Aachen – Access Institute. International partners directly involved were the Institute of Physics Riga (Latvia), the Coventry University (UK) and the company RGS Development B.V. Broek op Langedijk (The Netherlands).

Liquid metal technologies provide a surprising bandwidth in fundamental and applied research. Regarding energy conversion and storage they allow for high-temperature energy conversion, efficient and compact solar power receivers, advanced thermal storage units, new high current density batteries or carbon dioxide free production of hydrogen as chemical energy carrier. In the sense of sustainability and resource efficiency these technologies are of vital importance for the production and casting of steel and light metals, the separation of inclusions and the controlled dispersion of particles in melts for, e.g., new metal matrix nano-composites or solar-grade silicon. Further on liquid metals are used as targets in modern neutron sources, coolants of fusion and fission reactors but also in advanced laboratory experiments to investigate the origin and behavior of stellar and planetary phenomena in geo- and astrophysics. Characteristic for LIMTECH was the interlock of experimental and numerical activities in and among the 14 sub-projects additionally supported by the YIG. In addition, the YIG formed the core of a dedicated PhD program, which has been complemented by all research partners. A special focus of the Alliance was on cooperation with industry and the related transfer activities. Worth mentioning as well were the seed grants for which a certain budget was flexibly focused on new ideas and developments. A few of them developed very well up to the level of new projects with new partners.

LIMTECH provided a strong basis for further research projects, e.g. on European level. The international activities were steadily growing. For example, LIMTECH was co-organizer of the 10th PAMIR International Conference on Fundamental and Applied MHD held in Cagliari (Italy) on June 20-24, 2016, a leading conference in the field being organized every three years. Of particular interest became the International LIMTECH Young Scientist Award, which was for the first time given in 2016 to Dr. Jiang Wang (Shanghai University, China). In 2017 the winner was Dr. Kaspars Dadzis (Institute of Crystal Growth Berlin, Germany) who presented his award lecture at the present final LIMTECH colloquium.

In summary, the alliance LIMTECH was a very dynamic and successful type of consortium stimulating substantially the national collaboration but to an even larger extent significantly enhanced the visibility and recognition of liquid metal R&D activities in the worldwide community.

The editors would like to thank the authors for their contributions and for their cooperation during the editorial process. We are as well grateful to the reviewers for their constructive criticism and support. On behalf of all LIMTECH members and involved institutions we thank the Helmholtz Association for the substantial funding of LIMTECH during the past 5 years. We sincerely acknowledge the panel members of the initial and mid-term evaluation of our LIMTECH alliance for their constructive comments and hints, as well as the Scientific Advisory Board of the alliance (Prof. J. Etay (Grenoble), Prof. K. Pericleous (London), Dr. W. von Ammon (Ach, Austria)) for their continuous support in the scientific and structural development of LIMTECH during the funding period. Last but not least we are grateful to Dr. Gerd Mutschke who did most of the work for editing the present IOP issue and organizing the final LIMTECH colloquium.

Gunter Gerbeth and Robert Stieglitz
Colloquium Co-chairs and Co-editors

IOP Conf. Series: Materials Science and Engineering **228** (2017) 011002 doi:10.1088/1757-899X/228/1/011002

Peer review statement

All papers published in this volume of *IOP Conference Series: Materials Science and Engineering* have been peer reviewed through processes administered by the proceedings Editors. Reviews were conducted by expert referees to the professional and scientific standards expected of a proceedings journal published by IOP Publishing.

IOP Conf. Series: Materials Science and Engineering **228** (2017) 012001 doi:10.1088/1757-899X/228/1/012001

Thermo-hydraulic flow in a sudden expansion

W Jaeger[1,4], T Schumm[2], M Niemann[3], W Hering[1], R Stieglitz[1], F Magagnato[2], B Frohnapfel[2], J Fröhlich[3]

[1]Institute of Fusion and Reactor Technology, Karlsruhe Institute of Technology (KIT), Kaiserstraße 12, 76131 Karlsruhe, Germany
[2]Institute of Fluid Mechanics, Karlsruhe Institute of Technology (KIT), Kaiserstraße 10, 76131 Karlsruhe, Germany
[3]Institute of Fluid Mechanics, Technische Universität Dresden, 01062 Dresden, Germany
[4]wadim.jaeger@kit.edu

Abstract. The paper deals with the turbulent flow of liquid metal directed upwards in a vertical channel featuring a backward-facing step. The vertical wall behind the step is heated at various rates thereby inducing forced and mixed convection. Due to the low Prandtl number of liquid metal flow a data basis for this technically relevant flow type did not exist so far. Here, DNS and LES results are presented to provide detailed information about the statistics of the turbulent motion, budgets of turbulent kinetic energy and other quantities. This information is then further used to develop suitable statistical turbulence models capable of properly covering this flow and similar ones, i.e. forced, mixed and free convection of liquid metals. Finally, the paper reports on the construction of an experiment conceived for exactly the same configuration as simulated, with the purpose of close cross validation between the different approaches.

1. Introduction

The thermo-hydraulic behaviour of liquid metal flowing through a vertical backward facing step (BFS) is the subject of this study. The flow separation due to sudden changes of the geometry like the BFS is of crucial importance for the thermo-hydraulic characterisation of various components like heat exchanger, thermal storage containers, manifolds and flow collectors. Such a sudden geometry change paired with an anisotropic heating profile causes flow stratification, stagnation points, re-circulation zones, etc. The resulting unsteady, heterogeneous temperature profile can have a negative impact on the life-time and performance of the aforementioned components, so that reliable prediction of such flows is pressingly needed.

The flow over a BFS has been investigated frequently in the past for fluids with a Prandtl number around unity, like water or air. For low Prandtl number fluids like liquid metals (LM), the behaviour differs considerably. The high thermal conductivity of liquid metals induces a separation between the thermal and the viscous boundary layer scale. As a result, the thermal boundary layer is much thicker than the viscous one, making heat conduction the dominant form of heat transfer. Furthermore, the application of LM in engineering is mostly related to turbulent forced convection introducing additional multiscale flow structures for the velocity and temperature field. In addition, buoyancy influences the structure of the flow. Taking all these boundary conditions into account, the turbulent transport of momentum and energy is challenging, because it necessitates an anisotropic approach for the prediction of the velocity and temperature field. Due to the differences between fluids with low Prandtl number and those around unity a direct transfer of knowledge from one to the other exhibits several deficits. Hence, detailed simulations and experiments with BFS and liquid metal flows are conducted and planed in the near future to address the above mentioned challenges.

A generic BFS is sketched in Figure 1. Iso-thermal investigations are available [1][2], while studies with imposed heat flux [3] are less frequent and studies with mixed convection are very rare [4]. From the technical perspective the following particular problems are to be addressed, among others: identification of the different flow regimes, the transition from forced to mixed to free convection, flow patterns like thermal striping and stratification. From a technical point of view, the prediction of results as a function of dimensionless quantities, like the Péclet number ($Pe = Re \cdot Pr$), the Grashof number ($Gr = g \cdot \beta \cdot (T_s - T_\infty) \cdot D^3/\nu^2$), the Richardson number ($Ri = Gr/Re^2$) and the Stanton number ($St = Nu/Re$) are desirable.

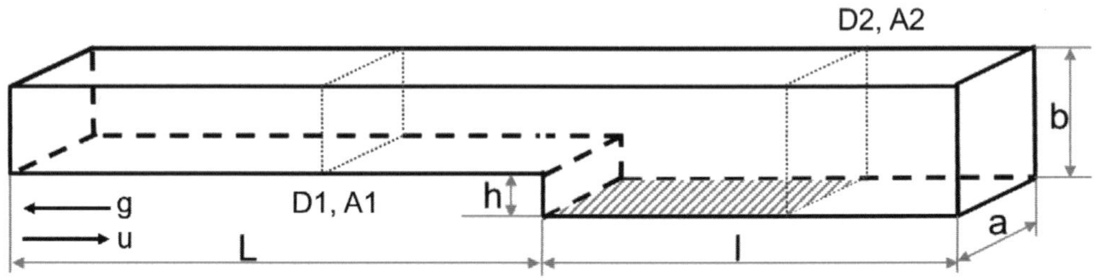

Figure 1. Generic sketch of the backward facing step (BFS) set-up (drawing not to scale). Note that gravity points to the left in this sketch, not downward, i.e. the mean flow is directed upwards.

2. Turbulence-resolving simulations

2.1. Simulation method

This section describes results obtained from simulations representing turbulent flow and buoyancy forces by the unsteady Navier-Stokes equations using the Boussinesq approximation. These equations are solved by a staggered Finite-Volume Method of second order accuracy in space and time [5]. For the lower Reynolds numbers Direct Numerical Simulations (DNS) are conducted resolving all turbulent scales. For the larger Reynolds numbers Large-Eddy Simulations (LES) are performed by means of the σ-model of Nicoud *et al.* [6] to represent the subgrid scales (sgs). Due to the low Prandtl number the smooth temperature field is always fully resolved by the grid employed, and modelling is only required for the dissipation of momentum. A more detailed account on the numerical methods involved can be found in [7] and [8].

2.2. Configuration and grids

For all simulations the configuration is as depicted in Figure 1 and with the inflow condition imposed at section A1. The inflow signal at each grid point of this plane is generated by a separate simulation of the fully developed turbulent flow through a channel of the same cross section and at the same Reynolds number as the main simulation. Most of the simulations are conducted with spanwise periodicity of the solution, the usual approach for DNS and LES in the case of statistically homogeneous conditions in the respective direction. This allows averaging in spanwise direction in addition to time averaging and reduces computational requirements substantially. Two further simulations are conducted assuming sidewalls to assess the impact of secondary flows on the obtained results. Their impact is larger the shorter the distance is between them, so that a width of a = h has been chosen to provide information about what could be termed a case of large comparatively alteration with respect to the spanwise homogeneous case. This distance is smaller than planned for the experiments discussed below to create a kind of upper bound for the impact of this feature. A no-slip condition is imposed on all solid walls and a convective outflow condition in the outlet plane at a distance of l = 30 h behind the step is considered, with a heated section of 20h in streamwise extension. The temperature is set to T_{in} = 150°C at the inlet and a transport equation is solved for the non-dimensional temperature rise $\theta = (T-T_{in})/\Delta T$. All solid walls are assumed adiabatic, except the wall behind the step, where constant heat flux $\dot{q}$ is imposed. The characteristic temperature difference ΔT is obtained from the imposed heat flux density $|\dot{q}| = \lambda \cdot \Delta T/h$. For all cases a Prandtl number of Pr = 0.0088 is assumed corresponding to that of liquid sodium at the chosen inlet temperature. Several cases for the Reynolds and Richardson numbers based on the step height h and the bulk velocity at the inlet U_b have been computed. The flow domain is discretized with Cartesian grids refined towards the walls and stepwise adapted for each case to resolve all boundary layers. Such an approach results in 40 up to 650 million cells, depending on the physical parameter magnitude.

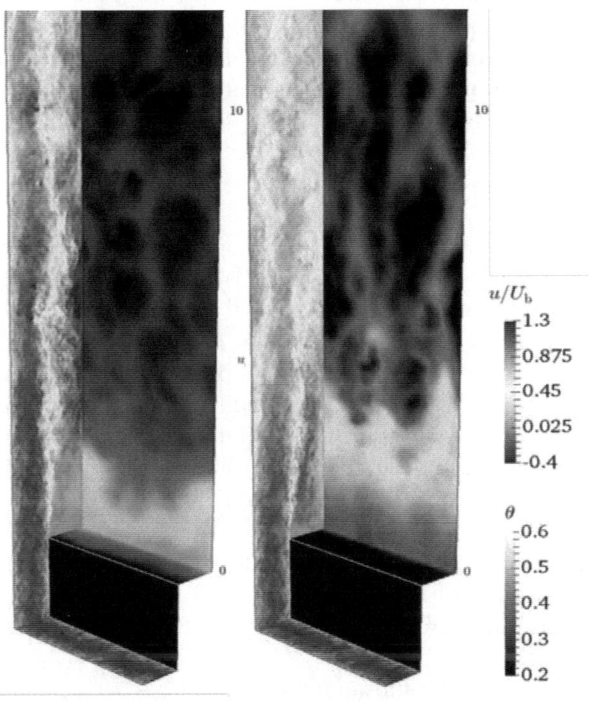

Figure 2. Overview of instantaneous LES dimensionless velocities and temperatures for Re_h = 20000 with Ri = 0.0 (left) and Ri = 0.2 (right). The streamwise velocity is shown on the inlet plane and at the periodic boundary. The step and the heated wall are colored by instantaneous temperature. Note the different scales.

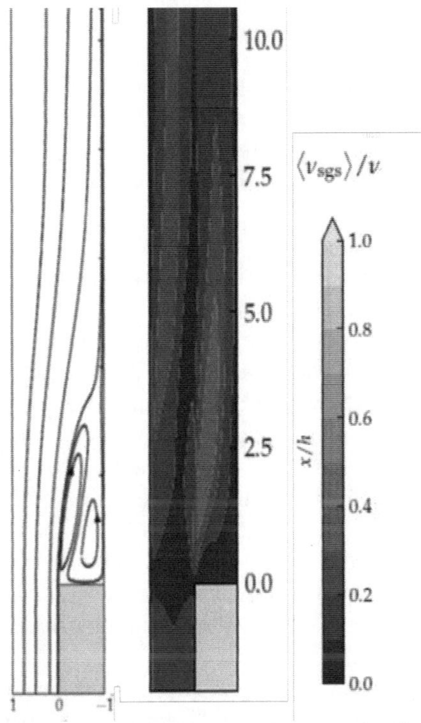

Figure 3. Streamlines of the main flow (left) and average subgrid-scale viscosity compared to the molecular viscosity (right) for the flow with Re_h = 20000 and Ri = 0.2.

2.3. Results for spanwise homogeneous flow

The numerical method has been validated with a DNS of a non-isothermal air flow over a BFS for Re_h = 4805, ER = b/(b-h) = 1.5 from [3], and the impact of temperature-dependent fluid properties of liquid sodium have been found to be marginal [7]. In [8], the flow of LM has been simulated with the same expansion rate for Ri = 0.338, comparing it to the non-buoyant flow with Ri = 0.0. Detailed one-point statistics, Nusselt number and wall friction profiles have been provided. These data are then used for model assessment in [9]. Selected results appear in Figure 6 - Figure 8 below.

The experiment described in Section 4. is supposed to be run with larger expansion ratio, ER = 2, so that DNS with this value and $Re_h = 10000$ have been performed varying the Richardson number from Ri = 0.0 to Ri = 0.4. On one hand, these data are used to assist in the construction of the experiment, but most of all they are aiming at model validation [9] as reported below in Section 3.3. and 3.4. For these cases, not only one-point statistics of velocity and temperature are determined, but also two-point correlations and all budget terms for the fluctuating quantities.

For even larger Reynolds numbers, a DNS demands excessive computational requirements, so that LES have been executed for $Re_h = 20000$ and $Re_h = 40000$ on grids of 420 to 549 million cells. Again, all one-point statistics and budget terms have been determined. An impression of these simulations is provided by Figure 2 and Figure 3, where instantaneous flow data for $Re_h = 20000$ are shown together with the streamlines of the mean flow for Ri = 0.2. It is visible that the upward buoyancy force accelerates the flow along the wall behind the step entraining the fluid towards this side resulting in a deceleration close to the straight wall. The ratio of the mean sgs viscosity to the molecular viscosity is reported as well. It is below 0.5 in most of the domain, indicating that the sgs model has an impact on the flow but does not dominate the dissipation. For the higher Reynolds number, the values are about twice as large (not shown here), so that in this case the model contributes sizably, as to be expected. Furthermore, the ratio is slightly larger for Ri = 0.0 than for Ri = 0.2 since the gradients of the mean flow are stronger without buoyancy which deviates the mean flow towards the heated wall.

2.4. Results for the step with sidewalls

An overview of the flow with sidewall at $Re_h = 10000$ without buoyancy is illustrated in the left picture of Figure 4 showing streamlines of the mean flow starting close to the wall upstream of the step and points between the channel wall (dark green) and points in the center (dark brown). Secondary flow of the first kind results from curvature of the mean flow behind the step. This secondary flow entrains streamlines close to the sidewalls towards the channel center within the recirculation zone and outwards in the reattachment region. On top of this effect, secondary flow of the second kind also affects the velocity field, as visible, for instance, in the inflow section where one of the depicted streamlines near the side wall is advected away from the heated wall. In the mixed convection case, addressed in the right picture of Figure 4, secondary flow of the first kind is also present in the recirculation zone but has no visible effect on the streamlines near reattachment. Here, the selected streamlines concentrate in the channel center at the heated wall, as a result of both the development of the wall jet and the effect of the secondary flow affecting the mean flow in this part of the domain.

The mean wall shear stress at the heated wall illustrates the substantial difference between the two regimes. A clear dependence on the distance from the side wall is visible for Ri = 0.0 and thick boundary layers at the side walls are present downstream of reattachment. For Ri = 0.2 the wall shear stress on the heated wall varies less in the channel center in spanwise direction and thinner boundary layers at the sidewalls are found.

Figure 5 shows profiles of the mean streamwise velocity component in the channel center for the two Richardson numbers addressed in Figure 4. Furthermore, these data are compared to the corresponding values for the cases with the same Reynolds number and spanwise periodicity. For both Richardson numbers, the difference between the cases with and without sidewalls are moderate. The main difference is a slightly higher velocity in the center for the case with sidewalls, as expected, since the bulk flow rate is assumed the same in both cases. This somewhat shifts the center of gravity of the velocity profile but does not introduce qualitative changes. The second graph in Figure 5 shows the turbulent kinetic energy. Here, the same observations can be made. Far downstream of the step, around x/h = 12 and 15, however, the results with and without sidewalls differ by a factor of 2 and more, which is substantial, and results from the differences in the secondary flow and the attenuating effect of the additional walls. All three normal stresses (not shown here) contribute to this behavior.

For the mixed convection cases, the decrease downstream of reattachment is smaller than for Ri = 0. In the recirculation zone substantial deviations between periodic case and side wall case are observed near the heated wall resulting from the substantial difference in the secondary flow.

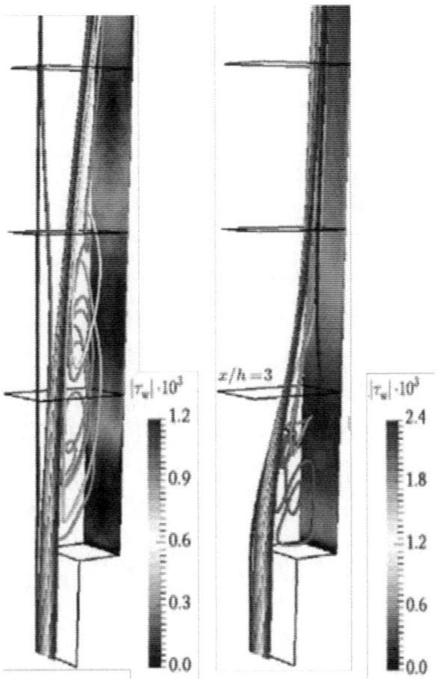

Figure 4. Selected streamlines and absolute value of the wall shear stress at the heated wall for the flow over the step with side walls with Ri = 0.0 (left) and Ri = 0.2 (right).

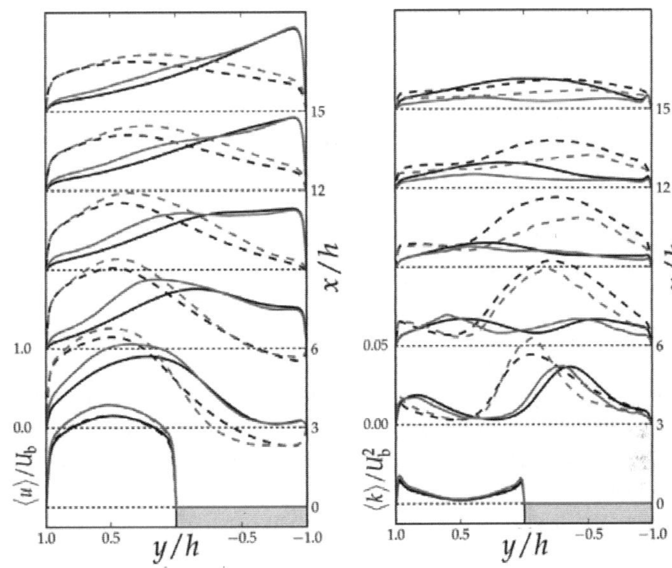

Figure 5. Mean streamwise velocity and turbulent kinetic energy at the channel center for the cases with side walls (light blue) in comparison with results of the periodic cases (black). Dashed lines: Ri = 0.0. Solid lines: Ri = 0.2.

3. Turbulence and heat flux modelling

3.1. Validation of turbulence and heat flux models for the BFS

The turbulent flow over a BFS is a well-known generic test case for turbulence models, as it features an adverse pressure gradient, flow separation and the development of a boundary layer. Additionally, due to the complex flow structure, the Reynolds analogy, describing the similarity between the viscous and the thermal field, is not valid [10]. Also for liquid metals, the Reynolds analogy cannot be assumed, as due to the high thermal diffusivity (α_t) a constant relation between the Reynolds shear stresses and turbulent thermal diffusion is not given. Consequently, the correct prediction of the heat transfer poses a challenge and needs special attention in terms of heat flux modelling.

Investigations of the heat transfer downstream of a BFS revealed, that the expected reattachment length is a good criterion to select turbulence models. During the investigation, two different cases with changing reattachment lengths are considered.

One major finding of previous investigations is that a lower expansion ratio, ER, features smaller reattachment lengths and vice versa [11][12][13][14]. Within the present investigation this finding is confirmed, however, for the case with ER = 2, a larger reattachment length than the one reported by [14] is found. The difference is attributed to the different boundary layer conditions at the separation point. From the turbulence modelling point of view, the estimation of the reattachment length seems to be important for the choice of the turbulence model. Linear k-ε-models (model of Abe, Kondoh and Nagano [15] – AKN – and the one of Launder Sharma [16] in conjunction with the Yap correction term [17]– LSY) seem to be better suited for predicting the flow structure at smaller reattachment lengths. Contrary, the k-ω based turbulence models (model of Wilcox [18] – SST – and Hellsten [19]), as well as the non-linear v^2-f model of [20], in combination with Durbin's realisability constraints [21], (V2F) show a better agreement with the data of [22][23] for the simulations featuring a larger reattachment length. Therefore, the results of the different simulations are discussed separately. The turbulent heat fluxes are modelled with the Single-Gradient-Diffusion-Hypotheses (SGDH), where the turbulent diffusion of heat (α_t), is computed by two different models, firstly the local correlation of Kays [24] and secondly the two equation heat transfer model of Manservisi and Menghini [25]. The AKN model is coupled with the two equation model of [25] and is referred to as MM, while all other models use the Kays correlation for the computation of the turbulent heat fluxes.

3.2. *Forced convective flow over a BFS with ER = 1.5*

The turbulent flow over a BFS with ER = 1.5 and Re_h = 5000 is discussed. As mentioned above, the flow features a short reattachment length (x/h = 6.9), shown in Figure 6. It can be noticed, that the AKN model and the LSY model show a good agreement with the data of [8] in respect to the reattachment length, whereas the other models over-estimate x. The differences in the reattachment length can be attributed to the discrepancies in the shear stress profiles predicted by the different models as illustrated in Figure 7. A good agreement between the numerical data of [8] and all RANS results is achieved for the mean velocity profiles, as well as for the turbulent kinetic energy, similar as in [26]. For the LSY model, the fluid experiences the step at a later position, resulting in a sharp edge in the velocity at x/ h = 1.0, as discussed in [26].

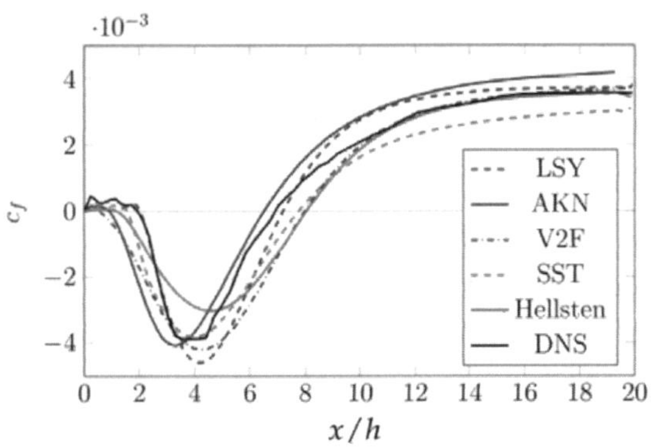

Figure 6. Skin friction distribution downstream of the separation point for Re_h = 5000.

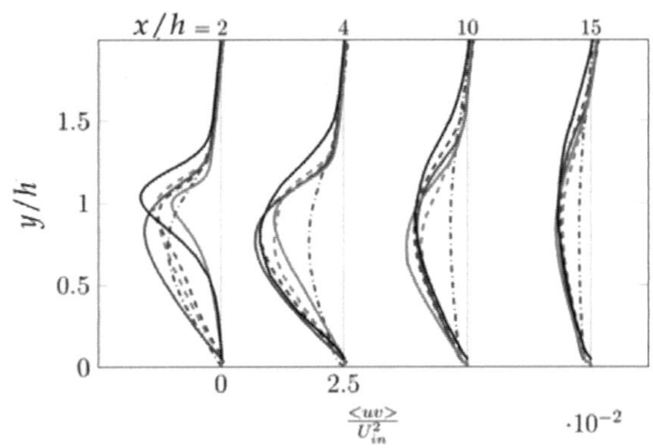

Figure 7. Comparison of the Reynolds shear stress at several positions downstream of the separation point (legend as in Figure 6).

A good prediction of the momentum field is only a pre-requisite for a correct thermal field according to the Reynolds analogy deficit in low Prandtl number fluids. This statement is supported by the large scatter in the Stanton number profiles predicted by the different models shown in Figure 8. The LSY model in combination with the Kays correlation provides the best prediction compared to the data of [8], see also [9], [26]. The other models, using the same correlation for the turbulent heat fluxes, exhibit reduced Stanton number profile. The V2F model predicts a significantly lower Stanton profile than the DNS data, as shown in Figure 4. This behavior is expected, as the Kays correlation is based on the correct prediction of the eddy viscosity. Thus, the considerable under-estimation of the shear stresses shown in Figure 7 implies lower turbulent heat fluxes compared to the other models. The MM model is not capable to predict the corner eddy in the velocity field, which influences the thermal prediction close to the step, indicated by a lower St. This shortcoming leads to an overestimation of Stanton number for the region of the recirculation zone. After the reattachment point the computed Stanton number converges towards the one of the DNS. That demonstrates, that the correct predictions of the flow physics after the separation point, like the existence of the corner eddy, is of great importance for the computation of the heat transfer, also for fluids with a very large thermal diffusion. To understand the contribution of the eddy diffusivity of heat to the total thermal transport at the given Reynolds and Prandtl number, the ratio of turbulent and molecular diffusion is illustrated in Figure 9. In the near wall region, the effect of turbulent diffusivity is very small and can be neglected. Further above, nearly half a step height away from the wall, its influence is increasing up to 50 % of the molecular diffusion at x/h=10. Further downstream of the reattachment point, the area affected by turbulent diffusion is expanding and the peak is shifting away from the heated wall towards the duct center.

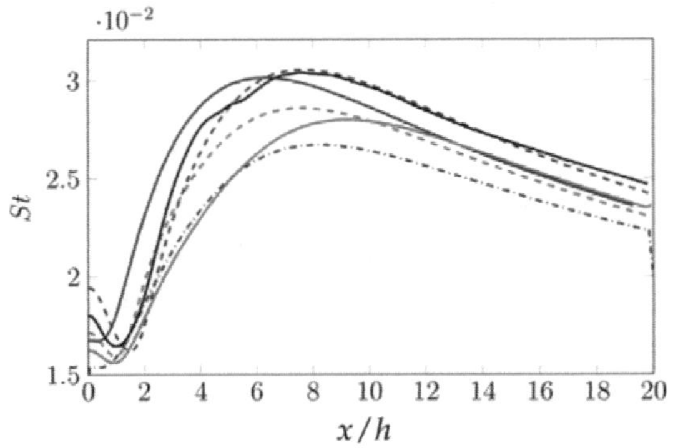

Figure 8. Stanton number distribution downstream of the separation point for $Re_h = 5000$ (legend as in Figure 6).

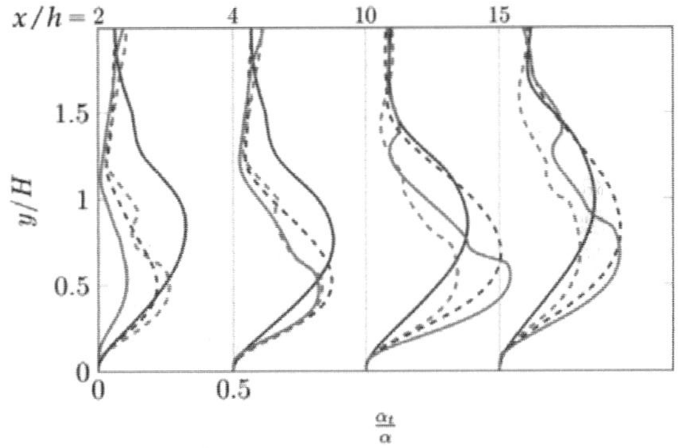

Figure 9. Comparison of the ratio of eddy to molecular diffusivity of heat at several positions downstream the separation point (legend as in Figure 6).

3.3. Forced convective flow over a BFS with ER = 2.0

As mentioned, the reattachment length for a turbulent flow over a BFS increases, compared to the previous case. Therefore, the k-ω models show a better agreement with the DNS data of [22], [23] in terms of the skin friction distribution, see Figure 10. Surprisingly, for $Re_h = 10000$, nearly all models over-predict the heat transfer within the recirculation zone. This difference is caused by the low inflow Reynolds number together with the high expansion ratio, leading to reduced shear stresses downstream the separation point and thus reduced turbulent heat fluxes compared to the previous case. The reduced contribution of the turbulent heat fluxes on the global heat transfer is supported by a simulation assuming that the thermal diffusivity is zero, shown in Figure 11. There, the Stanton number profile collapses with DNS data up to x/h = 3. Further downstream, the influence of the turbulent heat fluxes is enhanced and the temperature transport within the fluid increases. Thus, the Stanton number of the DNS data grows further.

3.4. Buoyancy aided mixed convective flow over a BFS with ER=2.0

The influence of buoyancy is investigated by means of RANS simulations for the turbulent flow over a BFS at $Re_h = 10000$ with ER = 2 [27]. The goal is to verify fast and simple methods for complex flows and thermal structures. Therefore, only the linear model of LSY in conjunction with the Kays correlation is used. Similar to investigations of [33], who reported an increase in heat transfer for buoyancy-aided mixed convection for liquid sodium with Pr = 0.005, a larger heat transfer is found for the turbulent flow of sodium over a BFS with a slightly larger molecular Prandtl number of Pr = 0.0088 as shown in Figure 12. The investigation reveals that buoyancy contributes in accelerating the fluid close to the heater. The incoming flow is diverted towards the heated wall more strongly for the mixed convection cases than for the forced convection cases. Subsequently, the flow impinges on the heater earlier with increasing influence of buoyancy. The impingement leads to a higher heat transfer as shown in Figure 12.

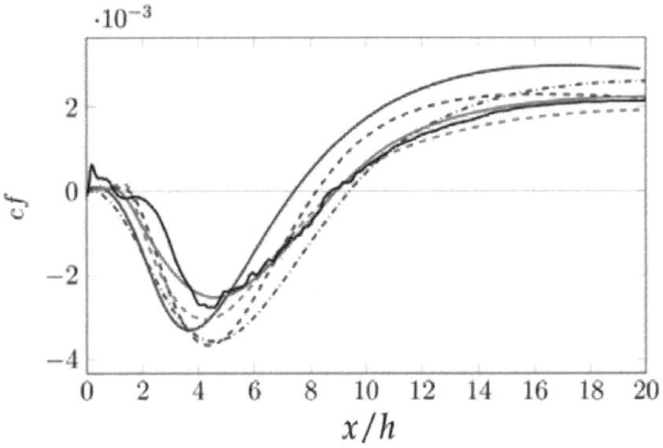

Figure 10. Skin friction distribution downstream of the separation point for Re_h = 10000 (legend as in Figure 6).

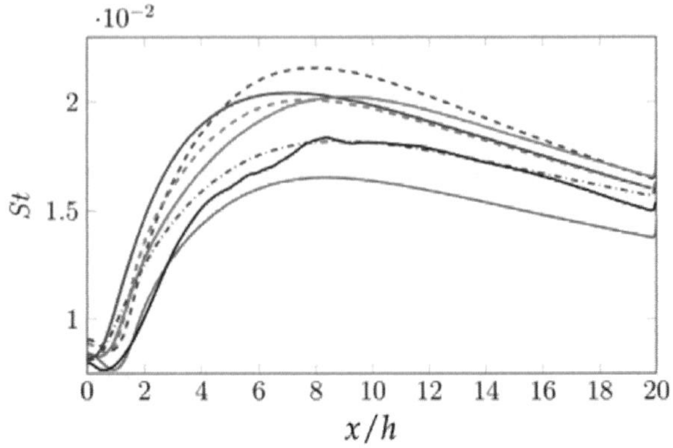

Figure 11. Stanton number distribution downstream of the separation point for Re_h = 10000 (legend as in Figure 6 - the green line corresponds to the simulation with $\alpha_t = 0$).

Towards the end of the heater, a decrease in c_f is found for Ri > 0.2. This decrease is attributed to the strong increase in turbulent shear stresses, which counter-acts the buoyancy force and decelerates the fluid, thus reducing the near wall velocity gradients as reported in [9], [26]. The influence of eddy diffusion of heat increases further downstream, as the turbulence is generated by the high shear stresses. Nevertheless, the acceleration of near wall velocity is responsible for the enhanced heat transfer found for increasing buoyancy.

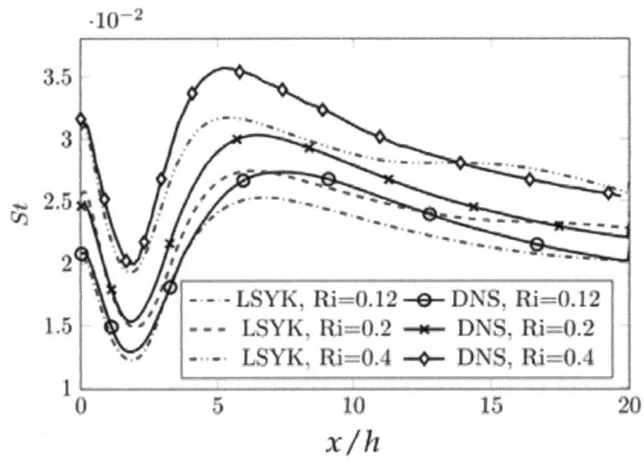

Figure 12: Comparison of the Stanton number distribution downstream of the separation point for Re_h = 10000 at different Richardson numbers with Ref. [23].

4. Preparation of a BFS experiment

4.1. The BFS design

To support the numerical investigations an experiment set-up for the BFS is currently under construction. The BFS test section is divided into four parts, as depicted in Figure 13. The first section is the inlet section. It is basically the transition region from circular cross section to rectangular cross section. Furthermore, it contains perforated plates for flow rectification. The second section is the rectangular part of the BFS. A commercially available duct with rounded corners is used. The main reason is that, first of all, for engineering applications commercially available components would be used rather than expensive single-part products. Second, the ducts with rounded corners have only one welded seem. A construction of a rectangular duct with sharp corners made from simple steel plates would require several welded seems. This is of disadvantage for the dimensional accuracy due to the large heat input. Furthermore, wire and sinker EDM are very expensive for the envisaged lengths of ducts. The commercially available ducts proof to be very accurate and the welded seem satisfies the requirements for, e.g., safety. Detailed studies are performed to investigate the difference between channels with sharp 90° corners and rounded corners [28]. The outcome is that no difference in the general behavior exists between the two flows. In the present case, the rectangular section of the BFS has a length of 3600 mm and an internal cross section of 90 x 40 mm. The outer radius of the corners is 2 times the wall thickness, which is 5 mm. With this cross section, the undisturbed length of the rectangular section is around 65 hydraulic diameters and equals x/h = 72. Studies related to the entrance length effect of liquid metal cooled rod bundles and circular ducts show that such an entrance length is sufficient to provide a hydraulically developed flow at the step [29]. The third section is the quadratic section of the BFS. It is 2000 mm long and has an internal cross section of 90 x 90 mm, which equals to x/h = 40. The wall thickness and the radius of the rounded corners are identical to the rectangular section. The inside surface and the welded seem will be mechanically treated to guarantee a homogeneous surface. The last section, section four is the outlet section, which is the transition from a quadratic cross section to a circular cross section. The presented approach limits the number of welds, which may have a negative impact on the flow.

To minimize the thermal losses during the operation, the BFS is equipped with a trace heating system and with a thermal insulation. The thermal insulation is made of a 100 mm thick rock wool layer (approximately thermal conductivity = 0.0038 W/m·K)

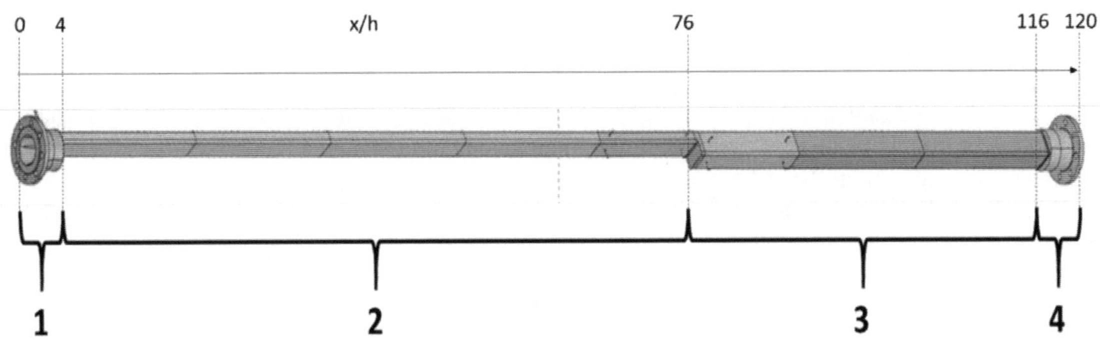

Figure 13. BFS test section (1 - inlet section, 2 - rectangular section, 3 - quadratic section, 4 - outlet section)

For the present BFS temperature and velocity measurements will be taken globally and at discrete local positions. First of all, the temperature of the heated wall, indicated by the red shaded area in Figure 1, and the fluid temperature above the heated section at the centerline are measured using sheated thermocouples. With the knowledge of the temperature dependent thermo-physical properties and the given heat flux density the heat transfer coefficient and also the Nusselt number can be derived. Second, traversable permanent magnet probe sensors (PMP) are foreseen to measure the temperature and the velocity at pre-defined positions, to identify the point of reattachment of the flow after the separation. Furthermore, the provision of temperature and velocity field data aims to provide input for CFD, LES and DNS simulation. Depending on the configuration of the permanent magnetic probe the turbulent heat fluxes on a local level can be evaluated. A detailed description of permanent magnet probe sensors can be found in [30][31][32].

Given the KASOLA corner data, the developed BFS test section is capable to cover the following parameter range: $5000 \leq Re \leq 10^6$; $0.005 \leq Pr \leq 0.008$; $25 \leq Pe \leq 8500$. Depending on the heating scheme and the pump operation, a Richardson number in the order of 1.0 can be realized. Thereby, the entire Richardson and Péclet number regime from purely buoyancy driven flow to pure convective flow can be covered. Additionally, the developed design allows for an exact determination of the transitional domains.

4.2. The KASOLA facility

The BFS test section will be integrated into the KASOLA (KArlsruhe SOdium LAboratory) facility, being illustrated in Figure 14.

The KASOLA facility is a flexible, medium size, loop type experimental facility. The facility is sheltered by a cylindrical steel containment with a height of 12 m and a diameter of 8 m. The only exception is the sodium storage tank, which is located in an underground storage outside the steel containment. The storage tank has a sodium volume of 7 m³. The KASOLA facility has a maximal electrical consumption of 2 MW (trace heating, experimental heating, pump, etc). The basic loop contains the following components: Two test section ports, a magneto-hydrodynamic pump, the sodium-air heat exchanger, a magnetic flow meter, an expansion tank and the sodium precipitator. The developed length of the basic loop equals 37 m. The piping between the components is made of DN 100 stainless steel pipes. During operation roughly 1 m³ of sodium is in the loop. Furthermore, the basic loop is equipped with several valves for operational and safety purposes. In a by-pass line to the basic loop, a second flow meter, Coriolis type, and a cold trap for sodium purification are installed. The whole facility, basic and by-pass loop are enveloped with a trace heating system to prevent the sodium from freezing. The sodium storage tank and the expansion tank are under an Argon atmosphere. To remove the energy dissipated into the sodium an air cooled heat-exchanger is integrated and in the cold trap three separate air cooling circuits are installed.

For a maximum of operational and experimental flexibility, two test section ports are installed. The first port is foreseen to house experiments related to duct and pipe flow like the BFS. With a total free length of six meter the installed test section allows additional flexibility for the study of different flow phenomena. The second test port is related to the investigation of thermal storages. A thermocline storage tank with a movable separating plate will be installed to investigate the feasibility of sodium as a heat transfer and storage fluid for concentrated solar power applications. The tank has a volume of about 0.5 m³ and a storage capacity of 50 kWh. The KASOLA configuration allows a simultaneous operation of both test loops.

The magneto-hydrodynamic pump can deliver a maximal volume flow rate 150 m³/h (37.65 kg/s) and a maximal relative pressure head of 6 bar. The maximal volume flow rate corresponds to a Reynolds number of approximately 10^6 for the quadratic test section of the BFS. The operational temperature of the facility is in the range of 150 to 550°C. For the sodium cooling, the sodium-air heat exchanger has a maximal thermal duty of 970 kW. The sodium-air heat exchanger can be operated on the air side with an air flow rate of 4.4 kg/s and a maximal temperature increase of 210°C. Because KASOLA is foreseen to operate anywhere between 150 and 550°C an expansion tank is installed at the top of the basic sodium loop.

The operation of the facility is realised by two automatic Programmable Logic Controller units. One is for the general operation of the loop, while the other is connected to the trace heating system. In addition, an automatic shut-down system is installed.

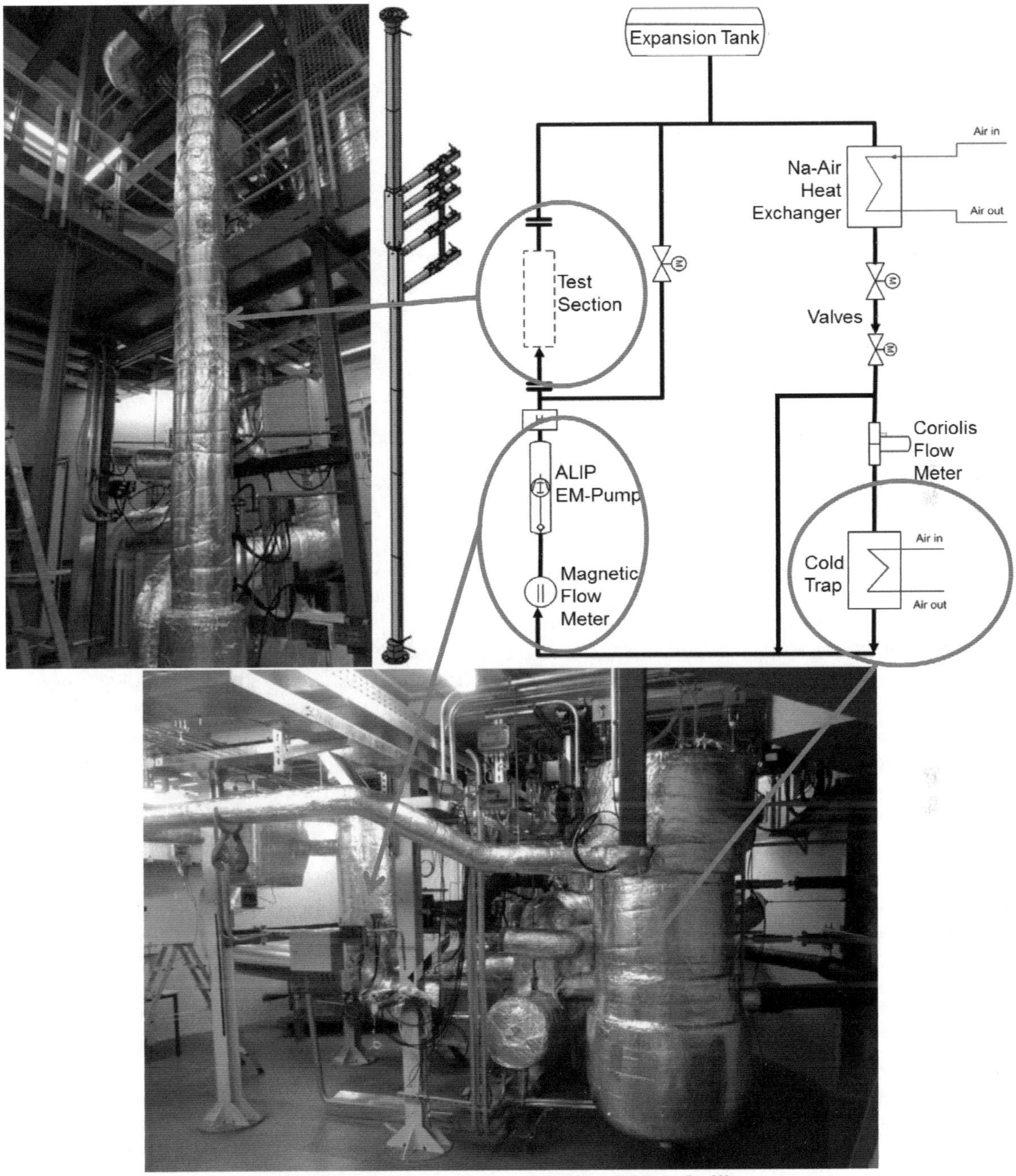

Figure 14. Impressions of the KASOLA facility

The KASOLA loop is connected to an Argon system. This system is used to avoid sodium oxidation, to maintain a pre-defined system pressure and to assist during the filling and evacuation of the main components of the KASOLA facility include:

- Storage tank: 8 m³ with 7 m³ of sodium and argon cover;
- Expansion tank: 0.3 m³ and argon cover;

- Stainless steel piping system: DN 100, PN 16, up to 550°C;
- Magneto-hydrodynamic pump: 75 kW, 150 m³/h (37.65 kg/s);
- Cold trap: Sodium purification;
- Sodium precipitator: Protection against sodium leakage into the argon system;
- Trace heating: 40 kW for the loop, 32 kW for the storage tank;
- Thermal insulation: Rock wool 100 mm piping, 200 mm thermal storage;
- Magnetic flow meter: Operation;
- Coriolis flow meter: Calibration and control:
- Sodium-air heat exchanger: ~ 1 MW:
- 2 test section ports: 1 for BFS, 1 for thermal storage;
- Containment: Steel cylinder, 12 m high, 8 m in diameter;

The KASOLA facility is currently in the commissioning phase (until August 2017). First general experiments are foreseen with the basic loop (until March 2018) and experiments with the BFS are foreseen to take place afterwards (beginning April 2018).

5. Summary

This article describes numerical and technical developments to investigate the vertical backward facing step (BFS) problem for a low Prandtl number medium such as liquid metals, appearing in several technical applications.

In this context a quasi-exact solution of the flow has been developed, verified and validated by means of a turbulence resolved modelling of the flow problem based on a Direct numerical simulation (DNS) or a corresponding DNS to serve as a reference solution. In this context not only the mean fluid wall interface temperatures are developed to provide an indication for the test-sectional set-up of the BFS experiments, but rather to describe the exact transport quantities for the momentum and temperature field wall normal to the heated section along the entire heated length to identify stagnation points and recirculation domains peculiar to any technical design. Since this approach demands excessive computational power, this solution formed the pre-requisite for a numerical computations relying on turbulence models providing a prediction of technically relevant parameters as the fluid/wall interface temperature (or Nusselt number) at a substantially faster computational speed. Several turbulence models have been analysed and compared in the budget terms to the DNS data provided. Thereby, it turns out no superior turbulence models is currently available to describe all potential developing flow and temperature distributions adequately. Most essential is however, that already the momentum field is correctly depicted to execute a reliable prediction of the thermal field, which excludes some of the conventionally used turbulence models.

Based on the obtained numerical data a BFS test-sectional lay-out with a corresponding instrumentation has been developed to be integrated in the KASOLA facility aiming either to verify the model calculations or to elucidate drawbacks of currently existing models requiring a more complete model description. Moreover, the set-up in the KASOLA facility is not only targeting a model validation but also to provide a data basis for parameter regimes also in the near future hardly numerically accessible. Unfortunately, due to technical constraints the successful operation could not be achieved within the project duration.

Acknowledgments

The financial support of the Helmholtz Alliance "Liquid Metal Technologies – LIMTECH" is gratefully acknowledged.

References

[1] Le H, Moin P and Kim J 1997 *J. Fluid Mech.* **330** 349-74
[2] Barri M and Andersson H I 2011 *J. Fluid Mech.* **665** 382-417

[3] Avancha R V R and Pletcher R H 2002 *Int. J. Heat Fluid Flow* **23** 601-14
[4] Abu-Mulaweh H I, Chen T S and Armaly B F 2002 *Int. J. Heat Fluid Flow* **23** 758-65
[5] Kempe T and Fröhlich J 2012 *J. Comput. Phys.* **231** 3663–3684
[6] Nicoud F, Toda H B, Cabrit O, Bose S and Lee J 2011 *Phys. Fluids* **23** 085106
[7] Niemann M and Fröhlich J 2014 *Proc. Appl. Math. Mech.* **14** 659-660
[8] Niemann M and Fröhlich J 2016 *Int. J. Heat Mass Transfer* **101** 1237-50
[9] Schumm T, Niemann M, Marocco L, Magagnato F, Frohnapfel B and Fröhlich J 2015 *Proc. of 8th Int. Symp. on Turbulence, Heat and Mass Transfer* vol **1** ed K Hanjalic et al (New York, Begell House Ind.)
[10] Vogel J C and Eaton J K 1985 *J. Heat Transfer* **107(4)** 922-9
[11] Eaton J K and Johnston J P 1980 Rept MD-39, Dept of Mech Eng, Stanford University, Stanford, CA
[12] Durst F and Tropea C 1981 *Proc. 3Rd Symp Turb Shear Flows* (Davis)
[13] Adams EW and Johnston J P 1988 *Exp. Fluid* **6** 493-9
[14] Nadge P M and Govardhan R N 2014 *Exp. Fluids* **55(1)** 1657
[15] Abe A, Kondoh T and Nagano Y 1994 *Int. J. Heat Mass Transfer* **37** 139-51
[16] Launder B E and Sharma B I 1974 *Letters in Heat and Mass Transfer* **1** 131-7
[17] Yap C R 1987, Ph.D. Thesis (Manchester University England)
[18] Menter F 1994 *AIAA J* **32** 1598-605
[19] Hellsten A 2005 *AIAA J* **43** 1857-69
[20] Davidson L, Nielsen P and SveningssonA 2003, *Proc. of 8th Int. Symp. on Turbulence, Heat and Mass Transfer* **4** 577-84
[21] Durbin P A 1996 *Int. J. Heat Fluid Flow* **17** 89-90
[22] Niemann M and Fröhlich J 2015 *Proc. Direct and large-Eddy Simulation* **10**
[23] Niemann M and Fröhlich J 2016 *Proc. Int. Symp on Eng. Turb. Mod. And Meas.* **11**
[24] Kays W M 1994 *J. Heat Transfer* **116** 284-95
[25] Manservisi S and Menghini F 2014 *Int. J. Heat Mass Transfer* **69** 312-326
[26] Schumm T, Frohnapfel B and Marocco L 2016 *J. Phys.: Conf. Ser.***745** 032051
[27] Schumm T, Frohnapfel B and Marocco L 2017 *Heat Mass Transfer* (in press)
[28] Jaeger W, Schaub Hahn T, Hering W, Otic I, Shams A, Oder J and Tiselj I 2017 *Proc Int. Topical Meeting on Nuclear Reactor Thermal Hydraulics* **17**
[29] Jaeger W 2017 *Heat Mass Transfer* **53(5)** 1667-84
[30] von Weissenfluh T 1985 *Int. J. Heat Mass Transfer* **28(8)** 1563-75
[31] Knebel J U and Krebs L 1994 *Exp. Therm Fluid Sci.* **8** 135-48
[32] Kappulla R, Sigg B, Horanyi S and Hudina M 2000 *Exp. Therm Fluid Sci.* **20** 115-36

IOP Conf. Series: Materials Science and Engineering **228** (2017) 012002 doi:10.1088/1757-899X/228/1/012002

Magnetic field dynamos and magnetically triggered flow instabilities

F Stefani[1], T Albrecht[2], R Arlt[3], M Christen[4], A Gailitis[5], M Gellert[3], A Giesecke[1], O Goepfert[6], J Herault[7], O N Kirillov[8], G. Mamatsashvili[1], J Priede[9], G Rüdiger[3], M Seilmayer[1], A Tilgner[6] and T Vogt[1]

[1] Helmholtz-Zentrum Dresden – Rossendorf, Bautzner Landstraße 400, D-01328 Dresden, Germany
[2] Department of Mechanical and Aerospace Engineering, Monash University, VIC 3800, Australia
[3] Leibniz-Institut für Astrophysik Potsdam, An der Sternwarte 16, D-14482 Potsdam, Germany
[4] Technische Universität Dresden, Institut für Energietechnik, D-01062 Dresden, Germany
[5] Institute of Physics, University of Latvia, LV-2169 Salaspils, Miera iela 32, Latvia
[6] Institute of Geophysics, University of Göttingen, Friedrich-Hund-Platz 1, D-37077 Göttingen, Germany
[7] Laboratoire des Sciences du Numérique de Nantes (LS2N), CNRS, IMT Atlantique, Nantes, France
[8] Northumbria University, Mathematics, Physics and Electrical Engineering, Ellison Building, D219 Newcastle upon Tyne NE1 8ST, United Kingdom
[9] Flow Measurement Research Centre, Coventry University, UK

E-mail: F.Stefani@hzdr.de

Abstract. The project A2 of the LIMTECH Alliance aimed at a better understanding of those magnetohydrodynamic instabilities that are relevant for the generation and the action of cosmic magnetic fields. These comprise the hydromagnetic dynamo effect and various magnetically triggered flow instabilities, such as the magnetorotational instability and the Tayler instability. The project was intended to support the experimental capabilities to become available in the framework of the DREsden Sodium facility for DYNamo and thermohydraulic studies (DRESDYN). An associated starting grant was focused on the dimensioning of a liquid metal experiment on the newly found magnetic destabilization of rotating flows with positive shear. In this survey paper, the main results of these two projects are summarized.

1. Introduction

Magnetic fields of planets, stars and galaxies are generated by the homogeneous dynamo effect [1, 2, 3]. Once produced, cosmic magnetic fields can play an active role in cosmic structure formation via various magnetically triggered flow instabilities, such as the magnetorotational instability (MRI) [4] and the current-driven Tayler instability (TI) [5].

Complementary to the decades-long theoretical and numerical efforts to understand these fundamental magnetohydrodynamic effects, the last years have seen great progress in dedicated experimental investigations [6, 7, 8]. After the pioneering Riga and Karlsruhe dynamo experiments [9, 10, 11], it was in particular the rich dynamics observed in the French von Kármán Sodium (VKS) experiment [12] that provoked much interest throughout the dynamo community. The observed reversals, excursions, bursts, hemispherical fields etc. inspired new activities to understand the essential physics behind the corresponding planetary phenomena [13, 14, 15, 16].

For magnetically triggered flow instabilities, the situation is more subtle. Interesting results had been obtained in a liquid sodium spherical Couette experiment in Maryland in form of coherent velocity/magnetic field fluctuations showing up in a parameter region reminiscent of MRI [17], as well as in the GaInSn Taylor-Couette (TC) experiment in Princeton which provided evidence for slow magneto-Coriolis waves [18] and a free-Shercliff layer instability [19]. Despite these achievements, both experiments have corroborated the intricacies of demonstrating the standard version of MRI (SMRI) for which a purely axial field is applied. For liquid metal TC flows, in particular, the complications result from the compromising effect of axial boundaries on the flow structure at those high Reynolds numbers ($> 10^6$) that are necessary to attain magnetic Reynolds numbers of the order of 10.

More conclusive, albeit less ambitious, was the experimental demonstration of two special types of the MRI, which arise when applying helical or purely azimuthal magnetic fields to the rotating flow. These instabilities have been coined helical MRI (HMRI) and azimuthal MRI (AMRI), respectively. As was first shown by Hollerbach and Rüdiger [20], the essentially inductionless, axisymmetric ($m = 0$) HMRI scales with the Reynolds and Hartmann number rather than with magnetic Reynolds and Lundquist number as SMRI with which it is monotonically connected, though [20, 21]. It is this scaling behaviour that makes experimental investigations of HMRI much easier than those of SMRI. Indeed, first evidence of HMRI occurring in the predicted parameter regions of Hartmann number with roughly correct eigenfrequencies was demonstrated in the PROMISE experiment at Helmholtz-Zentrum Dresden-Rossendorf (HZDR) [22, 23]. In 2009, an improved version of this experiment - using split end-rings installed at the top and bottom of the cylinder in order to minimize the global effects of Ekman pumping - allowed to characterize HMRI by a number of parameter variations, generally in good agreement with numerical predictions [24].

Nearly at the same time, Hollerbach et al. [25] identified AMRI as a second induction-less MRI version that appears for strongly dominant azimuthal fields in form of a non-axisymmetric ($m = 1$) perturbation. When relaxing the condition, implicit for AMRI and HMRI, that the azimuthal field should be current-free in the liquid, one enters the vast field of current-driven instabilities, which includes the Tayler instability (TI) [5]. This kink-type instability, whose ideal counterpart has long been known in plasma physics [26], is also discussed as a central mechanism of the non-linear Tayler-Spruit dynamo model for stellar magnetic fields [27, 28, 29]). While TI was experimentally observed prior to the start of the LIMTECH alliance [30], AMRI was first demonstrated within the funding period [31].

Despite – and partly inspired by – those experimental achievements, there are still a number of questions worth to be studied in laboratory. While realistic "bonsai" models of cosmic objects, with all dimensionless numbers matching those of planets or stars, are certainly outside the scope of laboratory feasibility [8], some new facilities still challenge the physical and technical limits of dynamo experiments. This certainly applies to the 3 m diameter spherical Couette experiment running at the University of Maryland [32, 33], to the 3 m diameter plasma dynamo experiment in Madison [34, 35], as well as to the 2 m diameter precession dynamo experiment presently under construction at HZDR [36, 37]. One of the unwelcome characteristics of these "second generation" dynamo experiments is a higher uncertainty of success. The Riga and Karlsruhe experiments were quite accurately described by kinematic dynamo codes and simplified saturation models [38, 39, 40], and even the unexpected VKS dynamo results were understood once the effect of the high permeability disks was accounted for properly

[41, 42, 43, 44]. By contrast, the outcomes of the experiments in Maryland, Madison, and Dresden are much harder to predict. In either case, this uncertainty results from the ambition to construct a truly homogeneous dynamo, which is neither driven by pumps or propellers, nor influenced by guiding blades or gradients of magnetic permeability. This higher degree of freedom makes those flows prone to exhibiting medium-size flow structures and waves, whose dynamo capabilities are still under scrutiny.

On the MRI side, the "holy grail" of observing SMRI in the lab is yet to be found. One promising set-up is the relatively flat TC experiment in Princeton, which circumvents the Ekman-pumping induced distortion of the original TC flow profile by separately driving different rings of the lids. Another promising way is followed in the plasma experiment in Madison, where already reasonable radial velocity profiles have been produced by a near-wall $\mathbf{j} \times \mathbf{B}$ driving of argon and helium plasmas [35]. At HZDR, a more traditional path towards SMRI is pursued: the respective experiment contains a long sodium column between two rotating cylinders exposed to a strong axial magnetic field. Yet, this axial field is complemented by an azimuthal magnetic field, which permits progressing from the well-known regime of HMRI towards the limit of SMRI by increasing the Reynolds and Hartmann number and simultaneously decreasing the ratio of azimuthal to axial field.

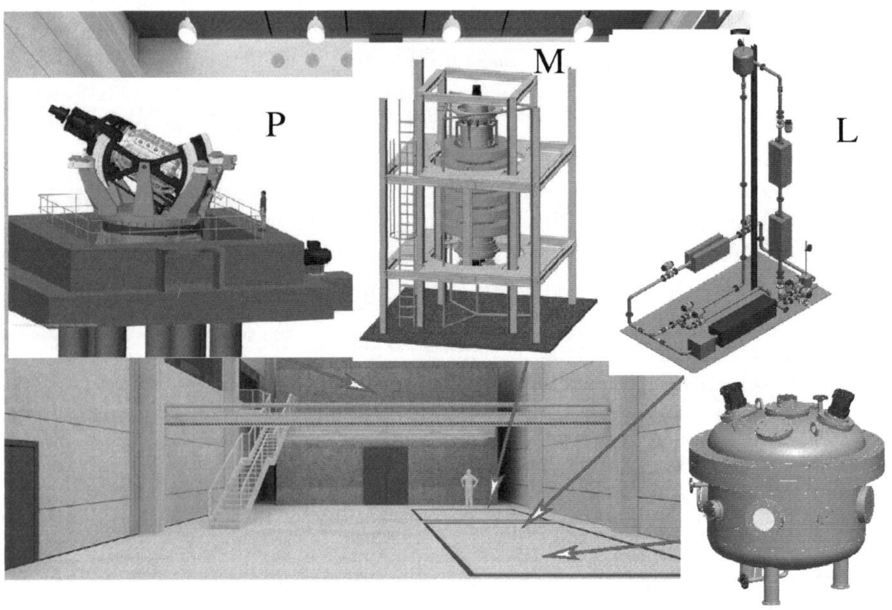

Figure 1. Interior of the central hall of the DRESDYN facility with the main planned experiments. Precession driven dynamo experiment (P) to be installed in the containment; Tayler-Couette experiment for the investigation of the magnetorotational and the Tayler instability (M); sodium loop (L); In-Service-Inspection experiment (I), a vessel of 1.2 m diameter with a number of flanges for testing various measuring techniques in liquid sodium. A further test stand for liquid metal batteries is also planned but not yet designed.

This paper summarizes research activities within the LIMTECH Alliance dedicated to the preparations of the precession-driven dynamo experiment and the large-scale MRI experiment. Both experiments are presently under construction in frame of the DRESDYN project at HZDR, which will include also a number of other experiments using liquid sodium as operation fluid (see Figure 1). On the dynamo side, activities included a number of numerical simulations of precession driven flows and their dynamo action in cylinders and cubes, as well as experimental work at a 1:6 down-scaled water experiment. They also comprised various activities to refurbish and recommission the Riga dynamo experiment, which – apart from having its own scientific goals – serves also for testing various measurement techniques to be applied in future DRESDYN experiments.

With regard to MRI/TI, the project supported the first demonstration of AMRI at the PROMISE facility at HZDR, and included theoretical and numerical work on various aspects of the interaction of rotating fluids and magnetic fields. These led to the characterization of a new magnetically triggered instability ("Super-AMRI") that destabilizes rotating flows with positive shear, to the "refutation" of Chandrasekhar's theorem for magnetized rotating flows, and to the establishment of a rigorous mathematical connection between the non-modal growth for purely hydrodynamic rotating flows and the growth rate of helical MRI.

2. Research related to experimental dynamos

In this section we summarize the main activities dedicated to existing and planned dynamo experiments. While the main focus was on various preparations of the DRESDYN precession driven dynamo, we will start with presenting the works related to the re-commissioning of the Riga dynamo.

2.1. Riga dynamo experiment

On 11 November 1999 the kinematic phase of magnetic-field self-excitation was shortly observed at the Riga dynamo facility before the experiment had to be stopped due to a minor leakage of liquid sodium [9]. After some repairs, in July 2000, a number of full runs clearly demonstrated both the kinematic and the saturated dynamo regime [10], thereby laying the basis for a comprehensive data base including growth rates, frequencies, and spatial structures of the magnetic eigenfield in dependence on the impeller's rotation rate [38]. Data of the kinematic regime was shown to be in very good agreement with numerical predictions [45], and even the saturation regime was reasonably understood by applying Lenz's rule to the specifics of this hydrodynamic dynamo [46].

After a series of experimental campaigns, which delivered quite reliable und reproducible results, the Riga dynamo was disassembled in order to replace and stabilize an inner cylinder that had been deformed during one run in 2010. The project A2 supported the refurbishment with the double aim of testing measurement techniques for the DRESDYN precession dynamo, and of preparing such a modification of the experiment that allows for observing new and non-trivial back-reaction effects of the self-excited magnetic field on the flow structure.

The latter was motivated by the numerical finding that a specific "de-optimization" of the flow field in the Riga dynamo could lead to a vacillation between two different states of the dynamo [47]. Actually, we set out from the hypothesis that a too high initial azimuthal component of the velocity might yield a subcritical Hopf bifurcation, just by virtue of a selective breaking of this component that provides a "re-optimization" of the velocity. Instead of such a "hard" subcritical Hopf bifurcation we found numerically a "soft" vacillation between two dynamo states with different kinetic and magnetic energies. For still larger magnetic Reynolds numbers even a transition to chaos was predicted.

Based on these findings, one of the tasks of the A2 project was to figure out how such scenarios could be realized in the Riga dynamo facility. We evaluated several technical provisions to increase the azimuthal velocity component beyond its optimal value, without completely destroying the flow structure that had been carefully optimized for the original Riga dynamo [45]. A hydraulic analysis provided some feasible shapes and pitch angles of the post-propeller vanes that indeed should produce the desired azimuthal velocities. Figure 2a shows the resulting geometries and the pitch parameters of the corresponding velocity profiles. However, before implementing this new vanes' configuration it was decided to first reassemble the Riga dynamo in its old form (with a stabilized second cylinder in order to prevent buckling) and to validate reproducibility of the former dynamo. Figure 2b shows the self-excited magnetic field at different radial positions within the dynamo as observed in the first experimental campaign after re-commissioning in June 2016. The propeller rotation rate was varied and fell below the critical approximetaly between 1000 and 1100 s. The final decay of the signal is due to the increase of the conductivity with rising fluid temperature. More details can be found in [48].

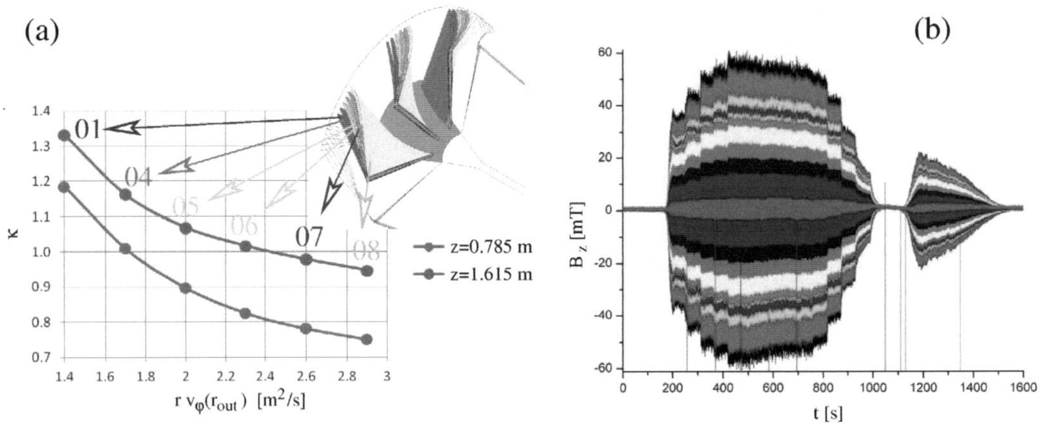

Figure 2. Riga dynamo experiment. (a) Targeted "de-optimization" of the flow profile behind the impeller by choosing vanes with varying pitch angles. The curves show the ratio $\kappa = \sqrt{\overline{v}_z^2/(2\overline{v}_\phi^2)}$ of mean axial to mean azimuthal velocity, in dependence on the swirl $rv_\varphi(r_{\rm out})$ of the flow at the outer radius of the vanes, for various investigated post-impeller vane geometries (see inset) and two distances z from the impeller. Vanes with high swirl, i.e. low κ, are considered for realising interesting back-reaction effects [47]. (b) Vertical magnetic fields measured at different radial positions (indicated by the different colours) in the upper port of the Riga dynamo during the first run after recommissioning (June 2016).

While the data are still under detailed analysis, we can conclude that the Riga dynamo experiment is now available for research into non-trivial back-reaction effects, and for testing measurement techniques for the DRESDYN precession dynamo.

2.2. Numerical and experimental results on precession-driven flows in cylinders

The largest installation in the framework of DRESDYN is a precession-driven liquid sodium experiment. Guided by early numerical estimations of the dynamo threshold of precession driven flows in cylinders [49, 37, 50] and cubes [51], we aimed at attaining a magnetic Reynolds number of $\mathrm{Rm} = \mu_0\sigma\Omega_{\rm rotation}R^2 \approx 700$, which is achievable with a cylinder of radius $R = 1\,\mathrm{m}$, rotating at an angular frequency of $\Omega_{\rm rotation} = 2\pi \times 10\,\mathrm{s}^{-1}$ (μ_0 is the magnetic permeability constant, σ the electrical conductivity of the fluid).

Design and construction of the precession dynamo experiment (see Figure 3) is the responsibility of SBS Bühnentechnik GmbH. A detailed shape optimization of the rotating vessel turned out to be necessary in order to cope with the huge stresses that result during precessing. The facility is presently under construction, and first pre-experiments with water are expected for late 2018.

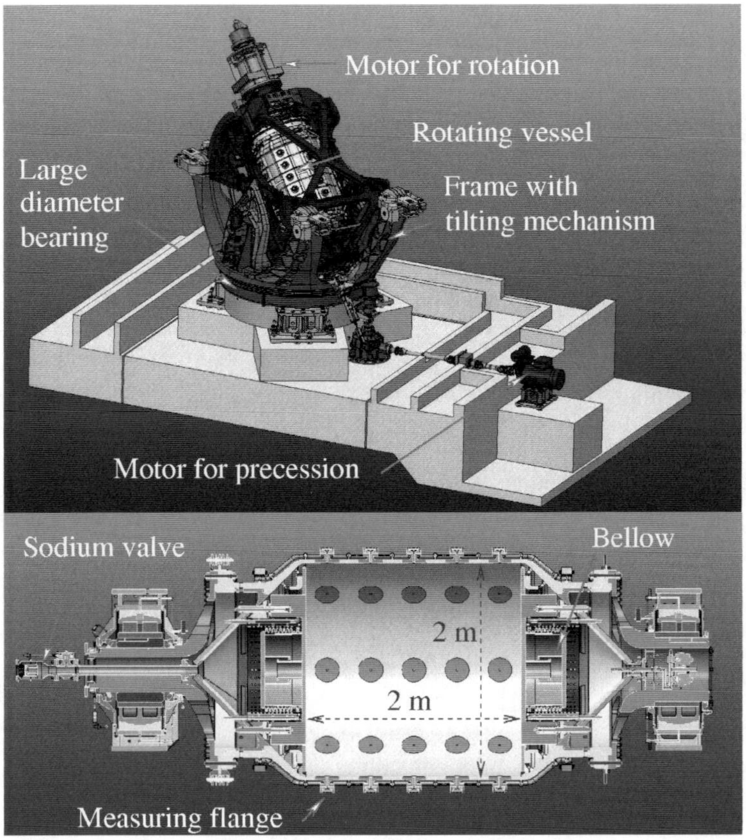

Figure 3. Design of the precession-driven dynamo experiment. Upper panel: entire system. Lower panel: details of the rotating vessel. Figure courtesy SBS Bühnentechnik GmbH.

In preparing this large-scale experiment, a 1:6 scaled water experiment with cylinder radius $R = 0.163$ m (see Figure 4a) was built and utilized in order to gain insight into the flow structure and the pressure field for varying Reynolds numbers and precession ratios (Poincaré numbers). Parallel to that, the spectral element code SEMTEX [52] was qualified and intensely used for precession-driven flows in cylinders. While such simulations are restricted to Reynolds numbers $\mathrm{Re} = \Omega_{\mathrm{rotation}} R^2/\nu \approx 10^4$ (ν is the kinematic viscosity of the fluid), the large experiment will reach approximately 10^8. The 1:6 water experiment reaches a number of 1.6×10^6 when run at 10 Hz, but can also be slowed down to a value of 10^4, thereby guaranteeing some overlap with numerical simulations.

A significant share of simulations and experiments was invested into estimating the pressure field, a crucial input for the (static and dynamic) strength evaluation of the vessel. While pure rotation with 10 Hz leads already to a centrifugal pressure of about 20 bar, it is the precession-driven pressure pulsation on the order of 10 bar that makes the (dynamic) strength validation so challenging.

Figure 4 illustrates the flow structure (c,d,e) and the corresponding pressure field (f,g,h) computed at Re = 6500 and a Poincaré number $\mathrm{Po} \equiv \Omega_{\mathrm{precession}}/\Omega_{\mathrm{rotation}} = 0.02$ for three angles between the rotation axis and the precession axis 45°, 90°, and 135°. The maximum pressure pulsations in dependence on Po are shown in Figure 4b, together with the corresponding values resulting from the small water experiment (note that all values are up-scaled to the conditions

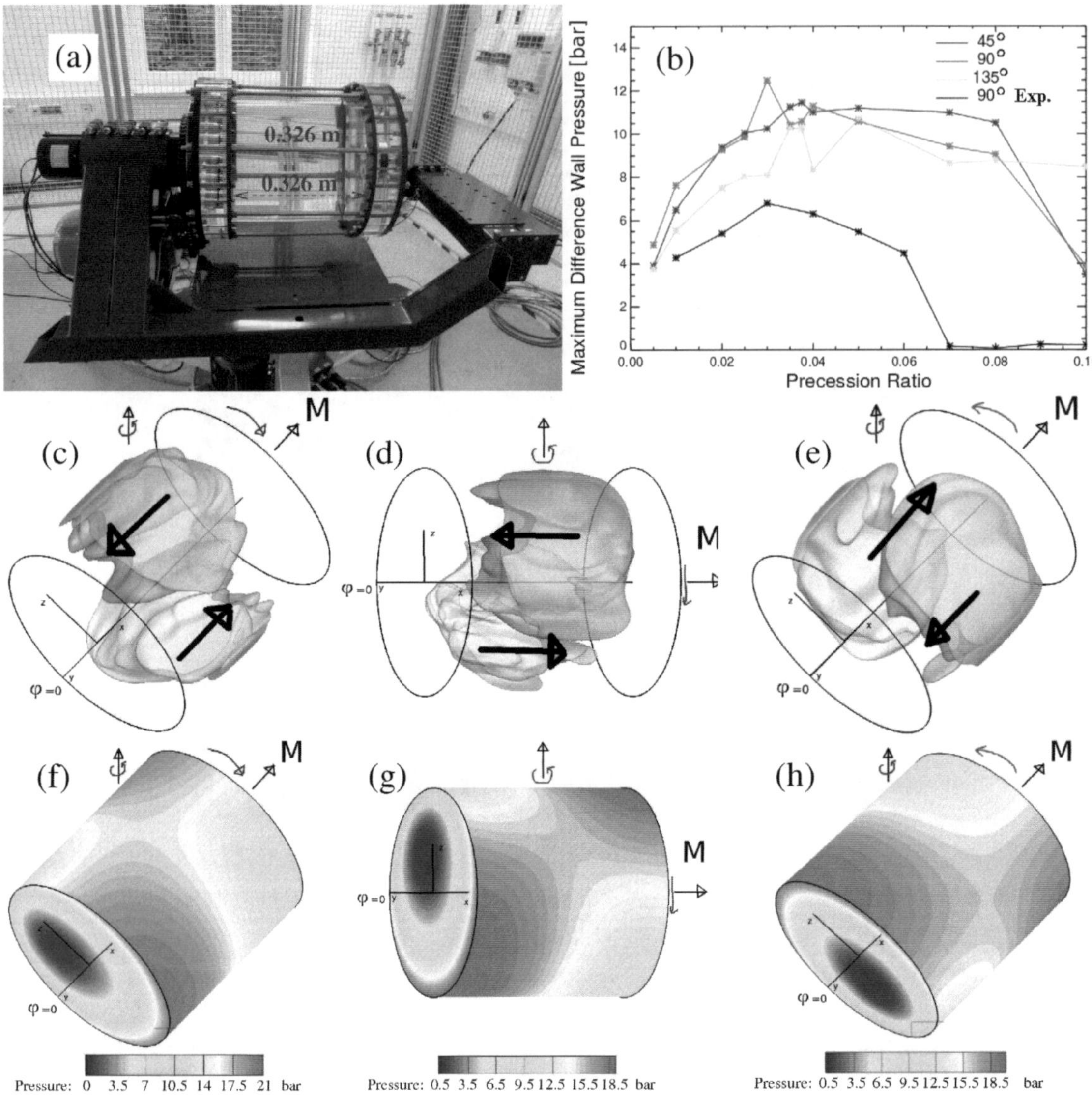

Figure 4. Experimental and numerical preparations for the large precession experiment. (a) 1:6 down-scaled water experiment. (b) Maximum pressure difference at the wall. The numerical curves, computed at Re $= 6500$, are for the three different precession angles 45° (blue), 90° (red), and 135° (green). The black curve gives the experimental data for 90° measured at Re $= 1.6 \times 10^6$. The middle row shows the numerically simulated axial velocity component v_z for the angles 45° (c), 90° (d), and 135° (e), the lower row (f,g,h) the corresponding pressure fields.

of the large experiment). Basically, at low Po the flow is dominated by the first ($m = 1$) Kelvin mode which is mirrored also by the increasing pressure pulsation. Approximately at Po $= 0.03$, higher azimuthal modes become relevant by drawing more and more energy from the forced $m = 1$ Kelvin mode by triadic resonances. The curve's kink at Po $= 0.07$ indicates a sudden transition of this quasi-laminar regime to a turbulent regime, which is also confirmed by the hysteretic behaviour of the motor power at this point [53]. There is an obvious difference between the simulated and experimental curves: the suppression of the $m = 1$ mode and the transition to turbulence is strongly delayed in the numerical case, which also shows significantly higher pressure pulsations.

Apart from this rather global feature, our numerical simulations have also characterized the onset of higher m-modes which usually are excited by triadic resonances with the forced $m = 1$ Kelvin modes [54].

2.3. Precession-driven dynamos in cylinders and cubes

Dynamo action of precession-driven flows is, unfortunately, a largely unsolved problem, despite a number of attempts to apply it to the dynamo of the Earth and other cosmic bodies [55, 56, 57, 58, 59]. The DRESDYN precession dynamo experiment was partly motivated by the optimistic numerical estimations obtained by Nore [49, 37] for the cylinder, and Krauze [51] for a cube; both pointed consistently to a critical Rm on the order of 700.

Unfortunately, this optimistic value is challenged when going to smaller values of the magnetic Prandtl number $\mathrm{Pm} = \mu_0 \sigma \nu$, i.e. to higher Reynolds numbers [61]. Here, even the structure of the hydrodynamic flow, and its dependence on the precession ratio and the tilt angle, is largely unknown. Recent simulations and experiments have revealed the occurrence of higher m-modes [60, 54] whose dynamo capabilities are still under scrutiny. Even less is known about the turbulent flow structure when crossing the critical precession ratio.

Typically, direct numerical simulations for precessing flows with correct no-slip boundary conditions work only until $\mathrm{Re} \approx 10^4$. A promising way of going to much larger values of Re is to replace no-slip boundary conditions by stress-free conditions, thereby avoiding the need to resolve the viscous boundary layers. Such an approach was pursued by Goepfert and Tilgner [62] who studied a precessing flow, and its dynamo action, in a cube. While this is by no means an astrophysically relevant geometry, one may naively expect the flow in a cube to resemble the flow in the largest sphere enclosed by the cube, with some dead water concentrated in the corners. The main hope is that there are some features common to precessing flows in all geometries, such as the appearance of triadic resonances, so that any geometry is useful as a model system.

The dynamo action of this flow depends quite sensitively, partly erratically, on the Reynolds and the Poincare numbers. Figure 5 illustrates the flow and the self-excited magnetic field for two different parameter sets. While the magnetic field for $\mathrm{Re} = 4 \times 10^3$ is concentrated at the boundaries (c), is appears to be wrapped around the central vortex for $\mathrm{Re} = 10^5$ (f).

3. Research related to magnetically triggered flow instabilities

We turn now to the field of magnetically triggered flow instabilities, and report the main experimental and theoretical results obtained within project A2.

3.1. Experimental demonstration of AMRI

After HMRI [22, 24] (and TI [30]) had been experimentally demonstrated prior to the start of the LIMTECH Alliance, the PROMISE facility was qualified for the experimental investigation of AMRI [31]. Since AMRI needs a minimum central current of about 10 kA (given the material parameters of GaInSn) the power supply for this current had to be enhanced to deliver 20 kA.

Figure 6 illustrates the dependence of various numerically and and experimentally obtained quantities on the central current (or the Hartmann number $\mathrm{Ha} = B_\phi(R_{\mathrm{in}}) R_{\mathrm{in}} (\sigma/(\rho\nu))^{1/2}$, where R_{in} is the inner radius of the Taylor-Couette cell, $B_\phi(R_{\mathrm{in}})$ the azimuthal field there, and ρ the density of the fluid). Figure 6a shows the growth rate obtained with a 1D stability code, for $\mathrm{Re} = 1480$, whose positive segment indicates the existence of AMRI in an interval approximately between 10 and 22 kA. Quite in correspondence with that, the lowermost curve ("Sim: ideal B") in Figure 6b shows the expected rms value of the axial velocity, under the assumption of a purely azimuthal applied magnetic field. But here is a subtlety: The curve with the highest

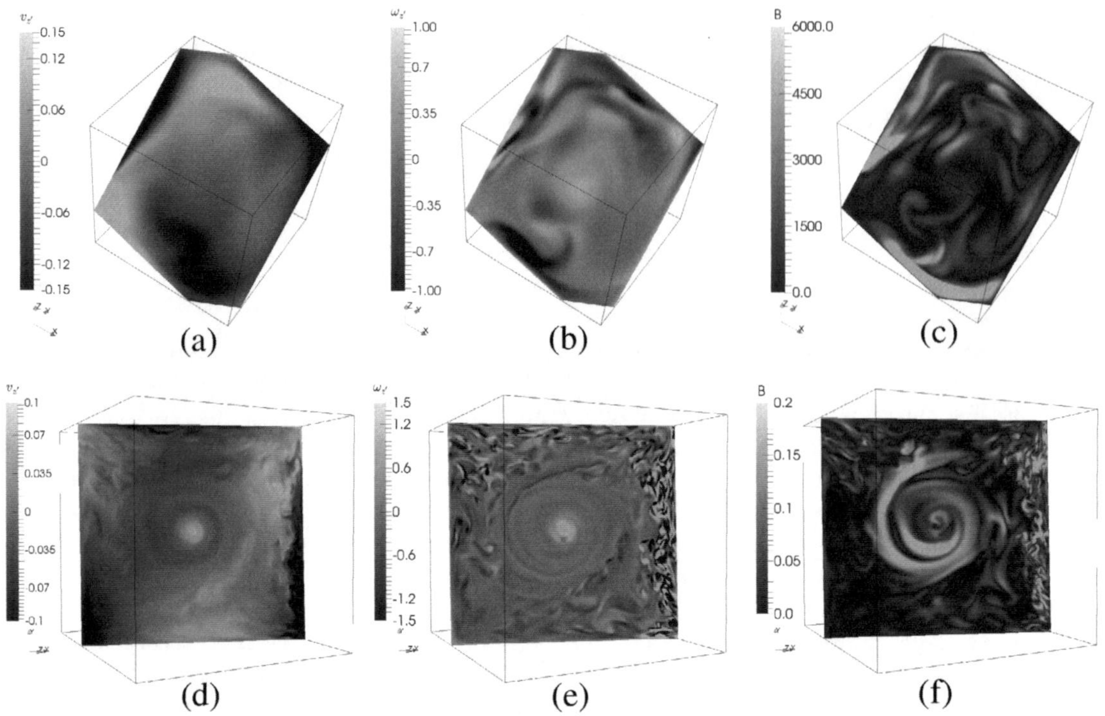

Figure 5. Two different kinds of dynamos in a precessing cube. Upper row: $\mathrm{Re} = 4\times 10^3, \overline{\mathrm{Po}} = -0.16$. Lower row: $\mathrm{Re} = 10^5, \overline{\mathrm{Po}} = -0.02$. Note the somewhat different definition of the Poincare number used here: $\overline{\mathrm{Po}} = \mathrm{Po}/(1+\mathrm{Po}\cos\alpha)$, with α denoting the angle between rotation and precession axis. The rotation axis is aligned with the z-axis, the precession axis lies at the time of visualization in the z-x-plane. Visualization of the velocity component parallel to the rotation axis of the fluid (a,d), of the total vorticity (b,e), and of the magnitude of the magnetic field (c,f) in the plane perpendicular to this rotation axis. The velocity is normalized with the product of the cube's edge lenght and the total angular frequency of the container about the z-axis. For more details see [62].

values ("Sim: real B") in Figure 6b gives the corresponding numerical rms values for the case that the real magnetic field of the experimental setting is assumed. This field deviates slightly from a pure B_ϕ, since the two leads from and to the power supply form a single-winding coil. Although the resulting deviation of the field is only in the region of a few per cent, the rms values are significantly enhanced, and show also a broadened range for AMRI. Interestingly, the experimental curve ("Experiment") shows a very similar shape, albeit with somewhat lower values.

The explanation of this phenomenon needs some peculiar symmetry considerations: Let us start with an infinitely long TC flow under the influence of a purely azimuthal magnetic field. As known from [25], this configuration has no preference for $m = 1$ or $m = -1$ modes, so that both together would be expected to form a standing wave. In a finite length TC cell, the Ekman pumping at the lids produces slight deviations from the perfect TC flow, which leads to some preference for either $m = 1$ or $m = -1$ modes in the upper and lower halves (see the simulation in Figure 6d). Adding now a second type of symmetry breaking, in form of an imperfect applied magnetic field, the original $m = \pm 1$ symmetry of the instability is to some extend restored. In other words, the upward and downward travelling modes that would be

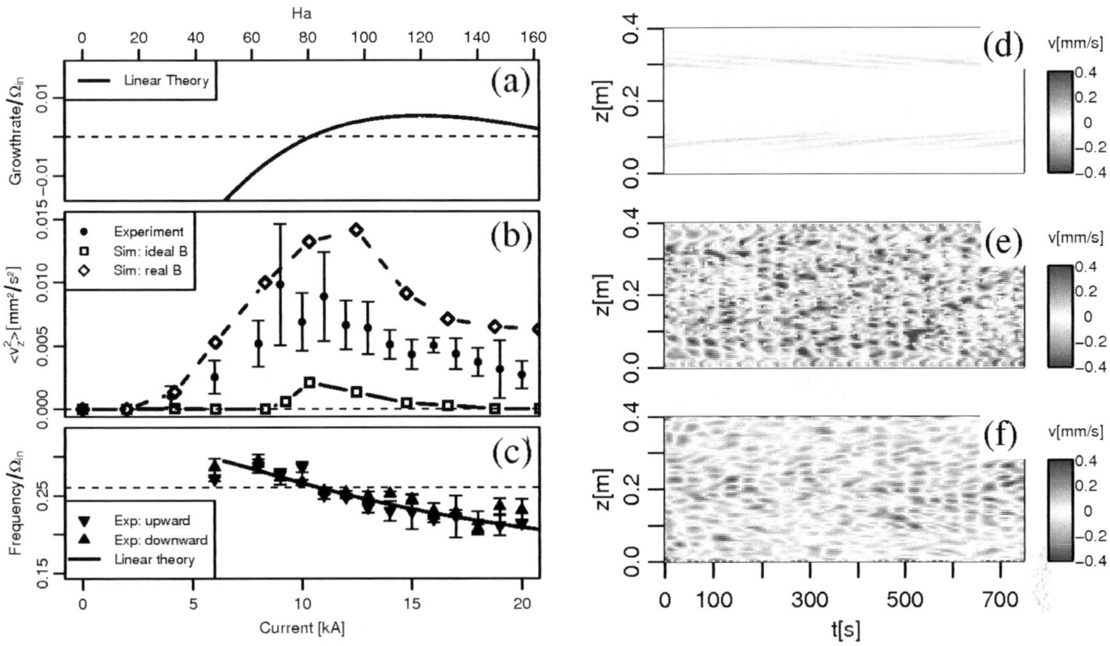

Figure 6. Results of the AMRI experiment. Left: Dependence of various quantities on Ha. (a) Numerically determined growth rate. (b) Mean squared velocity perturbation. (c) Angular drift frequency. In the frequency plot, "upward" and "downward" refer to the travel direction of the AMRI wave. Right: Velocity perturbation $v_z(m = 1, z, t)$ for $\mu := \Omega_{\rm out}/\Omega_{\rm in} = 0.26$, Re $= 1480$, and Ha $= 124$. (d) Simulation for ideal axisymmetric field. (e) Simulation for realistic field. (f) Experimental results. After [31].

expected to be concentrated in the upper and lower part, now interpenetrate each other (see simulation in Figure 6e) and populate now also the mid-height region, which effectively results in significantly enhanced rms values. This behaviour is indeed found in the experiment (Figure 6f).

While this simulated, and experimentally confirmed, effect of a double symmetry breaking on the AMRI is interesting in its own right, it was decided to improve the PROMISE facility in such a way that the azimuthal symmetry breaking is largely preserved. This is realized by a new system of wiring of the central current, comprising now a "pentagon" of 5 back-wires situated around the experiment. First experiments with this set-up show encouraging results, in particular transitions between AMRI and HMRI when adding an axial field to the azimuthal one.

3.2. The planned large-scale MRI/TI experiment

The second large-scale sodium experiment to be set-up within the DRESDYN project aims at investigating combinations of different versions of the MRI and the current-driven TI (see Figure 7). Basically, the set-up is designed as a TC experiment with 2 m fluid height, an inner radius $R_{\rm in} = 20$ cm and an outer radius $R_{\rm out} = 40$ cm. Rotating the inner cylinder at up to 20 Hz, we plan to reach Rm ~ 40, while the planned axial magnetic field $B_z = 120$ mT will correspond to a Lundquist number $S := {\rm Pm}^{1/2}{\rm Ha} \sim 8$. Both values are about twice the respective critical values for the onset of SMRI as they were derived in [64].

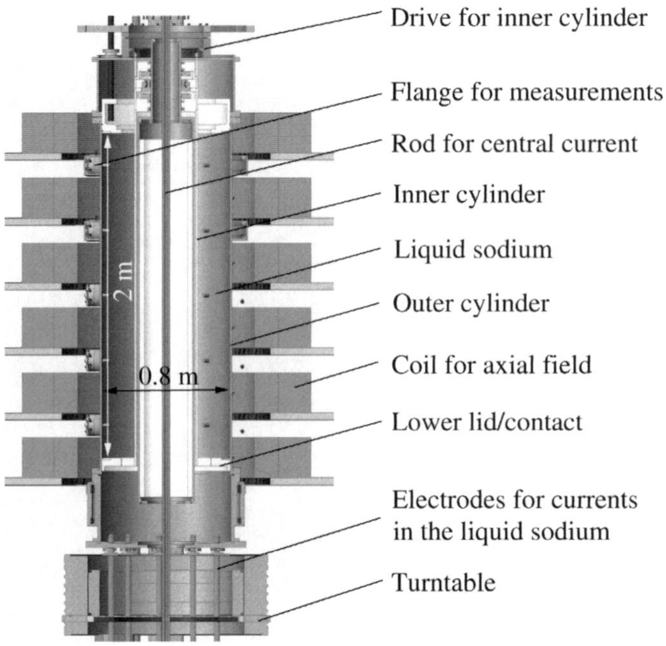

Figure 7. Design of the large-scale TC experiment for investigations of HMRI, AMRI, SMRI, TI and their combinations.

Below those critical values, we plan to investigate how HMRI approaches the limit of SMRI [20, 21]. To this end, we will use a strong central current, as it is already present in the PROMISE experiment [24, 31]. This insulated central current can be supplemented by another axial current guided through the (rotating) liquid sodium, which will further allow to investigate combinations of MRI and TI. Theoretical studies [65, 67, 68, 69] have shown that even a slight addition of current through the liquid extends the range of application of the helical and azimuthal MRI to Keplerian flow profiles.

3.3. Between HMRI, AMRI and TI: Some theoretical results

Shortly after the numerical revelation of HMRI by Hollerbach and Rüdiger [20], Liu et al. [71] had derived – in the framework of a short-wavelength (or WKB) approximation – two limits for the negative and positive shear between which HMRI should cease to exist. Expressed in terms of the Rossby number $\mathrm{Ro} \equiv r/(2\Omega)\partial\Omega/\partial r$, the lower Liu limit (LLL) $\mathrm{Ro}_{\mathrm{LLL}} = 2(1-\sqrt{2}) \approx -0.828$ has attracted a lot of interest [72, 73], in particular since it would prevent HMRI from working for astrophysically interesting Keplerian flows characterized by $\mathrm{Ro}_{\mathrm{Kepler}} = -0.75$. In contrast to this, the upper Liu limit $\mathrm{Ro}_{\mathrm{ULL}} = 2(1+\sqrt{2}) \approx 4.828$ for positive shear flows has been largely ignored, although positive shear flows are indeed relevant in astrophysics, for example in a $\pm 30^\circ$ strip of the solar tachocline. We note in passing that positive shear flows were usually considered perfectly stable (even under the action of vertical magnetic fields), so that Deguchi's recent discovery of a linear instability for very large Reynolds numbers came as a big surprise [74].

A first interesting result regarding the two Liu limits was published in [65]. This work dealt with the question, how HMRI (and AMRI) would be modified if the central current, which generates the azimuthal field component, was gradually complemented by some second axial current through the liquid. The respective weight of the two azimuthal field parts can be

IOP Conf. Series: Materials Science and Engineering **228** (2017) 012002 doi:10.1088/1757-899X/228/1/012002

quantified by the magnetic Rossby number $\mathrm{Rb} \equiv r^2/(2B_\phi)\partial(r^{-1}B_\phi)/\partial r$ which is constructed as a counterpart of Ro. A pure current-free field (as assumed for HMRI and AMRI [66]) gives $\mathrm{Rb} = -1$, while a pure current in the fluid (as for TI) corresponds to $\mathrm{Rb} = 0$. In [65], and more detailed in [67], it was shown that the LLL and the ULL are just the endpoints of one common stability curve (Figure 8) in the Ro − Rb plane which acquires the surprisingly simple analytical form

$$\mathrm{Rb} = -\frac{1}{8}\frac{(\mathrm{Ro}+2)^2}{\mathrm{Ro}+1}, \tag{1}$$

when both Re and Ha tend to infinity and either the ratio of azimuthal to axial field (for HMRI) or the shape of the perturbation (for AMRI) is optimized. The "metamorphosis" of the helical MRI, when changing the ratio of the axial currents within the fluid and on the axis, were also studied with a 1D stability code [68].

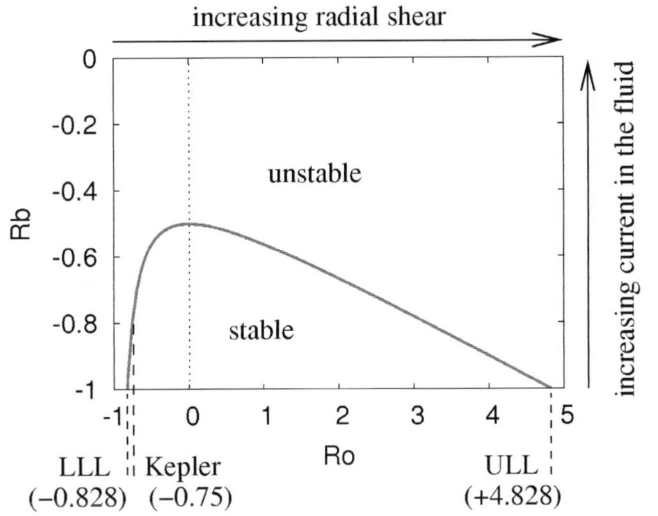

Figure 8. Stability chart in the Ro − Rb plane, for $\mathrm{Pm} = 0$ and Ha and Re tending to infinity. The Liu limits LLL and ULL apply only for $\mathrm{Rb} = -1$, while for $\mathrm{Rb} > -1$ shallower shear profiles can as well be destabilized (including Kepler rotation with $\mathrm{Ro} = -0.75$, starting at $\mathrm{Rb} = -0.78125$). The dotted line separates flows with negative shear (to the left) and positive shear (to the right).

The treatment of problems with variable Rb led also to the reconsideration of an old problem of magnetohydrodynamics, which is known as Chandrasekhar's theorem [75]. In the framework of ideal MHD this theorem states the stability of rotating flows of any radial dependence under the influence of an azimuthal magnetic field whose corresponding Alfvén velocity has the same amplitude and radial dependence as the rotation. Both in the WKB framework [67] and with a 1D-stability code [69] we showed that these Chandrasekhar solutions can be destabilized in non-ideal MHD.

We also mention the paper by Priede [70] that, treating a simplified pinch-type instability in a semi-infinite planar sheet of an inviscid incompressible liquid with a straight rigid edge, helped to clarify some differences between the instability mechanisms in highly resistive and well conducting fluids, resulting in different development times: magnetic response time for the former, and the much shorter Alfvén time for the latter.

3.4. Super-AMRI

Within the project, we have proved the existence of the upper Liu limit also for purely azimuthal magnetic fields, both in WKB approximation [76] and with a 1D stability code [77, 78]. The seed grant "Super-AMRI" was dedicated to find optimal parameters for a liquid metal TC experiment to show this "Super-AMRI", as we call it now. The main problem here is that such a TC experiment would need a rather thin gap in order to realize the enormous positive shear of Ro > 4.828, which leads to a very large critical value of the central current. For a liquid sodium experiment, both WKB [76] and 1D simulations with insulating boundary conditions [78] point to a critical current of about 80 kA, while a much lower (and experimentally more feasible) value of some 20 kA results for ideally conducting cylinder walls [78]. Figure 9 shows the stability maps in the Hartmann-Reynolds plane for $R_{\rm in}/R_{\rm out} = 0.75$ (a) and $R_{\rm in}/R_{\rm out} = 0.9$ (b) and two magnetic Prandtl numbers 10^{-5} (red) and 10^{-2} (blue), without (AMRI) and with (TI) current in the fluid, together with the simulated axial velocity component (c) for the case $R_{\rm in}/R_{\rm out} = 0.75$. The technically feasible combination of using sodium and copper walls is expected to lead to a critical current between 20 kA and 80 kA, but a detailed simulation for that case is still in progress.

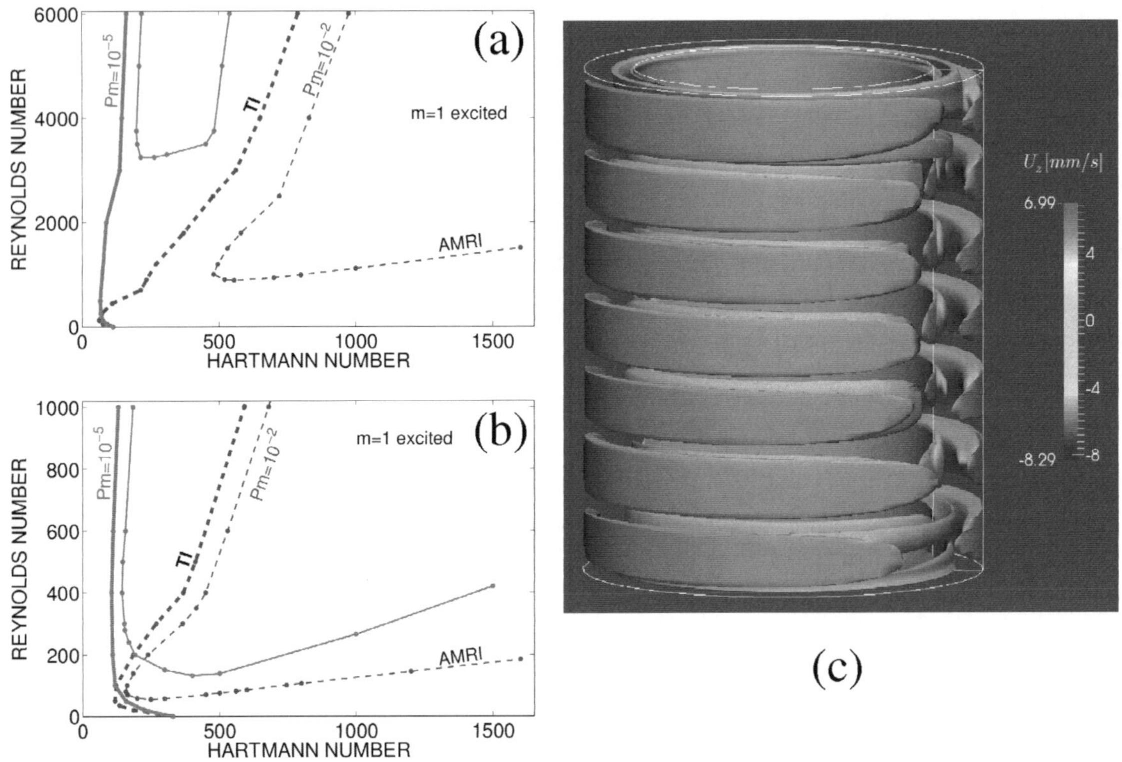

Figure 9. Azimuthal MRI for flows with positive shear (Super-AMRI): (a) Stability maps in the Hartmann-Reynolds plane for resting inner cylinder with $R_{\rm in}/R_{\rm out} = 0.75$ and two magnetic Prandtl numbers 10^{-5} (red) and 10^{-2} (blue), without (AMRI) and with (TI) current in the fluid. (b) The same for $R_{\rm in}/R_{\rm out} = 0.9$. (c) Simulated axial velocity component for $R_{\rm in}/R_{\rm out} = 0.75$.

Hence, it might be worthwhile to come back to the idea of "Super-HMRI", for which a 1D simulation is also pending. With view on the significant differences in the axial currents that are needed for the experimental observation of the negative-shear versions of HMRI (4 kA) and

AMRI (10 kA), we expect a similar reduction in the positive shear case that could reduce the technical efforts significantly.

3.5. Linking dissipation-induced instabilities and non-modal growth

While the two Liu limits for HMRI have been known for more than a decade [71], the physical reason behind them was never brought into question. In a recent paper [79], we have revealed a link between the modal growth *rate* of HMRI (from which the Liu limits follow) and the non-modal growth *factor* of the underlying purely hydrodynamical problem, which sheds some new light on the physical essence of both.

Non-modal (or transient) growth is typical for the time evolution of dynamical systems governed by non-normal operators. Due to the non-orthogonality of their eigenfunctions, an appropriately chosen initial state can experience a transient growth of its amplitude even if the (two or more) individual eigenfunctions of which it is composed have negative modal growth rates. Non-modal instabilities are known to play a key role in explaining the onset of turbulence in pipe flows [80].

As shown by Afshordy et al. [81], the non-modal growth factor for axisymmetric perturbations of purely hydrodynamic rotating flows can be expressed by the surprisingly simple equation $G = (1 + \mathrm{Ro})^{sgn(\mathrm{Ro})}$ which is illustrated by the red curve in Figure 10. Comparing this with the equation for the modal growth rate γ (green curve in Figure 10) of HMRI [67], we find the following link [79]:

$$\gamma = \frac{\mathrm{Ha}^2}{\mathrm{Re}} \left[\frac{(\mathrm{Ro}+2)^2}{8(\mathrm{Ro}+1)} - 1 \right] = \frac{\mathrm{Ha}^2}{\mathrm{Re}} \left[\frac{(G+1)^2}{8G} - 1 \right] . \tag{2}$$

For large $|Ro|$ this leads to a simple linear relation of γ and G: $\gamma \approx \mathrm{Ha}^2/\mathrm{Re}(G/8 - 3/4)$. Figure 10 illustrates this asymptotic connection. At the two Liu limits of HMRI, $\mathrm{Ro} = 2(1 \pm \sqrt{2})$, G acquires the same value $G = 1 + 2(1 + \sqrt{2}) \approx 5.828$.

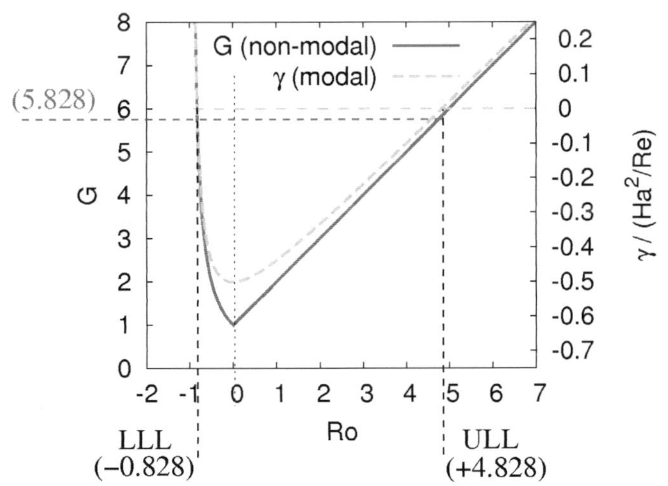

Figure 10. Non-modal growth factor G for purely hydrodynamic rotating flows (red curve), and normalized growth rate γ of helical MRI (green curve), in dependence on the Rossby number Ro. The two quantities are connected by Equation (2) and match asymptotically according to $\gamma \approx \mathrm{Ha}^2/\mathrm{Re}(G/8 - 3/4)$.

Keeping in mind that HMRI is a (double-)diffusive instability, we learn that it is inherently based on the non-modal growth of the underlying purely hydrodynamic flow. For a linear instability to become effective, it requires some dissipation to allow a coupling between different components of the eigenfunctions (actually between meridional and azimuthal flow perturbations, as shown in [82]). Once this coupling is established, the growth rate of the linear HMRI becomes (nearly) proportional to the non-modal growth factor of the underlying hydrodynamic flow.

It remains to be seen, whether similar links exist also in other fields of hydrodynamics. This is not obvious, as we have already experienced when analyzing the corresponding connection for AMRI [83] that is more subtle due to the shear-induced time-variation of the wavenumber of the $m = 1$ mode.

4. Summary and perspectives

Two large-scale liquid sodium experiments, on precession and on MRI/TI, are presently under construction at HZDR. For both of them, the project A2 of the LIMTECH Alliance has made theoretical and experimental contributions. The project has further supported re-commissioning of the Riga dynamo experiment which is now ready for studying new interesting back-reaction effects as well as for testing measurement techniques for the DRESDYN experiments. A particular achievement of the project is a deepened insight into the diffusive instabilities of rotating flows under the action of magnetic fields. As an outcome of those activities, the starting grant on "Super-AMRI" has delivered first parameter estimates for a new liquid sodium experiment on the magnetic destabilization of positive shear flows. Another result concerns the rigorous mathematical link between the non-normal growth factor of purely hydrodynamic rotating flows with the normal growth rate of the HMRI which gives – to the best of our knowledge – the first link between non-normal growth and diffusive (or dissipation-induced) instabilities. We should also mention here the recent astrophysical application of AMRI for explaining the angular momentum redistribution of post-main sequence low-mass stars [84].

Many problems related to the TI in liquid metals were investigated in close collaboration with the project B3 which was dedicated to applied problems of liquid metal batteries [85, 86, 87]. As one of the potentially dangerous instabilities, which would occur even if electrovortex flows and sloshing instabilities were completely suppressed, the TI was numerically treated with an OpenFOAM code enhanced by a Poisson solver for the electric potential and Biot-Savart's law for the induced magnetic field [85]. With this code it was possible to test various provisions for suppressing the TI and to understand its saturation mechanism [86]. An interesting by-product of these simulations was the detection of helicity oscillations [88] for small magnetic Prandtl numbers. Re-applying this result (obtained for liquid metal batteries) to a simple model of a Tayler-Spruit solar dynamo, we identified a mechanism that could empower the weak tidal forces of planets to synchronize the Hale cycle of the solar magnetic field [89, 90]. It is here where the synergies between basic research and liquid metal technologies, which were fostered by the LIMTECH Alliance, proved particularly useful.

Acknowledgments

This work was supported by Helmholtz-Gemeinschaft in frame of the LIMTECH Alliance. Close cooperation with Tom Weier and Norbert Weber (project B3) on various aspects of the Tayler instability is gratefully acknowledged.

References

[1] Jones C 2011 *Annu. Rev. Fluid Mech.* **43** 583
[2] Charbonneau P 2014 *Annu. Rev. Astron. Astrophys.* **52** 251
[3] Beck R 2009 *Astrophys. Space Sci. Trans.* **5** 43

[4] Balbus S A 2003 *Ann. Rev. Astron. Astrophys.* **41** 555
[5] Tayler R J 1973 *Mon. Not. R. Acad. Sci.* **161** 365
[6] Gailitis A, Lielausis O, Platacis E, Gerbeth G and Stefani F 2002 *Rev. Mod. Phys.* **74** 973
[7] Stefani F, Gailitis A and Gerbeth G 2008 *ZAMM* **88** 930
[8] Lathrop D P and Forest C B 2011 *Phys. Today* **64** 40
[9] Gailitis A et al 2000 *Phys. Rev. Lett.* **84** 4365
[10] Gailitis A et al 2001 *Phys. Rev. Lett.* **86** 3024
[11] Stieglitz R and Müller U 2000 *Phys. Fluids.* **13** 561
[12] Berhanu M et al 2010 *Eur. Phys. J. B* **77** 459
[13] Stefani F, Gerbeth G, Günther U and Xu M 2006 *Earth Planet. Sci. Lett.* **243** 828
[14] Sorriso-Valvo L, Stefani F, Carbone V, Nigro G, Lepreti F, Vecchio A and Veltri P 2007 *Phys. Earth Planet. Int.* **164** 197
[15] Petrelis F, Fauve S and Dormy E 2009 *Phys. Rev. Lett.* **102** 144503
[16] Benzi R and Pinton J F 2010 *Phys. Rev. Lett.* **105** 024501
[17] Sisan D et al 2004 *Phys. Rev. Lett.* **93** 114502
[18] Nornberg M D, Ji H, Schartman E, Roach A and Goodman J 2010 *Phys. Rev. Lett.* **104** 074501
[19] Roach A H, Spence E J, Gissinger C, Edlund E M, Sloboda P, Goodman J and Ji H 2012 *Phys. Rev. Lett.* **108** 154502
[20] Hollerbach R and Rüdiger G 2005 *Phys. Rev. Lett.* **95** 124501
[21] Kirillov O N and Stefani F 2010 *Astrophys. J.* **712** 52
[22] Stefani F, Gundrum T, Gerbeth G, Rüdiger G, Schultz M, Szklarski J and Hollerbach R 2006 *Phys. Rev. Lett.* **97** 184502
[23] Stefani F, Gundrum T, Gerbeth G, Rüdiger G, Szklarski J and Hollerbach R 2007 *New J. Phys.* **9** 295
[24] Stefani F, Gerbeth G, Gundrum T, Hollerbach R, Priede J, Rüdiger G and Szklarski J 2009 *Phys. Rev. E* **80** 066303
[25] Hollerbach R, Teeluck V and Rüdiger G 2010 *Phys. Rev. Lett.* **104** 044502
[26] Bergerson W F, Hannum D A, Hegna C C, Kendrick R D, Sarff J S and Forest C B 2006 *Phys. Rev. Lett.* **96** 015004
[27] Spruit H C 2002 *Astron. Astrophys.* **381** 923
[28] Gellert M, Rüdiger G and Elstner D 2008 *Astron. Astrophys.* **479** L33
[29] Rüdiger G, Schultz M and Gellert M 2011 *Astron. Nachr.* **332** 17
[30] Seilmayer M et al 2012 *Phys. Rev. Lett.* **108** 244501
[31] Seilmayer M et al 2014 *Phys. Rev. Lett.* **113** 024505
[32] Zimmermann D S, Triana S A, Nataf H-C and Lathrop D P 2010 *J. Geophys. Res. - Sol. Earth* **119** 4538
[33] Adams M M, Stone D R, Zimmerman D S and Lathrop D P 2015 *Prog. Earth Planet. Sci.* **29** 1
[34] Cooper C M et al 2014 *Phys. Plasmas* **21** 013505
[35] Weisberg D B, Peterson E, Milhone J, Endrizzi D, Cooper C, Désangles V, Khalzov I, Siller R and Forest C B *Phys. Plasmas* **24** 056502
[36] Stefani F, Eckert S, Gerbeth G, Giesecke A, Gundrum T, Steglich C, Weier T and Wustmann B 2012 *Magnetohydrodynamics* **48** 103
[37] Stefani F, Albrecht T, Gerbeth G, Giesecke A, Gundrum T, Herault J, Nore C and Steglich C 2015 *Magnetohydrodynamics* **51** 275
[38] Gailitis A, Gerbeth G, Gundrum T, Lielausis O, Platacis E and Stefani F *Comptes R. Phys.* **9** 721
[39] Tilgner A and Busse F H 2001 *Saturation mechanism in a model of the Karlsruhe dynamo* (Dynamo and Dynamics: A mathematic challenge) ed P Chossat, D Armbruster, I Oprea (Dordrecht: Springer) pp 109-116
[40] Rädler K-H and Brandenburg A 2003 *Phys. Rev. E* **67** 026401
[41] Giesecke A, Stefani F and Gerbeth G 2010 *Phys. Rev. Lett.* **104** 044503
[42] Giesecke A, Nore C, Stefani F, Gerbeth G, Léorat J, Herreman W, and Guermond J-L 2012 *New. J. Phys.* **14** 053005
[43] Nore C, Léorat J, Guermond J-L, 2015 *Phys. Rev. E* **91** 013008
[44] Nore C, Quiroz D C, Cappanera L, Guermond J-L 2016 *Europhys. Lett.* **114** 65002
[45] Stefani F, Gerbeth G and Gailitis A 1999 *Velocity profile optimization for the Riga dynamo experiment* (Transfer phenomena in magnetohydrodynamic and electroconducting flows) ed A Alemany, Ph Marty, J-P Marty (Dordrecht: Kluwer) pp 31-44
[46] Gailitis A, Lielausis O, Platacis E, Gerbeth G and Stefani F 2004 *Phys. Plasmas* **11** 2838
[47] Stefani F, Gailitis A and Gerbeth G 2011 *Astron. Nachr.* **332** 4
[48] Gailitis A and Lipsbergs G 2017 *Magnetohydrodynamics* **53** 349
[49] Nore C, Léorat L, Guermond J-L and Luddens F 2011 *Phys. Rev. E* **84** 016317

[50] Cappanera L, Guermond J-L, Léorat and Nore C 2016 *Phys. Rev. E* **93** 043113
[51] Krauze A 2010 *Magnetohydrodynamics* **46** 271
[52] Blackburn H M and Sherwin S J (2004) *J. Comp. Phys.* **197** 759
[53] Herault J, Gundrum T, Giesecke A and Stefani F 2015 *Phys. Fluids* **27** 124102
[54] Giesecke A, Albrecht T, Gundrum T, Herault J and Stefani F 2015 *New J. Phys.* **17** 113044
[55] Malkus W V R 1968 *Science* **160** 259
[56] Vanyo J P 2004 *Geophys. J. Int.* **158** 470
[57] Tilgner A 2005 *Phys. Fluids* **17** 034104
[58] Dwyer C A, Stevenson D J and Nimmo F 2011 *Nature* **479** 212
[59] Fu R R et al 2012 *Science* **338** 238
[60] Albrecht T, Blackburn H M, Lopez J M, Manasseh R and Meunier P 2015 *J Fluid Mech.* **778** R1
[61] Giesecke A, Albrecht T, Gerbeth G, Gundrum G and Stefani F 2015 *Magnetohydrodynamics* **51** 293
[62] Goepfert O and Tilgner A 2016 *New J. Phys.* **18** 103019
[63] Gans R 1970 *J. Fluid Mech.* **45** 111
[64] Rüdiger G and Shalybkov D 2003 *Phys. Rev. E* **67** 046312
[65] Kirillov O N and Stefani F 2013 *Phys. Rev. Lett.* **111** 061103
[66] Kirillov O N, Stefani F and Fukumoto Y 2012 *Astrophys. J.* **756** 83
[67] Kirillov O N, Stefani F and Fukumoto Y 2014 *J. Fluid Mech.* **760** 591
[68] Priede J 2015 *Phys. Rev. E* **91** 033014
[69] Rüdiger G, Schultz M, Stefani F and Mond M 2015 *Astrophys. J.* **811** 84
[70] Priede J 2017 *J. Fluid Mech.* **816** 705
[71] Liu W, Goodman J and Ji H 2006 *Astrophys. J.* **643** 306
[72] Rüdiger G and Hollerbach R 2007 *Phys. Rev. E* **76** 068301
[73] Priede J and Gerbeth G 2009 *Phys. Rev. E* **79** 046310
[74] Deguchi K 2017 *Phys. Rev. E* **95** 021102(R)
[75] Chandrasekhar S. 1956 *PNAS* **42** 273
[76] Stefani F and Kirillov O N 2015 *Phys. Rev E* **92** 051001(R)
[77] Rüdiger G, Schultz M, Gellert M and Stefani F 2016 *Phys. Fluids* **28** 014105
[78] Rüdiger G, Gellert M, Schultz M and Stefani F 2016 arXiv:1601.01964
[79] Mamatsashvili G and Stefani F 2016 *Phys. Rev. E* **94** 051203
[80] Trefethen A E, Trefethen L N and Schmid P J 1999 *Comput. Methods Appl. Mech. Engrg.* **1926** 413
[81] Afshordi N, Mukhopadhyay P and Narayan R 2005 *Astrophys. J.* **629** 373
[82] Priede J, Grants I and Gerbeth G 2007 *Phys. Rev. E* **75** 047303
[83] Mamatsashvili G and Stefani F 2017 *Magnetohydrodynamics* **53** 107
[84] Spada F, Gellert M, Arlt R and Deheuvels S 2016 *Astron. Astrophys.* **589** A23
[85] Weber N, Galindo V, Stefani F, Weier T and Wondrak T 2013 *New J. Phys.* **15** 043034
[86] Weber N, Galindo V, Stefani F and Weier T 2014 *J. Power Sources* **265** 166
[87] Weber N, Beckstein P, Herreman W, Horstmann G M, Nore C, Stefani F and Weier T 2017 *Phys. Fluids* **29** 054101
[88] Weber N, Galindo V, Stefani S, Weier T *New J. Phys.* **17** 113013
[89] Stefani F, Giesecke A, Weber N and Weier T 2016 *Solar Phys.* **291** 2197
[90] Stefani F, Galindo A, Giesecke A, Weber N and Weier T 2017 *Magnetohydrodynamics* **53** 169

Magnetically induced instabilities in duct flows

L Bühler[1], T Arlt[1], T Boeck[2], L Braiden[3], V Chowdhury[1], D Krasnov[2], C Mistrangelo[1], S Molokov[3], J Priede[3]

[1]Karlsruhe Institute of Technology, Postfach 3640, 76021 Karlsruhe
[2] Technische Universität Ilmenau, Postfach 100565, 98684 Ilmenau, Germany
[3]Applied Mathematics Research Centre, Coventry University, Coventry CV1 5FB, UK

leo.buehler@kit.edu

Abstract. The occurrence of magnetically induced instability in magnetohydrodynamic duct flows is studied for Hunt flow, where one pair of walls parallel to the magnetic field is electrically insulating and the Hartmann walls perpendicular to the field are electrically conducting. The onset of time-dependent flow patterns and their intensity depends on the strength of the magnetic field and on the flow rate in terms of the Hartmann and Reynolds numbers, respectively. The problem is studied by a complementary approach using laboratory experiments, linear stability analysis and high-resolution direct numerical simulations.

1. Introduction

Liquid metal flows in channels with arbitrary wall electric conductivity are of importance for engineering applications such as liquid metal blankets for nuclear fusion reactors with magnetic confinement of the plasma. Under the action of strong magnetic fields, liquid metal flows in electrically conducting ducts exhibit jet-like velocity profiles in the thin boundary layers along walls parallel to the magnetic field. The velocity in these so-called side layers may exceed several times the mean velocity in the channel core, depending on magnetic field strength and electrical conductance of the walls. This kind of MHD flow received attention in recent years for validation of numerical tools [1] and it is of fundamental interest since the jet-like velocity distributions are potentially unstable. It is known that the flow in these layers becomes unstable already at relatively small Reynolds numbers [2]. A linear stability analysis [2] predicts a critical Reynolds number for the onset of instabilities, which is almost one order of magnitude smaller than the one obtained experimentally [3] [4]. The discrepancy between theory and experiments was attributed in the latter references by the need of a non-linear stability analysis, or by the influence on the experimental data of 3D effects at the entrance of the magnet. However, a conclusive explanation has not been found yet.

Numerical and experimental investigations of instabilities in parallel boundary layers have received increasing attention in the last years due to their significant influence on heat and mass transfer in liquid metal duct flows and therefore on the performance of liquid metal devices and blankets for fusion reactors. It has been shown that an applied magnetic field tends to damp three-dimensional flow perturbations and supports the formation of two dimensional structures aligned with the magnetic field. The electromagnetic damping of instabilities is exploited e.g. by electromagnetic brakes in metallurgical casting processes.

The present work focuses on the study of the onset of time-dependent instabilities in liquid metal magnetohydrodynamic (MHD) duct flows and the transition of unstable flow patterns to fully developed

IOP Conf. Series: Materials Science and Engineering **228** (2017) 012003 doi:10.1088/1757-899X/228/1/012003

MHD turbulence. MHD flows in ducts with electrically conducting Hartmann walls and insulating side walls exhibit highest velocities in side layer jets (strong destabilisation) while Joule dissipation due to the conducting Hartmann walls (damping) should stabilise the flow. From this point of view this so-called Hunt flow [5] constitutes an ideal prototype of magnetically induced instabilities for fundamental research of the influence of these counteracting mechanisms. A sketch of fully established MHD flows in conducting rectangular ducts is shown in Figure 1.

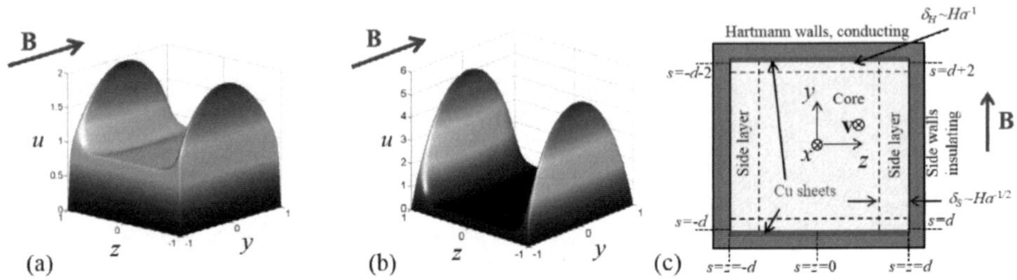

Figure 1. Fully developed velocity profiles in rectangular ducts with all walls conducting (a) and conducting Hartmann walls and insulating side walls (b), sketch of geometry and coordinates used for Hunt flow (c).

2. Problem description

We consider the stability of fully developed laminar duct flow of an electrically conducting incompressible viscous fluid under the influence of a uniform constant magnetic field. The flow is characterized by three non-dimensional groups, Hartmann number, Reynolds number, and wall conductance ratio,

$$Ha = LB\sqrt{\frac{\sigma}{\rho\nu}}, \quad Re = \frac{u_0 L}{\nu}, \quad \text{and } c = \frac{\sigma_w t_w}{\sigma L},$$

where L is a typical length scale of the problem, B the magnitude of the applied magnetic field, u_0 the average velocity, and ρ, ν, and σ, stand for the thermophysical properties of the fluid, its density, kinematic viscosity, and electric conductivity. Here, Ha is a dimensionless measure for the strength of the magnetic field, Re for velocity and c quantifies the conductance of the wall compared to that of the fluid. Electrically conducting walls are characterized by their electric conductivity σ_w and wall thickness t_w.

The flow is described by the balance of momentum and conservation of mass

$$Re\left(\frac{\partial \mathbf{u}}{\partial t} + (\mathbf{u}\cdot\nabla)\mathbf{u}\right) = -\nabla p + \nabla^2\mathbf{u} + Ha^2\mathbf{j}\times\mathbf{B}, \tag{1}$$

$$\nabla\cdot\mathbf{u} = 0, \tag{2}$$

by Ohm's law and an equation for electric potential

$$\mathbf{j} = -\nabla\phi + \mathbf{u}\times\mathbf{B}, \tag{3}$$

$$\nabla^2\phi = \nabla\cdot(\mathbf{u}\times\mathbf{B}). \tag{4}$$

The latter ensures conservation of charge, $\nabla\cdot\mathbf{j}=0$. Here $\mathbf{B}=\hat{\mathbf{y}}$, $\mathbf{u}$, $\mathbf{j}$, p, and ϕ stand for the magnetic field, velocity, current density, pressure and electric potential, scaled by B, u_0, $\sigma u_0 B$, $\rho\nu u_0/L$ and $u_0 BL$, respectively. The magnetic field is assumed constant and not modified by the flow (low R_m limit).

For conducting walls, equations (3) have to be solved also inside the wall with conditions at the fluid wall interface, $\mathbf{u} = 0$, $\phi = \phi_w$, $\mathbf{j}\cdot\mathbf{n} = \mathbf{j}_w\cdot\mathbf{n}$. The external surface of the wall is insulating $\mathbf{j}_w\cdot\mathbf{n}=0$. For thin walls, the potential equation for the wall may be reduced to the so-called thin-wall condition that applies at the fluid-wall interface

$$\mathbf{j}\cdot\mathbf{n} = \nabla_w \cdot \left(c\nabla_w \phi_w\right), \tag{5}$$

where $\mathbf{n}$ denotes the inward unit normal and ∇_w stands for the projection of the gradient operator on the interface.

The geometry and coordinates used to describe Hunt flow in the present paper, are shown in Figure 1. Here x denotes the streamwise coordinate, the magnetic field is parallel to the non-conducting side walls and aligned with y. In the sketch all dimensions are scaled with the Hartmann length, i.e. the duct has nondimensional extensions $-1\leq y\leq 1$ and $-d\leq z\leq d$, where d stands for the aspect ratio. Electric potential data on the fluid-wall interface is discussed below along a circumferential coordinate s as shown in the sketch.

For strong magnetic fields ($Ha\gg 1$) the flow exhibits a core with essentially uniform velocity and thin Hartmann layers of thickness $\delta_H \sim Ha^{-1}$ in which the velocity drops to zero at the Hartmann walls. Side layers have a thickness $\delta_s \sim Ha^{-1/2}$ in which the high-velocity jets are located.

The stability of Hunt flow and its transition to time-dependent and eventually fully turbulent conditions has been analyzed for various combinations of Ha and Re, for different aspect ratios d, and for different wall conductivities c. Results have been obtained by experiments performed in the MEKKA laboratory at the Karlsruhe Institute of Technology, by a linear stability analysis and numerical simulations in the weakly nonlinear regime, and by high-resolution direct numerical simulations for states that might approach full turbulence.

3. Experiments

The design of the employed liquid metal loop and details of the test section are shown in Figure 2. Two lateral ducts serve as electromagnetic conduction pumps that feed the test section in the middle. This double loop configuration aims at realizing the most symmetric velocity distribution at the entrance of the central duct [6]. The entire loop is exposed to a uniform magnetic field (no 3D perturbing effects since the flow is completely inside the magnet), that yields Hartmann numbers up to Ha=2700. With a current of 2000A in both pumps it is possible to reach Reynolds numbers up to $Re\approx 6\cdot 10^4$.

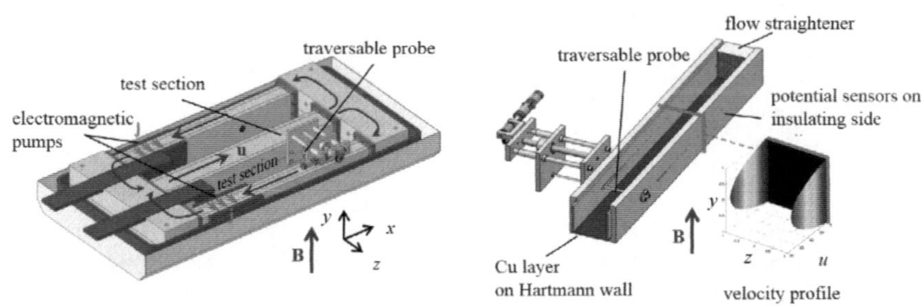

Figure 2. Liquid metal double loop with electromagnetic pumps, central test section, and details of the test section with copper layers on Hartmann walls. The entire loop is placed in a uniform magnetic field.

The test section has two electrically insulating walls parallel to the magnetic field. Electrically conducting Hartmann walls at a distance $2L$=60mm are realized by lamination of non-conducting walls with thin foils of copper (t_w=50μm) yielding c=0.03. Distribution of electric potential along the side-walls and along one Hartmann wall is measured at a number of electrodes. A traversable probe is used to measure profiles of transverse potential gradient within the fluid at y=0 for $-d<z<d$ to estimate the

axial velocity $u(y=0,z)\approx\Delta\phi/\Delta z$ between the tips of the probe over their distance Δz=1.6mm. The distance between flow straightener at the entrance and the instrumented cross-section is about 15L which should be long enough that the flow becomes fully developed far upstream of the instrumented cross-section [7] [6]. GaInSn is used as model fluid so that experiments can be performed at room temperature. Good electrical contact between fluid and conducting Hartmann walls has been achieved by first cleaning the metallic surfaces using hydrochloric acid and by rubbing GaInSn into the copper [8]. This procedure avoids contact resistance between the fluid and the copper surface and yields consistent results [9].

Potential signals consist of a mean part, a fluctuating part and noise, $\Delta\phi=\Delta\overline{\phi}+\Delta\phi'$+noise. The mean potentials on the duct walls are recorded by a Beckhoff KL3312 system and potentials at the tips of the traversable probe by a Prema 8017 nanovolt meter with resolution of 100nV. A multichannel, in-house developed system with an amplification factor of 5000 has been used to investigate time-dependent signals of potential differences of the order of 10^{-6} V. After a moderate pre-amplification an analog high-pass filter removes the mean part $\Delta\overline{\phi}$ with a cut-off frequency of 2Hz, then the time-dependent signals are strongly amplified, and a digital low-pass filter with a cut-off frequency of 95Hz is applied to remove high-frequency noise.

Instability of Hunt-type flow was investigated experimentally first for square ducts with aspect ratio d=1 for Hartmann numbers $500 \le Ha \le 2000$ and Reynolds numbers $Re \lesssim 30000$. One example of potential and velocity data is shown in Figure 3. We observe a linear variation of ϕ along the Hartmann wall and parabolic-like behaviour along the side walls. The agreement of measured wall potential data (symbols) with results of an exact analytical solution [10] is perfect for small Re. For higher Re, data of potential at the side walls shows slightly smaller magnitude which is a clear indication that the flow rates carried by the jects become smaller as a result of unsteady or eventually turbulent flow in the side layers. This is confirmed by measurements of potential gradients at the movable probe as indication of velocity $u(y=0,z)\approx\Delta\phi/\Delta z$. Results shown in Figure 3 agree again well with the laminar analytical predictions for moderate Re. Instabilities at Re=15637 lead to an increased momentum transfer, reduction of maximum velocity in the jets and thickening of the layers.

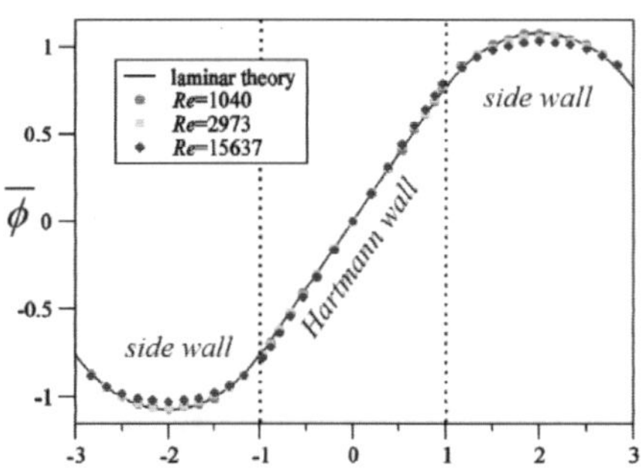

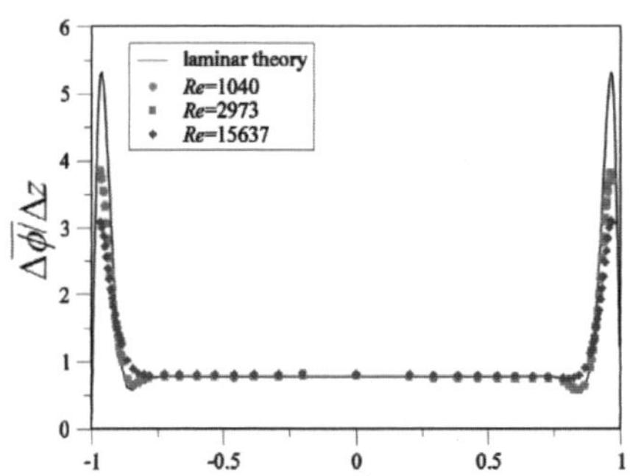

Figure 3. Mean potential distributions along the wall of a square duct and velocity profiles $\overline{u}(y=0,z)\approx\Delta\overline{\phi}/\Delta z$ for Ha=1000.

A systematic investigation for the onset of instabilities was performed in the following way: for a given strength of the magnetic field, i.e. for fixed Ha, the flow rate in terms of Re was increased while potential signals obtained at the probe and from walls were recorded. The onset of unstable flow i.e. the critical Reynolds number Re_c was obtained when measured potential fluctuations clearly emerge from the electromagnetic noise of the measuring circuit.

Values of critical Reynolds numbers obtained by this method are shown in Figure 4. We observe first an apparently *stable laminar* regime when $Re<Re_{c,I}$. Above $Re_{c,I}$ the flow is time-dependent and perturbations have measurable amplitude. These instabilities are located close to the side walls while

the core of the flow is practically still not affected by the instability. This flow regime is indicated in the figure as *unstable regime I*. Increasing *Re* further, leads to the sudden occurrence of perturbations with amplitudes that are one or two orders of magnitude larger. This transition into another *unstable regime II* is also indicated in the diagram. In *unstable regime II* the core flow is also affected and perturbations become visible on the Hartmann walls. The *unstable regime 0* was detected only for higher *Ha*. Here, instabilities occur at much smaller values of *Re*. By increasing *Re* in this regime the flow becomes apparently laminar again before a transition to *unstable regime I* occurs.

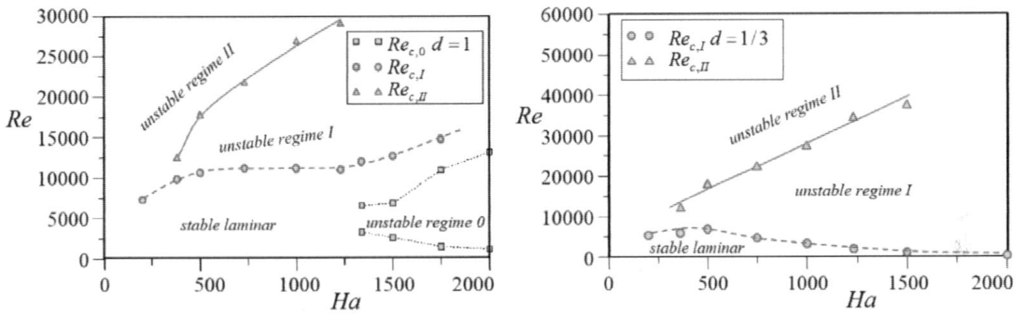

Figure 4. Map of flow regimes as a function of *Ha* for a square duct *d*=1 and for aspect ratio *d*=1/3.

Time-dependent potential gradient data has been recorded simultaneously along a line on a side wall. Two particular results are shown as an example in Figure 5. The coordinate *t* stands for time, scaled by L/u_0. Alternatively, assuming Taylor's hypothesis, we may interpret *t* as axial coordinate $x=-ct$, where *c* stands for the velocity at which perturbations move downstream. For *Re*=17506 we observe small-scale structures. Transition to large-scale patterns and higher modes along *y* are observed at *Re*=26599.

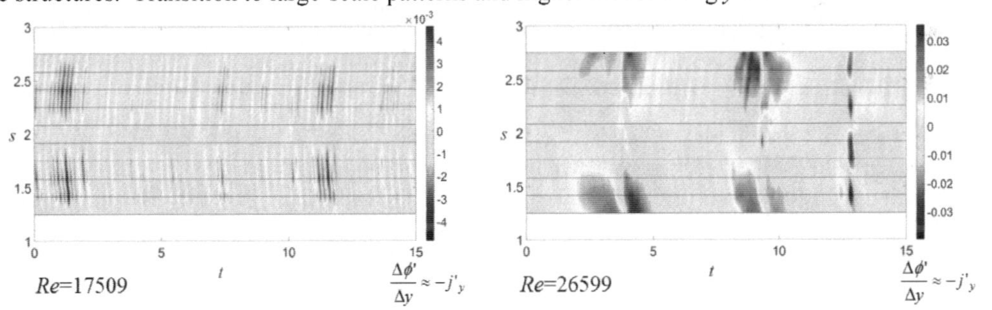

Figure 5. Data of $\Delta\phi'/\Delta y(y,t)$ along the side wall at z=1 for *Ha*=1000, showing small-scale structures for *Re*=17506 and transition to large-scale patterns and higher modes along *s* (i.e. along *y*) at *Re*=26599 and aspect ratio *d*=1.

In a second series of experiments, the aspect ratio of the duct was changed to $d=1/3$ and measurements were performed for Hartmann numbers $500 \leq Ha \leq 2000$ and Reynolds numbers $500 \lesssim Re \lesssim 37000$. Time averaged results for the velocity distribution in transverse direction are presented in Figure 6 for two values of *Ha* and various *Re*. For the smallest *Re*, good agreement between measurements and laminar theory can be observed. With increasing *Re* the maximum jet velocity decreases and the jets become thicker which indicates unstable flow. This decrease of velocity occurs for *Ha*=500 in both jets, causing a symmetric velocity profile. For *Ha*=2000 the velocity profile is non-symmetric for $Re \geq 5054$. As a result the jet at $z=0.3$ becomes unstable much earlier, while the other one at $z=-0.3$ does not show any thickening until $Re=25782$. A map of flow regimes for *d*=1/3 showing transitions between *stable laminar* flow, unstable flow with very small fluctuations (*unstable regime I*), and unstable flow with large amplitude (*unstable regime II*) is displayed in Figure 4. Further results and more details can be found in [11], [12]. Finally it is worth to notice that even when the flow is strongly time-dependent, the pressure drop measured in the experiment still follows the laminar predictions [12].

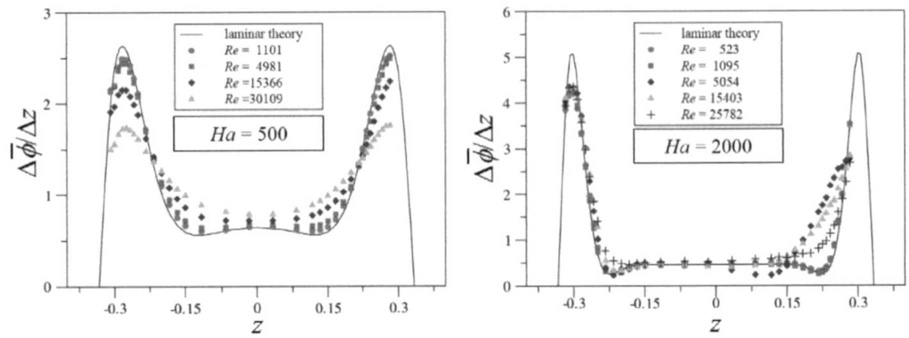

Figure 6. Mean velocity profiles $\bar{u}(y=0,z)\approx\Delta\bar{\phi}/\Delta z$ for Hunt flow with aspect ratio d=1/3.

4. Stability analysis and weakly nonlinear numerical simulations

A linear stability analysis was performed to determine theoretically the critical Reynolds number for the onset of instabilities. Direct numerical simulations were conducted for confirmation of the analysis and for further investigation of the nonlinear unstable flow behaviour in a certain range above the stability limit.

For the linear stability analysis pressure is eliminated by taking the curl of (1), the vorticity $\omega=\nabla\times\mathbf{v}$ is introduced, and the velocity is defined through a vector potential ψ as $\mathbf{v}=\nabla\times\psi$ that satisfies identically mass conservation (2). The equations are linearized assuming a fully developed steady state with infinitesimally small perturbations of all variables that are assumed to have an axial wave number k.

$$(\omega,\psi,\phi)(x,y,z,t)=(\bar{\omega},\bar{\psi},\bar{\phi})(y,z)+(\hat{\omega},\hat{\psi},\hat{\phi})(y,z)\exp(ik(x-st)). \quad (6)$$

The real part of the complex variable s denotes the phase velocity and the imaginary part vanishes at marginal stability. The resulting eigenvalue problem is obtained by discretization with a Chebychev collocation method and solved using the Fortran LAPACK package. For efficient performance of the solution method, the problem is considered only in a quarter of the duct assuming appropriate symmetry conditions as shown in Table 1. A detailed formulation of the problem and results can be found in [13], [14] and [15].

mode	*y*-symmetry	*z*-symmetry
I	even	even
II	odd	even
III	even	odd
IV	odd	odd

Table 1: Symmetries of vorticity perturbations in magnetic field direction [13]

The results presented in the following have been obtained for Hunt flow in ducts with Hartmann walls of arbitrary conductivity c. Figure 7 shows the critical Reynolds number Re_c as a function of Ha for the least stable modes for several values of c. Increasing the magnetic field for well conducting walls (c=1) has first a destabilizing effect at small values of Ha. After passing a local minimum (e.g. $Re_c \approx$ 1700 at $Ha \approx 11$ for $c = 1$), Re_c increases until mode I becomes the most unstable one (e.g. at $Ha \approx 46$ for $c = 1$). This mode shows strong decrease of Re_c with increasing Ha. If $cHa \gg 1$, Re_c shows only weak dependence on Ha. A similar behavior has been reported from previous studies [2] [13]. The dependence of the critical wavenumber k_c on *Ha* is shown in Figure 7. At small *Ha*, k_c shows only moderate variation with *Ha*. But for higher Hartmann numbers, when mode I becomes most unstable, the critical wavenumber k_c increases as $k_c \sim Ha^{1/2}$ for $Ha \to \infty$. This shows that the wave length $\lambda_c = 2\pi/k_c \sim Ha^{-1/2} \sim \delta_s$ becomes proportional to the thickness of the side layers, which is a clear indication that the instability is triggered by the physics in the side layers.

Analyses for different aspect ratios have been also performed. A presentation of all results is out of scope of the present paper. Further details can be found in [14]

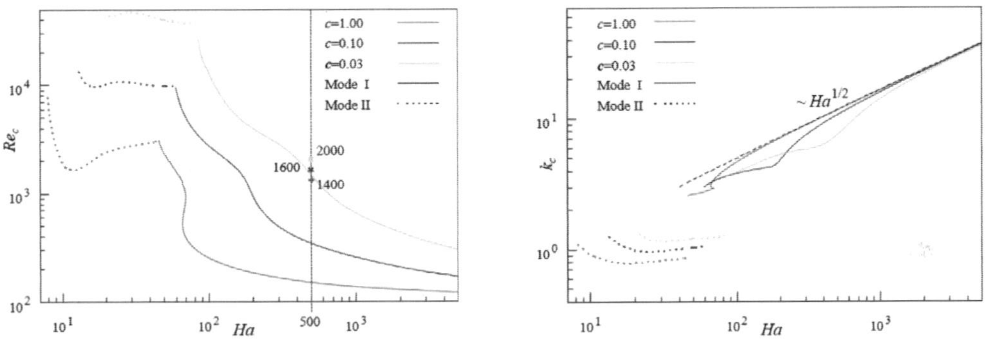

Figure 7. Hunt flow in a square duct. Critical Reynolds number Re_c and wave number k_c as a function of *Ha* for different wall conductance ratios *c*. Symbols at *Ha*=500 in left figure denote parameters used later in numerical simulations.

Results obtained by the linear stability analysis have been verified with numerical simulations for parameters close to the critical Reynolds number $Re_c(Ha)$ using OpenFoam [16]. A wall conductance ratio *c*=0.03, as present in the experiment, was chosen. For *Ha*=500 the linear stability analysis yields $Re_c \approx 1490$. A number of numerical simulations were performed for $Re<Re_c$ and for $Re>Re_c$ as indicated by symbols in Figure 7. For $Re=1400<Re_c$ the flow is stable and stationary with no variations of variables along the streamwise direction. This can be seen from Figure 9 in which the axial component of velocity near the side wall at *y*=0, *z*=0.96 is displayed. The agreement with the exact analytical solution is very good and deviations are smaller than 0.3%. When the Reynolds number is increased to $Re=1600>Re_c$ one can observe small wavy perturbations (Figure 9), which confirms that the flow is already unstable. However, the magnitude of perturbations is still very small and the maximum value of turbulent kinetic energy is $E_{kin} = \overline{u'^2} = O(10^{-6})$. Patterns of axial velocity perturbations *u'* at *Ha*=500 obtained at the stability threshold by the linear theory and at slightly supercritical *Re* by OpenFOAM are quite similar as displayed in Figure 8.

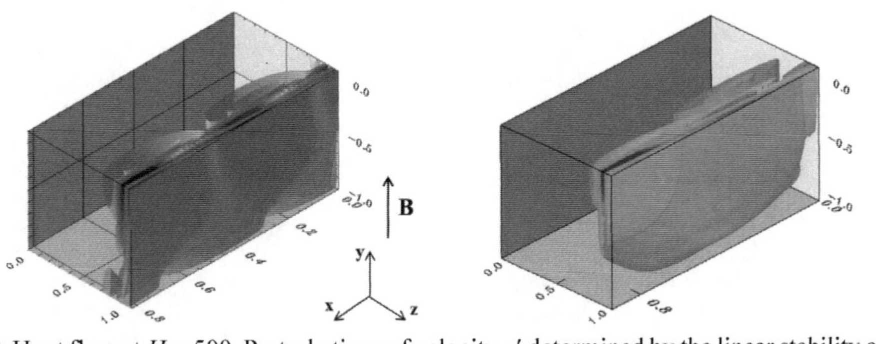

Figure 8. Hunt flow at *Ha*=500. Perturbations of velocity *u'* determined by the linear stability analysis at $Re_c \approx 1500$ (left) and OpenFOAM (right) at *Re*=1600. Only a quarter of the duct is shown.

When the Reynolds number is further increased up to Re=2000 the magnitude of perturbations becomes much stronger (Figure 9), and E_{kin} reaches values up to $E_{kin}=O(10^{-1})$. Finally we may conclude that both, the linear stability analysis and the numerical simulations, predict the onset of time-dependent motion at nearly the same critical Reynolds number. In the experiment, however, first occurrence of instabilities at Ha=500 could be detected only for $Re>10^4$. This apparent disagreement might be explained by the fact that the used experimental hardware has difficulties to detect small perturbations with frequencies below 2Hz while the numerical predictions and the stability analysis yield values close to 1.4Hz at the onset of instability.

Perturbations of potential gradients $\partial_y\phi'$ are visualized on the side wall in Figure 10 for Ha=500 and Re=2000. Results obtained in experiments show qualitative agreement (Figure 5) [12].

The present paper can give only a brief overview about the stability analysis and weakly nonlinear numerical simulations. Further results and details can be found e.g. in [14], [15], [17].

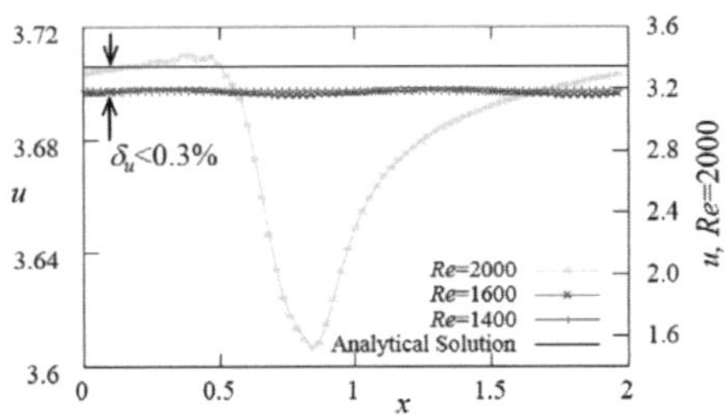

Figure 9. Numerical results for streamwise velocity $u(x,z=0.96)$ for different Re at Ha=500. Note: The axis for Re=2000 is shown at the right side.

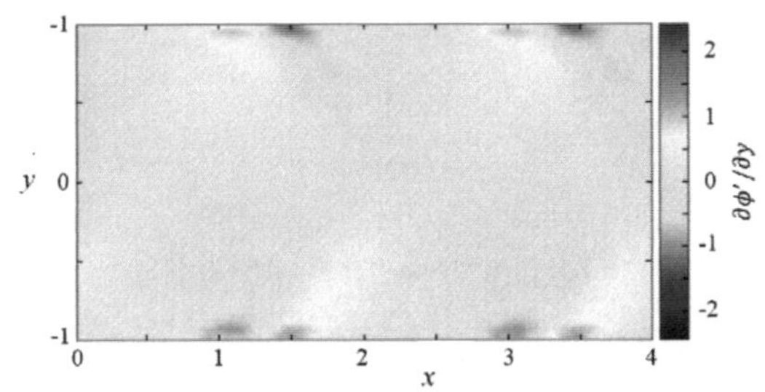

Figure 10. Numerical results for $\partial_y\phi'$ at the side wall for Re=2000 at Ha=500.

5. High-resolution direct numerical simulations

The linear stability analyses and numerical simulations with OpenFOAM focused on the onset of instabilities and the behavior at fairly low Reynolds numbers. Direct numerical simulations (DNS) at high Reynolds numbers approaching those realized in experiments have been performed with a finite-difference code for the rectangular duct geometry [18]. The computational grid is based on non-equidistant grid spacing which can be clustered near walls by one-dimensional mappings for the y- and z-directions. Thanks to this structured grid layout, the code is parallelized by one-dimensional domain decomposition, and the Poisson equations for pressure and electric potential are solved very efficiently by fast Fourier transforms and the two-dimensional cyclic reduction method implemented in the FISHPACK library [19]. For this reason, numerical resolutions of up to 2048×384×384 grid points could be realized.

DNS were performed for the ideal Hunt flow with perfectly conduction Hartmann walls and insulating side walls. The computational domain is assumed periodic in the streamwise direction. Relatively low values of the Hartmann number are considered first, and the changes of the flow with the Reynolds number Re are explored. This is done by systematically increasing or decreasing Re between successive simulations.

For the moderate magnetic fields, i.e. for Ha=100, uniformly spaced columnar vortices appear at the side walls as Re is increased to 500 from the laminar state (see [20]). For Re=1000 these vortices become larger and their spacing wider. When Re is increased further to 1400, large vortical structures appear in localized patches with leading and trailing vortices of smaller size. In all these cases, the flow appears symmetric with respect to the horizontal mid-plane. Concerning the left-right symmetry, the vortices form a staggered pattern, which maintains a constant phase shift. For slightly higher Re of about 1600 the regular vortical structures are replaced by so-called jet detachments, where the jets are lifted away from the wall into the core region. By that, strong perturbations are generated which also affect the opposing side layers and lead to similar detachments there. The detachments are associated with a significant mean turbulent kinetic energy (TKE) of up to 10% of the energy of the mean flow. By contrast, the columnar vortices attached to the side walls at lower *Re* are generally quite weak with a mean TKE below 0.1%. When the Reynolds number is increased further, the mean TKE decreases slowly, and the jet detachments become less pronounced. At the highest *Re*=10000 considered for *Ha*=100, the side wall regions exhibit turbulent bands with small-scale vortices, and the core region fluctuates less than in the presence of detachments. When the Reynolds number is systematically decreased from such a high value, the jet detachments reappear and increase in strength down to *Re*=2000. However, detachments do not disappear when *Re* is further reduced. Instead, they persist down to *Re*=200 although the mean TKE is gradually decreasing. This hysteresis indicates a complex behavior with a multiplicity of different states. In this respect, the behavior resembles subcritical transition in certain other shear flows although the linear stability is presumably supercritical in the present case.

Flows at higher Hartmann numbers up to *Ha*=1000 have been also explored for increasing *Re*. Different states are shown in Figure 11 as instantaneous velocity snapshots, visualized in the horizontal plane *y*=0. Unevenly spaced groups of weak detachments appear already at *Re*=1000. With increasing *Re*, there is a progression to detachments that spread into the core. At *Re*=10000 the level of turbulence in the core region is strongest. It diminishes with increasing *Re*, and at *Re*=50000 there are turbulent bands at the side layers and the core is weakly fluctuating. This change of flow regimes is also illustrated in Figure 12 for *Re*=1000 – 50000 by instantaneous distributions of streamwise velocity. We observe that the strongest flow distortions occur at *Re*=10000.

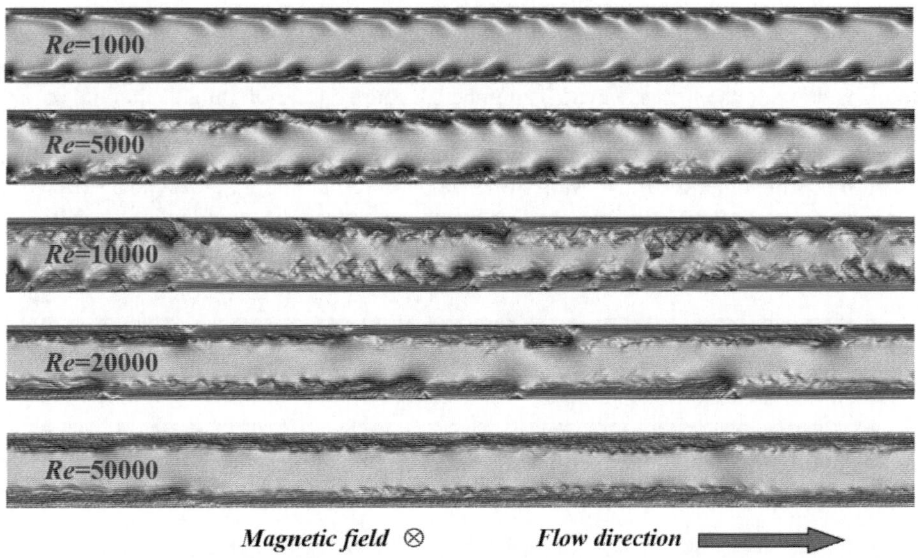

Figure 11. Hunt flow in a square duct at *Ha*=1000 and *Re*=1000 – 50000. Instantaneous patterns of streamwise velocity are visualized in the mid-plane at *y*=0.

Near *Re*=1000 also multiple states have been observed [19]. Especially for this study a series of simulations has been conducted at *Ha*=1000 with decreasing *Re*, similar to [20]. Several target values in the range 500>*Re*>100 were checked. In all cases the flow at *Re*=1000 with jet detachments (Figure 11) was used as initial state. These simulations reveal the interesting observation that no weak structures, such as small-scale vortices [2], have been obtained. Instead, the flow domain remains populated with detached jets of gradually decreasing TKE while decreasing *Re*. The observed behavior of hysteresis corresponds again to approximately the same range *Re* = 200...1000 and at *Re*=100 the flow becomes fully laminar as in the low-*Ha* case. The latter observation perfectly confirms the results of a linear stability analysis of Hunt flow, where the critical Reynolds number Re_c=112 has been predicted for $Ha \gg 1$ [13].

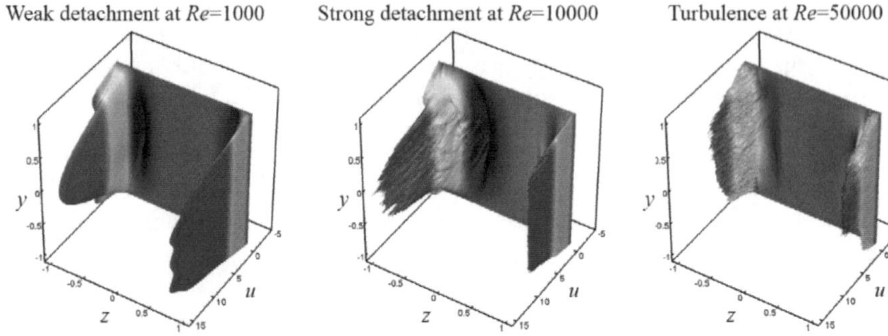

Figure 12. Instantaneous velocity profiles of Hunt flow in a square duct for *Ha*=1000.

Usually magnetic fields are known to suppress turbulent or time-dependent motion. However, in the present case of Hunt flow, the magnetically induced jets are a source of instability. In order to judge the stabilising and destabilizing mechanisms caused by the Lorentz force, reduced settings are postulated: the flow is driven by the balance between pressure gradient and streamwise constant Lorentz force obtained by axial averaging of turbulent Hunt flow. This force which is applied as source term in the momentum equation for hydrodynamic simulations is able to produce the Hunt jets that become unstable and turbulent. As can be seen in Figure 13 and Figure 14, the side layers then show very strong turbulent fluctuations down to the lowest Reynolds numbers. Figure 14 also highlights that the core region is far more turbulent and that the velocity in the side layers is reduced. The reason for this behavior lies in the fact that the strong correlation of fluctuations along magnetic field lines and associated Joule damping is absent in this hypothetical flow. Moreover, we observe that the time-dependent flow regimes cease only at extremely low $Re \approx 10$. We thereby confirm that the magnetic damping effect on individual eddies is essential for the observed transitional flow states.

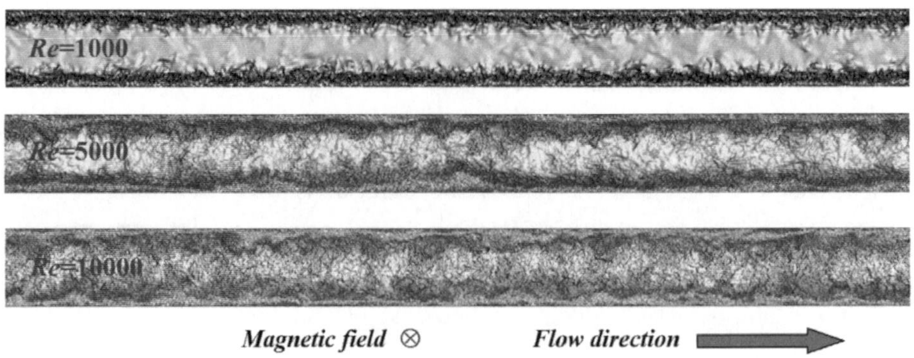

Figure 13. Numerical simulations of Hunt flow in a square duct at *Ha*=1000 and *Re*=1000 – 10000 with Lorentz force computed from the streamwise-averaged flow. Instantaneous patterns of streamwise velocity are visualized in the mid-plane at y=0.

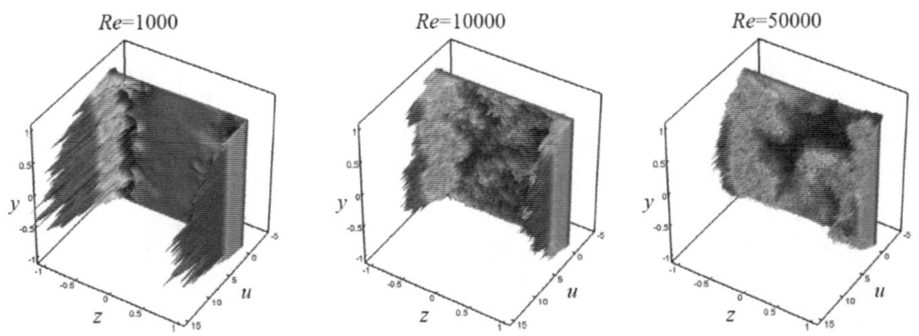

Figure 14. Instantaneous velocity profiles of Hunt flow in a square duct at *Ha*=1000 when the Lorentz force density is computed from the streamwise-averaged flow.

Direct numerical simulations have shown that transition in Hunt flow proceeds through different stages and some of them may be difficult to detect in experiments on account of the very small amplitude of fluctuations and the occurrence of multiple states. A direct comparison between simulations and experiments is also problematic due to the use of periodic boundary conditions for the computations. Spatial simulations with details of a flow straightener as inlet condition may seem more appropriate but they are computationally very expensive when *Ha* and *Re* are large

6. Conclusions

Magnetically induced instabilities in liquid metal duct flows have been studied for Hunt flow by a linear stability analysis, numerical simulations close to the onset of time-dependent motion, and by direct numerical simulations for high Reynolds numbers. The onset of time-dependent flow patterns and their intensity depends on the strength of the magnetic field and on the flow rate in terms of the Hartmann number *Ha* and Reynolds number *Re*. Numerical simulations show that unstable structures close to the stability limit have very small amplitude and occur in a fairly low frequency range that makes their experimental detection under laboratory conditions extremely challenging. This fact may explain difficulties in past and present comparisons of theoretical results and experimental observations. High-resolution direct numerical simulations give detailed insight into the rich variety of completely different flow patterns that might even coexist for the same set of parameters.

Acknowledgement

Financial support of this research by the German Helmholtz Association in the frame of the Helmholtz-Alliance LIMTECH is gratefully acknowledged. Computer resources were provided by the computing centers of Coventry University, Forschungszentrum Jülich (NIC) and TU Ilmenau.

References

[1] Smolentsev S, Badia S, Bhattacharyay R, Bühler L, Chen L, Huang Q, Jin H-G, Krasnov D, Lee D-W, Mas de les Valls E, Mistrangelo C, Munipalli R, Ni M-J, Pashkevich D, Patel A, Pulugundla G, Satyamurthy P, Snegirev A, Sviridov V, Swain P, Zhou T and Zikanov O 2015 *Fusion Engineering and Design* **100** 65-72

[2] Ting A L, Walker J S, Moon T J, Reed C B and Picologlou B F 1991 *Int. J. Engng. Sci.* **29** 939-948

[3] Reed C B and Picologlou B F 1989 *Fusion Technology* **15** 705-715

[4] Burr U, Barleon L, Müller U and Tsinober A B 2000 *Journal of Fluid Mechanics* **406** 247-279,

[5] Hunt J C R 1965 *Journal of Fluid Mechanics* **21** 577-590

[6] Bühler L, Mistrangelo C und Koehly C 2012 *IEEE Transactions on Plasma Science* **40** 590-595

[7] Gelfgat Y M, Dorofeev V S and Shcherbinin E V 1973 *Magnetohydrodynamics* **7** 26-29

[8] Morley N B, Burris J, Cadwallader L C and Nornberg M D 2008 *Review of Scientific Instruments* **79** 056107

[9] Chowdhury V, Bühler L and Mistrangelo C 2014 *Fusion Engineering and Design* **89** 1299-1303

[10] Müller U and Bühler L 2001 *Magnetofluiddynamics in Channels and Containers* (Wien, New York: Springer)

[11] Chowdhury V, Bühler L and Mistrangelo C 2016, in *Proceedings of the 10th PAMIR Int. Conference Fundamental and Applied MHD, Cagliari - Italy, June 20 - 24*, pp. 145-149

[12] Chowdhury V, PhD Thesis, Karlsruhe Institute of Technology, 2016

[13] Priede J, Aleksandrova S and Molokov S 2010 *Journal of Fluid Mechanics* **649** 115-134

[14] Arlt T, Priede J and Bühler L 2017 *Magnetohydrodynamics* **53** 35-44

[15] Priede J, Arlt T and Bühler L 2016 *Journal of Fluid Mechanics* **788** 120-146

[16] Mistrangelo C and Bühler L 2011 *Fusion Science and Technology* **60** 798-803

[17] Arlt T, Priede J and Bühler L 2017 *Journal of Fluid Mechanics,* in press

[18] Krasnov D, Zikanov O and Boeck T 2011 *Computers and Fluids* **50** 46-59

[19] Krasnov D, Boeck T, Braiden L, Molokov S and Bühler L 2016 Karlsruhe Institute of Technology, KIT-SR 7713, ISBN 978-3-7315-0562-4

[20] Braiden L, Krasnov D, Molokov S, Boeck T and Bühler L 2016 *Europhysics Letters* **115** 44002

IOP Conf. Series: Materials Science and Engineering **228** (2017) 012004 doi:10.1088/1757-899X/228/1/012004

Instabilities in electrically driven rotating MHD layers

C Mistrangelo and L Bühler

Karlsruhe Institute of Technology, P.O.B. 3640, 76021 Karlsruhe, Germany

chiara.mistrangelo@kit.edu

Abstract. Flows of electrically conducting fluids exposed to intense magnetic fields exhibit a common feature i.e. the formation of uniform cores in which electromagnetic forces are dominant. Cores are separated from each other by thin layers that extend along magnetic field lines. Across these parallel layers strong gradients of flow variables are present, which can lead to the onset of instabilities and non-linear flow transitions.

In this work we investigate dynamics and stability issues of rotating parallel layers driven by electromagnetic forces caused by the interaction of injected electric currents with an applied magnetic field. The geometry considered consists of two coaxial circular electrodes used for current injection. They are placed in parallel electrically insulating planes perpendicular to a uniform magnetic field. The basic axisymmetric steady state flow, characterized by a rotating velocity jet confined in a parallel layer that connects the rims of the electrodes, is rather well understood. By increasing the driving current above a critical value the basic flow becomes unstable and undergoes a sequence of supercritical bifurcations.

1. Introduction

Electrically driven magnetohydrodynamic (MHD) flows have been investigated numerically for the experimental configuration used in [1] and a test-section has been manufactured and instrumented to study MHD phenomena observed in the simulations. Electric current is injected between two coaxial, parallel circular electrodes, which are perpendicular to an externally applied uniform magnetic field. As a result of the MHD interaction of current and magnetic field, an azimuthal flow originates that is mainly confined in an annular fluid layer, which connects the rims of the electrodes. A similar type of flow can appear also in large scale phenomena typical of geo- and astrophysics [2]. The aim of the present study is getting an overview of unstable flow patterns that occur when the current injected across the liquid metal layer is increased above a critical value [3] [4].

In the past experiments and analytical studies have been carried out to investigate MHD flows, where the fluid is set in motion by electromagnetic forces generated by current injected between a pair of concentric electrodes on the bottom of an electrically insulating container [5] [6] [7]. These studies can be regarded as the electrical analogon of those performed in [8]. The established flow consists of a rotating annular core and two shear layers, aligned with the magnetic field. Here the current flows in axial direction before closing through the Hartmann layers that form along surfaces perpendicular to the magnetic field. The width δ of the internal layers reduces by increasing the magnetic field strength as $\delta \sim Ha^{-1/2}$. The Hartmann number Ha gives a non-dimensional measure for the strength of the imposed magnetic field. This flow distribution differs from the one described in this work, where cores are stagnant and the induced stationary rotating motion consists of jets confined in an internal parallel layer. Here the maximum jet-velocity increases with Ha and the flow in the layer can become unstable.

MHD flows created by injecting current through circular electrodes of finite diameter, as considered in the present study, are discussed in [1]. An approximate steady state asymptotic solution is developed and compared with experiments. For high Hartmann numbers the current is confined to the cylinder of fluid between the electrodes and when reducing the magnetic field the electric resistance of the fluid decreases due to the spread of current streamlines across the enlarged internal layer. The established azimuthal velocity distribution results from the balance between viscous and electromagnetic Lorentz forces in the layer. Discrepancies between experiments and asymptotic solution are explained by suggesting the occurrence of an axial-radial secondary flow. The contribution of the latter is not taken into account in [1]. A more rigorous asymptotic solution for the description of a steady, electrically driven shear layer is given in [9]. Experiments in [1] were carried out in a range of parameter where the flow is expected to be stable. Local electric potential was measured by using a potential probe and velocities by a Pitot tube. An attempt to investigate experimentally instabilities that occur in this type of electrically driven flows can be found in [10] [11], where instantaneous velocities are measured by means of hot-film anemometry. Experiments were reproducible only for sufficiently intense magnetic fields and data could be obtained for a limited range of parameters, $300 < Ha < 650$. Considerable deviations between experiments and theoretical studies [9] were observed and explained by a deterioration of measurements due to the influence of magnetic field on heat transfer properties of the hot-film sensor and the occurrence of contributions from the radial velocity. Malcolm [11] finds that for large magnetic fields the critical current for the onset of instabilities and the corresponding Reynolds number vary as $I_c \sim Ha^{4/3}$ and $Re_c \sim Ha^{1/3}$, respectively. Empirical observations point out that instability sets in as wave-like flow with steady amplitude.

Available theoretical studies of electrically driven MHD flows neglect the contribution of radial and axial velocities. In our analysis we consider this additional superimposed secondary circulation, the magnitude of which can be used to estimate the degree of anisotropy of the established MHD flow. For the described MHD flows only very few experimental results are available. Hence an experimental program has been proposed to thoroughly investigate instabilities in electrically driven rotating layers.

2. Formulation of the problem

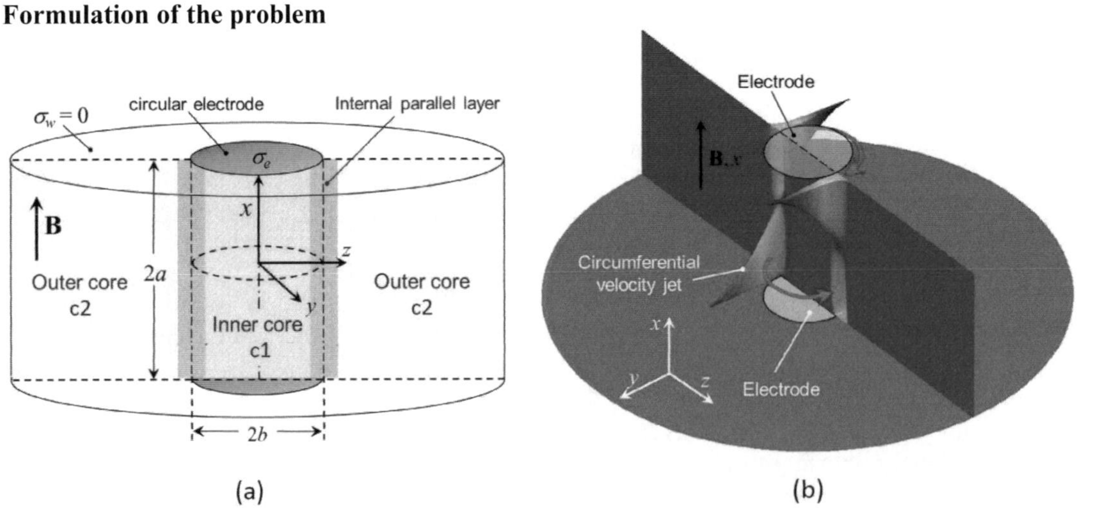

Figure 1. (a) Sketch of the geometry used for the simulations and main typical flow regions: an internal layer separates an inner and an outer core (c1 and c2). (b) Velocity distribution in case of a stable, laminar flow.

The geometry considered is schematically depicted in Figure 1a. We investigate the liquid metal flow in an electrically insulating cylindrical container of height $2a$ with two perfectly conducting electrodes embedded in the top and bottom walls, when a uniform magnetic field $B\hat{\mathbf{x}}$ is applied aligned with the axis of the container. The electrodes are kept at constant electric potential so that a potential difference $\Delta\phi$ is present between them. Half of the distance a between the electrodes has been chosen as characteristic size of the problem. The electrode rim is located at $b/a = 0.5$, where b is the electrode radius. The imposed potential difference $\Delta\phi$ drives a current between the electrodes and the resulting electromagnetic forces induce a primary azimuthal flow that is practically located inside a thin parallel layer originating from the electrode rims. The shear layer is caused by the discontinuity in the electrical boundary conditions on the top and bottom of the cylinder.

If $\Delta\phi$ is smaller than a critical value a steady state solution establishes and the layer takes the shape of a velocity jet travelling in circumferential direction. The 3D velocity distribution for a stable MHD flow is depicted in Figure 1b.

The established MHD flow can be mathematically described by equations for conservation of momentum, mass and charge that in non-dimensional form can be written as

$$\frac{1}{N}\left[\frac{\partial \mathbf{v}}{\partial t}+(\mathbf{v}\cdot\nabla)\mathbf{v}\right]=-\nabla p+\frac{1}{Ha^2}\nabla^2\mathbf{v}+\mathbf{j}\times\mathbf{B}, \tag{1}$$

$$\nabla\cdot\mathbf{v}=0, \quad \nabla\cdot\mathbf{j}=0, \tag{2}$$

where the current density $\mathbf{j}$ is expressed by Ohm's law

$$\mathbf{j}=-\nabla\phi+\mathbf{v}\times\mathbf{B}. \tag{3}$$

The variables $\mathbf{v}$, p, $\mathbf{j}$, $\mathbf{B}$ and ϕ indicate velocity, pressure, current density, magnetic field and electric potential normalized by u_0, $\sigma u_0 B^2 a$, $\sigma u_0 B$, B, and $u_0 Ba$, respectively. Here B is the magnitude of the magnetic field, and u_0 a typical velocity. The electrical conductivity σ, the density ρ and the kinematic viscosity ν of the fluid remain constant. It is assumed that the magnetic field induced by the flow is negligible compared to the imposed field. This assumption is called inductionless approximation and it is valid for small magnetic Reynolds numbers, $R_m = \mu\sigma a u_0 \ll 1$, where μ is the magnetic permeability [12].

The dimensionless groups in (1) are the interaction parameter N and the Hartmann number Ha:

$$N=\frac{\sigma a B^2}{\rho u_0}, \qquad Ha=aB\sqrt{\frac{\sigma}{\rho\nu}}. \tag{4}$$

The former one describes the ratio between electromagnetic and inertia forces, the latter gives a non-dimensional measure for the strength of the magnetic field. Its square quantifies the ratio of electromagnetic to viscous forces. The magnitude of the potential difference $\Delta\phi$ between the electrodes is expressed in dimensionless form by the Reynolds number Re

$$Re=\frac{\Delta\phi}{2}\frac{1}{B\nu}. \tag{5}$$

As electric boundary conditions fixed potential values are set at the electrodes, while the external surface of the cylinder is electrically insulated, i.e. $\partial_n\phi = 0$. The no slip condition, $\mathbf{v} = 0$, is applied to all walls.

3. Numerical results

Data from numerical simulations are discussed for a constant Hartmann number, $Ha = 588$, and various Reynolds numbers. The selected Hartmann number has been chosen since for this magnetic field strength experimental and theoretical data for stationary conditions are available in literature [9].

The code used for the numerical simulations has been obtained by introducing the Lorentz force term and the equations determining electric current density and potential into the finite volume open source code OpenFOAM.

3.1. Steady solution

For $\Delta\phi < 140\mu V$ ($Re < 500$) a steady state solution is found, which is characterized by two counter rotating layers (see Figure 1b) mainly confined in an annular region that separates an internal and an external core as schematically displayed in Figure 1a. The fluid is stagnant both in the internal and in the external cores and it moves with a jet-like profile inside the parallel layer. The 3D velocity distribution for a stable laminar MHD flow at $Ha = 588$ and $Re = 358$ is depicted in Figure 1b.

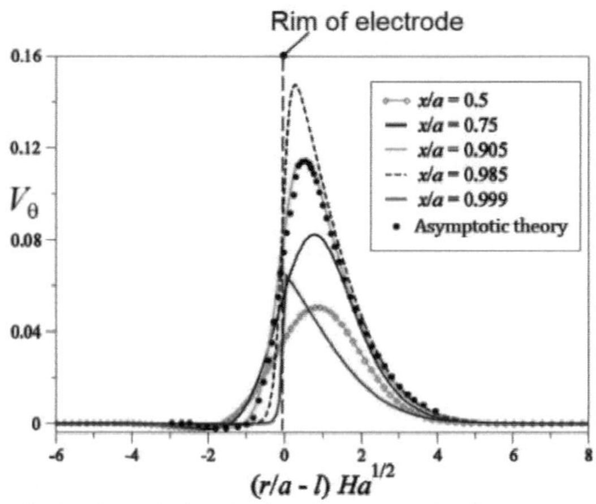

Figure 2. Radial profile of scaled azimuthal velocity for a steady flow at *Ha* = 588, at different axial locations. The velocity profile at $x/a = 0.905$ agrees well with the available asymptotic solution [9].

In Figure 2 the scaled circumferential velocity $V_\theta = v_\theta Ha^{1/2} bl(\sigma\rho\nu)^{1/2}/I$ is plotted along the normalized radial direction at various axial positions x/a for a stable MHD flow at $Ha = 588$. The total driving current density is named I. The velocity is scaled according to [9]. The predicted velocity profile at $x/a = 0.905$ reproduces well the one from the asymptotic solution indicated by dots, as given in [9]. This suggests that, for the potential difference considered, inertia effects are negligible compared to electromagnetic and viscous forces, and the secondary flow in axial direction is much weaker than the circumferential flow, as assumed in the cited theoretical study. The magnitude of the circumferential velocity increases linearly from zero at the symmetry plane to a maximum value when approaching the Hartmann walls perpendicular to **B**. In the Hartmann layers the velocity reduces with a strong gradient over a distance of $O(Ha^{-1})$ to satisfy the no slip condition at the wall.

The layer takes the shape of a velocity jet travelling in circumferential direction, whose thickness δ remains constants when it rotates, as visualized in Figure 3. Here the potential ϕ and the axial vorticity component ω_x are plotted on a plane perpendicular to the magnetic field, at an axial coordinate $x/a = 0.5$, at half distance between electrode and symmetry plane. Across the internal layer, in radial direction, the electric potential drops from the inner core value ϕ_{c1}, which varies linearly along x, to $\phi_{c2} = 0$ in the outer core. The strong radial potential gradient primarily balances the induced electric field, and the potential difference ϕ_{c1}-ϕ_{c2} through the layer is responsible for the formation of the rotating jet. The vorticity profile along the dashed line in the middle of the geometry at $z = 0$ (see Figure 3b) exhibits maxima and minima as marked by green stars. The occurrence of instabilities of the flow in the layers is associated with the presence of extrema in the vorticity field that correspond to inflection points in the velocity profile.

IOP Conf. Series: Materials Science and Engineering **228** (2017) 012004 doi:10.1088/1757-899X/228/1/012004

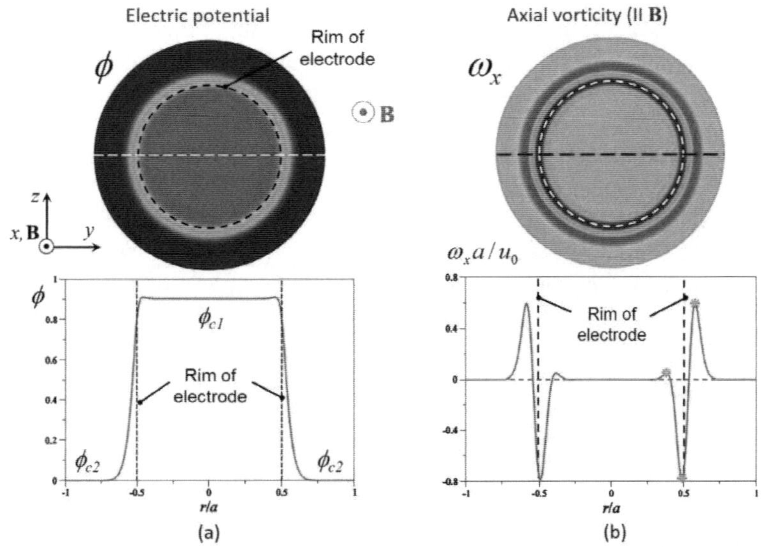

Figure 3. Results for a steady state flow at Ha = 588. (a) Electric potential distribution and (b) contours of axial vorticity on a plane at x/a = 0.5. Potential and vorticity are also plotted along the dashed line in the middle of the geometry.

3.2. Unstable flow regime

Simulations have been performed by increasing the applied potential difference, i.e. the Reynolds number, while holding the magnetic field constant, $Ha = 588$.

In Figure 4a the normalized time-averaged circumferential velocity is plotted along the scaled radial direction at x/a = 0.905 for various Reynolds numbers. By increasing Re the maximum of the azimuthal velocity reduces and shifts to higher radial positions. The negative velocity at the interface between inner core and layer progressively disappears, and a minimum occurs in the outer core. The time-averaged thickness δ of the internal layer becomes larger and the mean velocity in the jet reduces due to the increased transfer of momentum related to the occurrence of time-dependent fluctuations. This behavior is analogous to the one of unstable MHD flows in field-aligned boundary layers in ducts, although the physics behind the phenomena is different [13] [14] [15].

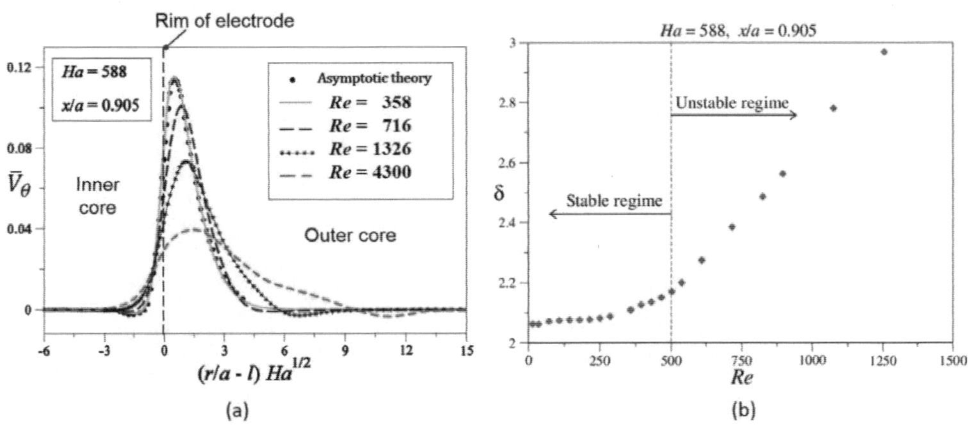

Figure 4. Flow at Ha = 588 and various Reynolds numbers. (a) Radial profile of scaled time-averaged azimuthal velocity at the axial location x/a = 0.905, for various applied potential differences (Re). (b) Thickness δ of the internal layer as a function of the Reynolds number.

IOP Conf. Series: Materials Science and Engineering **228** (2017) 012004 doi:10.1088/1757-899X/228/1/012004

In Figure 4b the thickness δ of the internal layer, estimated as the ratio between the integral of the velocity profile V_θ and its maximum $V_{\theta max}$, is plotted as a function of the Reynolds number. For 100μV<$\Delta\phi$<140μV, i.e. 350<Re<500, the flow is still steady and one could expect that δ remains constant. However, the internal layer thickens and flow structures extend to larger radial positions due to the occurrence of an increasing radial velocity. In the inertialess limit ($Re\rightarrow 0$) the thickness of the internal layers parallel to the magnetic field should scale as $Ha^{-1/2}$. In the present problem, when secondary flow and time dependent flow structures appear, the thickness δ of the layer can depend on the interaction parameter N, or a combination of N and Ha. This is due to the fact that radial-axial flow and unstable patterns are caused by inertial forces.

For Re>500 the flow becomes unstable in agreement with experiments in [11]. The instability sets in as sinusoidal oscillations and vortices develop, which have a main vorticity component ω_x in magnetic field direction.

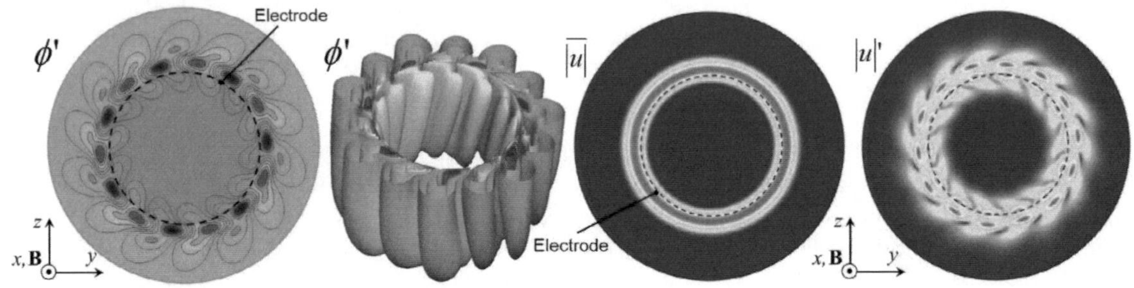

Figure 5. Example of unstable flow at Ha = 588 and Re = 540. Plots show contours of fluctuating part ϕ' of electric potential, magnitude of the mean velocity, fluctuations of the velocity magnitude, on a plane perpendicular to **B** at x/a = 0.5.

An example of unstable flow is visualized in Figure 5 for Re = 540. Here contours of the fluctuating part $\phi' = \phi - \bar{\phi}$ of electric potential, mean part and fluctuations of the magnitude of the velocity are plotted on a plane perpendicular to **B**, at x/a = 0.5. The unstable regime is characterized by a periodic vortex pattern and the instability takes the form of a chain of small vortices. The amplitude of the fluctuations decreases when moving from the Hartmann walls to the symmetry plane where they vanish. The 3D view of the fluctuations ϕ' of the electric potential in Figure 5 clearly highlights the three-dimensionality of the unstable structures. Perturbations are conveyed by the mean flow, $|u| = \overline{|u|} + |u|'$. Contours of axial vorticity are depicted in Figure 6 left.

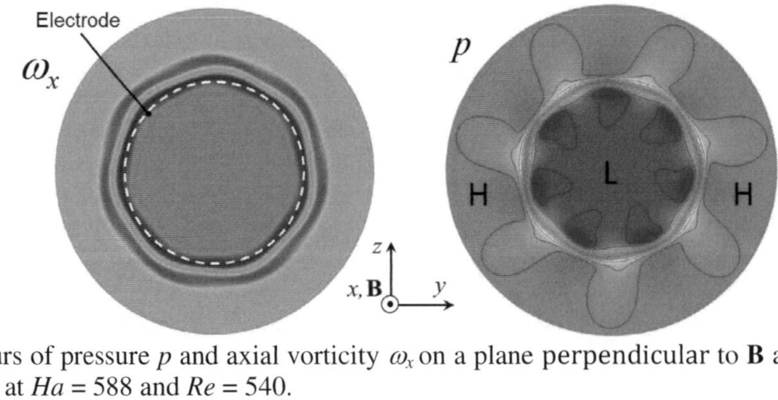

Figure 6. Contours of pressure p and axial vorticity ω_x on a plane perpendicular to **B** at x/a = 0.5 for the unstable flow at Ha = 588 and Re = 540.

An interesting flow feature that did not receive adequate attention in the past is the occurrence of a radial-axial flow superposed to the main azimuthal circulation. The occurrence of this secondary flow was already suggested in [1] as an explanation for discrepancies between experiments and asymptotic solution, but an exhaustive description of the phenomenon was not given. Preliminary numerical calculations [3] show that the secondary flow consists of a motion towards the electrodes in the internal part of the layer and towards the symmetry plane in the external portion. This results in a centripetal flow near the symmetry plane and a centrifugal flow at the interface between core and Hartmann layer. This secondary motion is caused by a balance of pressure, electromagnetic and inertial centrifugal forces. The pressure force points towards the inner core as visible in Figure 6 on the right, where pressure contours are depicted on a plane perpendicular to **B** at $x/a = 0.5$. The electromagnetic Lorentz force slows down the secondary flow and balances the viscous forces in layers.

4. Experimental set up

The conceptual design of the fabricated test-section is shown in Figure 7a. It consists of a vessel filled with the eutectic alloy GaInSn. On the electrically insulating top and bottom walls that are perpendicular to the imposed magnetic field (Hartmann walls) two circular coaxial copper electrodes with radius b are inserted flush with the internal surface of the plates. The axial length of the electrodes has been selected long enough to achieve sufficiently homogeneous current density at the fluid-solid interface. According to the numerical results, the diameter D of the outer cylinder has been defined such that the lateral wall does not interact with the layer or with unstable vortex patterns. Another constraint for the size of the vessel is represented by the dimension of the bore of the superconducting magnet at TUI (376 mm), where experiments will be performed. A reasonable choice is $D = 9b$.

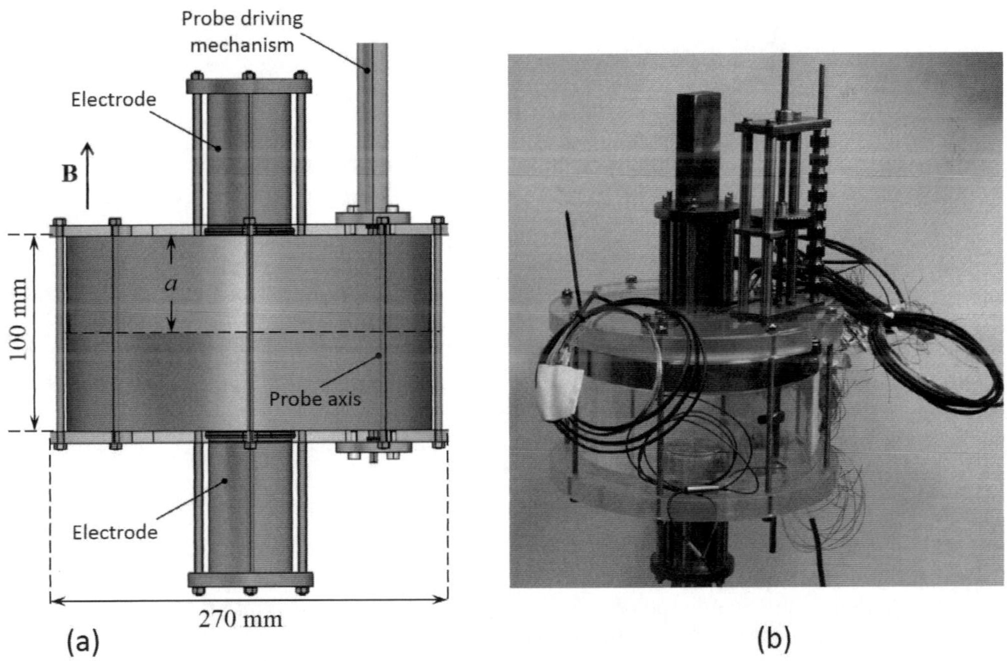

Figure 7. (a) Design and (b) picture of the manufactured and fully instrumented test-section

The test-section is designed according to a modular design by using flanges to connect different components. This permits an easy replacement of the cylindrical container in order to investigate the influence of the axial distance between the electrodes on onset and development of instabilities in the rotating fluid layer. This design flexibility allows also modifying the set-up of the instrumentation.

In order to preclude flow artifacts arising from symmetry deviations of the experimental configuration special care is required to ensure a precise positioning inside the magnet. The fluid vessel has to be placed concentrically inside the bore of the magnetic system.

4.1. Instrumentation

During the experimental campaign, values of local electric potential and potential differences are recorded by means of sensors located on Hartmann walls and by movable probes in the liquid metal. These data can be directly compared with numerical results. Spatial-temporal information about velocity field can be obtained by using ultrasound Doppler velocimetry (UDV) techniques. Temperature has also to be monitored at different locations on the test-section in order to calculate the physical properties of the liquid metal and to verify that experiments are carried out under isothermal operating conditions so that thermo-electric effects can be neglected. Figure 7b shows a picture of the fully instrumented test section and in Figure 8 the position of the various instruments is displayed.

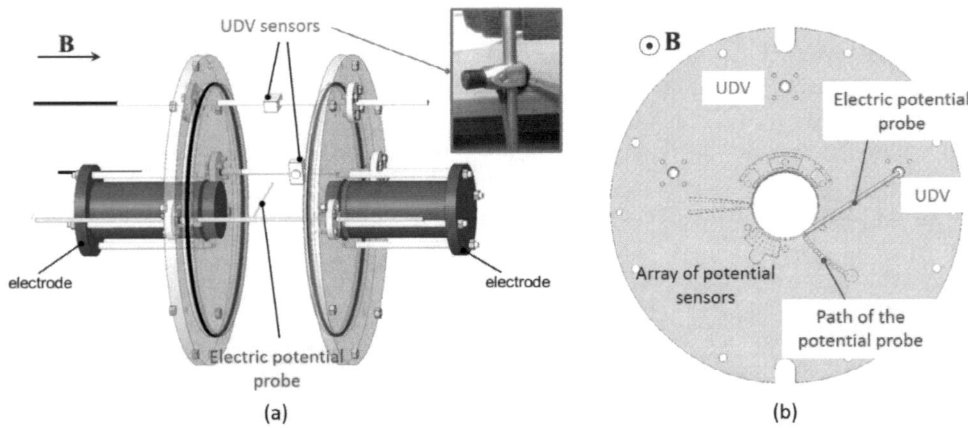

Figure 8. (a) Design of the test-section and available instrumentation: electric potential probe (red), UDV sensor (green). (b) Arrangement of electric potential sensors on the Hartmann wall of the container. The pins aligned along the path described by the movable probe are used to determine cross-correlation functions of potential fluctuations along magnetic field lines.

4.1.1. Electric potential probe

In order to determine local velocities in liquid metal flows electric potential differences, induced by the motion of an electrically conducting fluid in the imposed magnetic field, are measured by means of potential probes. Their operating principle is based on Ohm's law (3), through which the velocity components perpendicular to $\mathbf{B} = \hat{\mathbf{x}}$ can be expressed in non-dimensional form as

$$\mathbf{v}_{\perp} = \left(-\frac{\partial \phi}{\partial z} - j_z \right)\hat{\mathbf{y}} + \left(\frac{\partial \phi}{\partial y} + j_y \right)\hat{\mathbf{z}} \tag{6}$$

The local velocity is determined by potential gradients and by currents induced in the fluid. Therefore, the potential difference measured at the sensor tips can be directly interpreted in terms of velocity components only if the current density is very small or known [16].

In the planned experiments it is foreseen to employ a movable potential probe with four sensing tips in order to measure the intensity of time-dependent fluctuations of potential and potential gradients. A principle sketch of the probe is shown in Figure 9(a). Tips a and b are located on the same plane zy perpendicular to the magnetic field ($x_a = x_b$), while the central electrodes c are shifted along y and x with respect to tips a and b in order to measure fluctuations of the radial velocity. The body of the movable electric potential probe consists of a carbon rod the diameter of which should be minimized in order to reduce the obstruction of the flow by the sensor (Figure 9(b)). Due to circumferential symmetry of the flow it is not possible to record average profiles of radial velocity, since the mean potential is constant along a circumference. However, fluctuations of the radial velocity can be measured. Detection of the onset of these oscillations, compatibly with the accuracy of measuring instruments, could be an efficient method to identify the stability limit as a function of Reynolds and Hartmann numbers. Measurements of fluctuations of the much larger circumferential velocity require the average value to be filtered out before amplifying it.

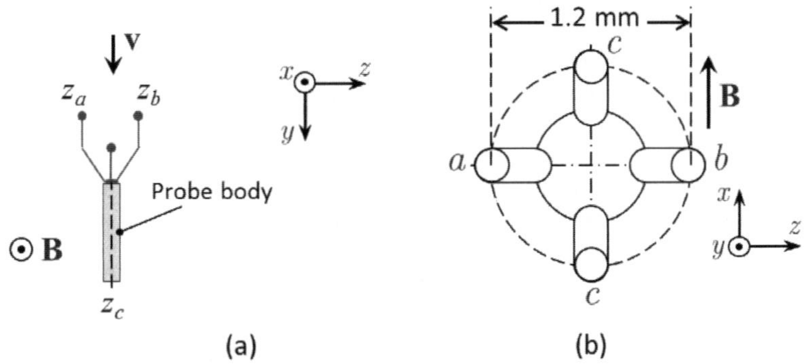

Figure 9. (a) Sketch of an electric potential probe to record local potential values and gradients. The probe has four sensing tips. (b) Frontal view of the probe.

4.1.2. Electric potential sensors on Hartmann wall

The distribution of potential gradient on walls perpendicular to the applied magnetic field can be directly related to the flow velocity. It gives information about flow structure at the interface between Hartmann layer and core of the flow if two conditions are satisfied. First there should be no variation of the electric potential across the thin Hartmann layer, i.e. the potential at the wall coincides with the value at the core-layer interface, and second the current density at the core-layer edge is negligible. Under these assumptions measurements of electric potential on the Hartmann walls allow obtaining an estimate of velocity components on a plane orthogonal to the magnetic field. Comparison with numerical simulations can support the proper physical interpretation of recorded quantities.

In order to measure the two dimensional distribution of the electric potential on a plane perpendicular to the magnetic field, an array of sensors (copper pins) has been installed on the Hartmann wall as shown in Figure 8b. They are in electrical contact with the liquid metal. This array permits to reconstruct an image of vortex patterns, i.e. to map the electric potential distribution at the core-Hartmann layer interface. The sensors are located along radii and regularly spaced along concentric circles. The spacing between the pins is selected so that higher spatial resolution is guaranteed. The proper distance is chosen according to numerical results that allow identifying small scale recirculating flow patterns. Signal correlations between points on the same circumference allow determining the travelling speed of the waves, the wave number and therefore the number of vortices forming the fluctuating patterns.

Electric potential sensors on the Hartmann wall are also located along a circle that corresponds to the path of the movable probe (see Figure 8b). In this way it is possible to define correlations between pairs of measured electric potential fluctuations recorded simultaneously at corresponding points along the same magnetic field line.

5. Conclusions and future work

Numerical simulations have been performed to study the stability of rotating liquid metal layers that are driven by MHD interaction of injected currents with an imposed magnetic field. Two coaxial electrodes are inserted in insulating parallel plates, which are perpendicular to an imposed magnetic field. The basic steady state flow consists of two counter rotating layers whose velocity vanishes at the symmetry plane. The maximum velocity is reached near the Hartmann layers. The radial profile of the azimuthal velocity exhibits inflection points and for sufficiently large driving current, the flow becomes unstable.

At the onset of instability the flow pattern is characterized by a chain of periodic small vortices. By increasing the Reynolds number the vortices merge together to give rise to larger structures.

An experimental test-section has been fabricated and it will be inserted in the superconducting magnet at the technical university of Ilmenau (TUI). The aim is both complementing the numerical results and comparing the measured data with the simulations. A stability map will be defined to identify the different flow regimes depending on Reynolds and Hartmann numbers.

During the experiments electric potential will be recorded on the Hartmann walls of the cylinder as well as inside the liquid metal by means of a movable probe.

Acknowledgement

Financial support of this research by the German Helmholtz Association in the frame of the Helmholtz-Alliance LIMTECH is gratefully acknowledged.

References

[1] J. C. R. Hunt and D. G. Malcolm 1968 *Journal of Fluid Mechanics* **33** 775

[2] R. Hollerbach 2000, in *Physics of Rotating Fluids*, vol. 549, C. Egbers and G. Pfister, Eds., Springer, pp. 295-316

[3] C. Mistrangelo and L. Bühler 2011, in *Proceedings of the 8th Pamir International Conference on Fundamental and Applied MHD, Borgo, Corsica September 5-9*, pp. 175-179

[4] C. Mistrangelo and L. Bühler 2015, in *6th International Symposium on Bifurcations and Instabilities In Fluid Dynamics, Paris, July 15-17*

[5] V. B. Levin 1980 *Magnetohydrodynamics* **16** 70

[6] A. A. Klyukin, Y. B. Kolesnikov and V. B. Levin 1980 *Magnetohydrodynamics* **16** 75

[7] A. A. Klyukin and V. B. Levin 1984 *Fluid Dynamics* **19** 815

[8] B. Lehnert 1955 *Proceedings of The Royal Society London A* **233** 299

[9] J. C. R. Hunt and K. Stewartson 1969 *Journal of Fluid Mechanics* **38** 225

[10] D. G. Malcolm 1969 *Nature* **224** 909

[11] D. G. Malcolm 1970 *Journal of Fluid Mechanics* **41** 531

[12] P. H. Roberts, The Whitefriars Press Ltd., London and Tonbridge, 1967.

[13] U. Burr, L. Barleon, U. Müller and A. B. Tsinober 2000 *Journal of Fluid Mechanics* **406** 247

[14] L. Bühler and S. Horanyi 2009 *Fusion Engineering and Design* **84** 518

[15] M. Kinet, B. Knaepen and S. Molokov 2009 *Physical Review Letters* **103**, 154501

[16] A. Kolin and F. Reiche 1954 *Journal of Applied Physics* **25** 409

IOP Conf. Series: Materials Science and Engineering **228** (2017) 012005 doi:10.1088/1757-899X/228/1/012005

Investigations of fluid flow effects on dendritic solidification: Consequences on fragmentation, macrosegregation and the influence of electromagnetic stirring

N Shevchenko[1,*], H Neumann-Heyme[1], C Pickmann[2], E Schaberger-Zimmermann[3], G Zimmermann[2], K Eckert[1], S Eckert[1]

[1] Helmholtz-Zentrum Dresden-Rossendorf (HZDR), 01328 Dresden, Germany

[2] ACCESS e.V., Intzestrasse 5, 52072 Aachen, Germany

[3] RWTH, Foundry Institute, Intzestrasse 5, 52072 Aachen, Germany

[*] n.shevchenko@hzdr.de

Abstract. Solidification experiments and numerical simulations have been performed to improve the understanding of the complex interrelation between melt flow and the formation of dendritic structures during solidification of Al-Cu and Ga-In alloys. Melt flow induces various effects on grain morphology primarily caused by convective transport of solute, such as a facilitation of the growth of primary trunks or lateral branches, dendrite remelting, fragmentation or freckle formation depending on the dendrite orientation, the flow direction and intensity. Within this project special interest was focused on fragmentation and segregation phenomena. Natural convection is caused by density variations within the solidifying alloys. Forced convection was produced by electromagnetic stirring. X-ray radioscopy was applied as a powerful tool for the visualization of dendritic growth and coarsening.

1. Introduction

Fluid flow plays an important role in solidification processes, since the convective transport of heat and solute influences the undercooling at the solidification front and thus determines the growth conditions of the solid phase decisively. The occurrence of a flow in a solidifying melt must be regarded as the normal case, since natural convection begins once there are significant differences in temperature or composition. In particular, thermo-solutal convection has been identified as the main reason for the development of solute-rich channels in the mushy zone [1-4]. The incidence of such segregation channels in fully-solidified castings has to be considered as a serious casting defect.

A sufficient understanding of the strongly coupled interaction between the melt flow and the solidification requires a detailed knowledge of the velocity field especially in the vicinity of the solidification front. During an upwards solidification solutal convection may arise from an unstable density stratification at the solid-liquid interface if the solute component rejected during the freezing of the primary crystals is lighter than the initial composition of the melt. In turn, the melt convection redistributes the solute concentration in boundary layer and mushy zone leading to dramatic changes in growth rate and direction and may provoke the development of segregation freckles [5-12]. Further effects concern the selection of the secondary branches of growing dendrites [9]. In general, the flow impact might be different for different regions of the mushy zone. The growth rate is increased by a removal of the solute, whereas an influx of solute-rich melt causes the formation of segregation freckles by a remelting of already solidified crystals.

Electromagnetic stirring by rotating or travelling magnetic fields can be used as an effective tool for flow control in solidifying metal alloys [6, 8, 13-18]. This research is also driven by the idea of a potential "repair" of the defects arising from the action of natural convection [6]. However, reaching this goal is not straightforward. By applying a traveling magnetic field Medina et al. [6] had to notice that the electromagnetic forces can control the positions of segregation channels and partly prevent the unsteadiness of buoyancy effects, but, they cannot fully eliminate the segregation effects.

It was shown recently, that the application of modulated AC magnetic fields offers considerable potential for optimizing the melt stirring [19-21]. Especially, the secondary flow can be organized in such a way that periodic reversals of the flow direction occur adjacent to the solidification front, which has been proven as an important method to achieve homogeneous, fine-grained, globular structures without flow-induced macrosegregation [18]. Moreover, it became apparent that a careful adjustment of the modulation parameters is required in order to create intense secondary flows with periodic reversals of the flow direction.

Nevertheless, the interaction between solidification and melt flow is rather complex and a further improvement of understanding the mechanisms is still needed. A better founded knowledge requires new diagnostic tools which allow *in-situ* observations of the process and provide high resolution data of the flow field, the concentration field and the solidified structure. Moreover, access to different length scales, as for example dendrite morphology and flow structure, is of particular interest.

During the last two decades new X-ray sources and innovative procedures for image analysis have dramatically improved the capabilities for visualization of solidification processes in metallic alloys [22-26]. The X-ray radiography technique enables real-time and *in-situ* observations of the solidification front with a spatial resolution of a few microns. For instance, Koster et al. [23] employed X-ray radioscopy for the investigation of melt convection effects in liquid Gallium - Indium alloys. Besides a reconstruction of the phase distribution inside the sample they demonstrated the capability to reproduce temperature fields and qualitative flow patterns. Shevchenko et al. [11] have shown that the occurrence of different segregation structures can be attributed to the occurrence of different patterns of melt convection adjacent to the solidification front and in the interdendritic region of the upper mushy zone.

The project "Magnetic Flow Control Solidification" of the Helmholtz Alliance "LIMTECH" was a collaborative work between ACCESS e.V., the Technical University Dresden (TUD) and the Helmholtz-Zentrum Dresden-Rossendorf (HZDR) and focused on the effects of melt convection on dendritic growth and on the question of whether effective control of microstructure formation can be achieved with the elaborate application of electromagnetic fields. The project should deliver substantial contributions for the development of innovative technologies to achieve casting products with superior mechanical properties. Solidification experiments have been conducted in Al-Cu and Ga-In alloys. Corresponding numerical simulations investigated the impact of diverse temperature and concentration fields on the morphology of dendritic structures. This article gives an overview of the main research activities in this project and is composed as follows: section 2 describes the experimental setups used by ACCESS and HZDR and gives a short description of the numerical scheme developed by TUD. Two phenomena formed the focus of the investigations, the fragmentation of primary dendrites and the development of segregation structures, since both processes have a significant influence on the mechanical properties of the solidified alloy. The fragmentation is the subject of section 3. Results from solidification experiments in Al-10wt%Cu and Ga-25wt%In are presented here. A numerical model was developed which is able to predict the fragmentation of dendrite sidearms during coarsening. This model was successfully validated with experimental results from studies on the coarsening behavior of Ga-In alloys. The role of natural convection on the formation of segregation freckles is considered in section 4, while the influence of an electromagnetically driven convection on fragmentation and segregation is discussed in section 5.

IOP Conf. Series: Materials Science and Engineering **228** (2017) 012005 doi:10.1088/1757-899X/228/1/012005

2. Experimental methods

This section describes the experimental setup and methods used by the partners ACCESS and HZDR. While the experiments performed by ACCESS considered the solidification behavior of non-refined and refined Al-10wt%Cu alloys, the activities by HZDR were exclusively focused on solidification experiments using Ga-In alloys.

2.1. ACCESS

Al-Cu samples were prepared from high-purity Al and Cu and were cast in rod-like steel moulds according to a final composition of Al-10.0 wt%Cu. For directional solidification experiments, cylindrical samples with a diameter of 8 mm and a length of 200 mm were processed using a Bridgman-Stockbarger furnace (see figure 1a) in an argon atmosphere. The samples were directionally solidified at a temperature gradient of 10 K/mm. Different velocities between 8.3 µm/s and 167 µm/s were adjusted by moving the furnace vertically upwards along the sample.

In-situ observations were realized in a quasi 2D setup using sheet-like samples of 50 mm in length, 5 mm in width and 150 µm in thickness. These samples were integrated in another Bridgman-Stockbarger type furnace. The corresponding setup is shown in the figures 1b and 1c. A temperature gradient along the sample of 11 K/mm was imposed by two heaters separated by an adiabatic gap. Directional solidification was controlled by a relative vertical motion of the sample with respect to the furnace using a stepping motor. The whole setup is mounted on a support structure between an X-ray source and an X-ray detector. A magnification of about 5 was chosen which provides a field of view of about 5 x 5 mm^2. That way, *in-situ* and real-time observation of the solidification process is realized with an acquisition rate of 1 frame every 2s and an effective pixel size of about 2 µm at the CCD sensor. More details of the experimental set-up are given in [27].

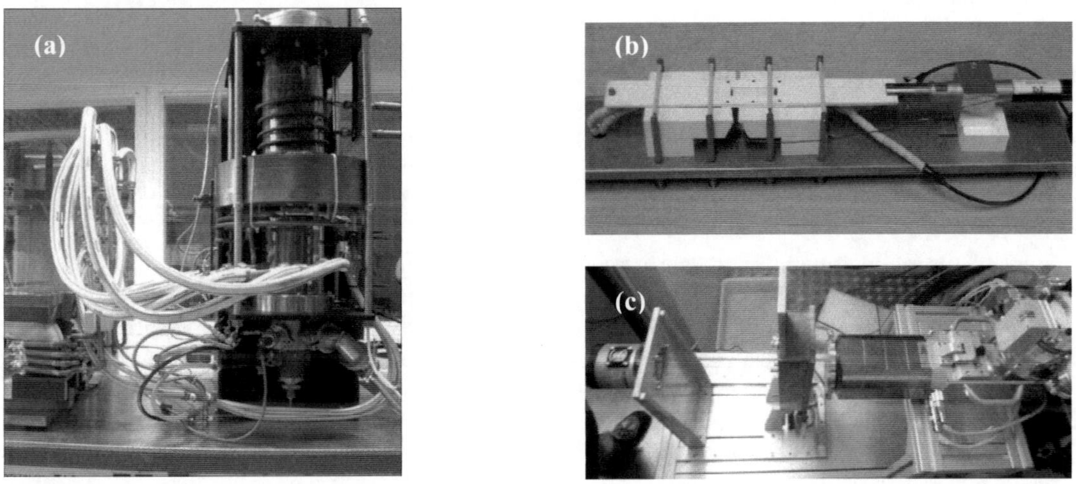

Figure 1. Scheme of the experimental setup at ACCESS: (a) Bridgman gradient furnace, (b) sample, integrated into a sample holder and adapted to the micro-Bridgman furnace unit (left) with linear actuator (right), (c) X-ray tube (right), micro-Bridgman furnace with active cooling in the copper housing (center) and CCD radiation detector (left).

2.2. X-ray laboratory at HZDR

A microfocus X-ray tube (XS225D from Phoenix|X-ray) was used for the solidification experiments carried out at Helmholtz-Zentrum Dresden-Rossendorf. The detailed description of the experimental setup can be found in [11, 28]. The majority of experiments was performed with a Ga–25wt%In alloy ($T_{Liquidus}$=25 °C, $T_{Solidus}$=15.3 °C) which was prepared from high-purity Ga (99.99%) and In (99.99%). It was melted in an furnace and filled into a 30 × 30 mm^2 Hele-Shaw quartz cell having a 150 µm gap parallel with respect to the direction of the X-ray beam (see figure 2a). The observation window was

IOP Conf. Series: Materials Science and Engineering **228** (2017) 012005 doi:10.1088/1757-899X/228/1/012005

approximately 9 × 12 mm^2. The images were captured with a scan rate of 50 half frames per second and integrated over a period of 1 s to reduce the noise level of single images. The tube voltage was 63 kV and the electrical current was 140 µA. The solidification cell was positioned between X-ray tube and detector as demonstrated in figure 2b. A linear array of Peltier elements on the lower edge of the solidification cell is operated as cooler. A simultaneous regulation of the Peltier cooler and the electric heater mounted at the upper part of the solidification cell enables controlling of the cooling rate and the temperature gradient across the cell during the experiment. A set of miniature K-type thermocouples were contacted to the outer surface of the cell to monitor the temperature. The electromagnetically driven flow was generated by a pumping device, which is represented by a rotating wheel consisting of two parallel discs with a set of permanent magnets (NdFeB) with alternating polarization at their inner sides. The top part of the solidification cell was placed between the disks. Rotation speeds of 30 and 80 revolutions per minute (rpm) were chosen for the magnetic wheel.

The analysis of the solidification front velocity and primary and secondary arm spacing follows the algorithm proposed by Boden et al. [29]. The concentration profiles were determined from the measurements of the local brightness in the images.

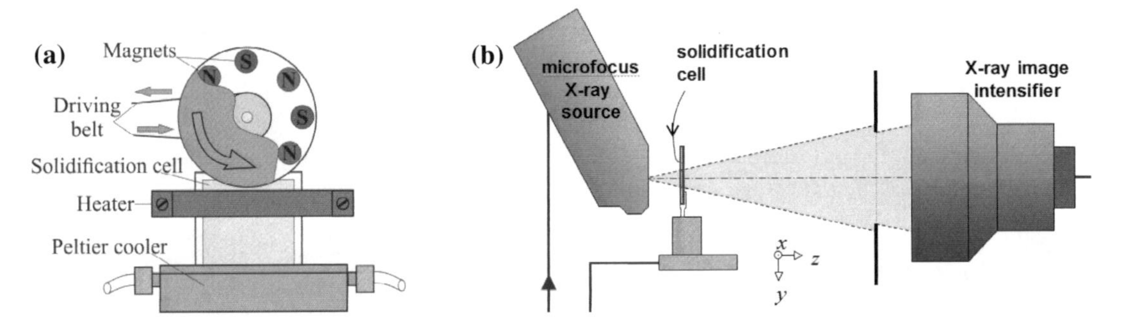

Figure 2. Experimental setup: (a) square Hele-Shaw solidification cell equipped with electric heater, Peltier elements and magnetic wheel for generation of the melt flow; (b) X-ray diagnostic system.

2.3. Beam-line ROBL at ESFR

Further visualization experiments were performed at the ROBL beam line (BM20) at the ESRF in Grenoble. The solidification cells used here for the solidification experiments were the same as already been employed for radiographic investigations carried out by means of a microfocus X-ray tube and described in the previous subsection. The solidification cell was exposed to a monochromatic X-ray beam characterized by an energy of 28.5 keV. Conventional transmission radiographs were obtained by means of a scintillator which provides a resolution of 2 µm. Further components for image acquisition were an optical magnifier and a PCO 2000 CCD camera with 2048 × 2048 pixels (pixel size: 0.34 × 0.34 µm^2). This equipment provides a field of view of about 700 × 700 µm^2. The distance between detector and sample was approximately 20 cm. In order to change the location of the observation window, the position of the solidification cell was manipulated with respect to the X-ray beam by a translation motor system with a minimal step < 10 µm. Images were acquired at exposure times ranging from 2 to 20 seconds. The exposure times were optimized in order to achieve an acceptable signal-to-noise ratio and a sufficient contrast for a specific process rate.

3. Fragmentation

The detachment of dendrite sidebranches is an important fragmentation mechanism and is considered as one of the major unresolved questions in the field of solidification. It is known that convection in a solidifying melt can promote the transition from a columnar to an equiaxed grain growth (CET). In particular, many studies have demonstrated that the application of electromagnetic stirring enhances the area of equiaxed grains and reduces the mean grain size (see e.g. [8, 14-17, 30-33]). It is widely accepted that flow-induced grain refinement and the CET in metallic alloys is triggered by the appearance of additional dendrite fragments originating from the columnar front. The mechanism for grain multiplication by melt convection is supposed to be complex. Hellawell et al. [34] suggested a multi-stage process including the generation and the transport of the fragments from the interdendritic area into the liquid region. In case of sufficient undercooling the fragments can survive and grow. In situations where the density of the primary crystals is greater than that of the liquid phase, sedimentation of the fragments and finally the blockage of the columnar front occur, if a critical concentration of equiaxed grains has been reached. However, none of these stages is fully understood until now.

In particular, two questions are discussed intensively in the literature: i) which mechanism governs the detachment of fragments from the dendritic skeleton and ii) how do the fragments reach the region ahead of the solidification front. Both issues have been investigated within our project. In the next section we will introduce a new theoretical model for improving the understanding of the dynamics of the pinch-off of dendritic side arms. This model was calibrated and validated by experiments performed at the ESRF synchrotron X-ray source in Grenoble (ROBL beamline, see section 3.2). Section 3.3 reports about experimental investigations carried out at ACCESS for studying the fragmentation in Al-10wt%Cu alloys during directional solidification.

3.1. Numerical sidearm model

A numerical model was developed within this project for predicting the evolution of important geometrical features of dendritic sidearms during coarsening and investigating the systematic effects of the cooling rate and basic geometrical parameters. The methodology which is described in [35] incorporates an axisymmetric phase-field model for capillarity-driven interface dynamics in a binary alloy. The model assumes the usual conditions for coarsening at vanishing melt undercooling where material transport is due to quasi-stationary diffusion through the liquid phase. In addition to the isothermal case, the effect of finite cooling rates is studied, where the solid volume varies over time.

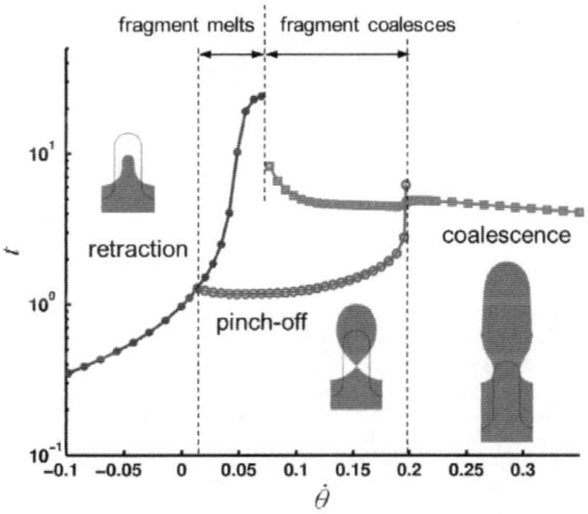

Figure 3. Durations of the different sidearm evolution scenarios as a function of cooling rate for some selected configuration; the initial and final evolution states are indicated by black and grey shapes respectively; all quantities are non-dimensionalized (for further details, see [35]).

The model was used to derive fundamental characteristics and limits of the pinching instability at the junction between a sidearm and its parent stem. It was shown that the evolution of the neck approaches the universal dynamics described in [36]. Moreover, the calculations revealed that the sidearm evolution is governed by its initial geometry and the rate of solidification. The durations and occurrence of different evolution scenarios, such as retraction or pinch-off, are shown in figure 3 as a function of the cooling rate. These results indicate that pinch-off only can only occur within a limited range of cooling rates and is in competition with retraction and coalescence. For further details we refer the reader to [35]. An application of the model to experimental data and quantitative comparison can be found in [38].

3.2. Fragmentation experiments at ROBL

The solidification experiments for validation of the sidearm model were conducted in two stages:

- Growth stage: The Ga–25wt%In alloy was cooled down over a period of 30 min at a cooling rate of 0.01 K/s. The cooling phase is dominated by dendritic growth.
- Isothermal stage: After completion of the cooling phase an isothermal state was maintained over a period of 150 min retaining a constant temperature difference of 15 °C between heater and cooler. The coarsening of the dendrite structure occurs during this stage.

For studying the temporal evolution of the side arm morphology during the isothermal holding stage we selected an observation window which is centered 6.5 mm above the bottom cooler (see figure 4). Figure 5a-c shows an enlarged image segment of figure 4. Here, the brightness gradients are plotted for a better identification of the contours of the solid structure. The picture sequence illustrates the evolution of the shape of the three sidearms (labelled 1, 2, 3) during the pinch-off process. The narrow neck is formed above the junction between the sidearm and the parent stem. Later, the sidearm 2 detaches at the neck and the resulting fragment coarsens into a spheroid (figure 5c).

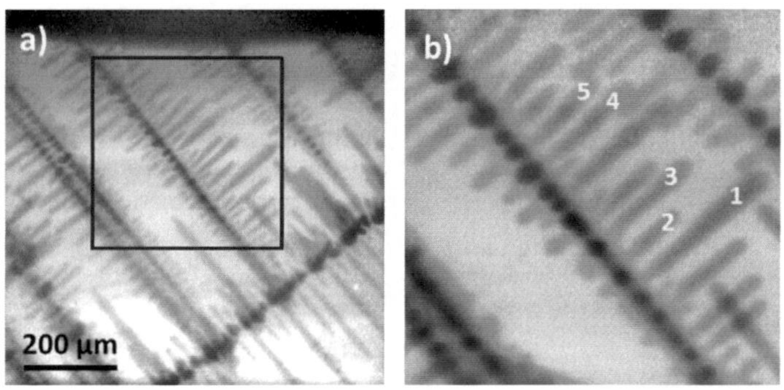

Figure 4. X-ray images showing the area of investigation: (a) dendritic structure with field of examination (blue box), (b) dendrite with five sidearms selected for quantitative measurements.

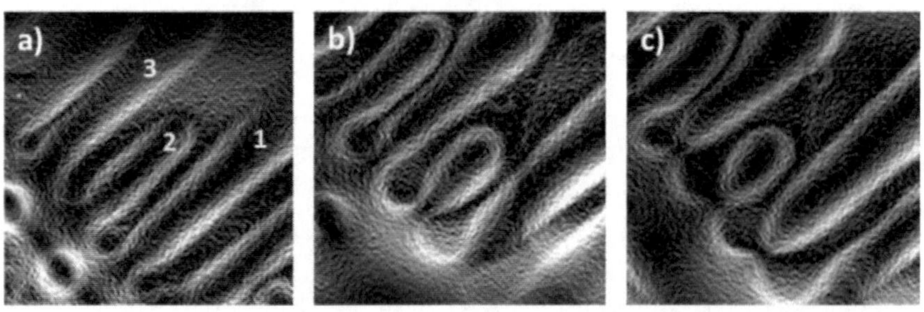

Figure 5. Images showing the development of secondary branches at different times during isothermal coarsening: (a) 5380 s, (b) 1670 s before- and (c) at the moment of pinch-off.

Measured values for the neck diameter as a function of time obtained from image processing are presented in figure 6. The same behavior of the neck diameter can be observed for all five side arms presented here. It shows the typical dynamics of capillarity-driven pinch-off, where the neck approaches a self-similar state during collapse [36]. The reduction of the neck follows the expected

$t^{1/3}$-law. Measurement uncertainties are responsible for the stronger decline of the neck diameter just before the pinch off. Likewise, the scatter of the data is caused by the limited contrast and image noise.

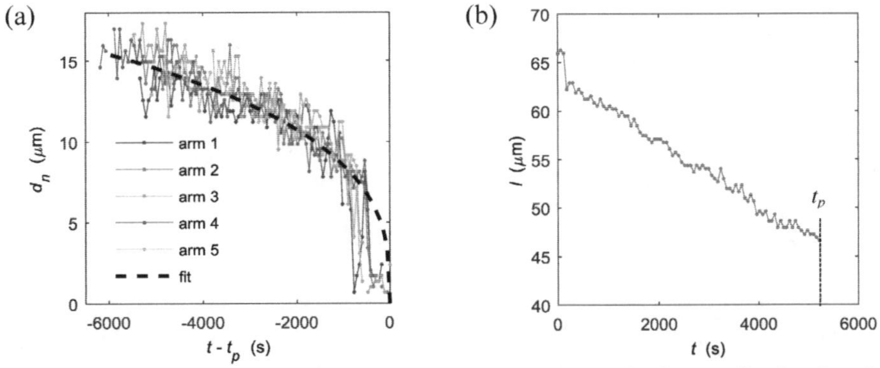

Figure 6. (a) Evolution of neck diameter for the sidearms labeled in figure 4b; the data is fitted to the pinching model in [36], where t_p is the pinch-off time. (b) Evolution of the length of the sidearm 2.

We demonstrate in [38] that the measured dynamics can be reproduced very well by means of numerical simulations using the axisymmetric sidearm model presented in the previous section. The corresponding temporal dynamics of the pinching processes have been utilized for determining the parameter Dd_0 (diffusion coefficient multiplied by the capillary length), which is a parameter of primary interest in modelling solidification and coarsening processes. This procedure has the potential to quantify the value for Dd_0 by *in-situ* observations followed by a rather simple theoretical analysis, which otherwise needs to be acquired from various delicate experiments.

3.3. Fragmentation in Al-10wt%Cu alloys

A series of solidification experiments investigated the formation of the grain structure in Al-10wt%Cu alloy during directional solidification. The experiments were carried out at ACCESS both in 3D bulk samples and in 2D sheet-like samples for visualization of grain growth and fragmentation using *in-situ* X-ray diagnostics. The specific alloy composition was chosen because of its special property that both the initial melt composition and the solidifying primary Al dendrites have almost identical densities. Therefore, gravity-related effects such as buoyancy or sedimentation acting on nucleated or fragmented solid particles in the melt are expected to be negligible.

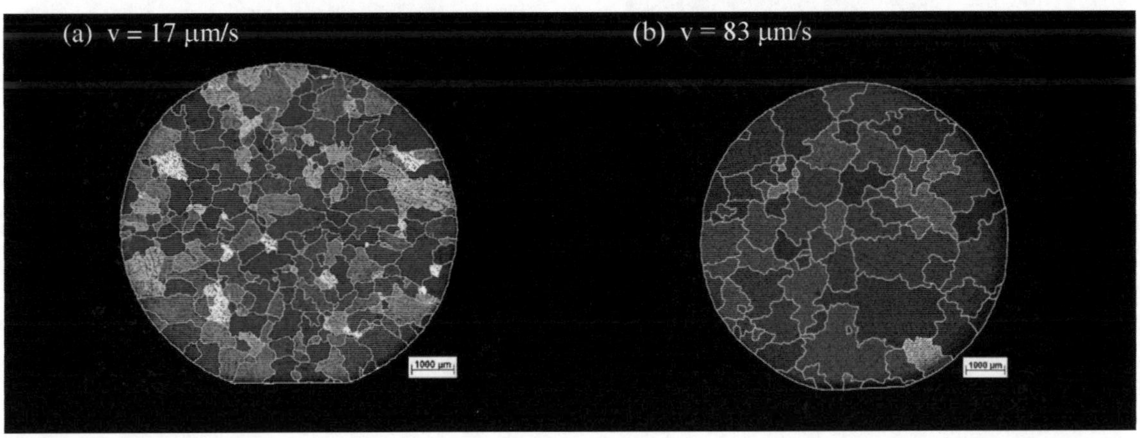

Figure 7. Grain structure in radial cross-section of the processed sample; the colors refer to the grain orientation.

IOP Conf. Series: Materials Science and Engineering **228** (2017) 012005 doi:10.1088/1757-899X/228/1/012005

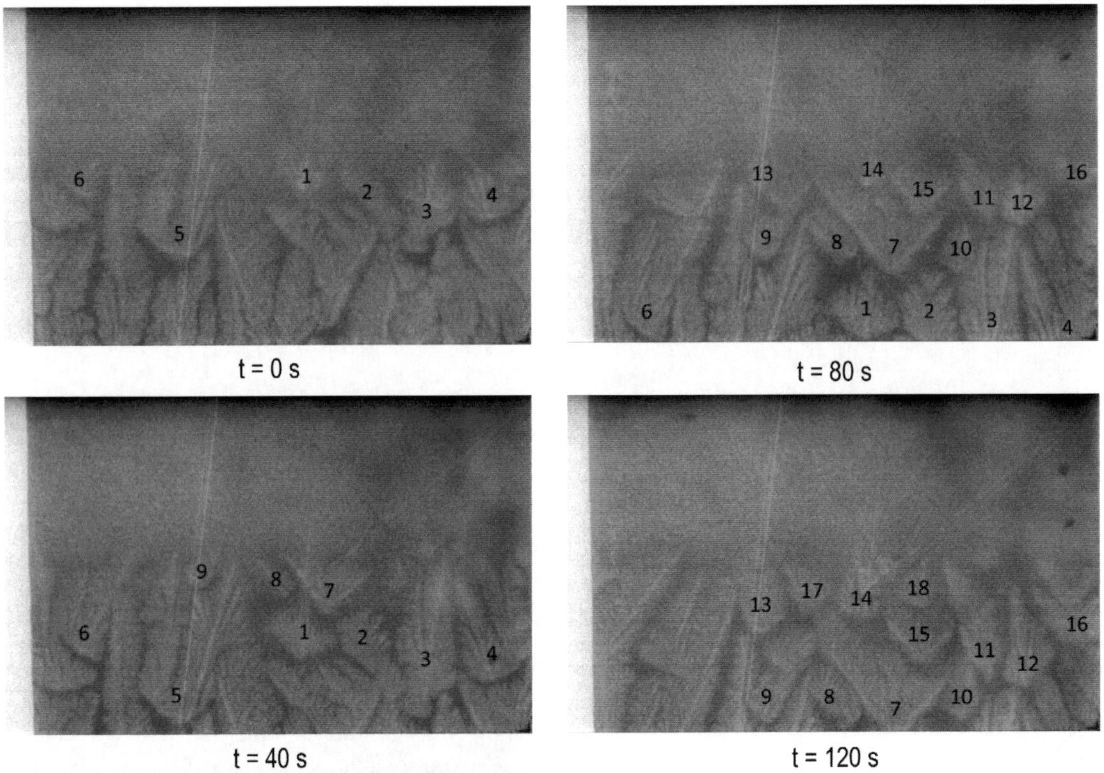

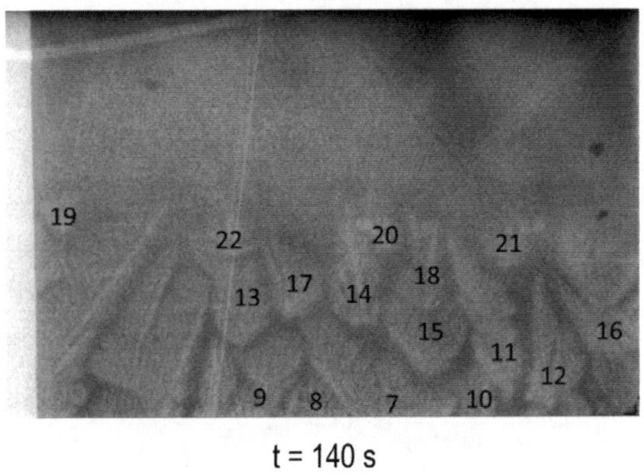

Figure 8. X-ray image sequence showing the directional solidification of a grain-refined Al-10wt%Cu alloy at G = 10 K/mm and V = 20 μm/s in an observation window moving with the solidification front.

The main result is that unexpected growth of equiaxed grains was found at low solidification velocities where normally columnar growth would be expected. By contrast we found that more columnar grains occur for higher solidification velocities (see [27] for details). Exemplary grain structures in radial cross-sections for different solidification velocities are presented in figure 7. For the lowest velocity many grains exist with different orientations (represented by the wide spectrum of different colors), whereas for higher velocities larger grains appear with a preferred orientation, see the narrower color spectrum compared to figure 7a. This surprising result can be explained by fragmentation. In-situ X-ray observations revealed that fragments continuously detach from the dendrite tip region. Within the solutal boundary layer at the solid-liquid interface they experience a moderate acceleration due to the higher solute concentration. If the resulting velocity of the fragments is larger than that of the solidification front, the fragments growing in the interdendritic liquid can leave the mushy zone. Buoyancy disappears as a driver for the motion of the fragments as soon as a fragment has left the solutal boundary layer. The fragments stay just ahead of the columnar front and form new equiaxed grains in the undercooled region [27]. This is the reason why the equiaxed grain structure appears at low solidification velocities, which is normally expected for much higher solidification velocities due to the classical columnar-to-equiaxed transition (CET).

Further solidification experiments were carried out in refined alloys. Here, 0.1wt%Al-Ti-B was added as grain refiner to the Al-10wt%Cu alloy. Directional solidification was achieved applying a temperature gradient of 10 K/mm and a solidification velocity of 20 μm/s. The results were compared to corresponding data obtained from non-refined alloys [27]. An exemplary sequence of images recorded at a time interval of 20 s is displayed in figure 8. Here, an ascending numbering of the grains was inserted for better identification. It can be seen that new grains are constantly being formed and that these grow subsequently. A fragment is usually the nucleus of the new grain. It is difficult to answer whether additional crystal nuclei are formed in the subcooled melt due to grain refinement. In order to detect such nuclei in their initial growth phase, the local and temporal resolution of the X-ray diagnostics used is not sufficient.

The nucleation rate can be estimated from figure 8. In the time interval $t = 40$ s to $t = 140$ s, about 15 new nuclei are produced (8 to 22). The respective nucleation rate of 0.15 nuclei/s is only slightly higher than that in non-grain refined alloy with 0.10 nuclei/s [27]. Thus, it can be concluded that fragmentation occurs in grain-refined as well as in non-grain-refined Al-10%wtCu alloys for the solidification parameters given here, and results in the same nucleation rates. Thus, the grain refining agent is not capable of sufficiently reducing the critical nucleation undercooling in order to generate further grains. As in the case of the non-grain-refined alloy, the vertical component of the buoyancy rate of the fragments is also estimated here before they are trapped by the solidification front. A movement of around 550 μm in a time interval of 40 s is typical. This means that an average buoyancy speed of about 14 μm/s can be determined. This value agrees very well with the value for the non-grain-refined alloy, which was between 10 μm/s and 16 μm/s. Thus, both alloys are not different in terms of fragmentation behavior.

4. Segregation

Thermosolutal convection has been identified as the main reason for the development of solute-rich channels in the mushy zone during the solidification of metallic alloys. The development of such kind of segregation channels was studied by means of X-ray radioscopy using the setup described in section 2.2. The Ga–25wt%In alloy was solidified directionally in vertical direction from the bottom of the solidification cell. Figure 9 presents a series of images documenting an exemplary solidification process. The solidification of this alloy composition leads to a growth of In dendrites. The restricted solubility of Ga in the In primary crystals causes an accumulation of rejected Ga and a formation of a solutal boundary layer as shown in figure 9a. The gallium-enriched melt is less dense as compared to the initial composition. Due to the unstable density stratification an intense solutal-driven convection occurs in form of rising plumes containing Ga-rich liquid. As a result, an arrangement of convection rolls is formed along the solidification front (see figure 9b). Further solute accumulation in the upper mushy zone intensifies the solutal convection. Figure 9c illustrates the formation of distinct plumes, whose intensity and lifetime are determined by the extent to which the plume can be fed by new Ga-rich melt in the mushy zone. Due to the continuity condition, the roll structures at the solidification front induce a convective transport of In-rich fluid towards the mushy zone, which promotes the growth of the dendrites at the edge of the plumes. The resulting solute is carried by the flow into the region below the ascending plume. The increasing Ga concentration lowers the local melting temperature and causes an inhibition of dendritic growth and even a re-melting of solid material. Finally, a segregation freckle (chimney) originates (see figures 9c and 9d). The flow structures observed in our experiments reveal a transient behavior. Parameters like the intensity and the position of the plumes are not constant over time. The unstable flow obviously affects the development of the segregation chimneys. A perturbation of the ascending plume appears in figure 9d. This enables the inflow of In-rich melt into the chimney (figure 9e) which results in an accelerated crystal growth. The consequence is a “self-healing” process, i.e. the chimney is filled in a short time by new dendrites and finally disappears (figure 9f).

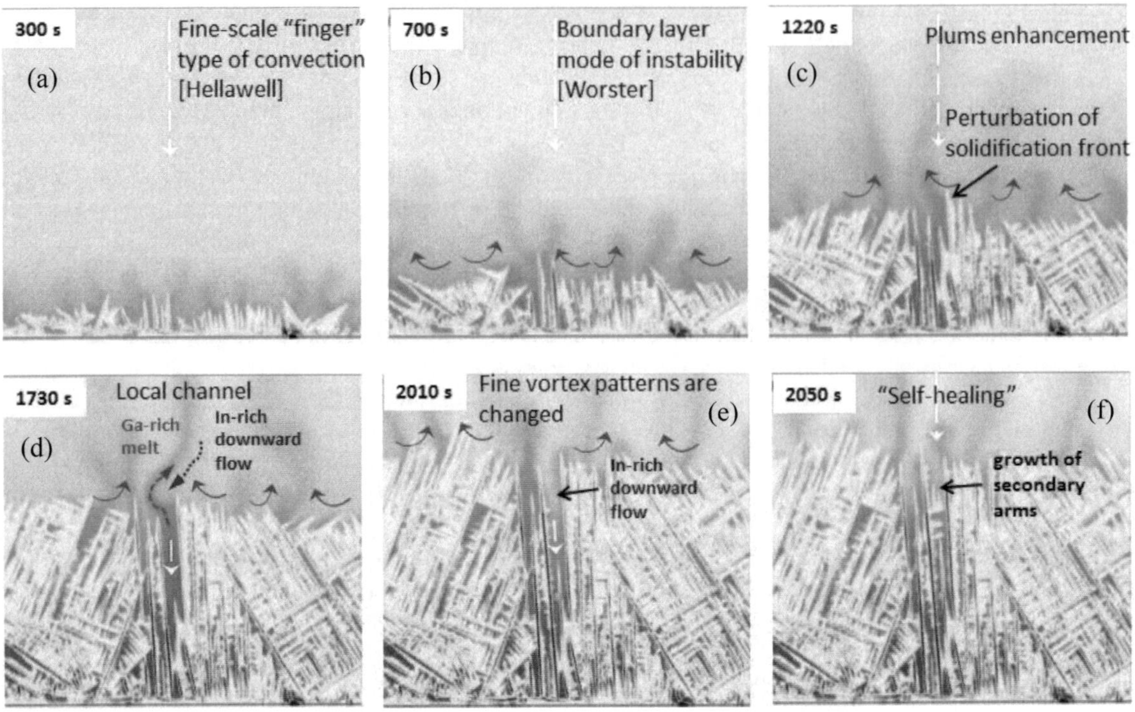

Figure 9. X-ray visualization showing an exemplary bottom-up solidification of a Ga-25wt%In alloy. The color scale represents the Ga concentration (red – Ga-rich melt, blue/green In dendrites).

The behavior described above raises the question how many of the emerging chimneys will survive and can be found in the solidified material. The long-term stability of these segregation channels and the role of the solute Rayleigh number in this context were investigated in [11]. This study showed that different segregation structures occur due to variations of the temperature gradients which in turn cause modifications of the melt flow pattern. While the stability of the chimneys is very much subject to fluctuations in the flow structure in case of low temperature gradients, the situation at higher temperature gradients is characterized by two dominating convection rolls in the liquid phase which are driven by a lateral temperature gradient and the convex shape of the solidification front. This configuration induces a continuous accumulation of solute in certain regions of the mushy zone and promotes the development of sustaining chimneys. The experimental data obtained by the X-ray radiography were also used for the validation of corresponding numerical simulations [12]. Both simulation and experiment show the complex interaction between the flow structure and the solidification dynamics. For instance, a "pulsing" behavior of the plumes and the occurrence of solute-enriched zones in the upper mushy zone beneath dominating plumes were found. Moreover, a 3-D model of freckle formation at a microstructural level was coupled with in the *in-situ* X-ray radiography to predict the mechanisms of freckle initiation and growth [37].

5. Electromagnetic stirring

5.1. Solidification of Al-10wt%Cu alloys under the influence of an RMF

Another series of solidification experiments were performed at ACCESS where a rotating magnetic field was additionally incorporated at the gradient furnace. Three coil pairs are dimensioned in such a way that homogeneous field strength is ensured over a height of 10 cm. The coil system is moved along with the furnace so that the magnetic field during solidification is always effective in the region of the solidification front. The frequency of the rotating magnetic field is constant at 50 Hz. The magnetic induction can be continuously adjusted from zero up to 13 mT allowing to generate a flow with different intensities in the solidifying melt. The experimental investigations were carried out for solidification velocities varying between 8.3 and 167 µm/s. Figure 10 compares cross sections of samples solidified at a solidification velocity of 16.7 µm/s and different magnetic field strengths of the RMF. At the solidification rate considered here the electromagnetically driven convection causes an increase of the number of grains. For the weaker excitation (B = 2 mT), the mean grain size and average grain size looks rather similar as without an external magnetic field. Based on the quite uniform color spectrum of the electrolytically etched sections for B = 2 mT we can infer small differences in the orientation of the dendrites. This is presumably due to the fact that the slow flow homogenizes the melt and promotes the growth of a few well-directed grains. The stronger excitation (B = 10 mT) generally leads to a clear increase in the mean grain count, as well as a correlation with a decrease in average grain size.

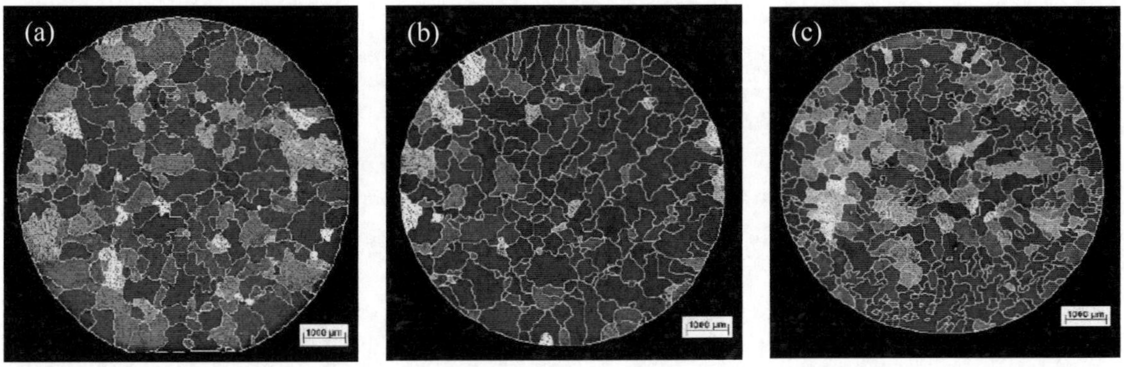

Figure 10. Cross sections of the samples solidified at V = 16.7 µm/s revealing the grain structure: (a) B = 0, (b) B = 2 mT, (c) B = 10 mT.

In general, in the case of low solidification rates (V = 16.7 µm/s), fragmentation seems to play the decisive role. Almost independently of the intensity of the flow in the melt, new grains are formed, which are rapidly overgrown by neighboring grains. Intensive flow leads to a growth of inclined grains, while an equiaxial grain structure remains predominant. With increasing solidification rates - starting at about 50 µm/s - a columnar grain structure predominantly forms. The number of grains decreases and the mean grain size increases. This effect occurs in the case of no induced flow (B = 0) and weak forced convection (B = 2 mT). With intensive flow (B = 10 mT) one finds many, often obliquely growing grains.

5.2. Visualization of the effect of electromagnetically driven convection by X-ray radiography

The X-ray radiography was used for an *in-situ* study of the effect of electromagnetic stirring during the solidification of a Ga-25wt%In alloy in a Hele-Shaw cell. The experimental setup was the same as in Chapter 4. The equipment was extended by a magnetic wheel, which allowed for a controlled excitation of a melt flow in the liquid phase.

Figure 11 shows a sequence of images demonstrating a growth process of In dendrites captured at different times. The development of ensembles of dendrites with different orientations is caused by the fact that no special preparations were made to control the nucleation at the bottom of the solidification cell. The effect of natural convection becomes obvious in figure 11a showing the situation before the magnetic wheel was switched on. Distinct solutal plumes dominate which are ejected at the solid-liquid interface since their density is lower compared with the bulk melt. The rising plumes are accompanied by a downward flow of In-rich melt in the intermediate regions. Accordingly, measurements of the local Ga concentration presented in figure 12a reveal an inhomogeneous horizontal concentration profile along the solidification front. Figures 11b and 11c are recorded at two different times after the magnetic wheel started to rotate counter-clockwise with 30 rpm (rotation per minute). The resulting induced flow is almost horizontal with respect to the solidification front with a velocity of about 0.3 mm/s.

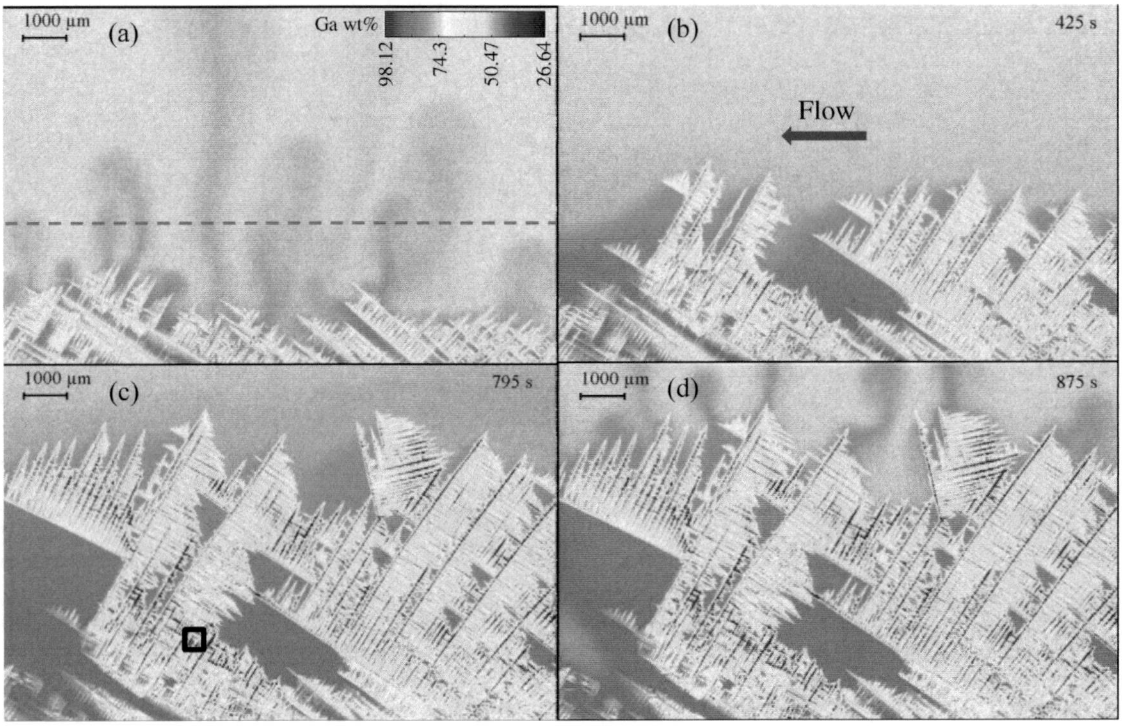

Figure 11. Image sequence showing the solidification of the Ga - 25wt%In alloy under the influence of forced convection at different times: (a) 155 s, (b) 425 s, (c) 795 s and (d) 875 s. Rotation rate of the magnetic wheel was 30 rpm.

In case of forced convection the solutal plumes are strongly damped and the final concentration profile along the solidification front becomes smooth (see the red curve in figure 12a). Slight increase of the Ga concentration on the left-hand side of the red curve is caused by blurred plumes running along the solidification front from right to left. The velocity of the solidification front decreases gradually from ~ 12.9 μm/s down to ~ 4.5 μm/s during its exposure to the driven flow (see figure 12b). This can be explained by a continuous increase of Ga concentration near the solidification front with progressing exposure time.

Another interesting effect can be observed during solidification: the uneven growth of primary dendrites at the beginning of the experiment leads to the formation of Ga-rich zones near the solidification front which develop into distinct segregation freckles as it can be seen on the left-hand side of figures 11b and 11c. Moreover, a competitive growth between secondary branches and primary dendrites leads to an increase of the primary arm spacing. Preferential growth of the secondary arms is observed at the upstream side of the dendrites. In contrast, high solute concentration at the downstream side strongly influences the formation of secondary arms: in case when dendrites are inclined against the upcoming flow the secondary arms at the downstream side are fully eliminated; in case when the dendrites are inclined towards the upcoming flow only few secondary arms develop and create tertiary arms. Furthermore, the inclination angles of the growing dendrites change by several degrees under the influence of the forced flow. A short time after the magnetic wheel is switched off the solute plumes re-appear (see figure 11d) and the solidification velocity increases again (see figure 12b).

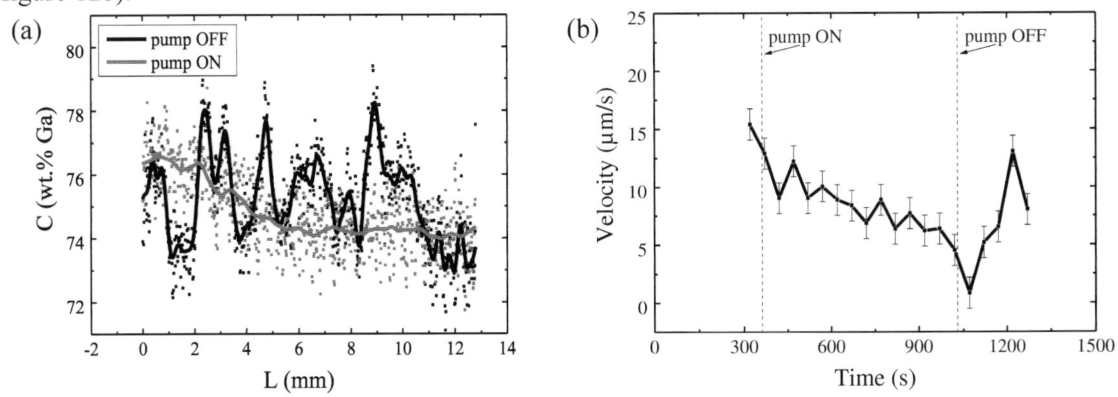

Figure 12. (a) Concentration profiles along the solidification front: 10 s before (black line) and 30 s after (red line) the initiation of the magnetic wheel. Measurements were carried out along the dashed line highlighted in figure 11a. (b) Velocity of the solidification front. The rotation rate of the magnetic wheel was 30 rpm.

Figure 13 shows another experiment carried out at a higher rotation rate (80 rpm, clockwise direction). The situation in figure 13a corresponds to the initial phase of the experiment when the magnetic wheel remains idle. As in the previous experiment the plumes are completely suppressed once the rotating magnetic wheel induces a significant forced flow (figures 13b and 13c). Figure 14a shows the corresponding Ga concentration profiles before (black) and after (red) the magnetic wheel was switched on. It can be seen that the forced flow straightens the Ga concentration along the solidification front considerably. The growth velocity of the solidification front decreases from a value of ~ 14.2 μm/s down to ~ 2.5 μm/s (see figure 14b). Such a fast and strong decrease can be explained by the accumulation of Ga-rich solute with time near the solidification front. A pronounced Ga-rich layer develops near the solidification front due to uneven growth of dendrites (see figure 13c). This leads to the evolution of distinct segregation freckles (see the right-hand side of figures 13b and 13c). The induced strong flow turns out to be an efficient carrier of In-rich material promoting the dendrite growth. As a result secondary arms grow preferably at the upstream side of the dendrites; while their development is impeded at the downstream side due to the high solute concentration. Increasing arm spacing due to competitive growth between secondary and primary arms is also observed. As in the previous experiment, the solute plumes re-appear again after the magnetic wheel is switched off (see Figure 13d).

Despite multiple similarities certain differences between the two experiments can be clearly identified: first, regions with higher In concentration are built up at higher rotation speeds (80 rpm) (see figures 11c and 13c for comparison). The average In concentration was measured in both experiments in the most In-rich areas marked as squares in the figures 11c and 13c (0.52×0.52 mm^2 size). A value of ~ 47.5wt% was found for 30 rpm, while a value of ~ 63.6wt% occurs in the second experiment at 80 rpm. Moreover, the effects of the reduction of the solidification front velocity and the flattening of the concentration profiles along the solidification front appear to be much stronger in case of the more intense forced flow.

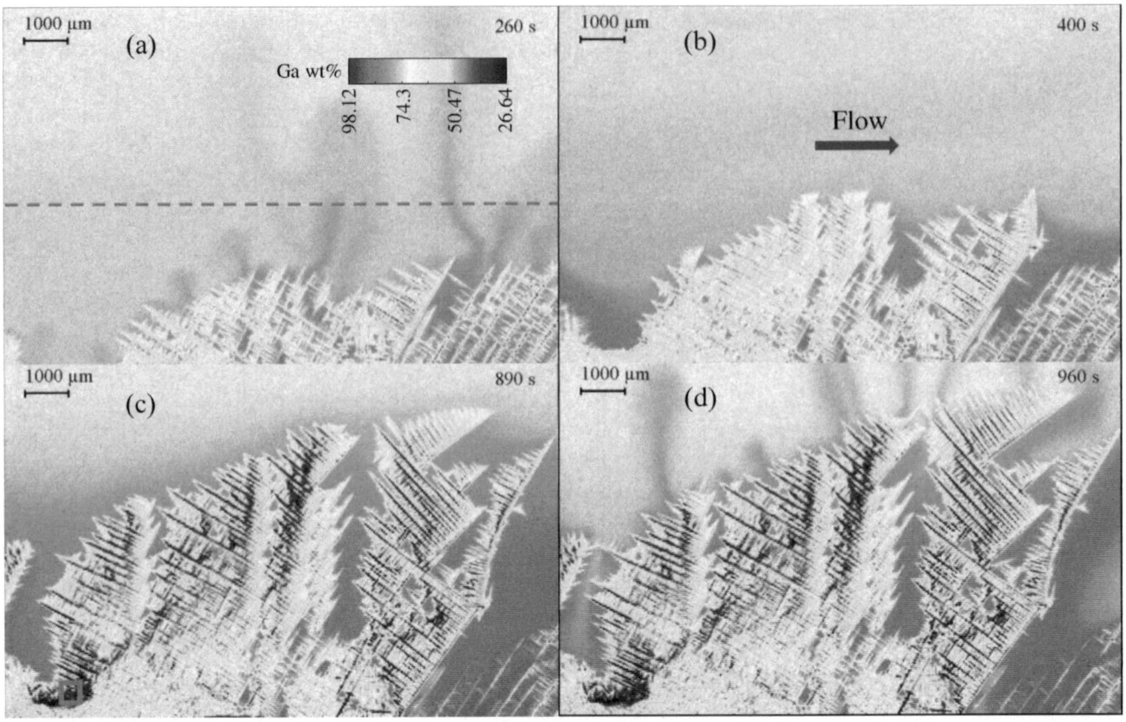

Figure 13. Image sequence showing the solidification of the Ga - 25wt%In alloy under the influence of forced convection at different times: (a) 260 s, (b) 400 s, (c) 890 s and (d) 960 s. Rotation rate of the magnetic wheel was 80 rpm.

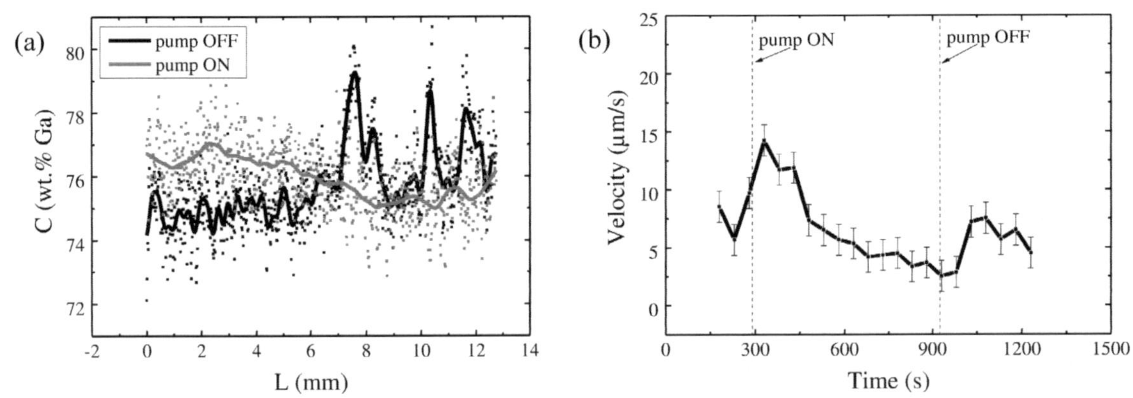

Figure 14. (a) Concentration profiles along the solidification front: 10 s before (black line) and 30 s after (red line) the initiation of the magnetic wheel. Measurements were carried out along the dashed line highlighted in figure 13a. (b) Velocity of the solidification front. The rotation rate of the magnetic wheel was 80 rpm.

6. Conclusions

We have presented experimental and numerical results that were obtained within the project "Magnetic Flow Control Solidification" of the Helmholtz Alliance "LIMTECH". The main goal of the project was to improve the understanding of the complex interactions between melt flow and solidification. In particular we have addressed several aspects in the context of grain refinement, dendrite fragmentation and segregation mechanisms. Various solidification experiments were performed for studying solidification processes under the influence of natural convection or electromagnetically driven flow. Directional solidification experiments with an Al-10wt%Cu alloy have revealed an unexpected equiaxed grain growth at low solidification velocities, instead of columnar growth. This exceptional behavior was conclusively explained by the specific conditions that affect formation and movement of dendrite fragments at the solidification front. Further activities have focused on the evolution of dendritic structures during coarsening. The combination of theoretical modeling and experiments performed at the ESFR synchrotron X-ray source in Grenoble has allowed to improve the understanding of the pinch-off of dendritic side arms and to obtain material information that is relevant for quantitative modeling. Our future work will include the systematic study of the evolution of dendritic morphologies at finite cooling rates and under the impact of melt flow.

Acknowledgements

The authors acknowledge the financial support by the German Helmholtz Association in form of the Helmholtz-Alliance "LIMTECH". The authors also gratefully acknowledge the Rossendorf Beam Line (material research) at ESRF (Grenoble, France) and the scientific and experimental support from Joerg Grenzer, Olga Keplinger and Carsten Baehtz.

References

[1] Hellawell A, Sarazin J R and Steube R S 1993 *Philosophical Transactions: Physical Sciences and Engineering* **345** 507
[2] Sample A K and Hellawell A 1984 *Metallurgical Transactions A* **15** 2163
[3] Tewari S N and Shah R 1992 *Metallurgical Transactions A* **23** 3383
[4] Bergman M I, Fearn D R, Bloxham J and Shannon M C 1997 *Metallurgical and Materials Transactions A* **28** 859
[5] Murakami K, Fujiyama T, Koike A and Okamoto T 1983 *Acta Metallurgica* **31** 1425
[6] Medina M, Du Terrail Y, Durand F and Fautrelle Y 2004 *Metallurgical and Materials Transactions B* **35** 743
[7] Boden S, Eckert S and G Gerbeth 2010 *Materials Letters* **64** 1340
[8] Willers B, Eckert S, Michel U, Haase I and Zouhar G 2005 *Materials Science and Engineering A* **402** 55
[9] Steinbach S and Ratke L 2007 *Metallurgical and Materials Transactions A* **38** 1388
[10] Zimmermann G, Weiss A and Mbaya Z 2005 *Materials Science and Engineering A* **413-14** 236
[11] Shevchenko N, Boden S, Gerbeth G and Eckert S 2013 *Metallurgical and Materials Transactions A* **44** 3797
[12] Saad A, Gandin C A, Bellet M, Shevchenko N and Eckert S 2015 *Metallurgical and Materials Transactions A* **46** 4886
[13] Eckert S, Nikrityuk P A, Willers B, Räbiger D, Shevchenko N, Neumann-Heyme H, Travnikov V, Odenbach S, Voigt A and Eckert K 2013 *Eur. Phys. J. Spec. Top.* **220** 123
[14] Roplekar J and Dantzig J 2001 *International Journal of Cast Metals Research* **14** 79
[15] Griffiths W D and McCartney D G 1996 *Materials Science and Engineering A* **47** 46
[16] Kovács J, Roósz A and Szőke J 2006 *Materials Science Forum* **508** 263
[17] Zimmermann G, Vitusevych V T and Sturz L 2010 *Materials Science Forum* **649** 249
[18] Willers B, Eckert S, Nikrityuk P A, Räbiger D, Dong J, Eckert K and Gerbeth G 2008 *Metallurgical and Materials Transactions B* **39** 304
[19] Kojima S, Ohnishi T, Mori T, Shiwaku K, Wakasugi I and Ohgarni M 1983 *66th Steelmaking Conf. Atlanta, GA, ISS-AIME, Warrendale* 127
[20] Wang X, Fautrelle Y, Etay J and Moreau R 2009 *Metallurgical and Materials Transactions B* **40** 82
[21] Eckert S, Nikrityuk P A, Räbiger D, Eckert K and Gerbeth G 2007 *Metallurgical and*

Materials Transactions B **38** 977
[22] Mathiesen R H, Arnberg L, Mo F, Weitkamp T and Snigirev A 1999 *Phys. Rev. Lett.* **83** 5062
[23] Koster J N, Seidel T and Derebail R 1997 *Journal of Fluid Mechanics* **343** 29
[24] Reinhart G, Mangelinck-Noël N, Nguyen-Thi H, Schenk T, Gastaldi J, Billia B, Pino P, Härtwig J and Baruchel J 2005 *Materials Science and Engineering A* **413** 384
[25] Yasuda H, Ohnaka I, Kawasaki K, Sugiyama A, Ohmichi T, Iwane J and Umetani K 2004 *Journal of Crystal Growth* **262** 645
[26] Liotti E, Lui A, Vincent R, Kumar S, Guo Z, Connolley T, Dolbnya I P, Hart M, Arnberg L, Mathiesen R H and Grant P S 2014 *Acta Materialia* **70** 228
[27] Zimmermann G, Pickmann C, Hamacher M, Schaberger-Zimmermann E, Eckert K and Eckert S 2017 *Acta Materialia* **126** 236
[28] Shevchenko N, Roshchupkina O, Sokolova O and Eckert S 2015 *Journal of Crystal Growth* **417** 1
[29] Boden S, Willers B, Eckert S and Gerbeth G 2008 *Metallurgical and Materials Transactions A* **39** 613
[30] Johnston W C, Kotler G R, Ohara S, Ashcom H V and Tiller W A 1965 *AIME MET SOC TRANS* **233** 1856
[31] Vives C 1989 *Metallurgical and Materials Transactions B* **20** 623
[32] Campanella T, Charbon C and Rapaz M 2004 *Metallurgical and Materials Transactions A* **35** 3201
[33] Liu S F, Liu L Y and Kang L G 2008 *Journal of Alloys and Compounds* **450** 546
[34] Hellawell A, Liu S and Lu S Z 1997 *JOM* **49** 18
[35] Neumann-Heyme H, Eckert K and Beckermann C 2015 *Physical Review E* **92** 060401(R)
[36] Aagesen L K, Johnson A E, Fife J L, Voorhees P W, Miksis M J, Poulsen S O, Lauridsen E M, Marone F and Stampanoni M 2011 *Acta Materialia* **59** 4922
[37] Karagadde S, Yuan L, Shevchenko N, Eckert S, and Lee P D 2014 *Acta Materialia* **79** 168
[38] Shevchenko N, Neumann-Heyme H, Lei Z, Eckert K, Keplinger O, Grenzer J, Beckermann C and Eckert S 2017 *Acta Materialia (submitted)*

IOP Conf. Series: Materials Science and Engineering **228** (2017) 012006 doi:10.1088/1757-899X/228/1/012006

Combined experimental and numerical analysis of a bubbly liquid metal flow

B Krull[1], E Strumpf[2], O Keplinger[2], N Shevchenko[2], J Fröhlich[1], S Eckert[2], G Gerbeth[2]

[1] Institut für Strömungsmechanik, TU Dresden, 01062 Dresden, Germany
[2] Helmholtz-Zentrum Dresden-Rossendorf, 01328 Dresden, Germany

E-mail: benjamin.krull@tu-dresden.de

Abstract. The paper proposes a combined experimental and numerical procedure for the investigation of bubbly liquid-metal flows. It describes the application to a model configuration consisting of a recirculating GaInSn flow driven by an argon bubble chain. The experimental methods involve X-ray measurements to detect the bubbles and UDV measurements to gain velocity information about the liquid metal. The chosen numerical method is an immersed boundary method extended to deformable bubbles. The model configuration includes typical phenomena occurring in industrial applications and allows insight into the physics of bubbly liquid-metal flows. It constitutes an attractive test case for assessing further experimental and numerical methods.

1. Introduction

Bubbly liquid metal flows occur in different flow regimes for a wide range of gas flow rates: single bubbles, bubble chains, swarms and plumes. In many industrial processes, gas bubbles are injected into liquid metals to promote the mixing of the melt, to control chemical reactions or to remove undesired inclusions. The injection of a large number of gas bubbles leads to complex flow conditions, the prediction of which is extremely challenging. In addition to the interaction with the surrounding liquid, the bubbles generally interact with each other or with container walls. They deform distinctly due to the local pressure field and induce a flow in the surrounding liquid metal, which, in turn, acts upon the bubble. Substantial work, experimental as well as numerical, has been carried out to improve the understanding of these flows in their great diversity [1–7]. The physical and numerical models, however, still lack sophistication as they need reliable experiments for validation. The detection of flow properties in opaque fluids requires more advanced measurement techniques such as neutron- or X-ray radiography, ultrasound techniques or inductive methods. This is also the major reason for the fact that to date there are only a few systematic experimental studies on the behavior of gas bubbles in liquid metals. Driven by this unsatisfactory situation, extensive research was carried out within the last ten years at the HZDR and the last five years in the LIMTECH alliance for the development and application of specific measurement techniques in liquid metals. The present paper focused on work performed within project A5 of this alliance which aimed to develop and qualify powerful and efficient numerical methods for an appropriate description of dispersed liquid metal bubbly flows. In parallel, suitable model experiments were realized to provide an experimental data base for code validation. The capabilities of both approaches will be demonstrated focusing on the specific case of a bubble chain rising in an open, narrow container filled with liquid metal. This configuration includes the most important phenomena occurring in complex bubbly flows being relevant for industrial applications. Moreover, the problem is simplified enough to facilitate experimental measurements and direct numerical simulations and in-depth physical analysis.

The paper is organized as follows: The experimental and numerical methods are introduced in section 2. Results for the specific case of a bubble chain in a flat container are presented and discussed in section 3. A summary and conclusions will be given in section 4.

2. Methods for investigating bubbly liquid metal flows

2.1. A setup for investigating single bubbles, bubble chains and small swarms

This paper focuses on argon bubbles in GaInSn in a rectangular, narrow container. Such a flat geometry fulfills two crucial requirements: On one hand, the numerical model resolving very small scales of the flow has to face the limited power of current high-performance computers. A narrow container simply facilitates a higher resolution of the calculation domain. On the other hand, the X-ray radioscopy was chosen as measuring technique which provides an in-situ visualization of the gas bubbles rising in the opaque liquid metal. The high attenuation of X-ray radiation by liquid metals provides a good contrast but restricts the thickness of the fluid container along the direction of the X-ray beam.

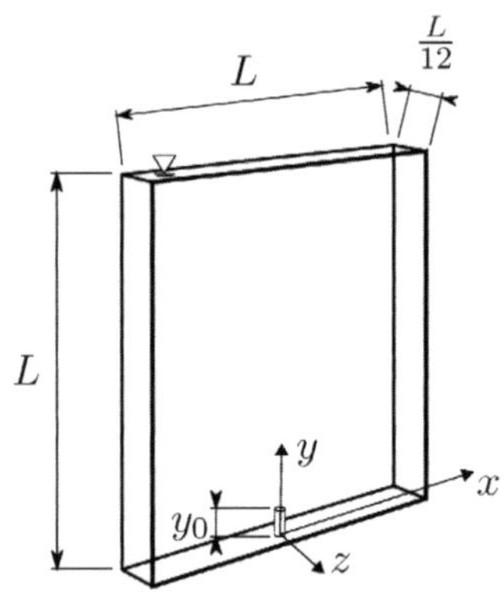

Figure 1. Sketch of the fluid domain with free surface and centric injection needle.

GaInSn is liquid at room temperature and, hence, highly suitable for low-cost experiments without heating. A compilation of thermophysical data of this alloy can be found in [8]. In the case considered here, argon gas was injected through a vertically oriented injection nozzle placed at the center of the bottom plate at a height, see figure 1. The vessel employed was made of acrylic glass with extensions 144 × 200 × 12 mm^3 and filled with GaInSn up to a height of 144 mm. Acrylic glass was chosen for the wall material because it is transparent and facilitates the X-ray and ultrasonic measurements due to its low X-ray absorption and a sound velocity comparable to the one of the liquid metal. The container walls have a thickness of 10 mm, except at the positions of the ultrasonic sensors where the wall thickness was reduced to 2 mm. The nozzle was a non-wetting medical injection needle (Sterican, Gauge 19) made of stainless steel with an outer diameter $D_a = 1.09$ mm and an inner diameter $D_i = 0.785$ mm with a bevel orifice facing in the x-direction placed at a height of about 5 mm. This simple arrangement allows the investigation of single bubbles, chains and small swarms.

IOP Conf. Series: Materials Science and Engineering **228** (2017) 012006 doi:10.1088/1757-899X/228/1/012006

2.2. Experimental methods

The applied experimental procedure combines the X-ray radioscopy to visualize the bubbles with the ultrasound Doppler velocimetry (UDV) to measure velocity profiles of the liquid metal flow. The following measurement equipment has shown to be suitable for investigating bubbly liquid metal flow.

For X-ray measurements, a high-power source was applied (ISOVOLT 450M1/25-55, GE Sensing & Inspection Technologies GmbH) operating with a maximum voltage of 320 kV and a current of 14 mA. A scintillation screen (SecureX HB, Applied Scintillation Technologies) was installed directly behind the vessel containing the liquid metal to convert the proportion of X-rays transmitted through the sample into visible light. The imaging system is completed by a custom-made lens system (Thalheim Optik) and a video camera. Two different types of video cameras were used. The first one was the CCD camera Pixelfly QE (PCO) operated with a frame rate of 23 fps. The second camera was a Pco.edge AIR (PCO) based on an sCMOS-sensor achieving about 100 fps. The images were captured by the Pixelfly camera with a resolution of 696 × 512 pixels, corresponding to a field of view of 90 mm × 150 mm and was used to capture the overall situation. The Pco.edge camera provided a resolution of 1728 × 1022 pixels, corresponding to a field of view of 60 mm × 100 mm and was used for the quantitative analysis. The exposure times (3 ms for Pixelfly and 5 ms for Pco.edge) were adjusted to achieve an acceptable signal-to-noise ratio and a sufficient contrast. Installing lead shields around the observation area was necessary to reduce the background noise arising from scattered radiation originating from regions outside the liquid metal volume and provided significantly increased image quality. Additional lead shielding was installed around the video camera to avoid radiation damage of the video sensors.

The quantitative analysis of the bubble sizes, positions and velocities was performed by off-line data processing using Matlab scripts for image analysis. After subtracting the mean dark current signal, a shading correction was applied using a mean reference image recorded for a situation without gas injection. Gaussian low-pass filtering was applied to reduce the image noise. Then, the intensity gradient was computed. The final image for analysis was obtained by summing up the smoothed image and the gradient image. Separation of individual bubbles and bubble clusters from the background was achieved by using a threshold value. All pixels with a brightness above the threshold value were added to the bubble area. The total image was further segmented to divide bubble clusters into individual bubbles using the binary image resulting from the previous step. This operation includes a watershed segmentation algorithm to distinguish between bubbles whose projections are close to another or even overlap. For each bubble the center of mass was calculated from its brightness distribution. The projected area of a gas bubble was obtained by counting pixels in the segmented zone. The equivalent bubble diameter was calculated from the projected bubble area. Finally, the derived quantities were converted from pixel units to metric units using the scaling of the image. This setup is similar to that applied in [9].

The UDV technique was applied to measure the velocity field in the continuous phase. This is restricted to zones without bubbles because of their blocking effect. A detailed description of the measuring principle can be found in [10, 11]. For the present study, specially manufactured sensor arrays (Richter Sensor & Transducer Technology) were used operating at an emission frequency of 8 MHz. The sensor array consists of 25 single piezo elements and has a total length of about 70 mm. Two adjacent elements were always activated simultaneously. Eight of these pairs of piezo elements were operated for recording vertical profiles of the vertical velocity with a time resolution of about 32 ms. The instrument DOP3010 (Signal Processing SA) was used for multiplexed excitation of the piezo elements and data acquisition. The overall time resolution for all eight measuring lines was about 4 frames per second, and a spatial resolution of about 2.4 mm was achieved in y-direction. The angle of divergence of the acoustic beam was approximately 5 °, the sensor array was attached to the bottom wall of the fluid container and the ultrasonic energy was transmitted through the acrylic glass wall using a water-soluble coupling gel. The following control parameters of the DOP3010 were applied: pulse repetition frequency of 220 ms, 64 emissions per profile and an internal amplification of 0 to 40 dB (slope). A value of 2730 m/s was assumed for the speed of sound within the liquid metal [12].

A gas flow control system (MKS Instruments Deutschland GmbH) was used to cover a wide range of gas flow rates from 10 to 680 sccm (standard cubic centimeters per minute, valid for nitrogen, 1 bar, 0 °). In the case of argon at room temperature 1 sccm corresponds to 1.07 cm^3/min or 17.8 mm^3/s.

2.3. Numerical method

The numerical simulations were performed using the multiphase code PRIME [13] with the extensions presented in [14] and [15]. It solves the incompressible Navier-Stokes equations on a staggered Cartesian grid by a finite-volume method and employs an immersed boundary method (IBM) for the coupling between fluids and bubbles. The bubbles are treated as deformable objects of axisymmetric shape. Their motion is decomposed into three parts: translation, rotation and change of shape. Translation and rotation are determined by the corresponding ordinary differential equations for linear and angular momentum, respectively, relating these to the force and torque exerted by the fluid onto the bubble.

The instantaneous shape of each individual bubble is represented by a finite series of spherical harmonics. The surface is described in spherical coordinates by the surface points $\mathbf{x}(\vartheta, \varphi) = \mathbf{x}_p + r(\vartheta, \varphi)(\cos\varphi \sin\vartheta, \sin\varphi \sin\vartheta, \cos\vartheta)^{\mathrm{T}}$ with $\vartheta \in [0, \pi]$, $\varphi \in [0, 2\pi)$, $\mathbf{x}_p$ the center point, and the radial approximation

$$r(\vartheta, \varphi) \approx \sum_{n=0}^{N} \sum_{m=-n}^{n} c_{nm} Y_n^m(\vartheta, \varphi) \,. \tag{1}$$

Here, Y_n^m identifies the spherical harmonic function of degree n and order m, c_{nm} the associated coefficients, and N the order of the approximation. In the present case, the series is restricted to an axisymmetric shape by reducing the second sum to the contribution $m = 0$. The representation equation (1) allows very accurate determination of geometrical properties such as surface curvature, etc., via appropriate analytical formulae for the spherical harmonic functions [16]. Furthermore, the bubble volume can be controlled exactly by controlling the coefficients in equation (1). The coefficients c_{nm} are computed by balancing the forces on the bubble surface minimizing the potential displacement energy [17]

$$W = \int_S [(p_i - p_o) - 2\sigma H]\, \delta n \, \mathrm{d}A \quad \rightarrow \quad \min \,. \tag{2}$$

This equation describes the balance of the forces by the inner pressure p_i, the outer pressure p_o and the surface tension effects proportional to the mean curvature H.

In equation (2), viscous stresses are neglected. On the bubble surface a no-slip condition is imposed, since the bubbles in the present liquid metal case have typically an oxide layer on the phase boundary. These assumptions are further discussed in [18] and [19].

Due to the narrow shape of the container considered in the present study, collisions of bubbles with walls, collisions between bubbles and bubble coalescence have to be taken into account. In the present approach, collisions are represented by the model of [20] and coalescence of bubbles is accounted for by the model proposed in [21]. These events, however, are relatively rare in the simulation reported below.

3. Application to a bubble chain

3.1. A bubble chain model problem

The configuration of a bubble chain was chosen for investigation within this study. This choice avoids dealing with the complex bubble swarms but allows investigating essential bubble-bubble interactions, like collisions, coalescence and the effect of bubble wakes. Bubble chains were already addressed in [22] covering low Galilei numbers. A bubble chain in a liquid metal relevant regime was studied in [21], but close validation was not possible due to a lack of detailed experimental data. The present work aims at filling this gap and providing such a validation and, beyond that, aims at providing new quantitative data on the physical phenomena. As mentioned above, the size of the container is L_x in the horizontal x-direction, L_y in the vertical y-direction, and a depth of L_z in the z-direction (figure 1). Bubbles of mean equivalent diameter d_{ref} are injected at the center of the bottom plane at height y_0 and rise upwards where they finally pierce through the free surface. The bubbles are introduced at a constant frequency f_b and are supposed to exhibit a Gaussian distribution of their diameter around the mean value d_{ref} with a standard derivation of σ_d. The present case features a rectangular container of size $(L_\mathrm{x}, L_\mathrm{y}, L_\mathrm{z}) = (1, 1, 1/12)L$ with $L = 144$ mm (figure 1). The bubbles are released with a frequency of $f_b = 30\,\mathrm{Hz}$. The mean bubble diameter is $d_{ref} = 6$ mm with the standard deviation $\sigma_d = 0.08\,d_{ref}$. The liquid metal is GaInSn with $\rho = 6\,440\,\mathrm{kg/m^3}$ and $\nu = 3.46 \cdot 10^{-7}\mathrm{m^2/s}$ [8] and the gas bubbles consist of argon with $\rho_b = 1.780\,\mathrm{kg/m^3}$ which is much smaller than the fluid density. The surface tension at the gas liquid interface is $\sigma = 0.553\,\mathrm{N/m}$. Hence, each bubble starts at $y_0 = 0.8\,d_{ref}$. The container size corresponds to $(L_\mathrm{x}, L_\mathrm{y}, L_\mathrm{z}) = (24, 24, 2)\,d_{ref}$.

Starting point is the container described in section 2.1 filled with GaInSn (figure 1). If one chooses the mean diameter d_{ref} as the reference length, the setting is then fully described by the container size ratios $L_\mathrm{x}/d_{ref}, L_\mathrm{y}/d_{ref}$ and L_z/d_{ref}, the non-dimensional initial bubble position y_0/d_{ref} and five more non-dimensional parameters related to gravitational driven flow

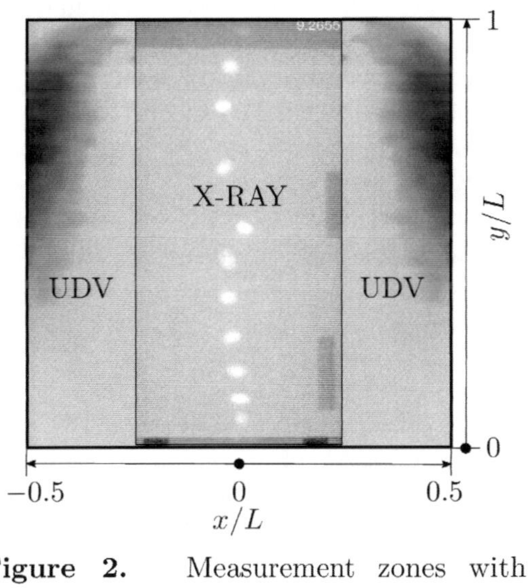

Figure 2. Measurement zones with snapshots of the respective measurement data: vertical velocity component obtained by UDV (left, right) and differences in density by X-ray (center).

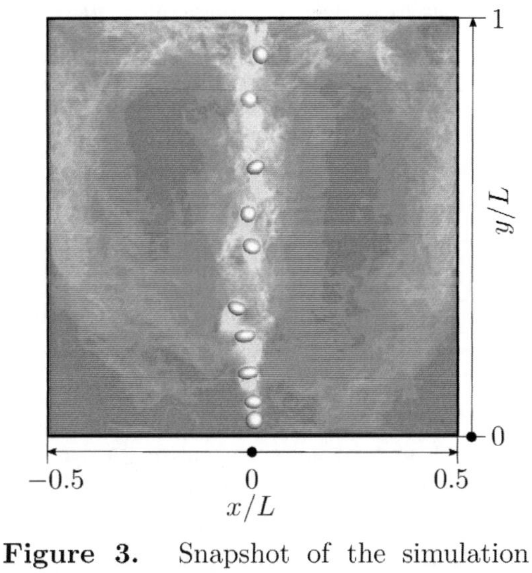

Figure 3. Snapshot of the simulation showing the bubble chain and the magnitude of the velocity field in the center plane.

and bubble injection:

$$\pi_{\Delta\rho} = \frac{\rho - \rho_b}{\rho} \tag{3}$$

$$Ga = \frac{g\, d_{ref}^3}{\nu^2} \tag{4}$$

$$Eo = \frac{\Delta\rho\, g\, d_{ref}^2}{\sigma} \tag{5}$$

$$\pi_{\sigma_d} = \frac{\sigma_d}{d_{ref}} \tag{6}$$

$$\pi_l = \sqrt{\frac{\pi_{\Delta\rho} g}{d_{ref}}}\frac{1}{f_b} \tag{7}$$

These parameters result from the Buckingham π-theorem applied to the governing physical variables, namely the gravitational acceleration g, the fluid density ρ, the fluid kinematic viscosity ν, the surface tension σ, the bubble density ρ_b, the mean bubble diameter d_{ref}, its standard derivation σ_d and the bubble release frequency f_b. The relative density difference $\pi_{\Delta\rho}$, the Galilei number Ga, the Eötvös number Eo and the relative standard deviation of the bubble diameter π_{σ_d} are well-known parameters. The value of $\sqrt{Ga}$ is a good *a priori* approximation for the bubble Reynolds number for single bubbles rising in stagnant fluids. The parameter π_l given in equation (7) is newly proposed here to complete the characterization of the physical situation. If l denotes a reference value of the center-center distance of two consecutive bubbles, the ratio $\pi_l = l/d_{ref}$ characterizes the interaction. Here, $l = u_c/f_b$ is derived from the characteristic buoyancy velocity

$$u_c = \sqrt{\pi_{\Delta\rho} g d_{ref}} \tag{8}$$

and the characteristic time scale $1/f_b$. Small values for π_l reflect high bubble frequencies with small bubble-bubble distances, with the result that interactions are more likely to occur. The limit $\pi_l = 1$ describes a bubble release with vanishing distance of the bubble surfaces, so that coalescence should have a sizable impact. In the limit of very large values of π_l, the bubbles rise without any bubble-bubble interaction. Note that the Morton number Mo can be computed from $Mo = Eo^3 Ga^{-2} \pi_{\Delta\rho}^{-2}$ and is therefore no independent parameter.

Thus, beyond geometrical parameters, the present case is defined by the following values of the dimensionless parameters defined by equation (3) to equation (7): $\pi_{\Delta\rho} = 1$ neglecting the bubble density, $\sqrt{Ga} = 4200$ as a first rough bubble Reynolds number approximation, $Eo = 4.3$ indicating bubble deformation, $\pi_{\sigma_d} = 0.08$ corresponding to a realistic bubble size distribution and $\pi_l = 1.35$ which indicates the possibility of bubble-bubble interactions.

3.2. Details for the experimental and numerical procedures

The methods described in section 2 were applied to the bubble chain model problem of section 3.1 in the following manner. Each experiment was performed in well-defined steps employing the methods described in section 2.2. First, an initial gas flow rate of 10 sccm was applied, resulting in a detachment of single bubbles at a frequency of about 1.5 bubbles per second. Then, data acquisition was started and the gas flow was increased up to the final value which was 200 sccm. This gas flow rate results in a bubble detachment frequency of approximately 27 Hz and equivalent bubble diameters of about 6 mm, which is close to the desired values. All experiments were conducted at room temperature and atmospheric pressure. Both measurement technologies were not be conducted simultaneously but were carried out subsequently under the same conditions. The analysis of the flow field in the continuous liquid phase by UDV focuses on regions on both sides of the bubble chain. The comparison with the numerical simulation relies on eight experiments conducted at a gas flow rate of 200 sccm, each averaging the velocity of the continuous phase over 5.3 seconds. Figure 2 contains exemplary results from both the X-ray and UDV measurements.

Employing the method of section 2.3, the simulations were conducted in dimensionless form on an equidistant grid with $(N_x, N_y, N_z) = (960, 960, 80)$ grid points, adding up to roughly $74 \cdot 10^6$ grid points in total. It resolves the mean bubble diameter d_{ref} with about 40 points. This ensures an acceptable resolution. A no-slip condition was imposed at the bottom wall and the side walls, the free surface was represented by a free-slip condition. Due to the complexity of the bubble detachment process, the nozzle itself was not resolved by the numerical model. Instead, a simplified detachment model was devised [21]. The bubbles are inserted at the bottom center at height $y_0 = 4.8$ mm assuming spherical shape, with the diameter selected according to a Gaussian distribution defined by the mean value d_{ref} and the variance σ_d and a frequency according to π_l. The initial velocity of the bubbles is set to the fluid velocity at this position which is justified due to the negligible bubble inertia. At the free surface the bubbles are removed from the simulation by just switching off the corresponding IBM force [21]. This is a very smooth way and an advantage of the IBM, as it avoids the representation of bubbles piercing through the surface which would be numerically very cumbersome. The time step was adjusted instantaneously during the simulation to maintain a CFL number equal to 0.8. Averaging was started after obtaining a constant value for the bubble mean velocity. The initial phase was excluded from averaging. The whole simulation time corresponds to 17.2 seconds in the experiment, the averaging period to 5.3 seconds. It is known that bubble chains and columns tend to form an s-shape which can suddenly shift to the other side [4, 23–25]. However, such a behavior was not observed in the present case.

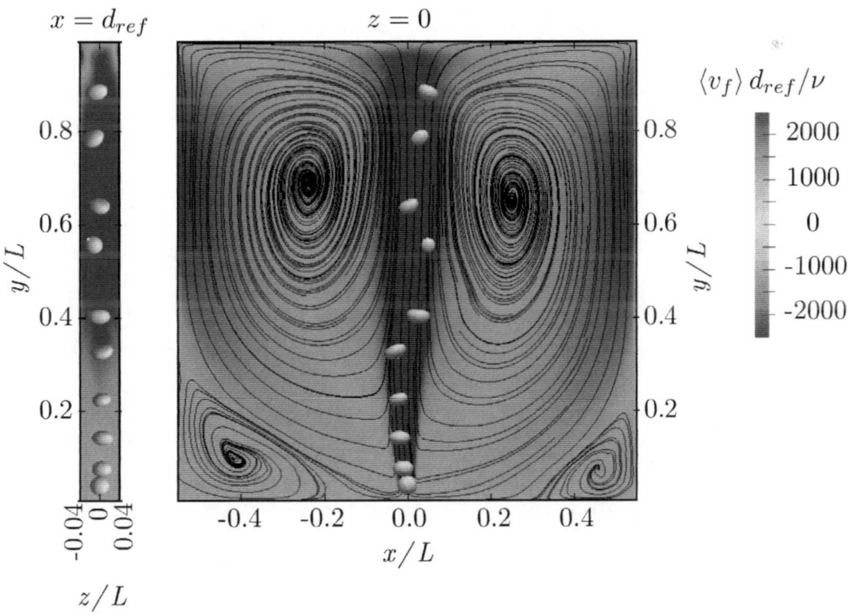

Figure 4. Averaged vertical fluid velocity and streamlines with instantaneous bubble constellation.

3.3. Results for the disperse phase

The center positions of the bubbles are tracked at discrete time steps. The resulting ensemble-averaged trajectory obtained from experiment and simulation are plotted in figure 5 and figure 7, scaled by the container height. Here, $\bar{\psi}$ denotes the average of a bubble property ψ related to

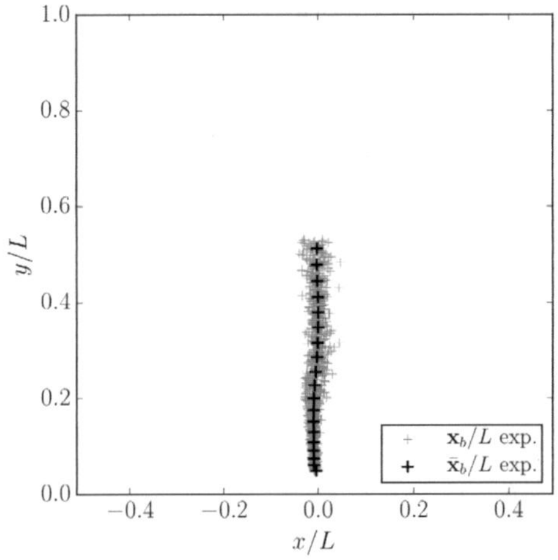

Figure 5. Experimentally determined bubble center positions and average trajectory.

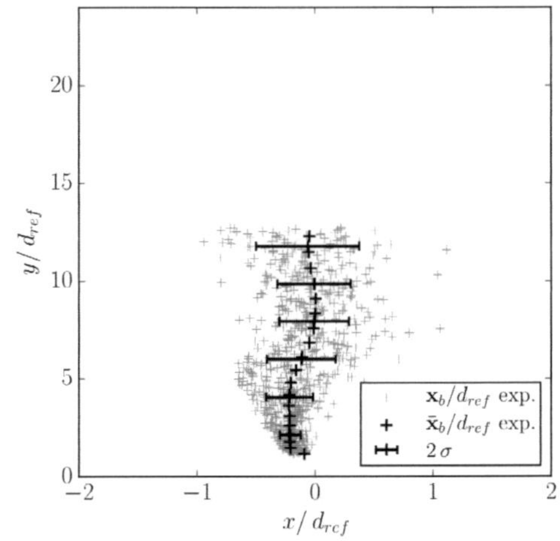

Figure 6. Experimentally determined bubble center positions, average and standard deviation. Same dataset as in figure 5 but scaled with d_{ref} instead of L, horizontally stretched.

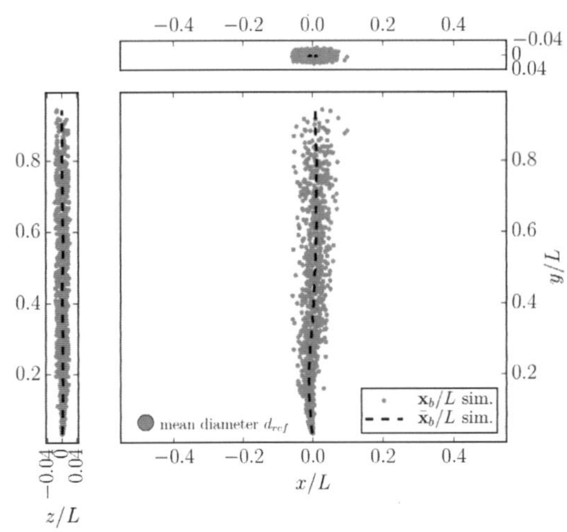

Figure 7. Numerically determined bubble center positions and average trajectory. Depiction covers the whole domain.

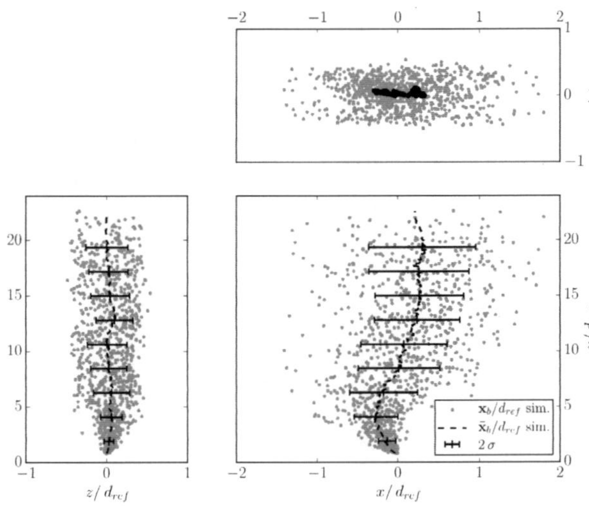

Figure 8. Numerically determined bubble center positions, average and standard deviation. Same dataset as in figure 7 but scaled with d_{ref} instead of L, horizontally stretched.

a certain elevation. It is obtained by subdividing the container vertically in equal parts and by computing the temporal average for each part independently. The same data were used to create more detailed plots, figure 6 and figure 8, horizontally stretched and scaled by the average bubble size. The standard deviation σ of a bubble property ψ with respect to the elevation average $\bar{\psi}$ is then computed by $\sigma^2 = \overline{\psi^2} - \bar{\psi}^2$. Note that there is a small horizontal shift between simulation and experimental data due to the angled injection nozzle. This shift was less than 2 mm and the position of the experimental trajectories were shifted accordingly for comparison.

The bubble centers stay within a band of $4d_{ref}$, whereas most of them occur within a band of $2d_{ref}$. The widening of the trajectories can be quantified by an undulation amplitude of about $d_{ref}/4$ and an increasing deviation σ (figure 6 and figure 8). The simulation results show the three-dimensional character of the bubble chain (figure 8). The slight s-shape of the chain is visible in both the experiment and in the simulation. Undulation and standard deviation are slightly larger in the simulation, but in general the experimental measurements confirm the numerical results.

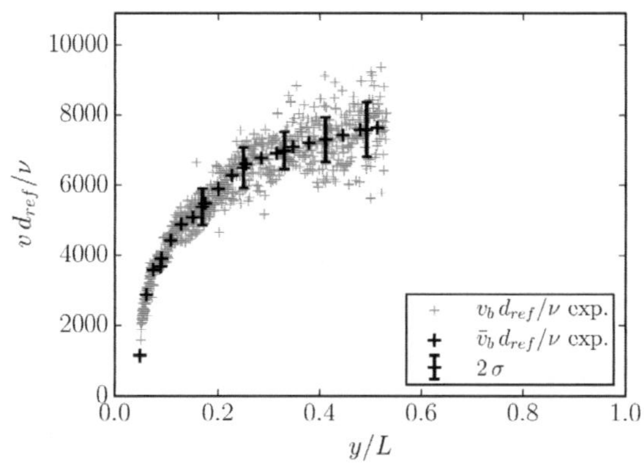

Figure 9. Experimentally determined absolute vertical bubble velocity against container height.

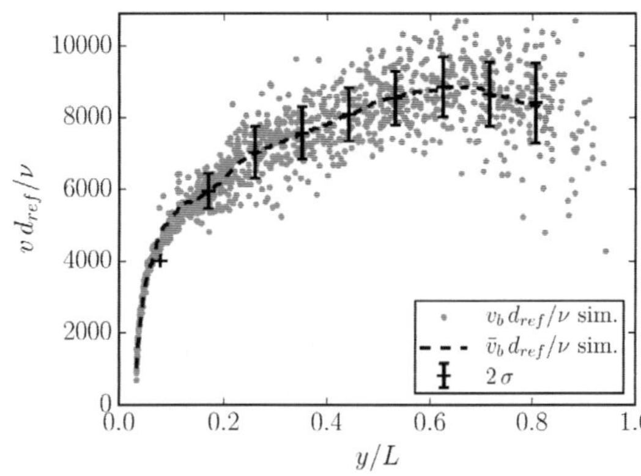

Figure 10. Numerically determined absolute vertical bubble velocity against container height.

Each bubble increases its velocity until a rising velocity of approximately $2\,u_c$. The scattering of the bubble velocity increases simultaneously with the scattering of the bubble position. Right after the injection, it is quite low. At a certain height however, located roughly at $y = 3\,d_{ref} = \frac{1}{8}L$, the variance of the rise velocity increases noticeably towards a constant value within the center region. Due to the low initial velocity, the void fraction close to the injection needle is about twice the void fraction in the developed center region (figure 11, figure 12). The experimental and numerical results coincide qualitatively, but the experimentally determined gas fraction is slightly higher. This could be caused by inaccuracies in the computation of the equivalent bubble diameter from the X-ray images. Such an error distorts the void fraction with third power. The void fraction distribution suggests, again, the existence of a center region bounded by a lower injection region and an upper close-to-surface region.

Such regions can be found also by analyzing bubble-bubble distances. Dominating distances between particles or bubbles can be identified by the discrete radial pair correlation function g_r [26, 27]. To account for the dimensions of the container, the analysis is done here in a two-dimensional manner. For a given distance value r, the value $g_r(r)$ is then evaluated by determining the number of bubble-bubble distances $r_{ij} = (|x_i - x_j|^2 + |y_i - y_j|^2)^{1/2}$ lying in the shell $\left(r - \frac{\Delta r}{2}, r + \frac{\Delta r}{2}\right]$ supplemented with a shell-specific scaling according to

$$g_r(r) = \frac{1}{N(N-1)V_{\text{shell}}/V} \sum_{i=1}^{N} \sum_{j=1}^{N} D\left(r_{ij} \in \left(r - \frac{\Delta r}{2}, r + \frac{\Delta r}{2}\right]\right) . \tag{9}$$

Here, N is the total number of bubbles at a given instant in time and V the volume of the container. The operator $D(x)$ is a discrete version of the Dirac distribution. It is equal to one if the statement x is true and zero otherwise.

The volume $V_{\text{shell}}(r, \Delta r)$ is the annulus volume $2\pi r \Delta r L_z$, regardless whether the considered distance is close to the boundary. The result for the radial pair correlation function is given in figure 13 and leads to the following conclusions. The correlation for the whole container shows a dominant peak at approximately $2\,d_{ref}$ which is equivalent to a surface-to-surface distance of $1\,d_{ref}$. Of course, it is much more revealing to search for subregions associated with characteristic correlations to detect regions of similar physics. Such regions can be found iteratively by checking the sensitivity of the associated correlation function to changes of the region size. Three distinct regions with different characteristic correlations can be found (figure 15). The distance of the bubbles in the injection zone is clearly dominated by one single peak resulting from the injection frequency. This distance is equal to the length scale l discussed above together with π_l. Hence, it corresponds exactly to the dimensionless length scale associated with the injection procedure:

$$\frac{r_{inj}}{d_{ref}} = \frac{l}{d_{ref}} = \pi_l = 1.35 \tag{10}$$

Hence, the radial pair correlation function is able to recover the value of the *a priori* known dimensionless injection bubble distance π_l. In the large center region, the core region, the dominant peak is at $2.3\,d_{ref}$ with a smaller one at $5\,d_{ref}$ probably caused by pairing scenarios. In the upper part, the close-to-surface region, the bubbles come closer to the free surface and are finally switched off in the simulation which leads to an unspecific correlation pattern.

Dominating directions of the connecting line between two bubbles can be identified by the discrete angular pair correlation function g_φ. If all particle distances are finite, it can be computed similarly to the radial pair correlation function, reading

$$g_\varphi(\varphi) = \frac{1}{N(N-1)V_{sector}/V} \sum_{i=1}^{N} \sum_{j=1}^{N} D\left(\varphi_{ij} \in \left(\varphi - \frac{\Delta\varphi}{2}, \varphi + \frac{\Delta\varphi}{2}\right]\right) . \tag{11}$$

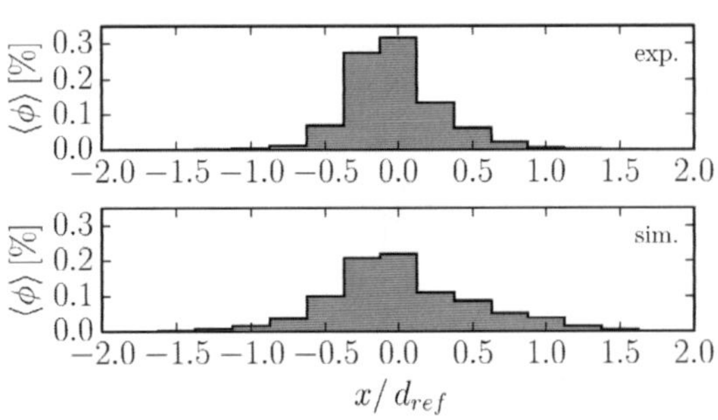

Figure 11. Horizontal distribution of time-averaged volume fraction $\langle\phi\rangle$.

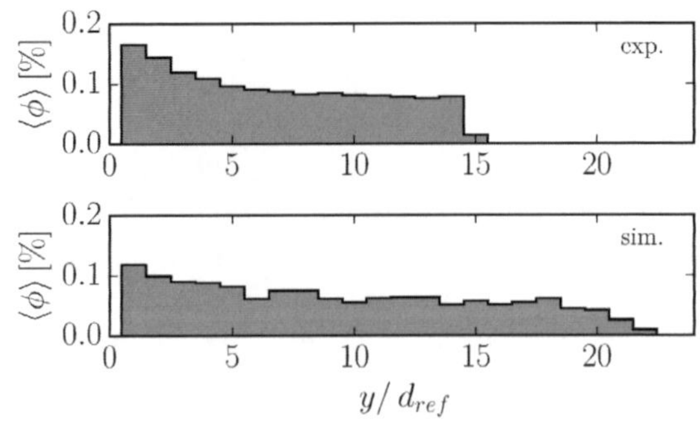

Figure 12. Vertical distribution of time-averaged volume fraction $\langle\phi\rangle$. The upper part of the container is not covered by the X-ray measurements.

Here, $V_{\text{sector}}(\varphi)$ is the volume of the angular sector bounded by $\varphi - \frac{\Delta\varphi}{2}$, $\varphi + \frac{\Delta\varphi}{2}$ and an outer radius $r_{\max}$, arbitrarily chosen as L. As before, it is computed in a two-dimensional manner as $V_{\text{sector}}(\Delta\varphi) = \frac{1}{2}r_{\max}^2\Delta\varphi L_z$. As expected, the bubbles are more or less aligned (figure 14).

3.4. Results for the continuous phase

The rising bubbles induce a recirculation flow of the liquid metal upstream in the center and downstream remote from the chain near the sidewalls. The velocity of the recirculating liquid metal was measured by UDV, as illustrated in figure 2. The simulated, time averaged magnitude of the fluid velocity is shown in figure 4, overlapped with a snapshot of the rising bubbles. Here, $\langle\psi\rangle$ denotes the temporal average. In figure 16, computational and experimental results for the vertical velocity of the liquid metal are compared at five equidistant container heights, i.e. at $(2, 3, 4, 5, 6)\frac{1}{8}L$ which corresponds to $(36, 54, 72, 90, 108)$ mm. The inner upstream flow generated

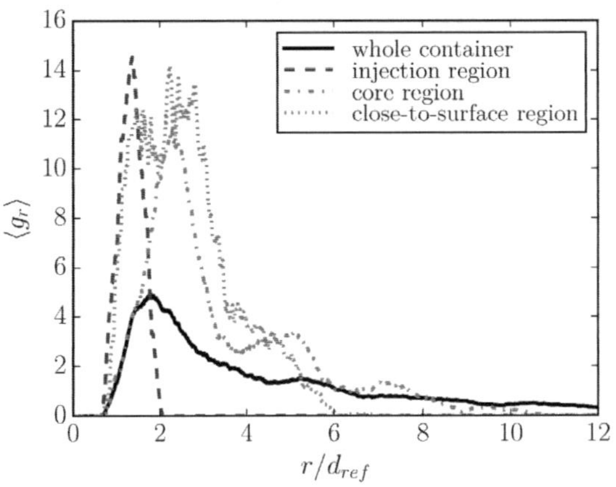

Figure 13. Numerically determined radial pair correlation function. The distribution for the injection region is scaled by 0.25.

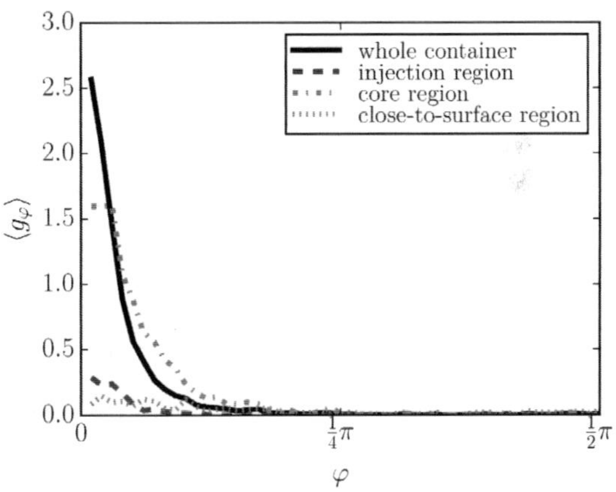

Figure 14. Numerically determined angular pair correlation function. The distribution for the whole container is scaled by 0.25.

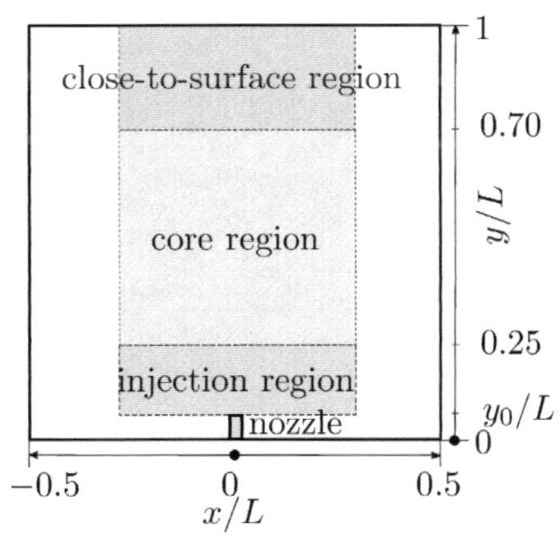

Figure 15. Regions with similar characteristics. The regions differ in terms of the pair correlation function, specific values for volume fraction and the deviation horizontal displacement and rise velocity.

by the rising bubbles is visible, as well as the outer downstream flow. The range of the velocity axis was limited to allow suitable comparison between the UDV data, recorded in the outer flow region, and the simulation results. The velocity gradient between the bubble chain and the walls is approximately constant $|\Delta v_f/\Delta x| \approx 4.7\,s^{-1} = 0.12\,u_c/d_{ref}$. Apparently, the mean flow is not fully symmetric with respect to the y-axis for simulation and experiment. This is not an error but a consequence of the persisting s-shape of the bubble chain discussed above. The differences between experimental and numerical velocities are small for the right side and tolerable for the left side of the container. In the experiment, the recirculation reaches slightly higher velocity values compaired to the simulation which suggests that the free-slip assumption might not be the optimal choice for free liquid metal surfaces. Apart from that, the numerical and the experimental velocity data are in good agreement. To understand the influence of the rising bubbles on the surrounding liquid metal, the turbulent kinetic energy provides further insight. An example is shown in figure 17 at half the container height. The bubbles produce large turbulent kinetic energy in the surrounding fluid. This energy is transported upwards close to the surface before it is carried downwards by the recirculating fluid near the side walls. Between these regions of increased turbulence, there are zones of low turbulent kinetic energy.

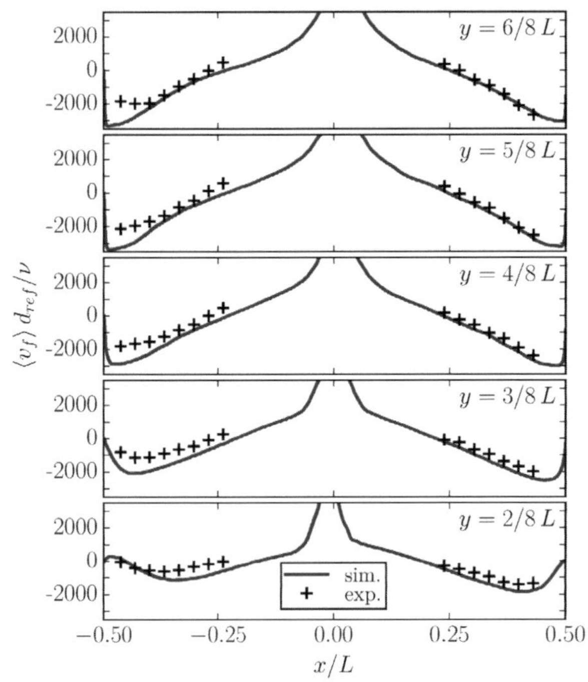

Figure 16. Profiles of vertical fluid velocity at five equidistant elevations derived from simulation and UDV measurement.

3.5. Results related to the interaction of disperse and continuous phase

The comparison between experimental and numerical results proves the reliability of the numerical approach. This allows a closer look at quantities that are inaccessible with present measurement technology. The three-dimensionality of the bubble shapes cannot be captured by high-speed X-ray measurements and the velocity of the liquid close to the bubble chain cannot be determined in a sufficient manner by UDV. The numerical approach, on the other

IOP Conf. Series: Materials Science and Engineering **228** (2017) 012006 doi:10.1088/1757-899X/228/1/012006

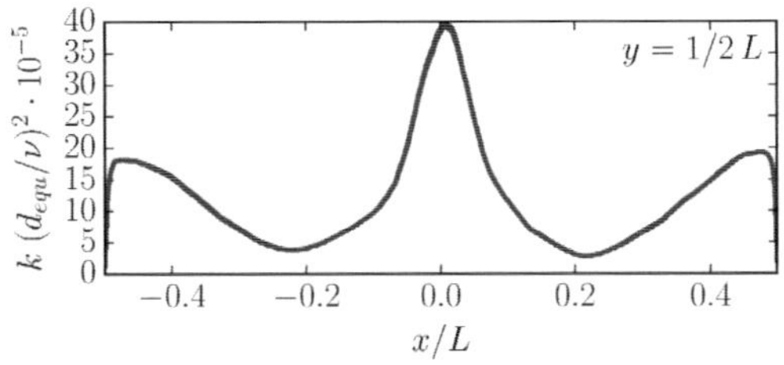

Figure 17. Turbulent kinetic energy from simulation data in the developed region exemplarily at $y = 0.5L$.

hand, includes a three-dimensional representation of each bubble with its individual shape and the entire velocity field. To derive a relative bubble velocity one has to determine a fluid velocity. Here, the fluid velocity v_f close to each bubble is determined by averaging the vertical fluid velocity component over a horizontal range around the bubble, $(\Delta x, \Delta z) = (2, 1)\, d_{ref}$, excluding the bubble itself. The notation $\tilde{\psi}$ is used here for such a local average combined with averaging over bubbles at a given height. This definition allows to compute the relative velocity $\bar{v}_{rel} = \bar{v}_b - \tilde{v}_f$ and the bubble Reynolds number $Re_b = \bar{v}_{rel} d_{ref}/\nu$. The result is shown in figure 18. After a very short distance, the relative velocity $\bar{v}_{rel}$ exhibits a fairly constant value in the injection region. It increases in the subsequent center region and saturates at around $3\,200\, d_{ref}/\nu$. Further up, both bubble and fluid velocity decay yielding a constant relative velocity. Obviously, the relation $Re_b \approx \sqrt{Ga}$, known to hold for rising single bubbles in quiescent fluid, is not met here as the present flow is substantially more complex. The evolution of the relative velocity finally approaches $\bar{v}_{rel} = 0.64\,\tilde{v}_f$ which corresponds to an abolute bubble velocity of $\bar{v}_b = 1.64\,\tilde{v}_f$.

Deformed bubbles can behave significantly different than spherical ones. The deformation can cause a separated wake behind the bubble which increases the drag noticeably [28]. It is

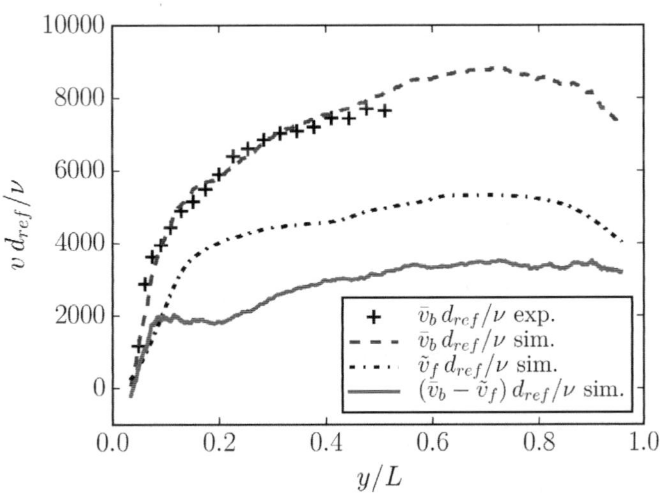

Figure 18. Absolute vertical bubble velocity, ambient fluid velocity and relative bubble velocity.

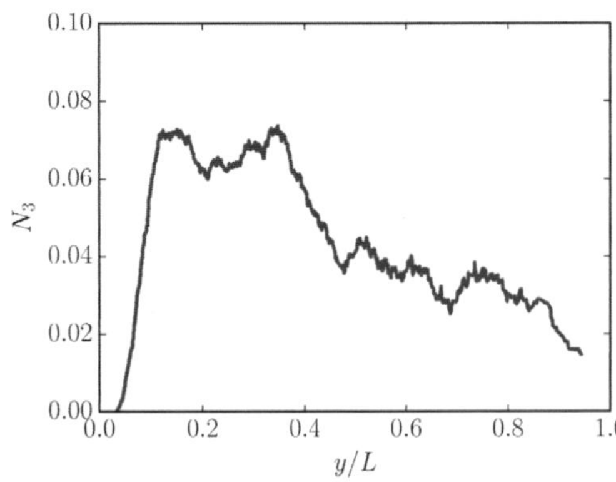

Figure 19. Averaged non-sphericity against elevation.

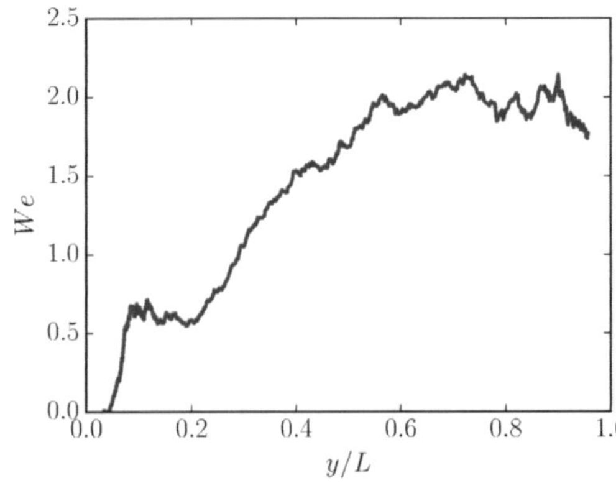

Figure 20. Averaged Weber number against elevation.

also known, that the near-wall behavior of spherical and non-spherical bubbles differs, so that deformable bubbles tend to drift to the center of channels [29, 30]. Therefore, deformation is an important quantity of interest. The three-dimensional deformation can be measured by the non-sphericity

$$N_3 = \frac{A - A_{veq}}{A} , \tag{12}$$

where A denotes the surface area of the bubble and A_{veq} the surface area of the volume-equivalent sphere. The experimental technique only provides two-dimensional projections of the shape which can be assessed with the non-circularity

$$N_2 = \frac{P - P_{aeq}}{P} , \tag{13}$$

where P denotes the perimeter of the projected bubble and P_{aeq} the perimeter of the area-equivalent circle. Both N_2 and N_3 are zero in case of a spherical bubble and larger than zero in case of deformation. For example, an oblate ellipsoid with an axial ratio of two exhibits a non-sphericity of $N_3 = 0.09$. The definitions in equation (12) and equation (13) where derived from definitions of sphericity and circularity [31]. It is obvious, that N_2 does not approximate N_3 in case of strong deformation in the direction of projection. Nevertheless, the comparison of both measures can be justified in the case of a symmetry axis orthogonal to the direction of projection and the assumption of minor deformations. For such cases, it can be shown mathematically, that the error $(N_2 - N_3)/N_3$ for an oblate ellipsoid with a deformation according to an axial ratio lower than two is below 5 %. Therefore, the non-circularity N_2 is a good approximation to the non-sphericity N_3 in the present case.

The deformation of a bubble is closely connected to the ratio of the square of the relative velocity and surface tension. This ratio is expressed by the Weber number $We = \rho_f \bar{v}_{rel}^2 d_{ref}/\sigma$. Large Weber numbers relate to large deformations in case of single bubbles [28]. As expected, this does not hold in the present case as visible in figure 19 together with figure 20. Moreover, the present simulation shows different results for the lower and the upper part of the container. It can be concluded from the comparison of figure 19 with figure 20, that the bubble acceleration is important here: The approach of the final relative velocity allows the bubbles to return to a more spherical shape. This change starts approximately at an elevation of $y/L = 0.4$. The final deformation of $N = 0.03 \pm 0.01$ in the core region is confirmed by the X-ray measurements.

Nevertheless, the re-deformation in the simulation occurs later than in the experiments. The final deformation, however, is predicted quite well.

4. Conclusions

The paper demonstrates the capability of the experimental and numerical methods for the investigation of bubbly liquid metal flows with deformable bubbles. A model configuration is proposed which is complex enough to contain industrially relevant phenomena and which is simple enough to allow detailed experimental and numerical investigation. Experimental and numerical results were found to be in very good agreement both for the disperse gas phase and for the continuous liquid metal phase. The physical phenomena are presented in detail, such as the connection between deformation and Weber number. In the future, the long-term behavior of the meandering and the influence of the distinct parameters could be studied. Apart from this, more complex configurations like swarms in wide containers or the influence of magnetic fields on coalescence could be investigated.

Along with the application of the method described so far and the physical evaluation of the results reported above further methodological achievements were made recently. An inconvenience of classical immersed boundary methods is the fact that vanishing mass of bubbles or particles can only be handled with difficulties [13]. The method in [14] provides a solution but does not allow a density exactly equal to zero and has comparatively severe stability limits. The problem was recently solved by devising a new formulation of the IBM introducing a better coupling to the fluid. It was presented in [32] for spherical particles and generalized recently. The second achievement is a new hybrid method for phase-resolving DNS of bubbly flows combining the advantages of front tracking and the above IBM. It bases on an equation for the coefficients c_{nm} in equation (1) directly, which guarantees a smooth and stable propagation of the bubble surface even in case of complex shapes, provided the surface point distribution is determined appropriately [33]. These recent developments will be used in future bubbly flow simulations.

Acknowledgements

Funding by the Helmholtz Alliance LIMTECH, project A5, and financial support by German Ministry of Economic Affairs and Energy under contract No. GRS-1501449 is gratefully acknowledged. Computing time was provided by ZIH, Dresden.

References

[1] Eckert S, Gerbeth G and Lielausis O 2000 *Int. J. Multiph. Flow* **26** 45–66

[2] Eckert S, Gerbeth G and Lielausis O 2000 *Int. J. Multiph. Flow* **26** 67–82

[3] Tomiyama A, Celata G P, Hosokawa S and Yoshida S 2002 *Int. J. Multiph. Flow* **28** 1497–1519

[4] Zhang X and Ahmadi G 2005 *Chem. Eng. Sci.* **60** 5089–5104

[5] Miao X, Lucas D, Ren Z, Eckert S and Gerbeth G 2013 *Int. J. Multiph. Flow* **48** 32–45

[6] Fröhlich J, Schwarz S, Heitkam S, Santarelli C, Zhang C, Vogt T, Boden S, Andruszkiewicz A, Eckert K, Odenbach S and Eckert S 2013 *Eur. Phys. J. Spec. Top.* **220** 167–183

[7] Zhang J and Ni M J 2014 *J. Comput. Phys.* **270** 345–365

[8] Plevachuk Y, Sklyarchuk V, Eckert S, Gerbeth G and Novakovic R 2014 *J. Chem. Eng. Data* **59** 757–763

[9] Timmel K, Shevchenko N, Röder M, Anderhuber M, Gardin P, Eckert S and Gerbeth G 2015 *Metall. Mater. Trans. Process Metall. Mater. Process. Sci. B* **46.2** 700–710

[10] Takeda Y 1991 *Nucl. Eng. Des.* **126** 277–284

[11] Eckert S, Cramer A and Gerbeth G 2007 *Velocity Measurement Techniques for Liquid Metal Flows* (Dordrecht: Springer Netherlands) pp 275–294
[12] Morley N B, Burris J, Cadwallader L C and Nornberg M D 2008 *Rev. Sci. Instrum.* **79** 3–6
[13] Kempe T and Fröhlich J 2012 *J. Comput. Phys.* **231** 3663–3684
[14] Schwarz S, Kempe T and Fröhlich J 2015 *J. Comput. Phys.* **281** 591–613
[15] Schwarz S, Kempe T and Fröhlich J 2016 *J. Comput. Phys.* **315** 124–149
[16] Garboczi E 2002 *Cem. Concr. Res.* **32** 1621–1638
[17] Landau L D and Lifšic E M 1991 *Hydrodynamik* 5th ed (Thun; Frankfurt am Main: Akademie-Verl.)
[18] Schwarz S and Fröhlich J 2014 *Int. J. Multiph. Flow* **62** 134–151
[19] Aland S, Schwarz S, Fröhlich J and Voigt A 2013 *Eur. Phys. J. Spec. Top.* **220** 185–194
[20] Glowinski R, Pan T W, Hesla T I and Joseph D D 1999 *Int. J. Multiph. Flow* **25** 755–794
[21] Schwarz S and Fröhlich J 2013 *Eur. Phys. J. Spec. Top.* **220** 195–205
[22] Ruzicka M C 2000 *Int. J. Multiph. Flow* **26** 1141–1181
[23] Delnoij E, Kuipers J and van Swaaij W 1997 *Chem. Eng. Sci.* **52** 3759–3772
[24] Delnoij E, Lammers F, Kuipers J and van Swaaij W 1997 *Chem. Eng. Sci.* **52** 1429–1458
[25] Sokolichin A, Eigenberger G and Lapin A 2004 *AIChE J.* **50** 24–45
[26] Bunner B and Tryggvason G 2003 *J. Fluid Mech.* **495** 77–118
[27] Santarelli C and Fröhlich J 2015 *Int. J. Multiph. Flow* **75** 174–193
[28] Loth E 2008 *Int. J. Multiph. Flow* **34** 523–546
[29] Lu J and Tryggvason G 2008 *Phys. Fluids* **20** 040701
[30] Dabiri S, Lu J and Tryggvason G 2013 *Phys. Fluids* **25** 102110
[31] Clift R, Grace J R and Weber M E 1978 *Bubbles, Drops, and Particles* (Mineola, NY: Dover Publ.)
[32] Tschisgale S, Kempe T and Fröhlich J 2017 *J. Comput. Phys.* **339** 432–452
[33] Krull B, Tschisgale S and Fröhlich J 2016 *Proc. Int. Conf. Multiph. Flow* (Florence, Italy)

IOP Conf. Series: Materials Science and Engineering **228** (2017) 012007 doi:10.1088/1757-899X/228/1/012007

Modelling and simulation of a copper slag cleaning process improved by electromagnetic stirring

H Yang[1], J Wolters[1,*], P Pischke[2], H Soltner[1], S Eckert[3], G Natour[1], J Fröhlich[4]

[1] Forschungszentrum Jülich, Central Institute for Engineering, Electronics and Analytics, ZEA-1, 52428 Jülich, Germany

[2] RWTH Aachen University, Institute of Heat and Mass Transfer, 52062 Aachen, Germany

[3] Helmholtzzentrum Dresden-Rossendorf (HZDR), Department of Magnetohydrodynamics, 01328 Dresden, Germany

[4] Technical University Dresden (TUD), Institute of Fluid Mechanics, 01062 Dresden, Germany

[*] E-mail: j.wolters@fz-juelich.de

Abstract. Electromagnetic stirring in a copper slag cleaning process aims at improving the recovery efficiency of the finely dispersed metallic materials from the waste. In the present study the multiphase problems involved in the slag cleaning process are numerically investigated. An Euler-Lagrange approach with advanced collision and coalescence modelling is employed. The corresponding methodologies are briefly introduced and discussed. Based on the implemented sub-models, the copper recovery is numerically investigated for operating parameters corresponding to industrial pilot plants.

1. Introduction

The recovery of valuable metallic material from non-ferrous industrial waste has attracted increasing attention in past decades. In the copper production industry, slags from submerged arc furnaces (SAF) still contain a remarkable amount of copper concentrate [1]. Single SAF furnaces generally produce 700,000 tons of slag per year, which represent a considerable economic recovery potential of the valuable copper material. As a result, a recovery furnace is conventionally arranged to recycle the remaining copper material from the SAF slag, allowing gravitational settling of copper material to the bottom layer due to the density difference (cf. Figure 1). In the slag two thirds of the total copper is present in the form of entrained matte droplets with radii ranging from several micrometers up to millimeters [2], whereas the remainder is dissolved in the slag in the form of copper oxide (Cu_2O). A strong direct current (DC) is applied to the slag between the top and the matte layer at the bottom. The electric current generates Joule heat to keep the slag molten and to reduce the level of dissolved copper in the form of Cu_2O through electrode reactions which segregate the dissolved copper near the anode. The segregated copper aggregates in the form of finely dispersed droplets of 3 to 50 μm diameter [3]. For these small-sized droplets, however, the gravitational settling is too slow for an efficient cleaning process. To increase the separation rate, a magnetic field orthogonal to the electric field is introduced, which generates considerable stirring of the slag. Stirring increases the probability of inter-droplet collisions and coalescence which accelerates the droplet growth and results in a higher settling velocity.

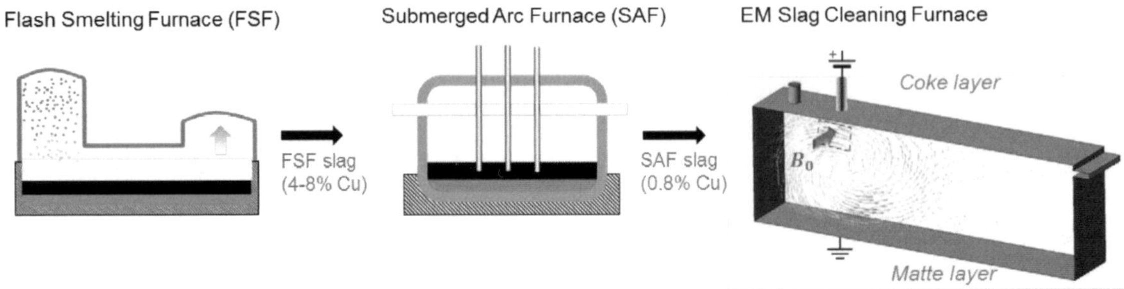

Figure 1. Flow diagram of an industrial copper production and slag cleaning process. The novel concept comprises a magnetic field in the recovery furnace to increase the recovery rate by electro-magnetic stirring.

The feasibility of the electromagnetic (EM) copper slag cleaning process was tested at the University of Chile on the laboratory scale [4]. The industrial scale pilot plant was realized by Aurubis AG and SMS group GmbH [5]. The experimental data showed a positive effect of the electromagnetic stirring on the copper recovery, but the relatively low economic efficiency based on the cost of energy have, among other things, so far prevented electromagnetic stirring from being employed commercially. The long-term goal of the present project is to advance the understanding of the complex physico-chemical mechanisms in the copper slag cleaning process and to optimize the industrial process based on CFD calculations. Current numerical work is carried out using the commercial code ANSYS FLUENT.

2. Numerical scheme

The numerical investigations of the EM copper slag cleaning process include multiphase modelling, in which the conducting slag is the continuous primary phase and the entrained liquid metal droplets constitute the dispersed secondary phase. The secondary phase features a small volume fraction but a massive number of droplets. Hence, the current simulations are performed with an Euler–Lagrangian scheme, where the primary phase is treated by solving the transport equations on a static grid and the secondary phase is treated as discrete phase in Lagrangian manner.

For the continuous phase, the laminar flow of the conducting slag is driven and stirred by the Lorentz force. The Navier-Stokes equation for an incompressible conducting fluid under the influence of an external electromagnetic field can be expressed as

$$\rho_f \left[\frac{\partial \boldsymbol{U}}{\partial t} + (\boldsymbol{U} \cdot \nabla)\boldsymbol{U}\right] = -\nabla p + \rho_f \boldsymbol{g} + \mu_f \nabla^2 \boldsymbol{U} + \boldsymbol{J_0} \times \boldsymbol{B_0}\,, \quad (1)$$

where $\boldsymbol{U}$ is the flow velocity, p is the pressure, $\boldsymbol{g}$ is the gravity, ρ_f and μ_f are the density and the viscosity of the slag, respectively. The term $\boldsymbol{J_0} \times \boldsymbol{B_0}$ is the Lorentz force density neglecting the induced fields, which are insignificant in comparison to the superposed fields $\boldsymbol{J_0}$ and $\boldsymbol{B_0}$ in our case.

For the dispersed phase the trajectory of a discrete droplet (or droplet parcel) is computed solving the force balance equation

$$m_d \frac{dU_d}{dt} = \boldsymbol{F_B} + \boldsymbol{F_D} + \boldsymbol{F_E} + \boldsymbol{F_{VM}} + \boldsymbol{F_L} + \boldsymbol{F_{EC}}, \tag{2}$$

where m_d and $\boldsymbol{U_d}$ represent the mass and velocity of the droplet, respectively. $\boldsymbol{F_B}$ is the buoyancy force, $\boldsymbol{F_D}$ is the drag force, $\boldsymbol{F_{VM}}$ is the virtual mass force, $\boldsymbol{F_L}$ is the lift force, $\boldsymbol{F_E}$ is the Lorentz force due to the higher conductivity of the droplets, and $\boldsymbol{F_{EC}}$ is the electro-capillary force. Detailed expressions for the individual forces can be found in [4] and [6]. In the stirring zone $\boldsymbol{F_D}$ and $\boldsymbol{F_E}$ are the dominant forces, while in the 'quite zone' the sedimentation of the droplets is dominated by $\boldsymbol{F_B}$ and $\boldsymbol{F_{EC}}$.

3. Collision and coalescence of liquid metal droplets in slag

3.1. Collision probability

The growth in droplet size as a result of inter-droplet collisions and coalescence is important in the slag cleaning process. In the present application the direct modelling of collisions for all individual droplets is prohibitive, since their large number would lead to unacceptable computational costs. This problem is avoided by a stochastic model, where the concept of parcels representing a certain number of droplets with the same properties is introduced. The most commonly used algorithm for collision modelling is a purely stochastic Monte Carlo algorithm which is based on the concept of the kinetic gas theory [7] and assumes that the droplets are homogeneously distributed in the collision volume. In this case, the collision probability depends on the mesh resolution, because the collision between droplets of two parcels is only possible when the centers of both parcels are located in the same cell of the grid for the continuous phase. The dependency can become significant for cases with relatively coarse grids as demonstrated by the present authors in [8]. The accuracy could be increased through refining the control volume mesh. However, it may be computationally very costly to introduce such a refinement in geometrically complex simulations and it is difficult to estimate the appropriate mesh-resolution for different cases in advance. To overcome mesh dependency problems of the purely stochastic collision algorithm, a new hybrid Lagrangian collision algorithm was proposed by Pischke et al. [9]. Here, the deterministic algorithm is transferred into a stochastic algorithm by redefining the collision probability assuming a normal distribution of droplets around the centers of the parcels.

To compare both algorithms a test case suggested by Schmidt [10] is considered featuring a cubic domain containing in-homogenously distributed droplets. Here, 10^8 droplets are assigned to a certain number of parcels, which are exponentially distributed in Y-direction and uniformly distributed in the other two directions. The radii of the droplets vary from 0 μm up to 50 μm and the velocities vary from 0 m/s to 100 m/s. The influence of the grid refinement with a certain number of cells in Y-direction is studied. To obtain statistically relevant results 20 independent runs were performed for each set-up with one time step of 0.1 ms.

Figure 2 shows the root-mean-square error between the average number of collisions predicted by different algorithms and the theoretical number. Compared with the purely stochastic algorithm, the hybrid algorithm is mesh-independent and results in less deviation for the same number of parcels. Due to this advantage, the hybrid algorithm was chosen and implemented for the present simulations.

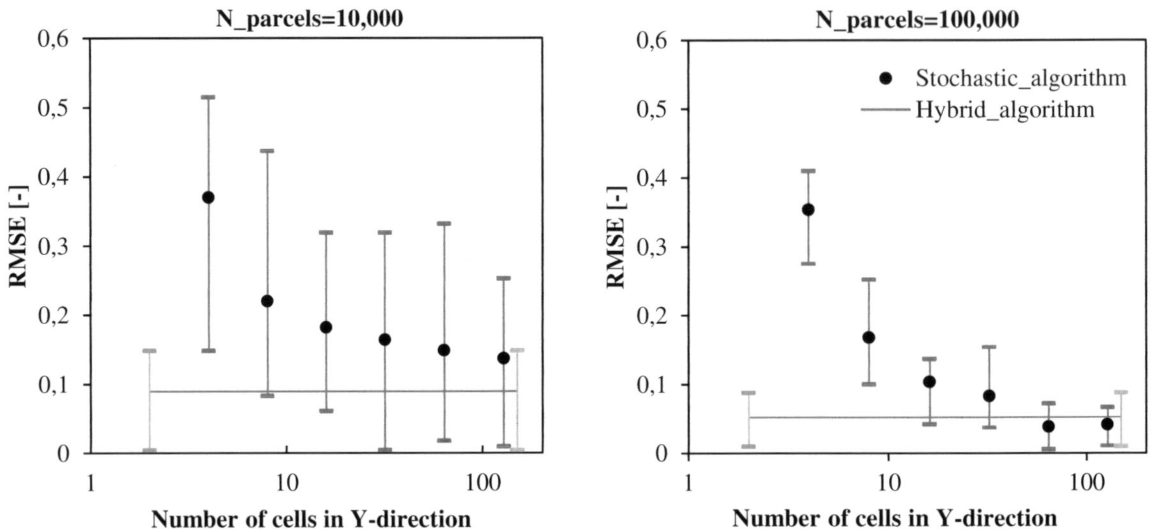

Figure 2. Root-mean-square error of the stochastic collision algorithm and the hybrid collision algorithm for an inhomogeneous benchmark case

3.2. Collision outcome

To predict the collision outcome in a stochastic collision modelling, a regime map is required. Generally, it uses two or more parameters to identify the boundary among different possible collision outcomes, here whether coalescence takes place or not. Ready-to-use regime maps, as suggested by Qian and Law [11], are typically established for inertia-driven collisions in a gas-liquid-system with the Weber number (We) as the main criterion. These regime maps, however, are physically not suitable for the present liquid-metal droplet and slag system. First, the collision of droplets in the slag cleaning process features very small Weber numbers ($We \ll 1$), which would always lead to droplet coalescence under the inertia-driven scheme. Second, the regime map has to account for the influence brought by the viscosity ratio between continuous phase and discrete phase. Since experimental results are difficult to obtain in such an opaque, high-temperature environment and consequently no such experimental data could be found by the present authors, the binary collision process between liquid metal droplets was theoretically and numerically studied in more detail to provide a suitable regime map for the collision outcome, taking into account the peculiarities of the slag cleaning process and the physical properties of liquid metal droplets in slags.

Based on our analysis in [12], instead of using a criterion relative to inertia-driven collisions, it was found to be physically more suitable to use the capillary number (Ca) as the main criterion. This includes the influencing factors of shear-driven collisions as encountered in the slag cleaning process. The important dimensionless parameters to characterize the process are

$$\mathrm{Ca} = \mu_f \frac{\dot{\gamma} R}{\sigma}, \quad \mathrm{Re} = \frac{\dot{\gamma} R^2}{\nu}, \quad \psi = \frac{\Delta y}{2R}, \quad \lambda = \frac{\mu_d}{\mu_f}, \quad h'(t) = \frac{h(t)}{R}, \quad t' = \dot{\gamma} t\ .$$

Here, R is the droplet diameter, $\dot{\gamma}$ is the shear rate, σ is surface tension, ν is the kinetic viscosity, μ_d and μ_f are dynamic viscosity for droplet and bulk phase, respectively, h(t) is the film thickness, ψ is the offset ratio of droplets in the direction perpendicular to the relative velocity.

To establish the regime map detailed numerical simulations of individual binary collisions of droplets of equal size were conducted using the volume-of-fluid (VOF) method with high mesh resolution. More details about the setup can be found in [13]. The droplets were positioned in a linear shear flow with a certain offset Δy in the direction of the velocity gradient (cf. Figure 3), so that they move on straight lines until they are close enough to interact.

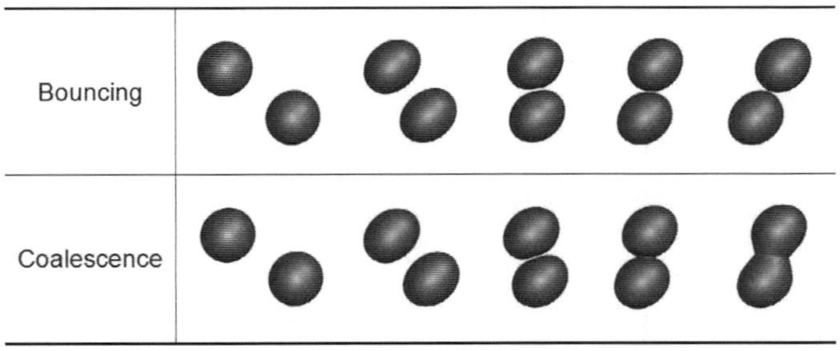

Figure 3. Coalescence and bouncing of two colliding droplets in a shear flow simulated with the VOF method.

At this point the confronting interfaces of the two droplets are flattened and the thin film of surrounding fluid starts to drain under the influence of the hydrodynamic forces exerted by the shear flow. Figure 4 exemplarily shows the dimensionless film thickness as a function of time during this process for different capillary numbers and offset ratios.

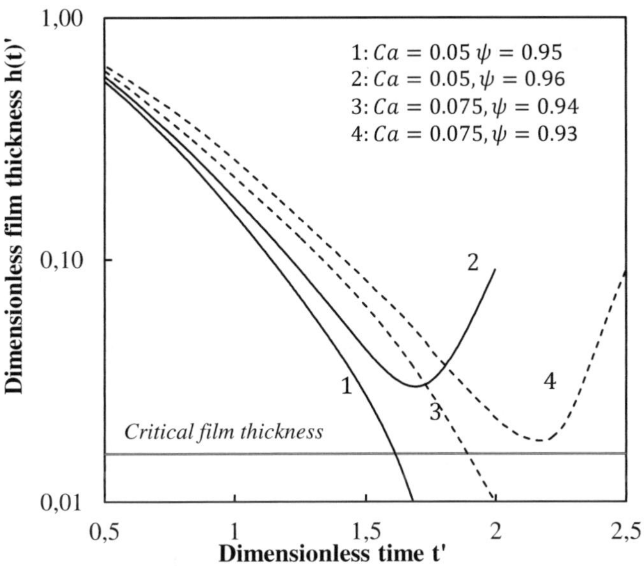

Figure 4. Dimensionless film thickness over time in simulations of binary collisions with $\lambda_R = 1$ resulting in coalescence or bouncing, depending on offset ratio ψ and capillary number Ca.

The collisions end up in coalescence if the critical value of the film thickness $h_c = \left(\frac{AR}{8\pi\sigma}\right)^{\frac{1}{3}}$ is achieved [14] and the inter-molecular force, becomes comparable with the surface tension force and leads to rupture of the film. Otherwise the droplets will slide along each other's surfaces and the collisions end up in grazed bouncing. A boundary curve separating the regimes of coalescence and bouncing in the plane spanned by the offset ratio and the capillary number can be fitted to the points identified as those with critical offset ratio and is presented in Figure 5 with $\lambda_R = 1$ designating the case of equal particle diameters. In the regime map the scaled capillary number $Ca\left(\frac{R}{\mu m}\right)^{0.84}$ is employed instead of the capillary number Ca itself with the empirical factor deduced from experiments by Hu et al. [15]. It is obvious that for larger droplets the film drainage takes more time than for smaller droplets, if the same capillary number is considered. For unequal-sized collisions an empirical reduction factor of coalescence efficiency $\left(\frac{4\lambda_R}{(1+\lambda_R)^2}\right)^{\kappa}$ in respect of the size ratio $\lambda_R = R_{smaller}/R_{larger}$ suggested by Mousa et al. [16] is employed to extend the regime map. The factor covers the effect that for a constant average radius the coalescence efficiency decreases if the difference in size increases. This decrease in coalescence efficiency is due to the fact that smaller droplets tend to better follow the curved streamline around the larger droplet in the Stokes regime. The regime map of Figure 5 was used in the simulations reported below.

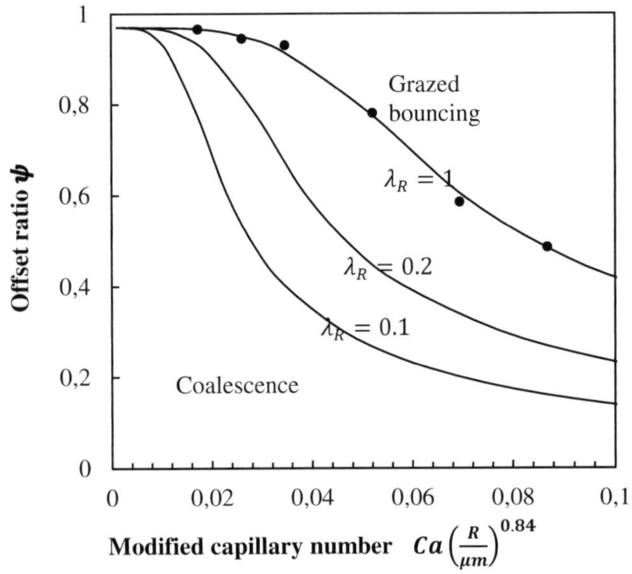

Figure 5. Capillary-number-based regime map for the collision result based on VOF simulations with R=0.284 µm.

4. Simulation of an industrial pilot plant

Between 2011 and 2012, tests on an industrial scale slag cleaning pilot plant were carried out in Hamburg, Germany. The schematic drawing of the furnace suggested by König et al. [5] is shown in Figure 6. The inner hearth of the furnace has overall dimensions of 2.778 m x 0.578 m x 0.94 m. It is filled with two molten material layers (a 0.14 m high matte layer and a 0.6 m high slag layer) which are covered by a coke layer of 0.2 m thickness. Two graphite electrodes acting as anodes are immersed in the coke layer and contact the slag. Electrode 2 serves only as an emergency electrode and is not considered in the present simulations. The furnace comprises two isolation layers of refractory material. The imposed magnetic field, which is located under the primary electrode, is generated by an electromagnet with a core cross section of 0.3 m x 0.3 m. The important operating parameters are shown in Table 1 and typical material properties are shown in Table 2. Applying the realistic operating parameters and the advanced collision sub-models introduced in the former sections, the preliminary simulations on the practical copper slag cleaning process were carried out. The first results of numerically predicting the copper recovery efficiency for an industrial scale pilot plant will be presented and discussed in this section. The enhanced reduction reaction of dissolved copper content by electromagnetic stirring is also taken into consideration. On the other hand, the heat transfer calculations are not included in the current stage and thus the average temperature of the slag is based on an internal report describing the experiments. Due to the high viscosity of the slag and the low characteristic velocity, the typical Reynolds number at the stirring zone of the investigating case is within the range of $\boldsymbol{O}(1)\sim\boldsymbol{O}(100)$, depending on the strength of the external magnetic fields. Since these values are below 1000, which is the typical boundary value for laminar flows in stirring tanks, the slag flow can be treated as laminar in the calculations.

IOP Conf. Series: Materials Science and Engineering **228** (2017) 012007 doi:10.1088/1757-899X/228/1/012007

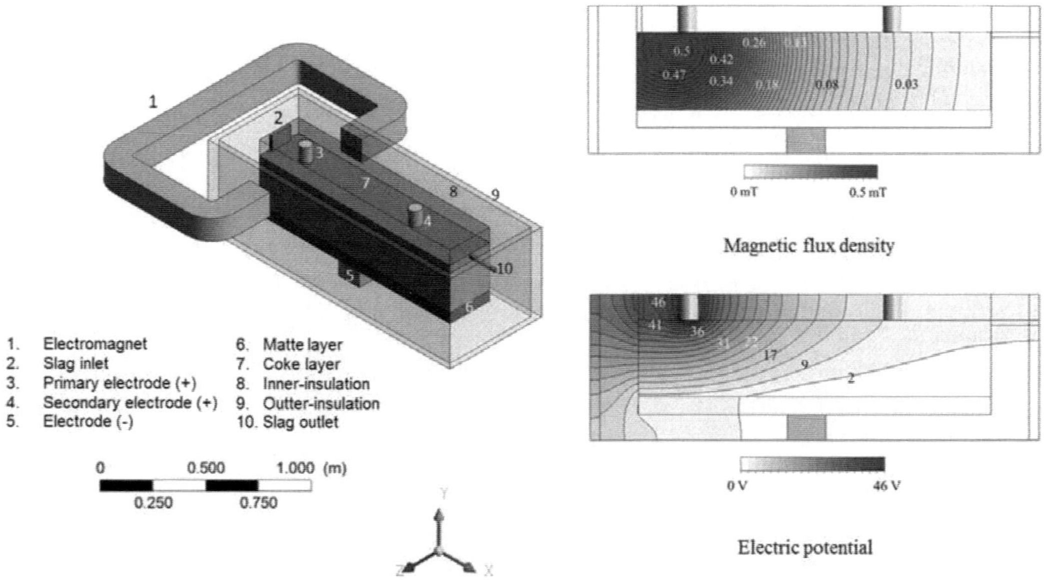

Figure 6. Schematic drawing of the slag cleaning furnace in the pilot-plant.

Table 1: Important material properties of the copper slag cleaning process.

	Density ($kg\,m^{-3}$)	Viscosity ($kg\,m^{-1}s^{-1}$)	Electric conductivity ($\Omega^{-1}m^{-1}$)	Resistance of metal/slag interface ($\Omega\,m^2$)
Slag	3600	0.1	80	$(0.7 \pm 0.2) \times 10^{-5}$
Matte	4950	0.005	5000	

Table 2: Operating parameters of the slag cleaning furnace in the pilot-plant.

Slag flow rate (kg/s)	Flow rate matte droplets (kg/s)	Electrode diameter (m)	Total current (kA)	Slag temperature (K)
1.667	0.044	0.15	3	1533

The injected slag has a copper concentration of 0.8%, of which 2/3 is in the form of matte droplets with the size distribution (50μm to 950 μm) suggested in [2]. The remaining copper is dissolved in the slag as Cu_2O, which can be segregated by the following reduction reaction near the coke layer:

$$2(Cu_2O)_{slag} + (C)_{coke} = 4(Cu)_{matte} + CO_2. \tag{3}$$

Here, $(Cu)_{matte}$ will further form the matte inclusion droplets with a diameter assumed uniform and equal to 50 μm. It is also assumed that the matte droplets and the matte generated by reaction have the same composition ($Cu_2S - FeS$). However, it is difficult to evaluate the generation of copper matte based on the reaction rate at an arbitrary time and location, since the concentration of Cu_2O and the slag temperature both keep changing along the reaction interface below the coke layer. An alternative method is to evaluate the change of dissolved copper concentration $\eta(Cu^+)$ at the inlet and outlet. The internal report by SMS group suggests that when the slag is continuously injected without electromagnetic stirring $\eta(Cu^+)$ can be treated as constant at the inlet $\eta_0(Cu^+) = 0.267\%$ and a 50% reduction can be expected at the outlet. The corresponding mass flow rate of reaction-generated matte Q_{matte} is

$$Q_{matte} = \Delta\eta_0(Cu^+) \cdot \frac{M(Cu_2S)}{M(Cu)} \cdot \frac{1}{0.25} \cdot Q_{slag} \quad , \tag{4}$$

where M represents the molar mass, Q_{slag} is the mass flow rate of slag, and the factor 0.25 reflects the proportion of copper sulphide in the matte inclusions. Due to the massive Joule heat released from the electrode the slag temperature near the primary electrode is much higher than in other parts of the furnace. Hence, it is assumed that the reduction reaction mainly takes place near the coke layer close to the primary electrode.

When the magnetic flux density (B) is applied the stirring velocity of slag is increased and a higher mass transfer coefficient to the reaction interface can be achieved. The change in $\eta(Cu^+)$ with respect to the mass transfer coefficient k_m can be expressed as [17]

$$\Delta\eta(Cu^+) = \eta_0(Cu^+)\left(1 - e^{-t \cdot k_m}\right). \tag{5}$$

The mass transfer coefficient k_m itself can be characterized by the Sherwood number (Sh). The concentration change $\Delta\eta(Cu^+)$ under different magnetic fields can be evaluated by the relationship between the stirring-improved mass transfer coefficient $k_m{}'$ and the mass transfer coefficient without stirring, $k_{m,0}$. When the Schmidt number (Sc) is assumed to be constant, the Ranz-Marshall equation describing the mass transfer can be expressed as [4]

$$\frac{k_m{}'}{k_{m,0}} \sim \frac{Sh'}{Sh_0} \sim \frac{\left(Re^{\frac{1}{2}}Sc^{\frac{1}{3}}\right)'}{\left(Re^{\frac{1}{2}}Sc^{\frac{1}{3}}\right)_0} \sim \left(\frac{U'}{U_0}\right)^{1/2} , \tag{6}$$

where U' is the average velocity of the slag near the coke layer and U_0 the slag velocity when no external magnetic field is imposed. The corresponding reaction rates at different magnetic fields can now be estimated by

$$Q'_{matte} = \left(1 - 0.5^{\sqrt{\frac{U'}{U_0}}}\right)\eta_0(Cu^+) \cdot \frac{M(Cu_2S)}{M(Cu)} \cdot \frac{1}{0.25} \cdot Q_{slag} \ , \tag{7}$$

with the corresponding values shown in **Table 3**.

Table 3: Generation rate of matte droplets with different magnetic fields.

Magnetic flux density (mT)	Average velocity near Electrode-1 (m/s)	$\eta(Cu^+)$ at outlet (%)	Generation rate of matte droplets (kg/s)
0	0.00074	0.1300	0.0111
5	0.01125	0.0188	0.0206
20	0.02826	0.0036	0.0219

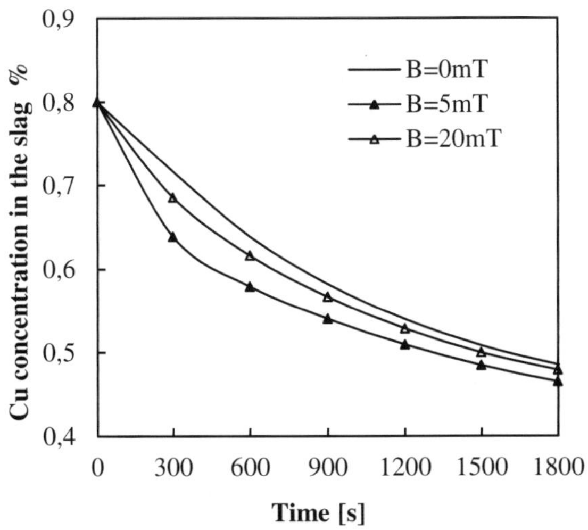

Figure 7. Level of copper content over operation time.

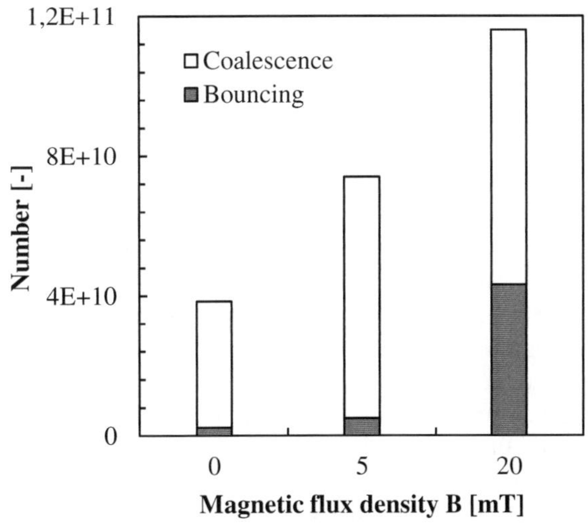

Figure 8. Outcomes of the inter-droplet collisions during 0.5 h for different magnetic fields.

The first simulations of the industrial pilot plant were performed with an operating time of t = 0.5 hour. During this time the slag with entrained copper droplets is continuously injected into the furnace while the reaction-generated droplets are simultaneously injected near the primary electrode. The concentration of remaining copper will reach a quasi-constant value after some time of operation beyond the interval covered. During the simulation the overall droplet population ($\sim 10^{10}$) was represented by $\sim 10^5$ parcels.

The temporal evolution of the copper content in the slag is shown in Figure 7 for different magnetic flux densities. As expected, compared to the case without electro-magnetic stirring the recovery of matte droplets is increased when introducing the magnetic field. The positive effect, however, decreases beyond certain strength of the magnetic field. This trend was also observed in the experiments. One reason for this is the increased drag force due to the stirred flow which makes settling even of the larger droplets resulting from coalescence more difficult and leads to a higher copper concentration remaining in the slag. On the other hand the increased stirring influences the collision outcome. Although the number of collisions is significantly increased with stronger magnetic fields and corresponding higher relative droplet velocities, the number of collisions resulting in coalescence does not increase any more or even decreases for higher velocities (Figure 8). According to the study in Section 3.2, coalescence as a result of shear-driven collisions is more likely for small impact velocities with similar size of the two droplets involved. As a result, the overall growth in droplets size due to inter-droplet coalescences is not significantly enhanced any more with stronger stirring, at least not enough to compensate the deceleration of sedimentation due to the highly disturbed flow field.

Based on the numerical results, optimization of the device and operating conditions is possible now. First of all the strength of electromagnetic stirring has to be adjusted to maximize the copper recovery rate. Moreover, the currently used C-electromagnet creates a widely spread magnetic field in the furnace causing a large stirring zone - in comparison with the length of the furnace - and compresses the settling zone. The setup of the electromagnet should therefore also be optimized.

A comparison between numerical and experimental results could not be conducted up to now, since some basic parameters and results of the experiments have to be further discussed with SMS group GmbH, such as, for example, the exact size distribution of the matte droplets entrained in industrial slags.

5. Conclusions and perspectives

A numerical approach to model an industrially relevant enhancement of a copper slag cleaning process through electromagnetic stirring was developed. An Euler-Lagrangian scheme for solving the multiphase problem was implemented, where slags are defined as continuous phase and liquid-metal droplets are treated as dispersed parcels. For the continuous phase, MHD (magneto-hydrodynamic) calculations on the conducting slag under the influence of orthogonally magnetic and electric fields were performed. For the dispersed phase, tracking of the liquid metal droplet was realized by solving the force balance accounting for the electromagnetic forces.

To predict the growth of the droplet size due to inter-droplet collisions and coalescence, a mesh-independent hybrid Lagrangian collision algorithm was introduced to calculate the collision probability, which has an excellent adaptability in both homogeneous and in-homogeneous cases. Moreover, to correctly predict the collision outcome in the stochastic collision modelling, a new regime map for the collision outcome was developed. It is based on a large number of VOF calculations of the binary collision process taking into account the peculiarities of the slag cleaning process and physical properties of liquid-metal droplets in slags. This regime map is available for use in subsequent now. Moreover, it points to the more general fact that inertia-based regime maps for collisions are not applicable in viscosity-dominated situations and need to be replaced by different ones.

The developed method was then applied to the setup of an industrial pilot plant investigating the efficiency of copper recovery under different imposed magnetic flux densities. Electromagnetic stirring using a relatively weak magnetic field of B = 5 mT significantly increases the generation of copper matte inclusions due to the reduction reaction near the coke layer. Moreover, the increased inter-droplet coalescence rate increases the droplet size and accelerates sedimentation of the droplets. Further enhancement of the magnetic field does not provide considerable improvements in the coalescence efficiency, though, but generates stronger disturbances in the flow field that lead to a decreased sedimentation rate of the droplets. As a result, simply increasing the strength of the imposed magnetic field fails to increase the copper recovery efficiency. Further validation will be performed.

The simulation tool is now ready for optimizing the industrial process, and more detailed calculations of the industrial copper slag cleaning process will be realized in the near future. For example, discontinuous electromagnetic stirring will be considered which seemed promising in the experiments.

Acknowledgments

The authors acknowledge the financial support from the German Helmholtz Association in the framework of the Helmholtz Alliance LIMTECH, Project A5, and like to thank Mr. König and Mr. Köneke from SMS group GmbH for providing information on the test facilities and for the fruitful discussion on the slag cleaning process.

References

[1] Schlesinger M and King M 2011 *Extractive metallurgy of copper 5th ed.* (UK: Elsevier)
[2] Warczok A and Riveros G 2003 *Yazawa International Symposium-Metallurgical and materials processing and technology* Volume **2** 417-429
[3] Degel R, Oterdoom H, Kunze J, Warczok A and Riveros G 2008 *Third International Platinum Conference 'Platinum in Transformation'* (Sun City, South Africa) 197-202
[4] Warczok A and Riveros G 2007 *Miner. Eng.* **20** 34-43
[5] König R, Degel R, Rose L, Schmidl J, Kadereit H and Specht A 2013 *TMS Annual Meeting & Exhibition* (San Antonio, USA)
[6] Choo R T C and Toguri J M 1992 *Canadian Metallurgical Quarterly* **31** 113-126
[7] O'Rourke P J 1981 Ph.D thesis Department of Mechanical and Aerospace Engineering Princeton University

[8] Yang H, Wolters J, Pischke P, Soltner H, Eckert S and Fröhlich J 2015 *8th International Conference on Electromagnetic Processing of Materials* (Cannes, France) 255-258
[9] Pischke P, Cordes D and Kneer R 2012 *Int. J. Multiphase Flow* **38** 1–16
[10] Schmidt D 2006 *J. Comput. Phys* **164** 62-80
[11] Qian J and Law C K 1991 *J. Fluid Mech.* **331** 59-80
[12] Yang H, Wolters J, Pischke P, Soltner H, Eckert S and Fröhlich J 2016 *9th International Conference on Multiphase Flow* (Firenze, Italy).
[13] Yang H, Wolters J, Pischke P, Soltner H, Eckert S and Fröhlich J 2017 *12th International Conference on CFD in Oil & Gas, Metallurgical and Process Industries, SINTEF* (Trondheim, Norway)
[14] Chesters A 1991 *Trans. IChemE* **69** 259-270
[15] Hu Y T, Pine D J and Leal L G 1989 *Phys. Fluids* **12** 484-489
[16] Mousa H and van de Ven T G M 1991 *Colloides Surfaces A* **60** 19-38
[17] Reddy R G, Prabhu V L and Mantha R 2003 *High Temp. Mater. Processes* **22** 25-33

IOP Conf. Series: Materials Science and Engineering **228** (2017) 012008 doi:10.1088/1757-899X/228/1/012008

Using Lorentz forces to control the distribution of bubbles in a vertical tube filled with liquid metal

S Heitkam[1], S Tschisgale[2], B Krull[2], T Wetzel[3], E Baake[4] and J Fröhlich[2]

[1] Technische Universität Dresden, Institut für Verfahrenstechnik und Umwelttechnik, D-01062 Dresden, Germany
[2] Technische Universität Dresden, Institut für Strömungsmechanik, D-01062 Dresden, Germany
[3] Institute of Thermal Process Engineering, Karlsruhe Institute of Technology, Kaiserstr. 12, Karlsruhe D-76131, Germany
[4] Institute of Electrotechnology, Leibniz University of Hanover, Wilhelm-Busch-Str. 4, Hannover D-30167, Germany

E-mail: sascha.heitkam@tu-dresden.de

Abstract. In this work, a method to increase the residence time of bubbles in tubes or pipes filled with liquid metal is investigated. Imposing a horizontal electric current and a perpendicular horizontal magnetic field generates an upward-directed Lorentz force. This force can counteract gravity and cause floating of bubbles. Even with homogeneous electric fields these float in the mean but fluctuate randomly within the swarm due to mutual interactions. In the present case the cylindrical shape of the container furthermore creates inhomogeneous electric currents and an inhomogeneous force distribution resulting in a macroscopic convection pattern stirring the bubbles and further homogenising the spatial distribution of the bubbles.

1. Introduction

An innovative technique for the production of hydrogen is the direct thermal cracking of methane into carbon and hydrogen. This can be achieved by introducing methane bubbles into tubes filled with liquid tin which serves as a heat transfer medium. The reaction takes place above $600°C$ but is much faster for higher temperatures. High temperatures, however, pose technical problems such as difficulties of handling the process and devising suitable materials for the construction of the apparatus. Another strategy is to work with somewhat lower temperatures but to increase the residence time of the bubbles by an appropriate amount to accommodate the slower reaction.

At KIT, an experimental facility including a bubble column reactor with molten tin as liquid medium has been developed and constructed in order to prove the feasibility of the concept. Several experimental campaigns have been run, during which parameters like gas volume flow rate, temperature and feed gas dilution were investigated [1, 2]. A major finding is the strong influence of the bubble residence time on the methane conversion. The residence time is limited by the strong acceleration of the gas bubbles in the liquid metal. On the other hand, this density difference is a major advantage of the process, as it facilitates the separation of the solid carbon produced. Therefore, additional measures which can resolve this conflict would be highly welcome.

To get an idea about the bubble dynamics in the experimental bubble column reactor additional numerical simulations were conducted in Hannover. In these studies, residence times for different volume flow rates and initial bubble sizes including slug flows were determined. To increase the efficiency, installations like packed beds, variation of multiple inlets including coalescence studies were done. Another aspect, the influence of an electromagnetic field with the potential of electromagnetic stirring to split bubbles into blebs, were numerically explored as well.

The option investigated in the present paper is to increase the well time by means of electromagnetic fields. In earlier simulations this was demonstrated for bubbles in liquid aluminum [3]. A horizontal electric and a perpendicular, horizontal magnetic field result in a Lorentz force pointing upwards and counteracting gravity. Further investigation of the floating state [4] revealed, that around each bubble, a swirling velocity field is induced. The interaction of these fields generates a global stirring motion homogenising the spatial distribution of the bubbles.

In the present work this concept is adapted to the bubble column reactor for hydrogen production investigated experimentally. Most importantly, a cylindrical domain is used here since this is an essential feature of the experiment. The cylindrical geometry of the electrodes results in an inhomogeneous distribution of the electric current thus resulting in an inhomogeneous Lorentz force and a macroscopic mean velocity field. Goal of this work is to evaluate the movement of liquid metal and bubbles in such an electromagnetic field and to provide first information about the suitability for an industrial process.

2. Numerical Method

2.1. *Euler-Lagrange approach*

The simulations were carried out with the in-house code PRIME [5]. It solves the incompressible Navier-Stokes equations

$$\frac{\partial \mathbf{u}}{\partial t} = -(\mathbf{u}\cdot\nabla)\mathbf{u} + \frac{1}{\rho_f}\nabla p + \nu\nabla^2\mathbf{u} + \frac{1}{\rho_f}\mathbf{j}\times\mathbf{B} + \mathbf{f}_{IBM} \tag{1}$$

$$\nabla\cdot\mathbf{u} = 0 \tag{2}$$

on a staggered Cartesian grid and employs a Runge-Kutta three-step low-storage method with pressure correction for time integration.

The bubbles are geometrically resolved and equations of motion for velocity and rotation are solved for each individual bubble. The bubbles are assumed mono-disperse and spherical, with a no-slip condition applied at their surface [6]. An immersed boundary method described below is used to couple the disperse and the continuous phase. In case of bubble collision, additional forces are added to the equation of motion of the bubbles, in order to model the physical collision process [7].

2.2. *Immersed boundary method*

Besides the bubbles, the immersed boundary method is also employed to represent the walls of the cylindrical fluid domain within a hexagonal Cartesian grid. This facilitates grid generation tremendously. Each embedded boundary Γ, whether bubble surface or container wall, is represented by a finite set of Lagrangian marker points evenly distributed over Γ to accomplish the coupling to the surrounding fluid. At the position of each marker point $\mathbf{x}_m \in \Gamma$ an artificial force $\mathbf{f}_{IBM}$ is added to the Navier-Stokes equation, as shown in (1). This source term is determined by the so-called direct forcing approach [8, 9, 10]

$$\bar{\mathbf{f}}_{IBM} = \frac{1}{\Delta t}\int_{t_{n-1}}^{t_n} \mathbf{f}_{IBM}\,\mathrm{d}t = \frac{\mathbf{u}_\Gamma^n - \tilde{\mathbf{u}}}{\Delta t}, \tag{3}$$

where $\bar{\mathbf{f}}_{IBM}$ is the average IBM force applied over the discrete time interval $\Delta t = t_n - t_{n-1}$. In equation (3) the velocity $\tilde{\mathbf{u}}$ is a shorthand for

$$\tilde{\mathbf{u}} = \mathbf{u}^{n-1} + \int_{t_{n-1}}^{t_n} \mathbf{rhs}\,\mathrm{d}t \quad , \tag{4}$$

with **rhs** representing the right-hand side of the momentum balance (1), excluding $\mathbf{f}_{IBM}$, so that the preliminary fluid velocity $\tilde{\mathbf{u}}$ is obtained without accounting for the effect of the immersed boundary.

Due to the no-slip condition the fluid velocity at a marker position is equal to the velocity of the interface, i.e. $\mathbf{u}_\Gamma^n = \mathbf{u}(\mathbf{x}_m, t^n)$. In case of the container walls $\mathbf{u}_\Gamma^n$ vanishes. In case of the bubbles, their surface is represented by 1259 Lagrangian marker points. At these points the no-slip boundary condition is imposed by $\bar{\mathbf{f}}_{IBM}$ in (1) and $\mathbf{u}_\Gamma^n$ obtained from the equation of motion of the bubble. The counteracting forces $-\bar{\mathbf{f}}_{IBM}$ at these markers are collected and introduced in the equations of motion of the bubble.

2.3. Magnetohydrodynamics

To represent magnetohydrodynamic effects in the simulation, an electric current density $\mathbf{j}$ and a magnetic field strength $\mathbf{B}$ have to be determined in (1). The magnetic susceptibility of liquid tin, methane and air are all close to zero. Thus, interfaces of these materials do not alter the magnetic field. Also the magnetic Reynolds number is low, so that the magnetic field $\mathbf{B} = B_0\mathbf{e}_y$ can be assumed constant in space and time.

The electric conductivity of liquid tin, σ_f, on the contrary, is much higher compared to methane, or hydrogen or air. Hence, bubbles of conductivity $\sigma_g << \sigma_f$ alter the electric current. Electric induction has to be taken into account as well. The electric current density, then is given by

$$\mathbf{j} = \sigma_{el}(\mathbf{u} \times \mathbf{B} - \nabla\phi) \quad , \tag{5}$$

and charge conservation $\nabla \cdot \mathbf{j} = 0$ results in a Poisson equation for the electric potential

$$\nabla^2\phi = \nabla \cdot (\mathbf{u} \times \mathbf{B}) \quad . \tag{6}$$

which is numerically solved. The boundary condition at the bubble interfaces are imposed by scaling the electric conductivity $\sigma_{el} = \alpha\sigma_f$ with $\alpha \in [0, 1]$, a phase marker function derived from a cut-cell algorithm [5]. At the surface of the tube, the marker points are employed to impose the correct value for the wall-normal electric current using the Neumann condition $\partial\phi/\partial n = j/\sigma_f$ with $j = j_0$ at one electrode, $j = -j_0$ at the other electrode and $j = 0$ otherwise. Once ϕ is known from (6), $\mathbf{j}$ is obtained from (5).

3. Setup

The material parameters for the simulation are derived from [11] and references therein. The considered temperature $1000°C$ is within the range $900°C$ to $1175°C$ used in the experiments at KIT. Higher temperatures allow for higher production rate but are more difficult to handle, because the magnet might have to be cooled and the material of the electrode has to withstand it.

Table 1. Parameters for the simulation. Material parameters are derived from [11] and references therein.

Parameter	Symbol	Value
Temperature	T	$1000°C$
Density	ρ_f	$6500\,kg/m^3$
Viscosity	ν_f	$1.1\,10^{-7}\,m^2/s$
El. conductivity	σ_f	$1.3\,10^6\,S/m^{-1}$
Tube diameter	D	$16\,mm$
Tube height	H	$16\,mm$
Bubble diameter	D_b	$2\,mm$
Number of bubbles	N_b	50
Surface tension	γ	$0.5\,N/m$
Mag. field strength	B_0	$0.1\,T$
El. current density	j_0	$0.5\,A/mm^2$

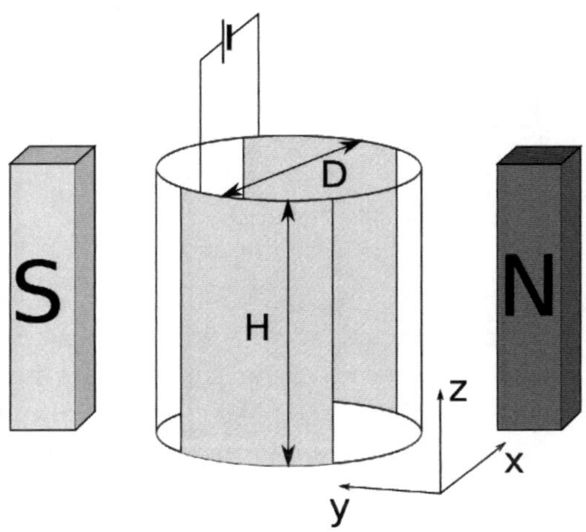

Figure 1. Setup of the investigation showing the tube with electrodes and magnets together with the coordinate system employed.

The geometrical setup is shown in figure 1. Top and bottom are non-conducting solid walls. The tube is filled with liquid tin, as in the experiments and has a diameter of 16 mm and a height of 16 mm as well. This is smaller than the experiment at KIT to limit computational resources. In an industrial process, the height and the diameter would be much larger, but important mechanisms can also be evaluated with this reduced size. Each of the electrodes covers the full height and 90 degrees of the circumference. The magnetic poles are placed perpendicularly to the electrodes.

At the initial state of the bubble-laden simulations 50 gas bubbles of 2 mm diameter were placed at random positions within the tube and allowed to freely propagate under the influence of the various forces.

IOP Conf. Series: Materials Science and Engineering **228** (2017) 012008 doi:10.1088/1757-899X/228/1/012008

4. Results

4.1. Static channel

First, computations in matlab® were conducted to determine the current density alone in the channel filled with stagnant material of constant conductivity, to understand the Lorentz force distribution in the present setup with curved electrodes. This is reported in figure 2a displaying the $x-$component of the current density. It shows significant minima close to the insulating parts of the tube surface. Close to the border of the electrodes strong maxima appear. This feature results from the electric boundary condition at the electrodes. For comparison, a case with constant voltage imposed at the electrodes, i.e. $\phi = \pm\phi_0$, was computed as well and is presented in figure 2b. For the following simulations, constant normal electric current at the

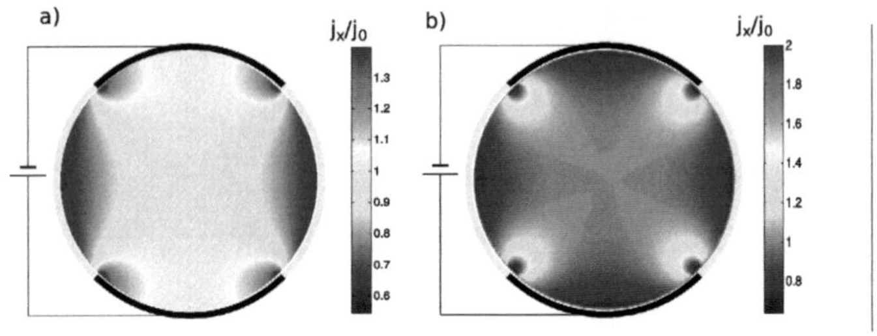

Figure 2. Theoretical current distribution in a cross section of the tube, filled with stagnant material of constant conductivity. The $x-$direction of the electric current is shown for two different boundary conditions at the electrodes. a) Constant normal current, b) constant potential. The $x-$direction is vertical in these graphs.

electrodes (Figure 2a) was imposed. Note, that the boundary condition only enforces the normal component, not the total value. At the apex of the electrode, the x-component equals the normal value, which is set by the boundary condition. At the rim, the electrode is tilted up to 45 degree. Thus, the local area of the electrode is larger than the corresponding projection of the electrode area perpendicular to the x-direction. Due to conservation of charge, the current density in x-direction has to be approximately $\sqrt(2) \approx 1.4$ times larger than the normal current at the electrode.

4.2. Single-phase flow

As a next step, the single-phase flow was computed with the code described above. The $x-$component of the electric current directly determines the vertical Lorentz force $f_{L,z} = j_x B_0$, so that the vertical velocity reflects the inhomogeneity of the electric current. The vertical velocity, averaged over the height of the tube, is shown in figure 3. Although the flow is not homogeneous in $z-$direction due to the top and bottom wall this figure provides an impression of the fluid flow generated by the Lorentz force with averaged values reaching $\langle w \rangle_z \approx 0.04$ m/s. This yields a Reynolds number in magnitude of

$$Re = \frac{\langle w \rangle_z D}{\nu_f} \approx 5000 \quad . \tag{7}$$

The velocity field was not averaged in time. The small irregularities and asymmetries visible in this graph reflect the fact that the instantaneous flow exhibits substantial turbulence. This issue is not discussed further here, because the introduction of bubbles changes the flow substantially.

Figure 4 shows the $x-$component of the electric current in the single-phase flow. It is obvious that this field is more homogeneous than the current density shown in figure 2a. This is for two

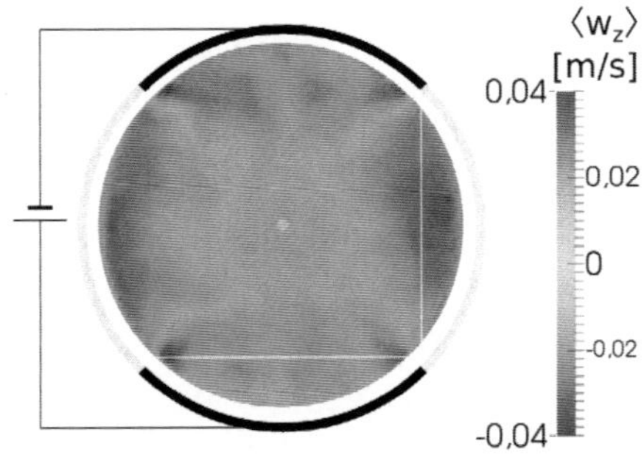

Figure 3. Upward velocity averaged over the height of the tube, $\langle w \rangle$.

reasons. First, the motion of the liquid metal is attenuated and homogenised by viscous effects. Second, the fluid moving in a direction different from the magnetic field itself generates electric current as described by (5). The load factor K relates the externally applied electric field j_0/σ_f to the induced electric field $\langle w \rangle_z B_0$

$$K = \frac{j_0}{\sigma_f} \frac{1}{\langle w \rangle_z B_0} \approx 0.96 \quad . \tag{8}$$

A load factor close to one means, that the induced electric currents are of the same magnitude as the applied ones. Regions of high electric current show high vertical Lorentz forces and, in turn, high vertical velocities. The vertical movement generates high induced electric fields counteracting the applied ones. This feed loop leads to more balanced distribution of j_x with smaller amplitude shown in figure 4.

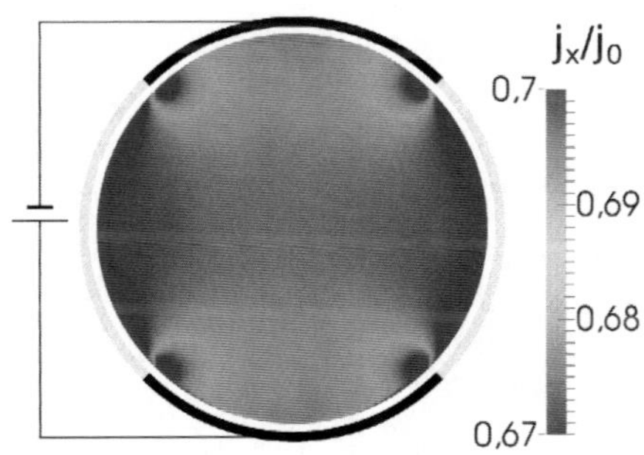

Figure 4. Normalised horizontal distribution of the $x-$component of the electric current density, averaged over the height of the tube.

4.3. Bubble distribution

Bubbles distributed in the tube experience several effects, such as buoyancy, drag force, pressure gradient and collision forces. Earlier simulations of the present authors of bubbles in a rectangular container [3] used that the vertical Lorentz force generates an upside-down pressure gradient $j_0 B_0$ that counteracts buoyancy $\rho_f g$ and causes floating of bubbles. For floating, the

Lorentz force has to overcompensate gravity by 1/3 [12] yielding

$$j_0 B_0 = \frac{4}{3} \rho_f g \quad . \tag{9}$$

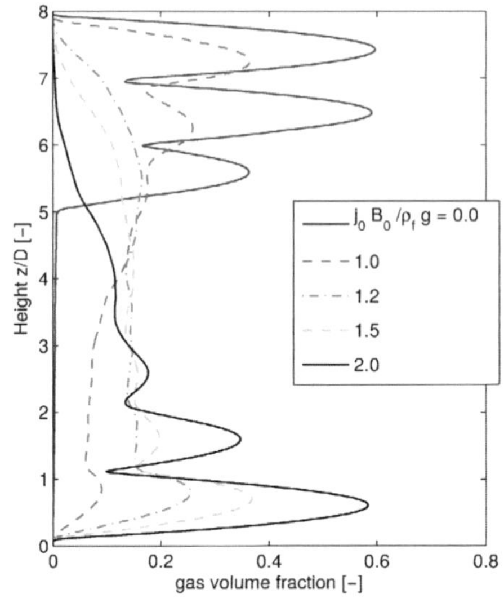

Figure 5. Mean bubble volume fraction over height obtained from averaging over time and horizontal directions in the considered tube with horizontal electric and magnetic fields. The latter is perpendicularly oriented to the electric field.

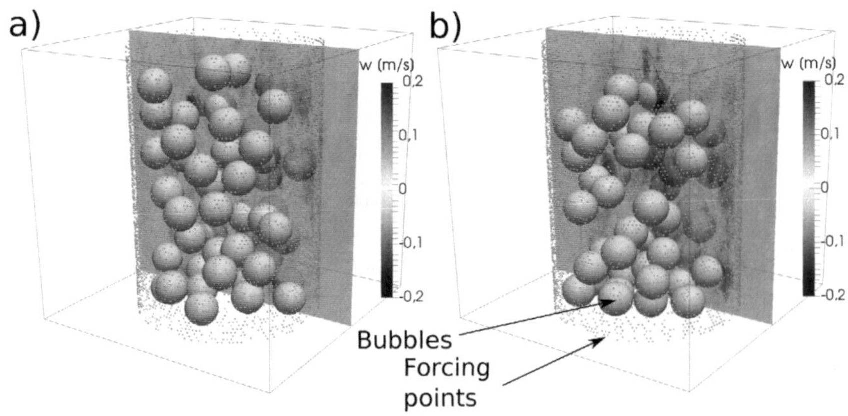

Figure 6. Snapshot of bubble distribution and vertical velocity component in a vertical plane off the axis. a) $j_0B_0/\rho_f g = 1.2$, b) $j_0B_0/\rho_f g = 1.5$. The small white points are forcing points on the cylindrical container surface serving to impose the boundary conditions. Only a small subset of these points is shown not to hide the interior of the domain.

As demonstrated in figure 5 this is also true for a round container. Without electromagnetic field the bubbles agglomerate at the top of the container. The undulating shape of the curve results from the fact that the top wall imposes a certain structure on the position of bubbles. With increasing field strength, the bubbles become mobilised, float and even collect at the bottom of the domain if the Lorentz force is substantially exceeding gravity. A stable state with mostly floating irregularly moving bubbles is achieved for $j_0 B_0/\rho_f g \approx 1.2$, corresponding to an overcompensation of gravity by about 20%.

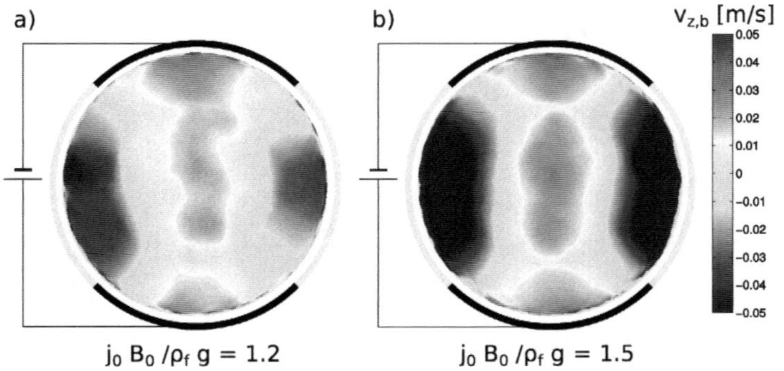

Figure 7. Vertical velocity of bubbles averaged over time and vertical direction for different values of applied electric field strength. a) $j_0 B_0/\rho_f g = 1.2$, b) $= 1.5$, respectively.

For the two cases depicted in figure 6a and 6b, the vertical velocity of the bubbles was averaged over height and in time yielding the data in figure 7. The resulting motion reflects the distribution of the Lorentz force and of the vertical fluid velocity. The fluid velocity of the single-phase flow exhibits a similar magnitude (cf. figure 3). This suggests, that the bubbles are not propelled by their interaction with the electromagnetic field but that they are transported by the macroscopic flow structure of the liquid metal.

5. Discussion

To reduce the complexity, the present simulations did not account for bubble deformation and coalescence. When the bubbles agglomerate, coalescence might occur, since presumably no surface active agent is present. However, in the floating state, collisions are less frequent and coalescence should be negligible due to the low contact times [13].

As pointed out in section 4.1 the structure of the inhomogeneity of the electric current highly depends on the electric boundary condition. In the present case, constant normal current is imposed at the walls. Constant potential at each electrode results in a more homogeneous current density and, hence, a more homogeneous Lorentz force in the empty tube. A constant potential boundary condition, however, has a major problem, because in cross sections with many bubbles the average resistance is higher resulting in lower electric current. This, in turn, reduces the Lorentz force near bubble clusters, so that no stable bubble distribution is to be expected. Consequently, in industrial applications a constant-current boundary condition has to be targeted. This is possible by splitting the electrode in parts imposing a constant current at each part. Measuring the corresponding potential could also allow measurement of the bubble distribution by impedance tomography.

The goal of the present study is to improve the generation process of hydrogen. The efficiency can be evaluated, roughly, by comparing the energy stored in the produced hydrogen, P_H, and the energy spent for imposing the electric current, P_j, giving $\eta = P_H/P_j$. The electric energy consumption per reactor volume depends on the electric current density. With a stronger magnetic field, a smaller electric current is necessary to achieve the floating state. However, magnetic field strengths above $B_0 = 0.1T$ are presumably difficult to handle in practical applications. This results in an electric power consumption of approximately $10^7 W/m^3$ per reactor volume. The production rate of hydrogen depends largely on the temperature and further parameters of the process and has yet to be assessed. It is clear, that the use of magnetohydrodynamic forces requires energy input which has to be justified by improved output of the process. This could be in the form of increased hydrogen reaction rates, but other processes might be considered as well, where the additional value is brought by the quality of the products, for example, so that the price for the energy is of lower importance. Further work is required to conclusively assess the potential of the "zero-gravity" option in liquid metal reacting bubbly flows.

Acknowledgements

The support of the Helmholtz Alliance LIMTECH via the project A5 and a Seed Grant in 2015, as well as the support of the German Research Foundation (HE 7529/1-1) is gratefully acknowledged. Computation time was provided by ZIH at TU Dresden.

References

[1] Geißler T, Plevan M, Abánades A, Heinzel A, Mehravaran K, Rathnam R, Rubbia C, Salmieri D, Stoppel L, Stückrad S, Weisenburger A, Wenninger H and Wetzel T 2015 *Int. J. Hydrogen Energy* **40** 14134–14146
[2] Geißler T, Abánades A, Heinzel A, Mehravaran K, Müller G, Rathnam R, Rubbia C, Salmieri D, Stoppel L, Stückrad S, Weisenburger A, Wenninger H and Wetzel T 2016 *Chem. Eng. J.* **299** 192–200
[3] Heitkam S, Schwarz S, Santarelli C and Fröhlich J 2013 *Eur. Phys. J. Special Topics* **220** 207–214
[4] Heitkam S 2014 *Manipulation of liquid metal foam with electromagnetic fields: a numerical study* Ph.D. thesis Institut für Strömungsmechanik, TU Dresden
[5] Kempe T and Fröhlich J 2012 *J. Comput. Phys* **231** 3663–3684
[6] Aland S, Schwarz S, Fröhlich J and Voigt A 2013 *J. Comput. Phys.* **220** 185–194
[7] Heitkam S, Sommer A, Drenckhan W and Fröhlich J 2017 *Journal of Physics: Condensed Matter* **29** 124005
[8] Mohd-Yusof J 1997 *Center for Turbulence Research. Ann. Res, Briefs. NASA Ames / Stanford University* 317–327
[9] Fadlun E A, Verzicco R, Orlandi P and Mohd-Yusof J 2000 *J. Comput. Phys.* **161** 35–60
[10] Tschisgale S, Kempe T and Fröhlich J 2017 *J. Comput. Phys.* **339** 432–452
[11] Sharafat S and Ghoniem N 2000 Summary of thermo-physical properties of Sn and compounds of Sn–H, Sn–O, Sn–C, Sn–Li, Sn–Si and comparison of properties of Sn, Sn–Li, Li, Pb–Li Tech. Rep. UCLA-UCMEP-00-31 University of California
[12] Lenov D and Kolin A 1954 *J. Chem. Phys.* **22** 683–688
[13] Schwarz S, Tschisgale S and Fröhlich J 2013 Modeling of bubble coalescence in phase-resolving simulations by an immersed boundary method Int. Conf. Multiphase Flows, Paper 750

IOP Conf. Series: Materials Science and Engineering **228** (2017) 012009 doi:10.1088/1757-899X/228/1/012009

Validation of X-ray radiography for characterization of gas bubbles in liquid metals

O Keplinger, N Shevchenko, S Eckert

Helmholtz-Zentrum Dresden-Rossendorf (HZDR), 01328 Dresden, Germany

E-mail: s.eckert@hzdr.de

Abstract. X-ray radiography has proved to be an efficient and powerful tool for the visualization of two-phase flows in non-transparent fluids, in particular in liquid metals. This paper presents a validation of the X-ray radiography by comparing measurements in water with corresponding results obtained by optical methods. For that purpose Ar bubbles were injected through a single orifice. The measurements results are compared in terms of bubble size, bubble shape and velocity. Furthermore, visualization experiments were performed in the eutectic alloy GaInSn where the image contrast between the liquid phase and the gas bubble is much stronger. Some obvious differences of the bubble dynamics in water and GaInSn are discussed.

1. Introduction

Liquid metal two-phase flows are an important part of many technical applications in metallurgy and continuous casting. Argon gas is injected during continuous casting in order to prevent clogging of the submerged entry nozzle and to separate undesired inclusions from the melt. Gas stirring is used in ladle metallurgy to homogenize the melt and to improve the cleanliness of the steel. On the other hand, injection of gas implicates many side effects as for example the generation of highly turbulent complex two-phase flows and creation of additional defects in final casted products. Effective control and optimization of this process requires a comprehensive understanding of the behavior of liquid metal two-phase flows. Many numerical and experimental studies of gas bubbles rising in water and transparent viscous liquids exist so far covering a broad range of Reynolds and Eötvös numbers (see for instance [1–9]. Many studies follow the approach to extrapolate the bubble behavior in liquid metals from experiments in water. However, strong differences in material properties such as density, viscosity and surface tension lead to discrepancies in essential non-dimensional parameters as the bubble Reynolds number, the Weber number or the Morton number. Therefore, direct experimental investigations in liquid metal two-phase flows become important and desirable. However, the value of experiments carried out in real liquid metal flows depends on the availability of trustworthy and efficient measurement techniques. Previous measurements in liquid metal bubbly flows were performed by means of ultrasound Doppler velocimetry (UDV) [10], by conductivity probes [11], by Local Lorentz force velocimetry (LLFV) [12] or by neutron radiography [13,14]. Most of these techniques have limitations. For example, ultrasonic methods meet problems with increasing gas content when many bubbles enter the measuring volume. Multiple echoes at the bubble interfaces may generate signal artifacts. Conductivity probes are intrusive and provide only local information. Recent studies have demonstrated the capability of X-ray radiography to be an efficient tool for the visualization of liquid metal two-phase flows [15–20]. X-ray radiography is a fully contactless method based on the absorption contrast between the liquid and gas phase. The weak point of this technique is the limitation of the sample thickness due to the high attenuation coefficients in liquid metals. Previous publications did not discuss the accuracy of the X-ray radiography for determining parameters like bubble size, bubble shape and bubble velocity in detail. The aim of this work is to demonstrate that the analysis of the X-ray radiography images provides accurate results for bubble size and shape. For this purpose we compare the results of the X-ray radiography with optical measurements performed in water. Further experiments were carried out in the eutectic GaInSn alloy.

2. Setup and methods

2.1. Experimental setup

Bubble visualization experiments were carried out at the X-ray laboratory at HZDR. The scheme of the setup is demonstrated in figure 1. A high power X-ray source (ISOVOLT 450M1/25-55 from GE Sensing & Inspection Technologies GmbH) operating with a maximum voltage of 320 kV and a current of 14 mA generates a divergent polychromatic X-ray beam. A scintillation screen (SecureX HB from Applied Scintillation Technologies) is attached to the surface of the container as shown in Figure 1. The non-absorbed part of the X-ray beam comes upon to this scintillation screen where its intensity is converted into visible light. The further imaging is completed with a lens system (Thalheim – Spezial - Optik) and a high-speed video camera (Pco.edge from PCO) equipped with a sCMOS-sensor. The images were captured with 100 frames per second (fps) and an exposure time of 3 milliseconds. The exposure time was optimized to achieve a good signal-to-noise ratio without causing bubble blurring due to their high rising velocities. The container is made of acrylic glass because the walls do not cause significant attenuation of the X-ray beam intensity. The container represents a rectangular tank with 12 mm gap and 144 mm width which was filled with water or with liquid GaInSn alloy up to a height of 144 mm. The eutectic GaInSn alloy is liquid at room temperature. Thermophysical properties of the alloy are reported in [21]. In our experiments the field of view was approximately 60×110 mm^2. The inert Ar gas was injected through a long bevel stainless steel orifice of 1.1 mm outer diameter (from Sterican®) positioned in the middle of the bottom part of the container. The Ar gas flow rate was 50 cm^3/min. A diffused light source was used for the optical measurements to minimize the light reflection at the bubble interface.

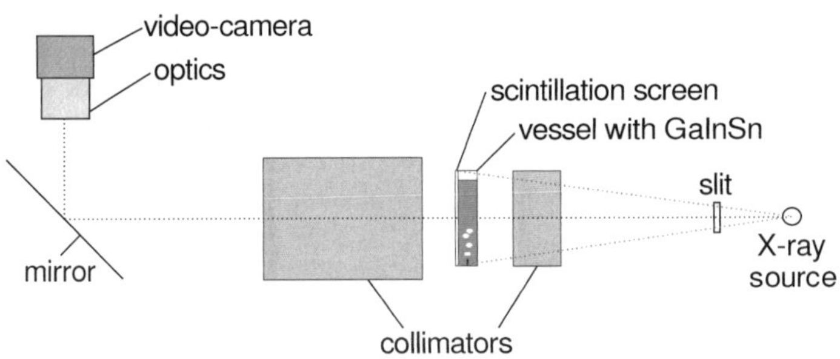

Figure 1. Scheme of the X-ray diagnostic setup

2.2. Image data processing

The quantitative analysis of the bubble dimensions, positions and velocities are performed by off-line data processing using Matlab scripts. The procedure of image processing is illustrated in figure 2 where an exemplary raw image is shown in figure 2a. Prior to the image analysis a shading correction is done by subtracting a mean reference image measured at zero gas flow rate (figure 2b). As a next step a Gaussian filter is applied to the images to reduce the noise signal (figure 2c). Further, a thresholding algorithm is applied to separate the individual bubbles and bubble clusters from the background. As a result, the images are converted to binary images where all pixels that belong to the bubbles are marked as 1, while the pixels marked as 0 correspond to the background (figure 2d). The corresponding parameters for the Gaussian filter applied to the optical measurements was chosen in such a way that the dimensions of the bright reflecting regions were reduced but no blurring of the bubble boarders was caused (i.e. the Gaussian filter was applied using small values of the variance σ^2). An appropriate threshold value was chosen which guarantees that all dark pixels in Figure 2c were counted as part of the bubbles. In turn, the parameters for the Gaussian filter and for the thresholding for GaInSn were obtained from a calibration measurement of two glass balls with 5 and 10 mm diameters surrounded by GaInSn at a zero gas flow rate. As a next step the binary images were analyzed using the function '*regionprops*' integrated into Matlab which allows to directly extract parameters like perimeter, area, center of masses, etc. Figure 2e presents the raw image with the determined bubble perimeters and their center of mass and Figure 2f shows the raw image with fitted ellipses and their main axis. The obtained parameters were converted from pixel to metric values using the image scaling. The equivalent bubble diameters are then calculated from the bubble projection area assuming a spherical bubble shape. Additionally a '*Simple tracker*' algorithm was implemented into the Matlab script. It allows to track the bubbles and, hence, to calculate the bubble velocities along their trajectories. All bubbles which move in close vicinity to each other or even overlap were excluded from further evaluation. The parameters derived from the X-ray radiography were compared to the parameters obtained from the optical measurements. Self-evidently, both measurements were performed in water applying the same process conditions.

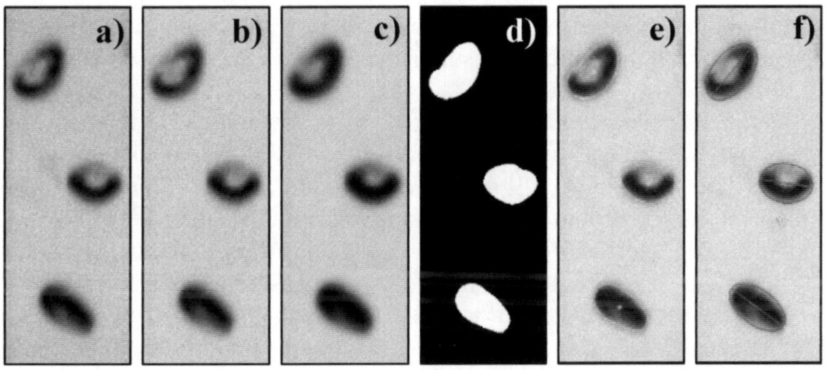

Figure 2. Image processing steps: a) raw image (cut from the full field of view), b) image with reduced reference image, c) gauss filtered image, d) binary image, e) raw image with delineated bubble interface and bubble centers, f) raw image with fitted ellipses and their main axis.

3. Results and discussion

3.1. Validation of X-ray radiography in water

Figure 3 displays snapshots of bubble chains rising in water. The optically captured image is shown in figure 3a while X-ray image can be seen in figure 3d. The gas bubbles can be clearly identified in Figure 3d even though the X-ray image demonstrates a rather weak contrast between the bubbles and the background. The subsequent bubble analysis is performed according to the algorithm described in the previous section. The parameters for the Gaussian filter applied at the X-ray measurements were taken over from the optical measurements. The only difference was the threshold value, which was chosen so that all the bright pixels in Figure 3d were assigned to the bubbles. Figures 3b and 3e illustrate the raw images with the bubble interface and the bubble centers. Figures 3c and 3f present the raw images with fitted ellipses and their main axis.

IOP Conf. Series: Materials Science and Engineering **228** (2017) 012009 doi:10.1088/1757-899X/228/1/012009

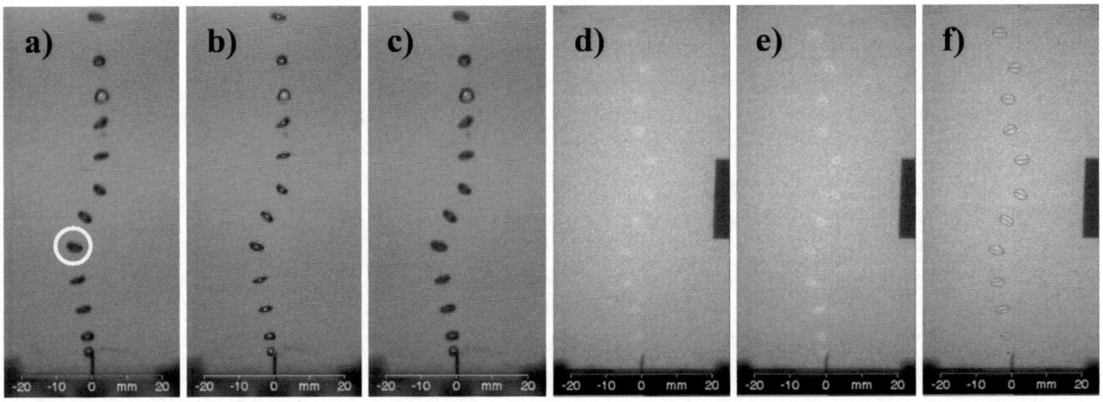

Figure 3. Snapshots of the bubble experiment in water measured at an Ar gas flow rate of 50 cm³/min by the optical method: (a) raw image, (b) raw image with delineated gas-liquid interface and bubble centers, (c) raw image with fitted ellipses and their main axis;
and by means of X-ray radiography: (d) raw image, (e) raw image with delineated gas-liquid interface and bubble centers, (f) raw image with fitted ellipses and their main axis.

The corresponding sauter mean bubble diameters were calculated from the bubble projections according to the formula $d = 2\sqrt{S/\pi}$), where S is the area covered by the bubble in the image. The results for bubble diameter, bubble velocity, etc. are shown in figure 4 where the blue and red data points correspond to the optical measurements and the X-ray radiography, respectively. The average bubble diameter derived from the optical measurements is ~3.4 ± 0.3 mm while a value of 3.25 ± 0.4 mm was found for the X-ray measurements. These results are in a very good agreement and correspond very well to the value calculated directly from the bubble detachment frequency taking into account a gas flow rate (50 cm³/min) and a number of 37 bubbles being ejected from the orifice per second (d ≈ 3.5 mm). Figure 4a displays the evolution of the bubble size along the height. Remarkable differences can be observed near the nozzle and at a height of approximately 37.5 mm. Moreover, the bubble sizes obtained from the optical measurements in the upper part of the container are slightly larger than the ones obtained from the X-ray measurements. These differences could be explained by two reasons. First, the X-ray beam has a Gaussian shape showing a maximum at a height of 60 mm height. Therefore, the whole container is not illuminated homogenously. Since the image thresholding is performed using a fixed value for the entire image the less illuminated bubbles provide a weaker signal and the algorithm might determine a too small bubble size. This problem explains the differences in the bubble size near the nozzle and in the upper part of the container near the free surface. Second, the interface between bubble and liquid is directly detected by the optical method whereas the X-ray radiography relies on the absorption contrast which is mainly determined by the local bubble cross-section along the X-ray beam direction. Therefore, it can happen that two bubbles of the same volume but different shapes (see the two examples on the left-hand side of figure 5a) will provide different X-ray signals. The corresponding X-ray intensity can be estimated using the Beer-Lambert law $I = I_0 e^{-\mu x}$, where I_0 is the primary beam intensity, μ is the X-ray absorption coefficient and x is the thickness of the liquid. Calculations of the X-ray intensity were carried out taking into account the values $I_0 = 100$ and $\mu = 0.1$ the latter is the water absorption coefficient for the X-ray energy of 320 KeV. As expected, the results show that bubbles with the larger thickness parallel to the X-ray beam (case A in figure 5) deliver a stronger X-ray signal compared to those with the smaller thickness (case B). A thresholding of both bubble signals using the same threshold value leads to a distinctly larger error in case B in comparison to case A. The consequence is that the total bubble size for case B will be underestimated. This effect might explain the differences in the bubble size at a height of approximately 37.5 mm. At that position the bubbles show significant increase of the size in the optical image (see the bubble highlighted in figure 3a by a circle). It is known from ultrasonic measurements that significant deformations of the bubble occur after detachment of the bubbles from the injector [22]. These fluctuations of the bubble shape lead to a flattening of the bubble in the vertical plane. The increased cross-section detected by the camera simulates a supposed but not real increase in the bubble size

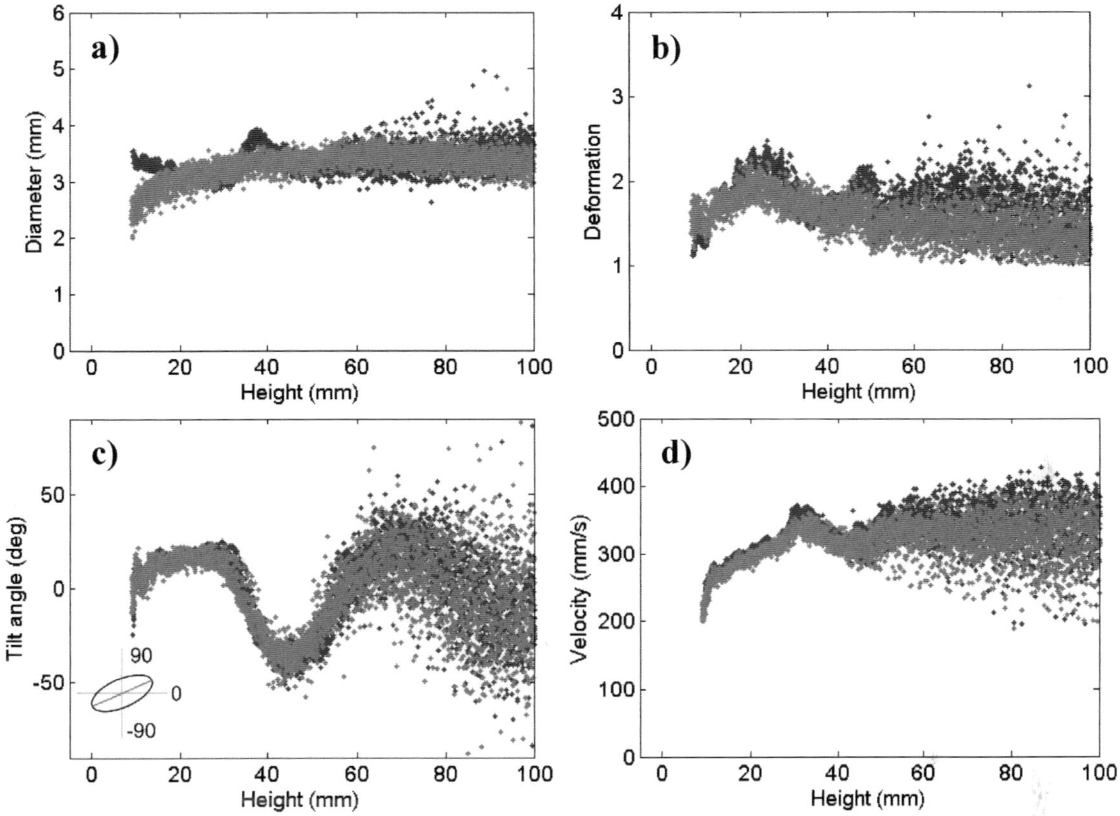

Figure 4. Bubble parameters obtained from optical measurements (blue) and X-ray radiography (red) in water: a) Bubble diameters, b) Bubble deformation, c) Bubble tilt angle, and d) Bubble velocity.

The bubble deformation is defined as the ratio between the length of the major and the minor axis of the ellipses. Corresponding results are shown in figure 4b. The bubble shape undergoes strong deformations along the whole bubble trajectory and reaches a maximum at the height of 25 ± 5 mm. Another deformation peak is observed at a height of 48 ± 2 mm. According to Bhaga [3] the bubble shape can be considered as an oblate ellipsoid with a wobbling surface. The bubble tilt angles (inclination angle of the major axis) and velocities are shown in figure 4c and 4d, respectively. The tilt angle reveals that the bubbles are detached from the nozzle with the main axis inclined positively to the horizontal line which is governed by the bevel shape of the injection nozzle. Then, the zig-zag motion of the bubble is starting fairly quickly. At a height of 29 mm the bubble orientation changes and the bubble tilt angle starts to decrease drastically till a turnaround point is reached at a height of 45 mm. Beyond the height of 68 mm the bubbles tilt angle data are strongly scattered but the zig-zag motion can be clearly identified. The data for the bubble velocity in figure 4d indicate a short acceleration phase just after detachment from the injection nozzle. The final velocity reaches values up to 400 mm/s with a pronounced peak at a height of 31 mm before the zig-zag motion appears. The scatter of the velocity data increases with increasing distance from the gas injection point.

Figure 5b illustrates the experimental X-ray signals from a bubble horizontal cross-section (black) at a height of ~25 mm above the bottom of the container. It becomes obvious that the X-ray signal in water is strongly affected by the noise. The corresponding signal-to-noise ratio calculated as $SNR = (A_{signal}/A_{noise})^2$, where A_{signal} is the signal amplitude and A_{noise} is the average noise amplitude, amounts to 1.3. Such a low signal-to-noise ratio is attributed to strong scattering of the X-rays in water. However, despite of the weak contrast between the bubble and the background and the related low signal-to-noise ratio the algorithm applied for the analysis of the X-ray images provides fairly accurate results with only slight discrepancy in the bubble size and shape (see figure 4). In conclusion, it can be successfully applied for the analysis of bubble rising in liquid metals where the higher X-ray attenuation of the metallic melt provides a better contrast and a better signal-to-noise ratio (see red line in figure 5b).

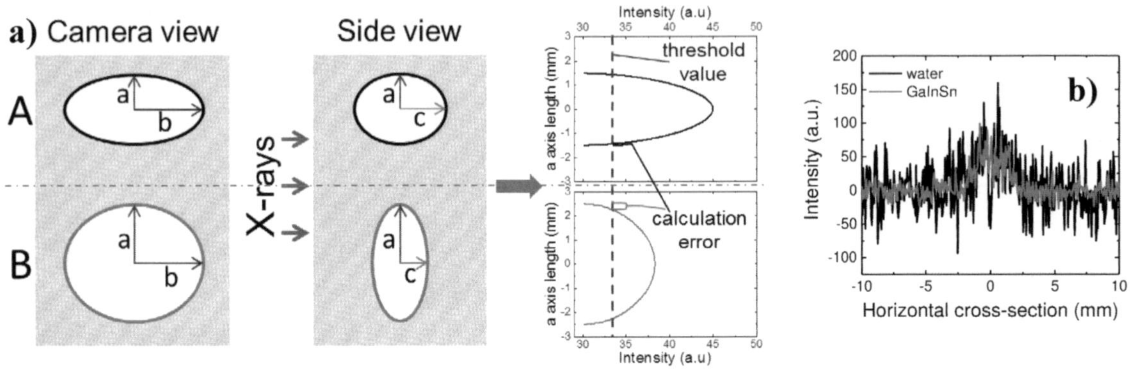

Figure 5. a) Camera view and side view of two bubbles of equivalent volume. Simulated X-ray intensity obtained from the case A (black) and case B (red) for the largest bubble cross-section parallel to the X-ray beam. b) X-ray signal from a horizontal cross-section of a bubble having a diameter of 3 mm diameter in water (black) and in GaInSn (red).

3.2. Comparison GaInSn versus water

This section aims to demonstrate the suitability of the X-ray radiography in the eutectic GaInSn alloy and to show some differences with respect to the bubble rising dynamics in water and GaInSn for a given Ar gas flow rate of 50 cm^3/min. Figure 6 presents the X-ray radiography image analysis for bubbles in GaInSn according to the image processing steps described in *section 2.2*: a) raw image, b) image with reduced reference image, c) gauss filtered image, d) binary image, e) raw image with bubble boundaries and bubble centers, f) raw image with fitted ellipses and their main axis. In comparison to the situation in water the horizontal cross-section of a 3 mm bubble in GaInSn shows a much better signal-to-noise ratio of 14.7 (see figure 5b). Such a large value is rather convenient for further data processing. As a result the analysis to be done for the experiments in GaInSn should provide much more accurate values for bubble size and shape as in water where the validation of the method was carried out.

The experiments in GaInSn were performed using the same gas injector and the same gas flow rates as in water. The most striking difference is that only a number of 6 bubbles are injected per second for the same gas flow rate from the orifice in GaInSn while a bubble detachment rate of 37 bubbles/s was observed in water. The average bubble diameter is 6.35 ± 0.4 mm which is twice as large as in water. The measured value is in a very good agreement with the value calculated from the gas flow rate and bubble detachment frequency (d ≈ 6.42 mm). The significant deviations with respect to the bubble number and size can be attributed to the strong differences in the surface tension and the wetting behavior at the injection nozzle [14].

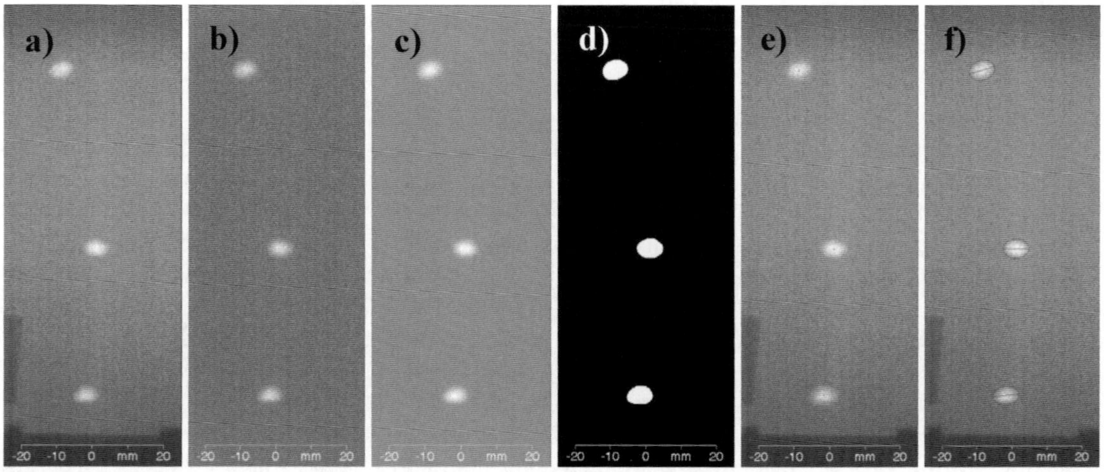

Figure 6. Image processing steps demonstrated for an exemplary image in GaInSn: a) raw image, b) image with reduced reference image, c) gauss filtered image, d) binary image, e) raw image with bubble boundaries and bubble centers, f) raw image with fitted ellipses and their main axis.

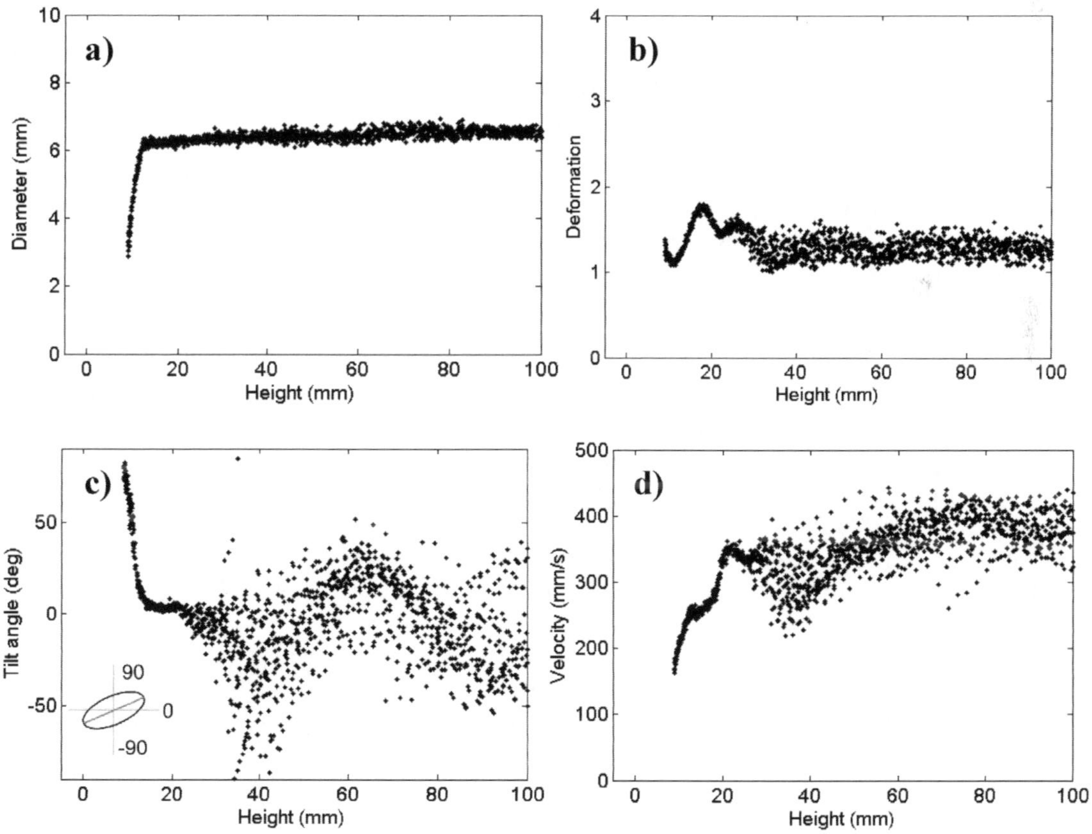

Figure 7. Bubble parameters determined by X-ray radiography in the eutectic GaInSn: a) Bubble diameter, b) bubble deformation, c) bubble tilt angle, d) bubble velocity.

Bubbles undergo the maximum deformation just after the detachment from the nozzle. The less scattering of the data points for the bubble deformation (see figures 4b and 7b) is also a direct result of the larger surface tension for GaInSn which prevents significant surface wobbling. During the initial stage of their rise the bubbles show a distinct deformation and can be described as ellipsoids but on their further trajectory the bubbles approach almost spherical shape: the deformation tends to become close to 1 at container heights above 30 mm. When leaving the nozzle the bubble main axis is almost parallel to the bottom of the container. Strong variations start beyond a height of 25 mm. At the positions below 10 mm the bubbles are still attached to the nozzle. This elongation of the bubble along the vertical direction explains the large tilt angles in figure 7c. The strong scattering of the tilt angle data at the upper part of the container is caused by the almost spherical bubble shape where all diagonals are almost equivalent. The almost spherical bubble shape impedes the accurate determination of the main axis by the Matlab algorithm. The bubble velocity in the GaInSn melt reaches values up to 400 mm/s with a pronounced peak at a height of 21 ± 1 mm.

As a summary, it can be concluded that the applicability of X-ray radiography for quantitative measurement of bubble parameters rising in a stagnant liquid was successfully validated by parallel optical measurements in water. Investigations in liquid metals benefit from a better signal-to-noise ratio due to better X-ray contrast. The experiments showed that for a chosen Ar gas flow rate the total number of injected bubbles is much larger in water than in GaInSn. Likewise, the bubble deformation is larger in water than in GaInSn. The bubbles in GaInSn tend to have a more spherical-like shape while the bubbles in water can be described as oblate ellipsoids with a wobbling surface. The main reason for these effects is the higher surface tension in GaInSn in comparison to water. The rising velocities of the gas bubble are found to be rather similar both water and GaInSn. The higher buoyancy in the liquid metal is obviously compensated by the increased drag force owing to the larger bubble size [10].

Acknowledgments

The research is supported by the German Helmholtz Association in form of the Helmholtz-Alliance "LIMTECH". We acknowledge the valuable technical assistance of Erik Strumpf during the measurement campaign.

References

[1] Wellek R M, Agrawal A K and Skelland A H P 1966 *AIChE J.* **12** 854–62
[2] Clift R, Grace J R and Weber M E 1978 *Bubbles, Drops, and Particles* (Academic Press)
[3] Bhaga D and Weber M E 1981 **105** 61–85
[4] Brücker C 1999 *Phys. Fluids* **11** 1781–96
[5] Valencia A, Cordova M and Ortega J 2002 *Int. Commun. Heat Mass Transf.* **29** 821–30
[6] de Vries A W G, Biesheuvel A and van Wijngaarden L 2002 *Int. J. Multiph. Flow* **28** 1823–35
[7] Kulkarni A A and Joshi J B 2005 *Ind. Eng. Chem. Res.* **44** 5873–931
[8] Smolianski A, Haario H and Luukka P 2008 *Appl. Math. Model.* **32** 641–59
[9] Legendre D, Zenit R and Velez-Cordero J R 2012 *Phys. Fluids* **24** 43303
[10] Zhang C, Eckert S and Gerbeth G 2005 *Int. J. Multiph. Flow* **31** 824–42
[11] Manera A, Ozar B, Paranjape S, Ishii M and Prasser H-M 2009 *Nucl. Eng. Des.* **239** 1718–24
[12] Lyu Z and Karcher C 2016 *Proc. 10th PAMIR Int. Conf. Fundam. Appl. MHD (Cagliary, Italy)* 243–6
[13] Mishima K, Hibiki T, Saito Y, Nishihara H, Tobita Y, Konishi K and Matsubayashi M 1999 *Nucl. Instruments Methods Phys. Res. A* **424** 229–34
[14] Saito Y, Mishima K, Tobita Y, Suzuki T, Matsubayashi M, Lim I C and Cha J E 2005 *Nucl. Instruments Methods Phys. Res. A* **542** 168–74
[15] Iguchi M, Chihara T, Takanashi N, Ogawa Y, Tokumitsu N and Morita Z 1995 *ISIJ Int.* **35** 1354–61
[16] Wang Z, Mukai K and Izu D 1999 *ISIJ Int.* **39** 154–63

[17] Shevchenko N, Boden S, Eckert S, Borin D, Heinze M and Odenbach S 2013 *Eur. Phys. J. Spec. Top.* **220** 63–77
[18] Fröhlich J, Schwarz S, Heitkam S, Santarelli C, Zhang C, Vogt T, Boden S, Andruszkiewicz A, Eckert K, Odenbach S and Eckert S 2013 *Eur. Phys. J. Spec. Top.* **220** 167–83
[19] Timmel K, Shevchenko N, Röder M, Anderhuber M, Gardin P, Eckert S and Gerbeth G 2015 *Metall. Mater. Trans. B* **46** 700–10
[20] Vogt T, Boden S, Andruszkiewicz A, Eckert K, Eckert S and Gerbeth G 2015 *Nucl. Eng. Des.* **294** 16–23
[21] Plevachuk Y, Sklyarchuk V, Eckert S, Gerbeth G and Novakovic R 2014 *J. Chem. Eng. Data* **59** 757–63
[22] Richter T 2017 Personal communication

IOP Conf. Series: Materials Science and Engineering **228** (2017) 012010 doi:10.1088/1757-899X/228/1/012010

Dynamics of liquid metal droplets and jets influenced by a strong axial magnetic field

D Hernández[1] and Ch Karcher[1]
[1]Institute for Thermodynamics and Fluid Mechanics
Technische Universität Ilmenau, 98684 Ilmenau, Germany
E-mail: daniel.hernandez@tu-ilmenau.de

Abstract. Non-contact electromagnetic control and shaping of liquid metal free surfaces is crucial in a number of high-temperature metallurgical processes like levitation melting and electromagnetic sealing, among others. Other examples are the electromagnetic bending or stabilization of liquid metal jets that frequently occur in casting or fusion applications. Within this context, we experimentally study the influence of strong axial magnetic fields on the dynamics of falling metal droplets and liquid metal jets. GaInSn in eutectic composition is used as test melt being liquid at room temperature. In the experiments, we use a cryogen-free superconducting magnet (CFM) providing steady homogeneous fields of up to 5 T and allowing a tilt angle between the falling melt and the magnet axis. We vary the magnetic flux density, the tilt angle, the liquid metal flow rate, and the diameter and material of the nozzle (electrically conducting/insulating). Hence, the experiments cover a parameter range of Hartmann numbers Ha, Reynolds numbers Re, and Weber numbers We within $0 < Ha < 440$, $340 < Re < 4500$, and $0.09 < We < 12.1$. As major results we find that under the influence of the strong magnetic field, droplet rotation ceases and the droplets are stretched in the field direction. Moreover, we observe that the jet breakup into droplets (spheroidization) is suppressed, and in the case of electrically conducting nozzles and tilt, the jets are bent towards the field axis.

Keywords: jet instability, magnetohydrodynamics, Lorentz force

1. Introduction

During the last decades, Electromagnetic Processing of Materials (EPM) have developed to a powerful engineering tool to influence or to control high-temperature processes of electrically conducting fluids like liquid metals (for an overview see [1]). In those processes, one exploits the fact that due to the interactions between a moving conductor and an externally applied magnetic field, the so-called Lorentz forces are induced. These forces can be used for stirring and melting processes, i.e. enhancing heat and mass transfer, as well as for the development of non-contact flow measurement techniques [2, 3, 4, 5]. In this paper, we focus on the effects of electromagnetic shaping and stabilization of liquid metal free surfaces [6, 7, 8, 9, 10]. In more detail we investigate the effect of an axial magnetic field on a liquid metal droplet and jet flow. A liquid jet which flows out from a nozzle to an ambient gas is inherently unstable. Plateu [11] has shown that, due to surface tension, an infinite jet breaks up into sections comparable to the initial circumference which have less total surface area than the original cylindrical jet. Rayleigh [12] showed that by neglecting the viscosity of the liquid, the ambient fluid, and gravity, the system is characterized by unstable perturbations of different wavelengths, for which the interfacial perturbations are amplified, i.e. the cylindrical shape becomes morphologically unstable. Rayleigh concluded that the maximum instability, in which perturbations grow faster, occurs at $\lambda_{crit} \approx 9r_0$, where r_0 is the radius of the original cylindrical jet. Favored by surface tension, λ_{crit} triggers the breakup of the liquid jet into droplets, and therefore defines the typical droplet size. Chandrasekhar [13] treated the general Navier-Stokes case and investigated the effect of the viscosity on the breakup. λ_{crit} can be shifted to larger or shorter wavelengths depending on the viscosity of the liquid. Chandrasekhar also investigated the effect of an axial magnetic field on this so-called capillary instability and concluded that a magnetic field of sufficient strength will stabilize the jet for varicose deformations of all wavelengths.

In this paper we validate experimentally the stabilizing effect predicted by Chandrasekhar for droplet and liquid metal jet flows under a strong axial magnetic field. The working fluid is GaInSn in eutectic composition and the steady and homogeneous axial magnetic field is produced by a 5 T cryogen-free superconducting magnet (CFM). We vary the diameter ($d_1 = 0.9$ mm, $d_2 = 1.55$ mm, $d_3 = 2.2$ mm) and material (electrically conducting/insulating) of the nozzle, the magnetic field strength ($B_0 = 0 \ldots 5$ T) and the liquid metal flow rate. We have focused at the beginning of this paper on the results using the biggest diameter, for which the stabilizing effect is more pronounced, covering three specific flow regimes: 1- droplet flow, 2- transition from droplet to continuous jet, and 3- continuous jet flow. Finally, we investigate the effect of electrically conducting and electrically insulating nozzles on a liquid metal jet, when the applied magnetic field is tilted by an angle of $\varphi = 5°$ with respect to gravity. In this case we vary the diameter of the nozzle while maintaining constant the Reynolds number and the magnitude of the applied magnetic field.

The structure of the paper is as follows: in section 2 we presents a brief introduction of droplet and jet flows with and without the influence of an axial magnetic field. Here, we review the definition of all dimensionless parameters that describe the physics involved. In section 3, the experimental set-up is described followed by section 4, which shows the main results. Finally, section 5 gives a summary of the principal conclusions.

2. Capillary instability of a liquid metal jet under a strong axial magnetic field

The flow regimes in droplet and liquid metal jet flows is a complex interaction between the effects of surface tension, inertia and friction. These effects are parameterized by the dynamic scaling ratios Reynolds Re, Weber We and Ohnesorge Oh numbers defined by the relations

$$Re = \frac{\rho u_0 d}{\mu}, \quad We = \frac{\rho u_0^2 d}{\gamma}, \quad Oh = \frac{\sqrt{We}}{Re} = \frac{\mu}{\sqrt{\rho \gamma d}}. \tag{1,2,3}$$

Here, u_0 is the characteristic velocity (average velocity inside the nozzle) and d is the characteristic length (nozzle diameter). ρ, μ and γ denote the density, the dynamic viscosity and surface tension of the liquid, respectively. Re and We represent the ratio of inertia to viscous forces and the ratio of inertia to surfaces tension forces, respectively. The breakup regimes of a given liquid jet flow depends on the corresponding values of Re and Oh. An example of a phase diagram Oh vs Re is depicted in [14]. Droplet flow occurs at low We and Oh numbers. For higher Oh and We, the flow regime of a liquid jet can be classified according to Lin [15] as: Rayleigh regime, first wind-induced regime, second wind-induced regime and atomization regime. Rayleigh and first wind-induced regime are characterized by low jet velocities, whereby the diameter of the droplets are larger than the jet diameter. For high speed jets, second wind-induced and atomization regime, the breakup is due to the growth of short-wavelength surface waves generating droplets smaller that the diameter of the jet.

Chandrasekhar [13] investigated the influence of an axial magnetic field on the capillary instability of an electrically conducting jet. Among others, he analyzed the cases in which the liquid is inviscid and shows a finite electrical conductivity, i.e. a finite resistivity. Here, the electrical resistivity η is defined as $\eta = (4\pi\mu_m\sigma)^{-1}$, where μ_m and σ are the magnetic permeability and electrical conductivity of the liquid, respectively. The effects of an axial magnetic field on the capillary instability of an electrically conducting jet are parameterized by the dimensionless group χ, Hartmann number Ha, and the electromagnetic interaction parameter N defined by the relations

$$\chi = \frac{\mu_m H^2}{4\pi\eta}\sqrt{\frac{r_0^3}{\rho\gamma}} = \frac{Ha^2}{4}\sqrt{\frac{l_\nu}{r_0}}, \quad Ha = \sqrt{NRe} = B_0 d\sqrt{\frac{\sigma}{\rho\nu}}, \quad N = \frac{\sigma B_0^2 d}{\rho u_0}. \qquad (4,5,6)$$

Here, $l_\nu = \nu^2\rho(\gamma)^{-1}$ is a viscous lengthscale according to Eggers [16] and H is the auxiliary magnetic field, where $B_0 = \mu_m H$. Ha and N represent the ratio of Lorentz forces to shear forces and the ratio of Lorentz forces to inertia forces, respectively. Ha is typical used for scaling the magnitude of B_0 in a given system. The main conclusion is that an axial magnetic field B_0 of sufficient strength will stabilize the jet for capillary instability (varicose deformations) of all wavelengths, whereby the wavelength λ_{crit} increases. For a liquid with zero viscosity and zero resistivity, the capillary instability can be totally suppressed. In the case of liquid metals, which in general have finite η, the stabilizing effect is present when the parameter χ exceeds the value of 6. Table 1 compares the values of χ for mercury, lithium, sodium and GaInSn for the same values of r_0 and B_0 like our experiments presented in section 4. According to table 1, we could expect a stronger suppression of capillary instability for hot liquids like lithium and sodium than for GaInSn or mercury under the influence of an axial magnetic field.

Liquid metal	ρ (kg/m^3)	ν (m^2/s)	σ (MS/m)	γ (N/m)	χ
Mercury @450°C[17]	$13.5 \cdot 10^3$	$1.15 \cdot 10^{-7}$	1	0.48	4.3
Lithium @450°C [17]	491	$7.1 \cdot 10^{-7}$	3.1	0.371	74.9
Sodium @450°C [17]	844	$3.0 \cdot 10^{-7}$	4.3	0.164	120.8
GaInSn @$\sim$ 26°C [18]	$6.33 \cdot 10^3$	$3.3 \cdot 10^{-7}$	3.3	0.585	17.8

Table 1. Summary of basic thermophysical and electrical properties for different liquid metals and the corresponding value of χ according to (4) for $r_0 = d/2 = 1.1$ mm and $B_0 = 3$ T like in our experiments.

3. Experimental set-up

The experimental set-up is depicted in figure 1. A syringe pump drives the liquid metal from the collecting vessel to the centering device at which the respective nozzle is fixed. Centering device and nozzle are located in the middle of a cryogen-free superconducting DC magnet (CFM) which produces an homogeneous axial magnetic field B_0. As a simple reciprocating pump, the syringe pump works in two modes. In the first one, it pumps GaInSn from the collecting vessel into its barrel. In the second one, it pumps the collected liquid into the centering device. The intake and outtake of GaInSn from and into the pump are controlled by a valve and the pressure in the tube is measured by a manometer. At the end of the centering device, which ensures an accurate angle between the applied magnetic field and the jet, the liquid metal flows through the nozzle and falls finally into a plastic funnel. At the outlet of the funnel the liquid metal then flows due to gravity to the collecting vessel. A high-speed camera placed on top of the magnet records the projection of the jet or droplet flow on a mirror. The light needed is given by a LED also placed in the middle of the magnet as well as the mirror.

The flow rate in the experiments for $d = 2.2$ mm lies between 11.6 ml/min and 154.5 ml/min covering a range of $340 < Re < 4500$ $(0.09 < We < 12.1)$. The velocity u_0 used in the dimensionless numbers is obtain by dividing the flow rate by the cross-section of the nozzle

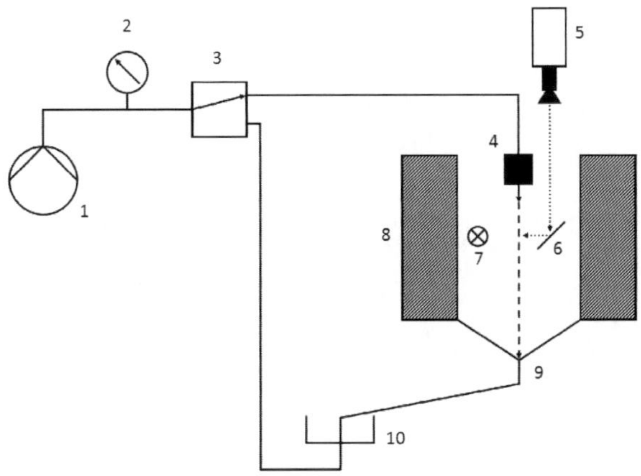

Figure 1. Sketch of the experimental set-up for the observation of the stabilizing effect of an axial magnetic field on liquid metal droplet and jet flows. The set-up is composed of: 1- mid pressure syringe pump, 2 - manometer, 3- valve, 4- centering device (brass), 5- high-speed camera (fps=231), 6- mirror, 7- LED, 8- cryogen-free super conducting DC magnet, 9- plastic funnel and 10- collecting vessel.

($\pi d^2/4$). In our experiments, we have selected the following Reynolds numbers which correspond to 3 different flow regimes:

- $Re = 340\,(We = 0.09)$: droplet flow
- $Re = 1690\,(We = 1.9)$: transition from droplet to continuous jet flow characterized by discontinuous short-length irregular jets
- $Re = 4500\,(We = 12.1)$: continuous jet flow (Rayleigh regime)

The magnitude of the applied magnetic field in the experiments starts at 0 T and it is gradually increased in 1 T steps until reaching 5 T. The magnetic field distribution in the center of the magnet, in which the liquid metal is observed, can be found in [19, 20]. In this area, the magnetic field is fairly homogeneous having about $\partial B_0/\partial r \approx 0.03$T/cm in radial direction and $\partial B_0/\partial z \approx 0.06$T/cm along the axis of the magnet. An homogeneous magnetic field produces no net electromagnetic force on droplets and/or jets. A high-speed camera is placed on top of the 5 T cryogen-free magnet and captures the dynamics of the flow at 231 frames per second. In our experimental set-up, the camera is capable to record droplet or jet flows up to a distance of about 7 cm away from the nozzle. The parameters taken into consideration are summarized in table 2.

Parameter	Value
Nozzle diameter d (mm)	0.9 , 1.55 , 2.2
Material of the nozzle	Brass (conducting) , PTFE (insulating)
Magnetic field B_0 (T)	0 , 1 , 2 , 3 , 4 , 5
Inclination of the magnet φ	0° , 5°
Re	340 , 1690 , 4500
We	0.09 , 1.9 , 12.1
Oh	$1.1 \cdot 10^{-3}$, $8.8 \cdot 10^{-4}$, $7.4 \cdot 10^{-4}$

Table 2. List of parameters.

4. Results

As explained in section 1, the results that will be presented at the beginning of this section correspond to the nozzle with diameter $d = 2.2$ mm in which $0 < N < 564.4$, $0 < Ha < 436.4$, and $0 < \chi < 49.5$. For smaller diameters, i.e. $d < 2.2$ mm at $Re = 340$ and $d < 1.55$ mm at $Re = 1690$, there was no evident stabilizing effect of the magnetic field even by $B_0 = 5$ T. Figure

Figure 2. Droplet formation at the outlet of the electrically conducting nozzle made of brass for $Re = 340$, $d = 2.2$ mm, $\varphi = 0°$, $B_0 = 0$ (upper sequence) and $B_0 = 5$ T (lower sequence).

2 presents a sequence of droplet flow ($Re = 340$) for 0 and 5 T flowing through the electrically conducting nozzle. It shows that the magnetic forces influence droplet formation just at the outlet of the nozzle. The strong axial magnetic field suppresses the growth of the droplet in radial direction, and therefore, it is forced to elongate streamwise. This same effect was also seen using the electrically insulating nozzle. However, the already deformed droplet maintains this form downstream until the bottom of the observable area. It appears that the influence of the magnetic field on the expected internal circulation patterns inside the droplet according to LeClair [21] is small and can not be identified based on the pictures taken by the high-speed camera. However, we have observed that $B_0 = 1$ T is sufficient for the droplets to align themselves with their axis into the direction of the magnetic field for every nozzle diameter. For a better understanding of this interaction between a liquid metal droplet under an axial magnetic field, we are currently preparing a numerical model of our experiments.

When the flow rate is increased up to $Re = 1690$, we observe that the capillary instability, spheroidization of the jet into droplets, is clearly suppressed. The discontinuous short-length jets have now characteristic top and bottom sharp tips which were already predicted by Eggers [22]. The shape of the jet segments with sharp tips appears to be caused by the complex interaction between the suppression of the radial velocity of the jet during the process of spheroidization under a strong axial magnetic field and the acceleration of the fall. Additionally, the breakup length is increased for $B_0 = 5$ T. We believe that with a magnetic field B_0 not much higher than the one considered in this paper, the capillary instability could be totally suppress, or at least sufficiently, resulting in long and continuous liquid metal jets at low Re numbers. However, for $Re = 4500$, there was no visible effect of the magnetic field as in this regime the instability occurs further than the observable area of the camera. The flow in this case is strictly axial, and therefore, co-aligned with the magnetic field without able to generate a Lorentz force density in the liquid. This is the same reason that we did not observe any comparable differences with electrically conducting and insulating nozzles as the velocity field in the nozzle is parallel to the magnetic field (figures 3 and 4). In the next section, as now the velocity and the applied magnetic field are no longer co-aligned with each other, we clearly see a notorious effect on the dynamics of the jet using an electrically conducting nozzle.

One of the problems or disadvantages of working with GaInSn is that it gradually reacts with the ambient air generating oxide that, after a long period of runs, it may accumulate inside the nozzle. This will cause unwanted deflections of the jet that are not related to the physics involved. An example of this situation can be seen in figure 3 for ($Re = 1690$, $N = 4.6$) and in figure 4 for ($Re = 1690$, $N = 4.6$) and for ($Re = 4500$, $N = 1.7$).

IOP Conf. Series: Materials Science and Engineering **228** (2017) 012010 doi:10.1088/1757-899X/228/1/012010

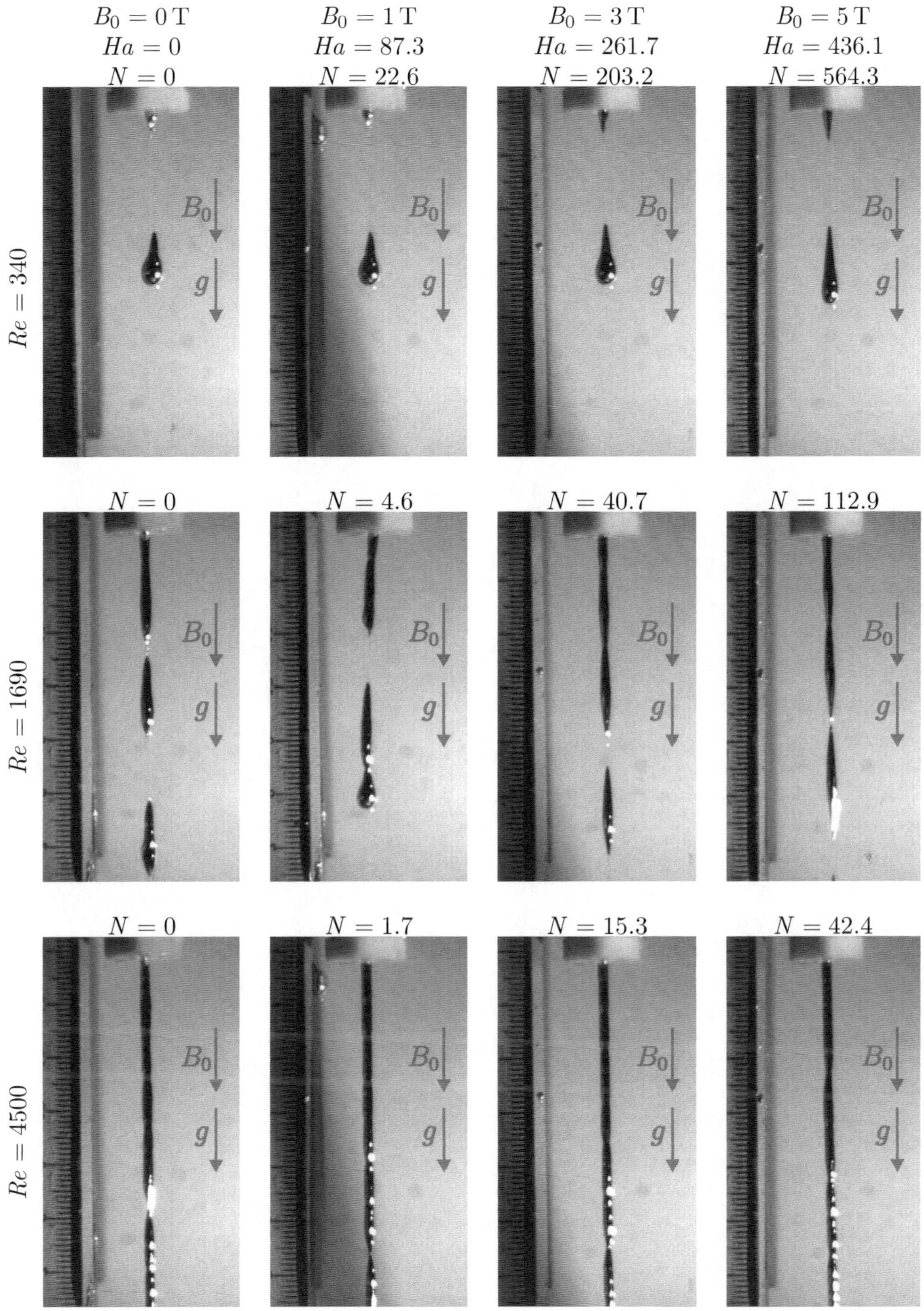

Figure 3. Single images of GaInSn droplet and jet flows falling from an electrically insulating nozzle. Re and Ha numbers are maintained constant in the horizontal and in the vertical axis, respectively. The direction of the applied magnetic field B_0 and gravity g are plotted as red and blue arrows, respectively. For ($Re = 1690$, $N = 4.6$), a non-physical bending of the jet occurs due to accumulation of GaInSn oxide at the outlet of the nozzle.

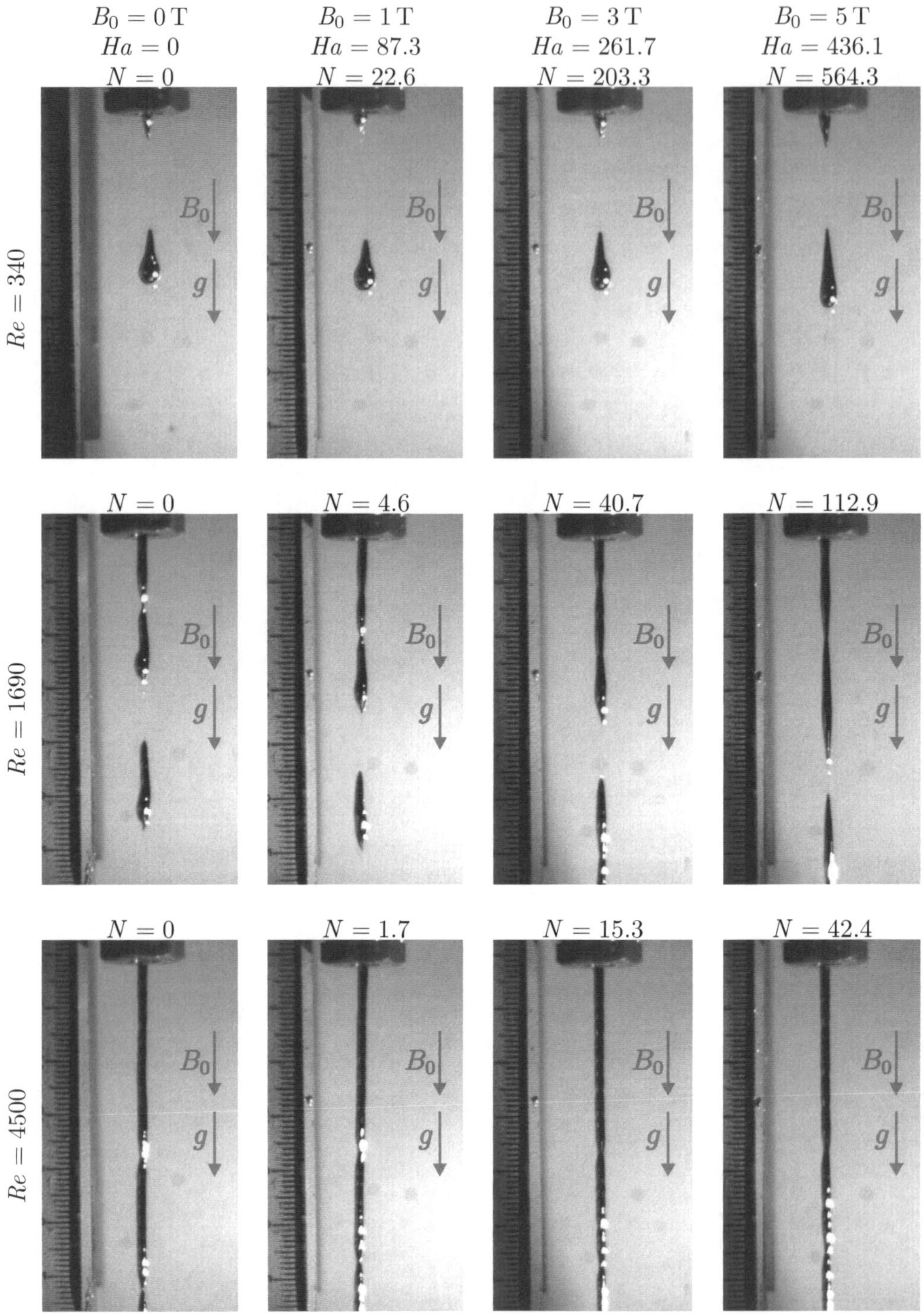

Figure 4. Single images of GaInSn droplet and jet flows falling from an electrically conducting nozzle made of brass. Re and Ha numbers are maintained constant in the horizontal and in the vertical axis, respectively. The direction of the applied magnetic field B_0 and gravity g are plotted as red and blue arrows, respectively. For ($Re = 1690$, $N = 4.6$) and ($Re = 4500$, $N = 1.7$), a non-physical bending of the jet occurs due to accumulation of GaInSn oxide at the outlet of the nozzle.

4.1. Tilted magnetic field

In this section, in contrast to the previous one, the magnetic field and gravity are not co-aligned but tilted 5° with each other. In this case we focus on $Re = 4500$ characterized by a continuous liquid jet (Rayleigh regime). Figure 5 presents the results of conducting and insulating nozzles by different nozzle diameters and with and without axial magnetic field. We observe that when the diameter d is increased while maintaining the same Re, i.e. N is increased, the effect of wall conductivity is more pronounced. At $B_0 = 5\,\mathrm{T}$, the liquid metal is strongly deflected into the direction of the applied magnetic field at the outlet of the nozzle, whereas in the insulating nozzle case no effect was observed. As now the flow inside the nozzle is not parallel to the magnetic field, eddy currents and flow-breaking Lorentz forces are generated in this area. However, it appears that in the insulating wall case, the induced Lorentz forces are too weak to affect the dynamics of the jet as the induced eddy currents are forced to close in a smaller area. In the case of conducting walls, the induced eddy currents and Lorentz force density are stronger due to the fact that the overall electric resistance of the system is minimized.

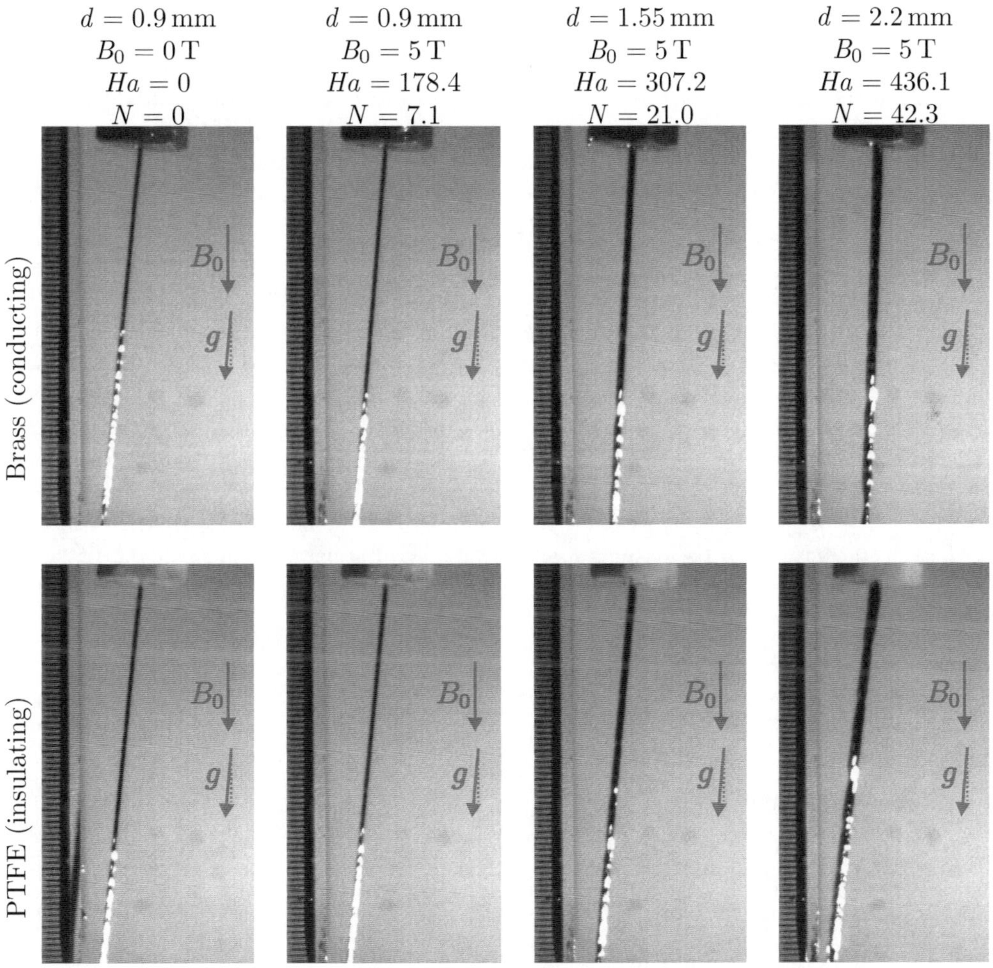

Figure 5. Sequence of pictures of GaInSn jets ($Re = 4500$) falling from an electrically conducting (top) and an electrically insulating nozzle (bottom). The direction of the applied magnetic field B_0 and gravity g are plotted as red and blue arrows, respectively. The magnetic field and gravity are tilted $\varphi = 5°$ with each other.

5. Conclusions

We experimentally investigated the effect of an axial streamwise magnetic field on the suppression of capillary instability of a liquid metal jet and its influence on a droplet flow. In this paper, we have focused on three different Reynolds numbers which correspond to three breakup regimes ($Re = 340$, $Re = 1690$, $Re = 4500$). We have presented the results for the biggest diameter ($d = 2.2\,\mathrm{mm}$) where the interaction between the flowing liquid metal and the magnetic field is more predominant. The first breakup regime corresponds to droplet and the third one to a continuous jet flow (Rayleigh regime). We have observed that at $Re = 1690$, the transition from droplets to a continuous jet flow occurs characterized by discontinuous short-length irregular structures. In this regime, the applied magnetic field of 3 T was sufficient to stabilize these segments showing characteristic top and bottom sharp tips, whereby the length of breakup is increased. In the case of droplet flow, the tips of the droplets are aligned vertically with the magnetic field even by the smallest magnetic field considered in this paper (1 T) and for 5 T, the droplet is deformed and elongated in the direction of the magnetic field. The strong axial magnetic field seems to suppress the radial growth of the droplet at the outlet of the nozzle. For continuous liquid jets ($Re = 4500$), there was no observable effect of the axial magnetic field on the flow for neither electrically conducting nor electrically insulating nozzles. However, when the magnetic field is not co-aligned but tilted 5° relative to gravity, the induced eddy currents and Lorentz forces are much stronger using the electrically conducting nozzle than the insulating one. The induced Lorentz forces are enough to deflect the liquid jet at the outlet of the electrically conducting nozzle into the direction of the magnetic field. Depending on the value of the interaction parameter N, the effect is less or more pronounced.

Acknowledgments

The authors acknowledge financial support from German Science Foundation within the Research Training Group Lorentz force under grant GRK 1567 and from Helmholtz Alliance LIMTECH within the seeding grant program in the Young Investigator Group. Moreover, the authors are thankful to student Jonas Ketterer for his experimental support. The authors gratefully acknowledge the fruitful discussions with Thomas Boeck (TU Ilmenau).

References

[1] Asai S 2012 *Electromagnetic processing of materials: Materials processing by using electric and magnetic functions* (*Fluid mechanics and its applications* vol 99) (Dordrecht: Springer)
[2] Priede J, Buchenau D and Gerbeth G 2009 *Magnetohydrodynamics* **45** 451–458
[3] Priede J, Buchenau D and Gerbeth G 2011 *Meas. Sci. Technol.* **22** 055402
[4] Stefani F, Gundrum T and Gerbeth G 2004 *Phys Rev E Stat Nonlin Soft Matter Phys* **70** 056306
[5] Thess A, Votyakov E and Kolesnikov Y 2006 *Phys. Rev. Lett.* **96** 164501
[6] Easter S, Bojarevics V and Pericleous K 2011 *J. Phys.: Conf. Series* **327** 012027
[7] Gailitis A 1966 *Magnetohydrodynamics* **2** 47–53
[8] Gerbeth G, Kaudze M and Gailitis A 1989 Deformation of an electrically conducting drop in a magnetic field *Liquid Metal Magnetohydrodynamics* (*Mechanics of fluids and transport processes* vol 10) ed Lielpeteris J and Moreau R (Dordrecht: Springer) pp 287–292
[9] Lielpeteris J (ed) 1989 *Liquid metal magnetohydrodynamics* (*Mechanics of fluids and transport processes* vol 10) (Dordrecht: Kluwer)
[10] Priede J 2011 *J. Fluid Mech.* **671** 399–416
[11] Plateau J A F 1873 *Statique expérimentale et théorique des liquides soumis aux seules forces moléculaires* (Paris: Gauthier-Villars)
[12] Rayleigh L 1879 *Proc. R. Soc. Lond.* **29** 71–97
[13] Chandrasekhar S 1981 *Hydrodynamic and hydromagnetic stability* dover ed., 1. publ., unabridged republ. of the 3. (1970) print. of the original clarendon pr. ed. 1961 ed (New York NY: Dover Publ)
[14] Schmidt P and Walzel P 1984 *Phys. Unserer Zeit* **15** 113–120
[15] Lin S P and Reitz R D 1998 *Annu. Rev. Fluid Mech.* **30** 85–105
[16] Eggers J and Dupont T F 1994 *J. Fluid Mech.* **262** 205

[17] 2008 *Thermophysical properties of materials for nuclear engineering: A tutorial and collection of data* (Vienna: International Atomic Energy Agency)
[18] Plevachuk Y, Sklyarchuk V, Eckert S, Gerbeth G and Novakovic R 2014 *J. Chem. Eng. Data* **59** 757–763
[19] Ziolkowski M, Brauer H, Halbedel B and Krieger U 2007 *16th Conference on the Computation of Electromagnetic Fields (COMPUMAG), Aachen, Germany* 871–872
[20] Hamann B, Schawohl J and Kraffert C 2006 *Information technology and electrical engineering - devices and systems, materials and technologies for the future : 51. IWK, Internationales Wissenschaftliches Kolloquium ; proceedings ; 11 - 15 September 2006* 489–490
[21] LeClair B P, Hamielec A E, Pruppacher H R and Hall W D 1972 *J. Atmos. Sci.* **29** 728–740
[22] Eggers J 1997 *Rev. Mod. Phys.* **69** 865–930

IOP Conf. Series: Materials Science and Engineering **228** (2017) 012011 doi:10.1088/1757-899X/228/1/012011

Transition between free, mixed and forced convection

W Jaeger[1,6], F Trimborn[3,5], M Niemann[4], V Saini[4,5], W Hering[2], R Stieglitz[2], B Pritz[3], J Fröhlich[4] and M Gabi[3]

[1]Institute of Fusion and Reactor Technology, Karlsruhe Institute of Technology (KIT), Kaiserstraße 12, 76131 Karlsruhe, Germany
[2]Institute for Neutron Physics and Reactor Technology, Karlsruhe Institute of Technology (KIT), Hermann-von-Helmholtz-Platz 1, 76344 Eggenstein-Leopoldshafen, Germany
[3]Institute of Fluid Machinery, Karlsruhe Institute of Technology (KIT), Kaiserstraße 12, 76131 Karlsruhe, Germany
[4]Institute of Fluid Mechanics, Technische Universität Dresden, 01062 Dresden, Germany
[5]Currently at: Laboratoire de Mécanique de Lille, Ecole Central de Lille, Boulevard Paul Langevin, 59655 Villeneuve d'Ascq Cédex, France
[6]wadim.jaeger@kit.edu

Abstract. In this contribution, numerical methods are discussed to predict the heat transfer to liquid metal flowing in rectangular flow channels. A correct representation of the thermo-hydraulic behaviour is necessary, because these numerical methods are used to perform design and safety studies of components with rectangular channels. Hence, it must be proven that simulation results are an adequate representation of the real conditions. Up to now, the majority of simulations are related to forced convection of liquid metals flowing in circular pipes or rod bundle, because these geometries represent most of the components in process engineering (e.g. piping, heat exchanger). Open questions related to liquid metal heat transfer, among others, is the behaviour during the transition of the heat transfer regimes. Therefore, this contribution aims to provide useful information related to the transition from forced to mixed and free convection, with the focus on a rectangular flow channel. The assessment of the thermo-hydraulic behaviour under transitional heat transfer regimes is pursued by means of system code simulations, RANS CFD simulations, LES and DNS, and experimental investigations. Thereby, each of the results will compared to the others. The comparison of external experimental data, DNS data, RANS data and system code simulation results shows that the global heat transfer can be consistently represented for forced convection in rectangular flow channels by these means. Furthermore, LES data is in agreement with RANS CFD results for different Richardson numbers with respect to temperature and velocity distribution. The agreement of the simulation results among each other and the hopefully successful validation by means of experimental data will fosters the confidence in the predicting capabilities of numerical methods, which can be applied to engineering application.

IOP Conf. Series: Materials Science and Engineering **228** (2017) 012011 doi:10.1088/1757-899X/228/1/012011

1. Introduction

Liquid metal flows are characterized by low molecular Prandtl numbers, which strongly influence the transport of heat from solid surfaces. As a result, they are capable to transfer quantities of thermal energy at low temperature differences to a large fraction by molecular diffusion. This allows operating thermodynamic processes on higher temperature levels attaining potentially higher overall efficiencies of the whole system. In order to optimize the design or evaluate the performance of such large scale facilities, local predictions are required. For the latter reliable experimental data are lacking, and hence one strategy is to obtain data from direct or partially resolved numerical simulations (DNS or LES) or data from simulations based on the Reynolds-averaged Navier-Stokes equations (RANS). Within the scope of the latter numerical methods, the RANS approach still seems to be the best compromise between required numerical effort and accuracy among the given methods, because the realization of DNS or even LES is still limited to rather simple geometries and specialized numerical codes. Experiments in liquid metals require a detailed knowledge of handling low molecular Prandtl number fluids. Safety aspects, purity of the material and chemical reactivity, e.g. the reactivity of liquid sodium or steel corrosion in Lead-Bismuth Eutectic (LBE) must be taken into account. The simultaneous measurement of local mean and fluctuating quantities with high accuracy and resolution to validate numerical methods is a further challenge. Detailed accounts on low Prandtl number thermal-hydraulics and related measurements can be found in the most recent publications of [1][2][3].

This article addresses this aspect. Because the usual approach based on a similarity assumption between the turbulent eddy viscosity and the turbulent thermal diffusivity leads to inaccurate results, more sophisticated turbulent heat flux models are developed, and their results are analyzed for validation purposes. Especially with respect to different convection regimes (forced, mixed and natural convection) and the transition from one regime to the other, the modelling of the turbulent heat fluxes has to incorporate the thermal-hydraulic peculiarities of liquid metal flows. Furthermore, the flow in a rectangular channel is characterized by the occurrence of the well-known secondary motions of the second kind, which show a notable impact on the heat transfer of already in higher Prandtl number flows. Thus, anisotropic modelling approaches for the closure of the turbulent momentum equations have to be investigated to provide an accurate basis for further modelling activities and investigations. Thereby, the subjects of the described task can be linked to the contributing organizations: Coordination and development of the experiment – KIT; Turbulent heat flux analysis by means of RANS CFD and LES – KIT; LES and DNS – TU Dresden.

2. Turbulent heat flux modelling

2.1. Analysis of the budget terms of turbulent heat fluxes

To account for most of the features and requirements of turbulent heat transfer in liquid metal flows, second-moment closures for the turbulent heat fluxes, based on the solution of three additional transport equations, seem to be the natural choice. This requires a detailed analysis of the main transport and destruction mechanisms of turbulent heat fluxes in dependence of the molecular Prandtl number Pr ($Pr = \eta \cdot c_p/k$), the Reynolds number ($Re = \rho \cdot u \cdot d/\eta$) and the Péclet number ($Pe_b = Pr \cdot Re_b$). The full transport equation of the turbulent heat fluxes $\overline{u_i\theta}$ in symbolic notation is given by:

$$\frac{\mathbf{D}\,\overline{u_i\theta}}{\mathbf{D}\,t} = P_{i\theta} + \Pi_{i\theta} + D^{v}_{i\theta} + D^{t}_{i\theta} - \varepsilon_{i\theta}\,, \tag{1}$$

where $P_{i\theta}$ represents the production of turbulent heat fluxes due to the mean flow, $\Pi_{i\theta}$ the temperature-pressure gradient correlation, $D^{v}_{i\theta}$ and $D^{t}_{i\theta}$ the molecular and turbulent diffusion and $\varepsilon_{i\theta}$ the dissipation term. The production term $P_{i\theta}$ needs no further modelling, which is one of the most important advantages of this level of closure. Each of the other terms, even the molecular diffusion term, needs further modelling strategies. The correlation term $\Pi_{i\theta}$ is often divided into the pressure-scrambling term $\phi_{i\theta}$ and the pressure diffusion term $D^{p}_{i\theta}$, which is only of minor importance far away from the wall in duct flows and therefore usually absorbed into the modelling of the turbulent diffusion term $D^{t}_{i\theta}$. To identify the behaviour of the main budget terms for different frictional Reynolds and Prandtl numbers, the DNS data by Abe et al. [4]are used; see Figure 1. Thereby, the frictional Reynolds number (Re_τ) is based on the frictional velocity u_τ.

For Prandtl numbers near unity, the correlation term $\Pi_{i\theta}$ is the main counterpart of the production term $P_{i\theta}$, whereas for lower Prandtl numbers the dissipation term $\varepsilon_{i\theta}$ mainly balances the production due to the mean flow. This may be one of the reasons, why most proposals from literature are mainly focussing on an appropriate modelling of the pressure-scrambling term $\phi_{i\theta}$ or $\Pi_{i\theta}$. Furthermore, only simple formulations for the dissipation rate $\varepsilon_{i\theta}$, which assume a vanishing value of the latter quantity in the core region of the flow, are incorporated. Therefore, in the present investigation, only proposals are investigated, which especially take into account the increasing or decreasing importance of these two budget terms in consideration of the molecular Prandtl number Pr.

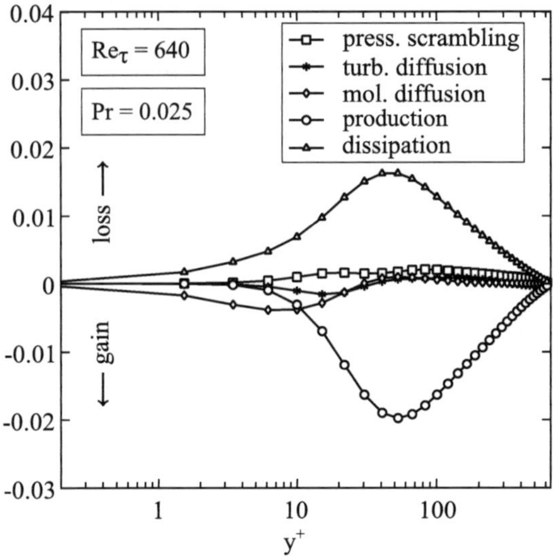

Figure 1. Analysis of the budget terms for a symmetrically heated planar channel flow and a molecular Prandtl number of Pr = 0.025 based on the DNS data of Abe et al. [4].

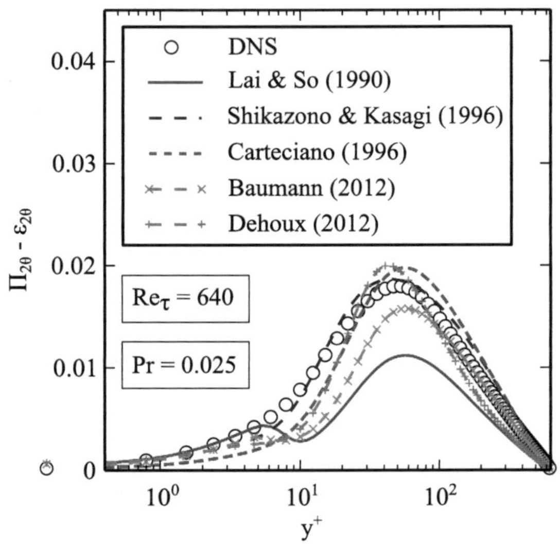

Figure 2. Comparison of different proposals for the difference of the pressure-scrambling $\Pi_{i\theta}$ and dissipation term $\varepsilon_{i\theta}$ based on the DNS data of Abe et al. [4].

As illustrated in Figure 2 for the wall-normal turbulent heat flux component, the best agreement for the prediction of the two budget terms can be achieved by the proposal of Shikazono & Kasagi [5]. One major disadvantage of this model is its limitation to rather simple channel flows, because the formulation of the correlation term $\Pi_{i\theta}$ is not frame invariant to rotation of the coordinate system. Therefore, it cannot be applied to complex geometries or flow types, e.g. square duct flows [5]. The model by Carteciano [6] introduces an additional term in the transport equations of the turbulent heat fluxes, which accounts for the effect of an increasing or decreasing importance of both of the budget terms in dependence of the Prandtl number Pr or turbulent Péclet number Pe_t, respectively. One shortcoming of this model is its violation of the asymptotic wall limiting behavior, which can be derived by a Taylor expansion series of the fluctuating quantities. This may be the result of its intended purpose; the numerical simulation of a hot sodium jet in different convection regimes [6]. The approach by Baumann [7] introduces rather empirical damping functions for both budget terms in dependence of the dimensionless wall-distance or the Prandtl number. It is thus not expected to be valid for a wide range of flow types. Dehoux [8] proposed a model for both budget terms based on the elliptic blending strategy, first published by Manceau & Hanjalić [9]. The model gives an acceptable agreement with the DNS data, but requires the user input of the time scale ratio in the homogeneous region of the flow (R_h). This value is usually not known a priori for more complex geometries and flow types considering buoyancy effects.

Due to its simplicity in comparison with the other proposals, the model by Carteciano [6] is selected for further investigations. The combination of a full second-moment closure in buoyancy-dominated flows for the turbulent Reynolds stresses and the turbulent heat fluxes requires at least the solution of 16 differential equations. For that reason, a further reduction of the numerical model by algebraic truncation is considered.

2.2. Derivation of a new algebraic turbulent heat flux model

For the derivation of a new algebraic turbulent heat flux model, several simplifications and assumptions have to be considered, which will only be briefly presented in this paper. The interested reader is referred to the work of Dol et al. [10] for a detailed discussion of the validity of the latter assumptions in buoyancy-dominated flows.

The first assumption is the so-called weak-equilibrium hypothesis, which is based on the formulation of a dimensionless anisotropic heat flux vector ζ_i, and implicates that the latter quantity is constant in space and time:

$$\zeta_i = \frac{\overline{u_i\theta}}{\sqrt{k}\sqrt{\overline{\theta'^2}}}, \qquad \text{and} \qquad \frac{\mathbf{D}\,\zeta_i}{\mathbf{D}\,t} - D_{\zeta_i} = 0. \tag{2}$$

The origin of this assumption can be traced to the work of Rodi [11], which discusses the truncation of the transport equation of the anisotropy of turbulent Reynolds stresses. The convective and diffusive transport can be expressed by means of the transport terms of the turbulent kinetic energy k and of the temperature variance $(\overline{\theta'^2})$:

$$\frac{1}{\sqrt{k}\sqrt{\overline{\theta'^2}}}\left(\frac{\mathbf{D}\,\overline{u_i\theta}}{\mathbf{D}\,t} - D_{i\theta}\right) = \frac{1}{2}\left[\frac{\overline{u_i\theta}}{k^{3/2}\sqrt{\overline{\theta'^2}}}\left(\frac{\mathbf{D}\,k}{\mathbf{D}\,t} - D_k\right) + \frac{\overline{u_i\theta}}{\overline{\theta'^2}^{3/2}\sqrt{k}}\left(\frac{\mathbf{D}\,\overline{\theta'^2}}{\mathbf{D}\,t} - D_{\theta\theta}\right)\right]. \tag{3}$$

In fully developed flows it is obvious, that the sum of the total diffusion of the turbulent kinetic energy and the temperature variance equals the following expression:

$$D_{i\theta} = \frac{1}{2}\overline{u_i\theta}\,(D_k + D_{\theta\theta}). \tag{4}$$

By introduction of the budget terms the formulation in equation (4) can be further modified to:

$$P_{i\theta} + G_{i\theta} + \Pi_{i\theta} - \varepsilon_{i\theta} - \frac{\overline{u_i\theta}}{2k}(P_k + G_k - \varepsilon) - \frac{\overline{u_i\theta}}{2\,\overline{\theta'^2}}(P_{\theta\theta} - \varepsilon_{\theta\theta}), \tag{5}$$

where $P_{i\theta}$, P_k and $P_{\theta\theta}$ represent the production terms, $G_{i\theta}$ and G_k the production due to buoyancy forces, $D_{i\theta}$, D_k and $D_{\theta\theta}$ the total diffusion terms and $\varepsilon_{i\theta}$, ε and $\varepsilon_{\theta\theta}$ the dissipation terms of the turbulent heat fluxes $\overline{u_i\theta}$, the turbulent kinetic energy k and the temperature variance $\overline{\theta'^2}$, respectively.

Assuming an equilibrium of the production and dissipation terms of the two latter quantities can further reduce this formulation:

$$P_k + G_k = \varepsilon, \qquad \text{and} \qquad P_{\theta\theta} = \varepsilon_{\theta\theta}. \tag{6}$$

The remaining production term $P_{i\theta}$ is exact and needs no further modelling. For the modelling of the pressure scrambling $\Pi_{i\theta}$ and dissipation term $\varepsilon_{i\theta}$ any model of the literature can be used. In the present investigation, a slightly modified version of the model by Carteciano [6] is used. A careful recalibration of the coefficients and the removal of the wall-reflection term, see Lai & So [11] for a detailed discussion of its necessity, leads to the following models for both budget terms:

$$\Pi_{i\theta} = -C_{1\theta}\frac{\varepsilon}{k}\overline{u_i\theta} + C_{2\theta}\,\overline{u_j\theta}\frac{\partial U_i}{\partial x_j} + C_{3\theta}\,\beta g_i\overline{\theta'^2}, \tag{7}$$

$$\varepsilon_{i\theta} = \frac{1+Pr}{2\sqrt{Pr}\sqrt{R}}\exp\left(-C_{5\theta}(Re_t+Pe_t)\right), \tag{8}$$

where the turbulent Reynolds and Péclet numbers are given by $Re_t = k^2/(\nu\varepsilon)$ and $Pe_t = Pr\, Re_t$. The time scale ratio $R = \tau_{th}/\tau_m$ represents the ratio of the thermal time scale $\tau_{th} = \overline{\theta'^2}/\varepsilon_{\theta\theta}$ and the mechanical time scale $\tau_m = k/\varepsilon$. For the present investigation, the two-equation model by So & Sommer [13] is used to calculate the time scale ratio *R,* which is validated for a wide range of molecular Prandtl numbers.

The final truncated algebraic heat flux model is therefore, given by the following expression:

$$\overline{u_i\theta} = -\left(\frac{k}{\varepsilon}\right)\frac{\overline{u_iu_j}\frac{\partial\Theta}{\partial x_j} + \alpha\,\overline{u_j\theta}\frac{\partial U_i}{\partial x_j} + \gamma\,\beta g_i\overline{\theta'^2}}{C_{1\theta} + \frac{1+Pr}{2\sqrt{Pr}\sqrt{R}}\exp\left(-C_{5\theta}(Re_t+Pe_t)\right)}, \tag{9}$$

where the parameters are given by $\alpha = (1 - C_{2\theta})$ and $\gamma = (1 - C_{3\theta})$. The complete set of coefficients of the present model is denoted in Table 1.

The present algebraic model is implicit in its formulation, which may result in numerical instabilities. Because the present simulations could be performed without any incident, the analysis of potential instabilities is postponed to further investigations. The present model is combined with a full second-moment closure for the turbulent Reynolds stresses based on the elliptic blending approach (RSM-EBM) by Dehoux [8]. This model is validated successfully for turbulent planar channel flows in different convection regimes.

Table 1. Coefficients of the present model.

Coefficient	Value
$C_{1\theta}$	3.6
$C_{2\theta}$	0.5
$C_{3\theta}$	0.5
$C_{4\theta}$	0.0
$C_{5\theta}$	0.0007/0.001

3. Results

3.1. *Turbulent planar channel flow*

The present algebraic model (AHM) is compared with available DNS and LES data of a turbulent channel flow from the literature [4],[14] at different Reynolds numbers Re_τ and at a molecular Prandtl number of Pr = 0.025. To analyze the influence of the coefficient for the dissipation term $\varepsilon_{i\theta}$, two slightly different values are investigated $C_{5\theta}$=0.0007 and $C_{5\theta}$=0.001.

As depicted in in Figure 3 and Figure 4, the present AHM exhibits a nearly perfect agreement in case of the dimensionless temperature profiles with the DNS data by Abe et al. [4] for different Re_τ. The results for both values of the coefficients are almost identical and only a slightly better prediction may be stated for $C_{5\theta}$=0.001. One of the most essential requirements for turbulent heat flux closures is an appropriate prediction of the Nusselt number *Nu*, which can be formulated as:

$$\mathrm{Nu} = \frac{2\delta\, \dot{q}_w}{\lambda\, (\overline{\Theta}_B - \overline{\Theta}_w)}, \tag{10}$$

where δ represents the half channel height, $\dot{q}_w$ the heat flux applied on the walls, λ the thermal conductivity and $(\overline{\Theta}_B - \overline{\Theta}_w)$ the difference between the bulk and wall temperature. An investigation of the Nusselt number is especially useful, because most design tools of large-scale facilities use this integral quantity, which allows depicting the global system performance in transient operational scenarios by means of one-dimensional analysis of the heat transfer. For the present algebraic model, a close agreement with the DNS and LES [4],[14] data over a wide range of bulk Reynolds numbers Re_b and bulk Péclet numbers $Pe_b = Pr\, Re_b$, respectively, is observed. The deviations are in the range of 10%.

Analyzing several flow quantities, the AHM exhibits the most satisfying agreement. Therefore, in a next step in the validation process, this model is applied to geometries that are more complex.

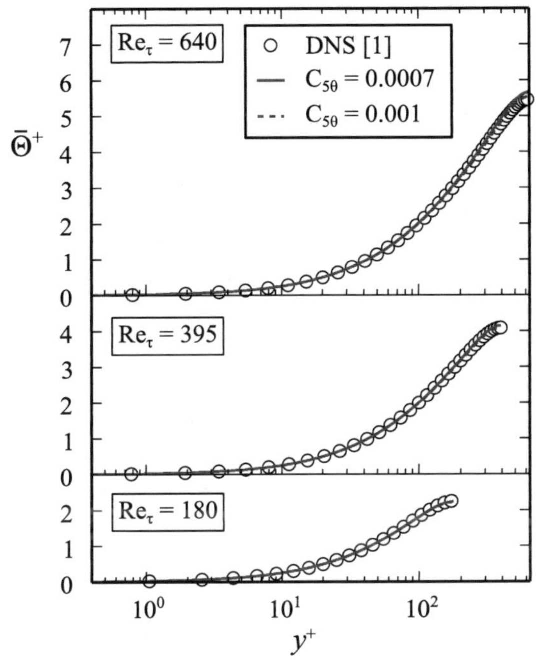

Figure 3. Dimensionless temperature profiles for the present AHM in comparison with the DNS data [4] at different Re_τ and $Pr = 0.025$.

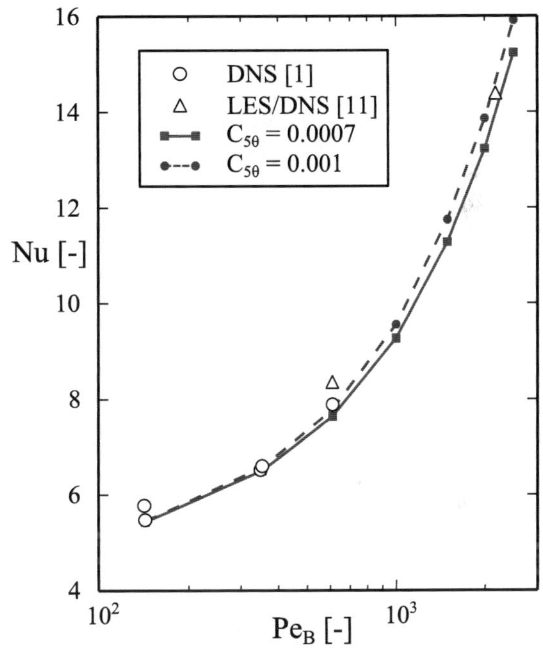

Figure 4. Comparison of the predicted Nusselt numbers for an increasing bulk Péclet number in comparison with DNS and LES data [4],[14].

3.2. Turbulent square duct flow

One of the main features of turbulent square duct flow is the intrinsic presence of secondary motions of the second kind, which are induced by differences in the turbulent normal and shear stresses. To the knowledge of the authors, presently no DNS data exists to investigate the turbulent heat transfer of liquid metals in turbulent square duct flows. Therefore, LES simulations are performed for different heating conditions and in different convection regimes. The results shown refer to a configuration for a constant, identical heating on one wall pair, whereas the other wall pair are assumed perfectly thermal insulating; see Figure 5. The molecular Prandtl and bulk Reynolds number are fixed at $Pr = 0.025$ and $Re_b = 7000$.

Figure 6 illustrates the results of the dimensionless velocity profiles of the LES and the RSM-EBM. It shows an acceptable agreement with the experimental data of Kawahara et al. [15] and the DNS data of Pinelli et al. [16]. Only a slight underestimation of the maximum of the velocity component in the cross-sectional plane is observed for the RANS approach. One major concern in the numerical model for liquid metal flows is the choice of a suitable domain size. As pointed out by Grötzbach [17] the largest turbulent structures are expected to occur in the temperature field. Therefore, it must be ensured that the domain size is sufficiently large to analyze the two-point correlations. The size of the domain and important parameters for the LES are given in Table 2.

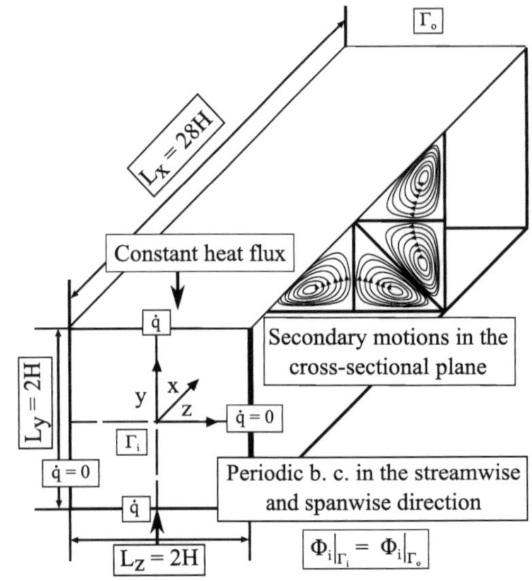

Figure 5. Sketch of the computation domain and definition of boundary conditions for the investigated turbulent square duct flow.

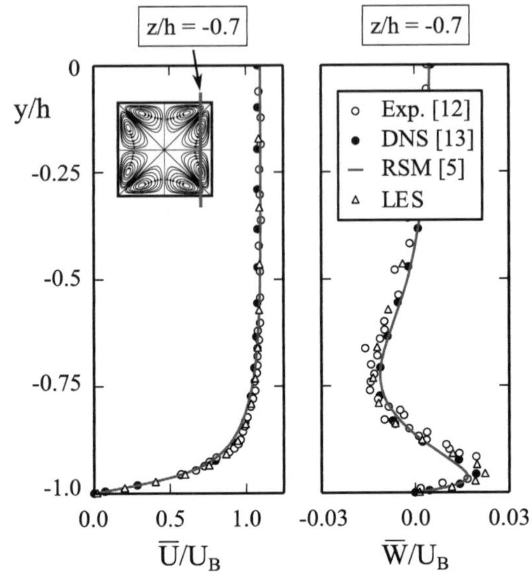

Figure 6. Comparison of the profiles of the streamwise and cross-sectional velocity component with exp. and DNS data [15],[16].

Table 2. Data sheet for the present LES.

Parameter	Symbol	Value
Bulk Reynolds number	$Re_b = U_b\, 2H/\nu$	7000
Mol. Prandtl number	Pr	0.025
Length of the domain	L_x	28H
Size of the cross-section	$L_y = L_z$	2H
No. of elements	N_x x N_y x N_z	245 x 120 x 120
Sampling time	$T_s\, U_b/H$	~8000
Spatial discretization	-	2nd order
SGS- model	-	Dynamic Smagorinsky
SGS-heat flux model	$Pr_{t,SGS}$	∞

For an acceptable resolution of the velocity field, one can assume a nearly full resolution of the temperature field for low molecular Prandtl numbers [17]. Therefore, the results of the present algebraic heat flux model can be compared with the results of the LES.

The dimensionless temperature profile $\overline{\Theta}^+$ is formulated based on the wall temperature at the wall bisectors $\overline{\Theta}_{wb}$ and the mean friction velocity averaged over the walls $\langle u_\tau \rangle$:

$$\overline{\Theta}^+ = \frac{\left(\Theta_{wb} - \overline{\Theta}\right)}{T_\tau}, \qquad \text{with} \qquad T_\tau = \frac{\dot{q}_w}{\rho c_p \langle u_\tau \rangle}. \tag{11}$$

where ρ and c_p represent the density and the specific heat capacity of the fluid, respectively.

This setup of the numerical model represents a limiting case of non-vanishing temperature fluctuations on the walls, which is opposite to the limiting case of zero wall temperature fluctuations in the planar channel flow. In reality, the magnitude of the fluctuations of the temperature field is largely dependent on the thermal activity ratio, describing the ratio between the thermal properties of the solid and fluid. Due to the unknown distribution of the temperature variance at the surface, a zero value is assumed for the AHM, which is also adopted by other researchers [18],[19].

Both, in terms of the dimensionless temperature a the turbulent heat flux, see Figure 7 and Figure 8, the results of the present AHM are in a good agreement with the LES data. The dimensionless temperature in the centre of the square duct is slightly underestimated by the present AHM. This can be related to a slight overestimation of the turbulent heat flux component in the y-direction of the duct. Due to the higher contribution of turbulence to the total heat fluxes, hot fluid is penetrating deeper into the core region of the duct.

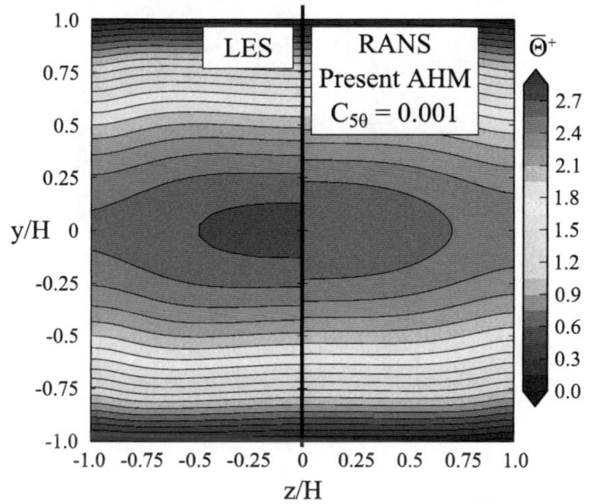

Figure 7. Comparison of the dimensionless temperature distribution in the cross-sectional plane of the duct.

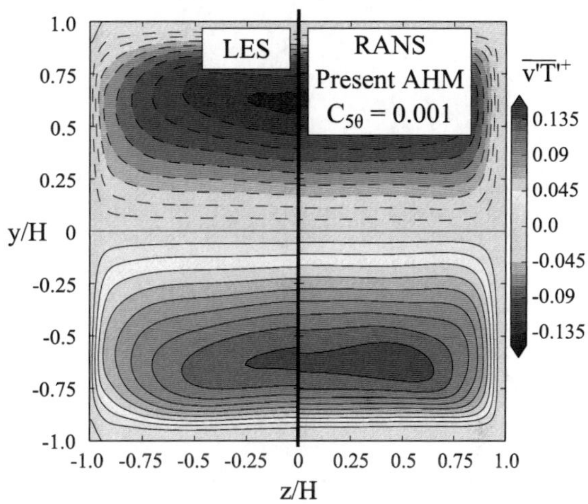

Figure 8. Comparison of the dimensionless turbulent heat flux in the cross-sectional plane of the duct.

To investigate the performance of the model within the mixed convection regime, additional LES for a fully developed square duct flow are performed at a Richardson number Ri = 0.09:

$$\mathrm{Ri} = \frac{\mathrm{Gr}}{\mathrm{Re}_b^2} = \frac{\beta \cdot g \cdot \dot{q}_w \cdot 2H^2}{\lambda \cdot u_b^{\,2}}, \tag{12}$$

with

$$\mathrm{Gr} = \frac{\beta \cdot g \cdot \dot{q}_w \cdot \rho^2 \cdot 2H^4}{\lambda \cdot \eta^2}, \tag{13}$$

where β represents the volumetric expansion coefficient, $\dot{q}_w$ the heat flux on the solid wall, H the wall distance, λ the thermal conductivity and u_b the bulk velocity.

Only the wall at the bottom is heated and all three other walls are assumed perfect thermally insulated. The gravity force g_i is applied in the streamwise directions. The size of the computational domain corresponds to that of the forced convective flow case.
As can be seen in Figure 9 the agreement between both simulations is acceptable for the streamwise velocity component even though the curvature of the isotachs into the corners is slightly underrated by the RANS approach. The magnitude of the secondary motions $\overline{S}/u_b$ is smaller for the RANS approach compared to the LES (see Figure 10), which can be referred to the underlying Reynolds stress model (RSM-EBM). Because secondary flow motion is a direct consequence of an appropriate prediction of the turbulent normal and shear stresses, further improvements should focus on the influence of buoyancy effects within the framework of the elliptic blending procedure.

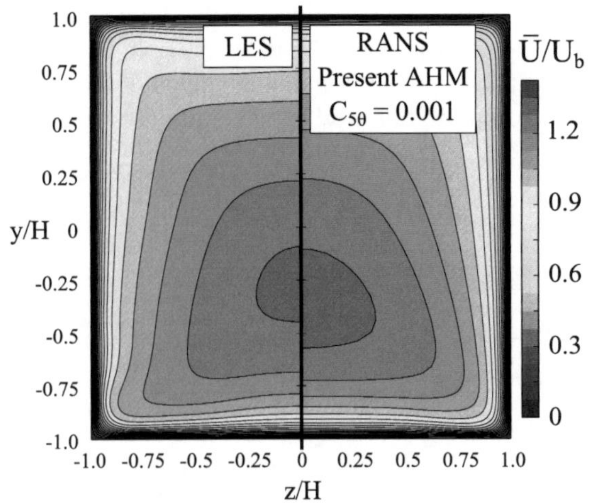

Figure 9. Comparison of the dimensionless streamwise velocity distribution in the cross-sectional plane of the duct.

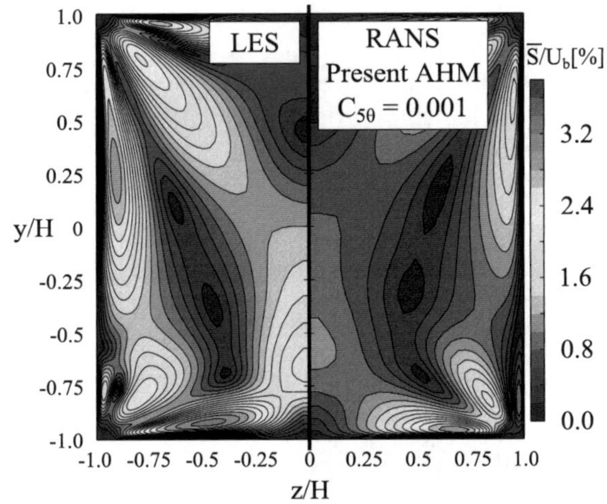

Figure 10. Comparison of the dimensionless magnitude of secondary motion in the cross-sectional plane of the duct.

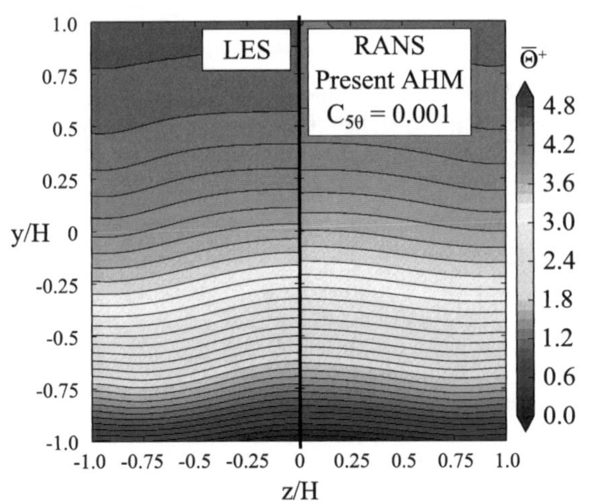

Figure 11. Comparison of the dimensionless temperature distribution in the cross-sectional plane of the duct.

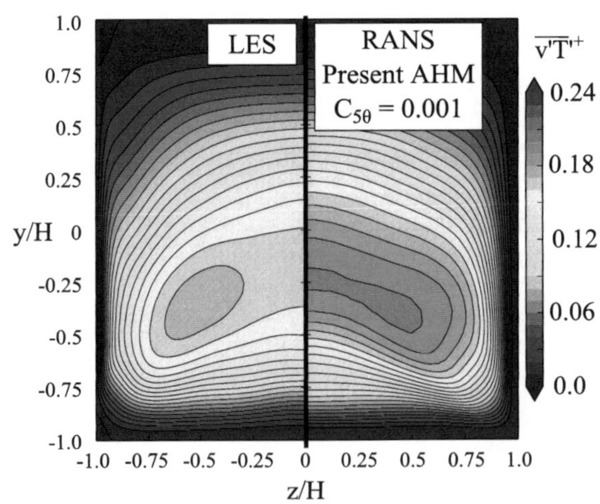

Figure 12. Comparison of the dimensionless turbulent heat flux in the y-direction within the cross-sectional plane of the duct.

With respect to the predicted dimensionless temperature profile the agreement is nearly perfect (see Figure 11) with only a minor deviations (underestimation) near the insulated wall at the top. Due to a slightly overestimation of the turbulent heat flux component in the y-direction of the AHM (see Figure 12), the dimensionless temperature profile is slightly shifted from the bottom to the top of the duct in comparison with the LES results.

As listed in Table 3 the AHM shows a close agreement with the LES for the Nusselt number in both convection regimes.

Table 3. Comparison of the predicted Nusselt numbers

Type	Heating condition	Ri [-]	Nu [-]
LES	symm.	0	5.91
RANS	symm.	0	5.86
LES	asymm.	0.09	3.75
RANS	asymm.	0.09	3.76

4. Direct Numerical Simulations

Very recently, Direct Numerical simulations of the laterally heated flow in a vertical square duct could be accomplished at TU Dresden. The Navier-Stokes equations with Boussinesq approximation have been solved using a Finite-Volume method of second order on staggered grids together with a Runge-Kutta scheme of second order in time [20][21]. Here, the reference length $H^*=L_y=L_z$ is used for normalization and definition of similarity numbers. The length of the channel is assumed to $L_x = 24H^*$and no-slip conditions are imposed at the walls. For the dimensionless over-temperature $\theta = (T-T_b)/\Delta T$, Dirichlet conditions $\theta(y = 0) = 1$ and $\theta(y = H) = 0$ are considered together with an adiabatic condition on the remaining sidewalls. The grid consists of 1024×166×166 cells, being uniform in x-direction, with the same distribution in y- and z-direction including refinement close to the walls. This grid resolves all the turbulent scales with the first grid point at the wall located at $y_1^+<0.4$.

Simulations are conducted for $Re^*_b = U_bH^*/\nu = 6200$, and $Ri^* = 0$ as well as $Ri^* = 0.1$ assuming liquid sodium with $Pr = 0.0088$, which is lower than the value used above. Averaging has been performed in time over about $1500H^*/U_b$ and $1600H^*/U_b$, respectively, as well as in x-direction, and the symmetry of the solution with respect to the plane $z/H = 0.5$ has been exploited for an additional averaging step. Simulation results in case of forced convection have been validated with the data of Ma et al. [22].

Figure 13 and 14 provide information on the mean streamwise flow (top row) as well as the secondary flow in the cross plane (second row). These latter show the magnitude of the secondary flow, termed $u\perp$ here, in terms of a contour plot as well as the corresponding streamlines of the secondary flow. It is obvious that under the prevailing conditions the pattern is substantially altered with respect to Ri=0. The graphs in the third row show the temperature which is very smooth, similar to Figure 11 above and which drives the axial flow on the left-hand side, i.e. close to y=0, upwards. As a result, the flow is accelerated near the hot wall and decelerated near the cold wall. Furthermore, the pattern of the secondary flow is substantially altered, with a pair of two vortices near the hot wall and a smaller pair near the cold wall and the stagnation point shifted towards the cold wall. The changes of the secondary flow pattern result from altered balances of the individual Reynolds stress components. The secondary flow transporting hot fluid towards the cold wall in the channel center creates enhanced fluctuations near the cold wall so that the turbulent kinetic energy is substantially increased in this region. The cross-plane turbulent heat fluxes are visualized in the last row with the color scale determined by the magnitude of this vector. It is visible that this quantity is larger by a factor of more than 3 and exhibits a pattern similar to the kinetic energy.

IOP Conf. Series: Materials Science and Engineering **228** (2017) 012011 doi:10.1088/1757-899X/228/1/012011

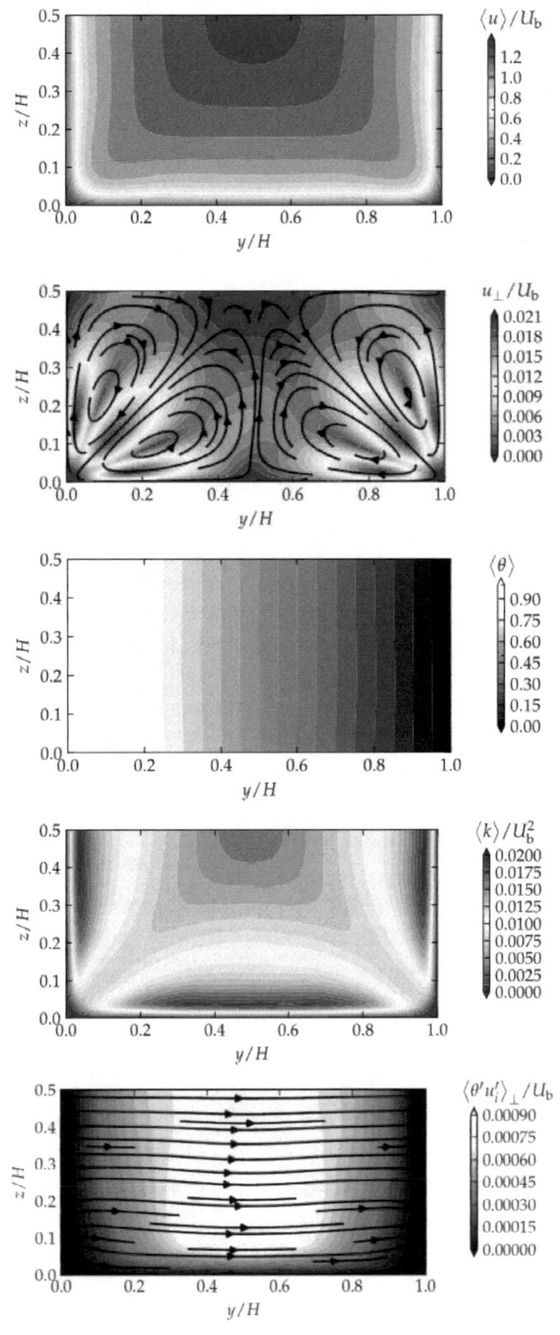

Figure 13. Mean streamwise velocity, secondary flow, temperature, turbulent kinetic energy, and cross-plane turbulent heat flux for Ri* = 0, i.e. forced convection.

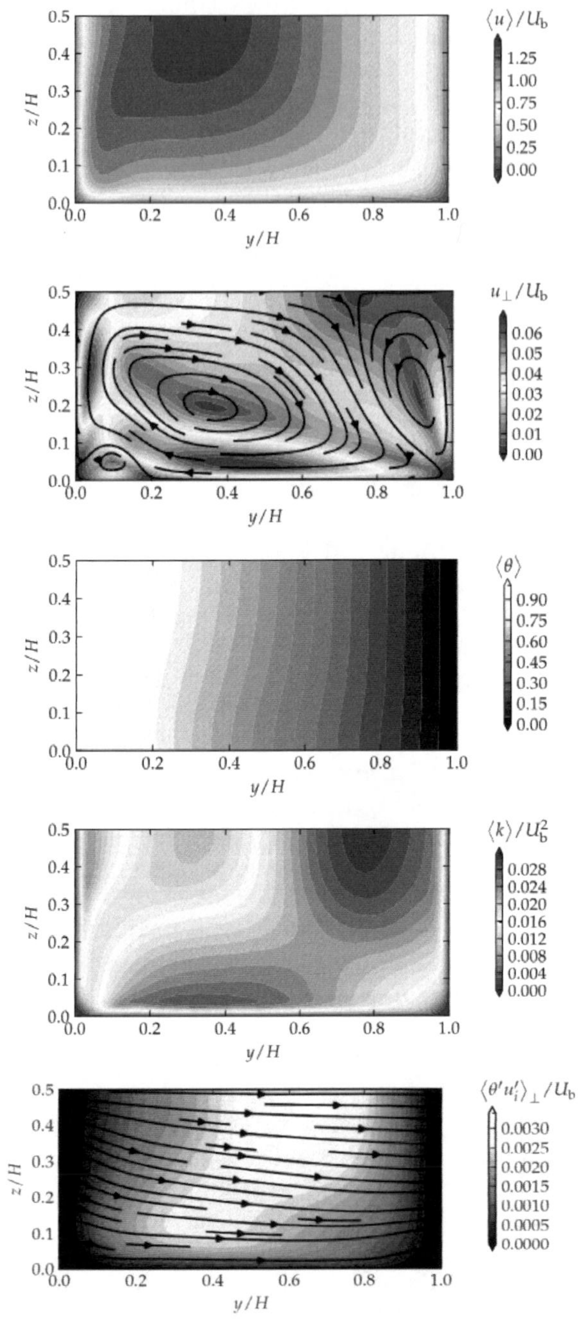

Figure 14. Mean streamwise velocity, secondary flow, temperature, turbulent kinetic energy, and cross-plane turbulent heat flux for Ri*=0.1 (the color scales adapted in case of different range).

From the DNS just described, all Reynolds stresses were determined, as well as all terms in their budget equations. These are now available for further analysis and can be used in modeling activities.

More evaluations and additional simulations with Ri*=0.02 and Ri*=0.5, together with LES for Re^*_b=40000 and Ri*=0, 0.02, and 0.1, will be the subject of subsequent investigations.

5. Comparison with experimental results

The above presented modelling approach and results exhibits one of the drawback related to CFD code modelling – the presence of a validated reference solution. Reference solution can be either obtained by quasi-exact direct numerical solutions or high resolution experimental data. But, the number of available and high quality experiments is rather sparse. Furthermore, these experiments are mainly related to forced convection, although a large fractions shows unintended buoyancy support. In Jaeger et al. [23] experiments are listed, which focus on quadratic and rectangular flow geometries operated with liquid metals. Mostly those studies provide empirical correlations for the Nusselt number as a function of the Péclet number, which are valid for rectangular ducts with an aspect ratio of less than 10 (width over depth) and for parallel plate-like ducts with an aspect ratio of larger or equal to 10. In addition, the available data for wall drag coefficients are compared to common friction factor correlations. It is shown that correlations for pipe flow conditions are applicable to quadratic and rectangular channel flows. These results are compared in Jaeger et al. [24] with results generated by turbulence models for RANS CFD in a benchmark.

For the benchmark, a rectangular duct with an aspect ratio of 10 is selected (width = 250 mm, depth = 25 mm). For the sake of completeness, DNS data from Kawamura for a parallel plate geometry are added to the comparison. The results for the Nusselt number and the wall drag coefficient are depicted in Figure 15 and Figure 16.

The benchmark shows, that a rectangular duct with an aspect ratio of 10 is very similar to a parallel plate, indicated by the similar results of the DNS data and all other data. The secondary motions caused by the corners lose their importance with increasing aspect ratio. From an engineering point of view, the differences between the modelling approaches are marginal. In the same way, the wall drag coefficients predicted with empirical models, RANS CFD and DNS show almost identical results.

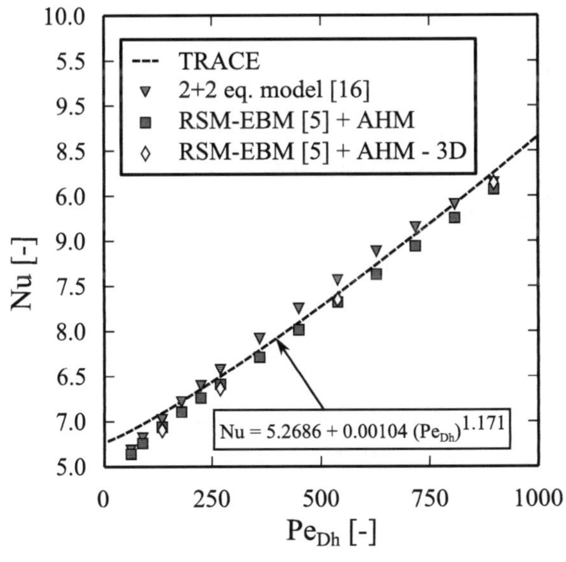

Figure 15. Comparison of numerical results originating from different modelling approaches for the heat transfer under forced convection.

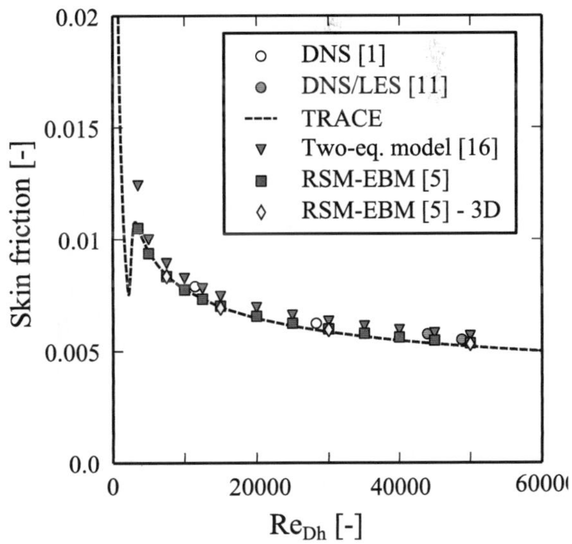

Figure 16. Comparison of numerical results originating from different modelling approaches for the wall drag coefficient under forced convection.

Even though the comparison of empirical models derived from and developed for experiments and the results generated with CFD and DNS codes shows identical qualitative trends and similar quantitative values, a conclusive statement concerning the prediction capabilities of numerical codes cannot be made.

The only way to proof the validity of the used models and approaches, for both empirical models and turbulent heat flux models, is by a direct comparison with experimental data. Therefore, experiments are under construction and currently in assemblage at the Institute for Neutron Physics and Reactor Technology at the KIT.

The first experiment foreseen is a rectangular flow channel with an aspect ratio of two. This rectangular duct will be integrated into the KASOLA facility. This duct consists of a 3600 mm long commercially available duct. The internal dimensions of this duct are 90 mm in width and 40 mm in depth. A detailed description of the KASOLA facility may be taken from [25].

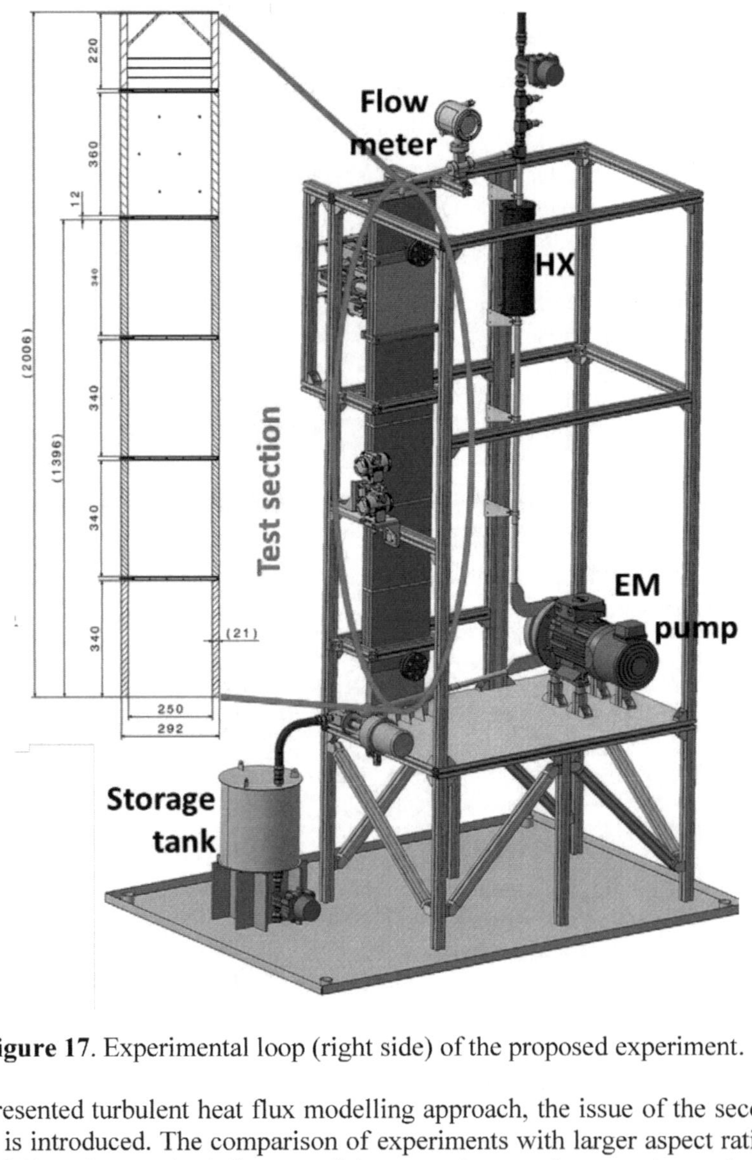

Figure 17. Experimental loop (right side) of the proposed experiment.

In the above presented turbulent heat flux modelling approach, the issue of the secondary motions due to the corners is introduced. The comparison of experiments with larger aspect ratios showed that for aspect ratios of 10 and more the influence of the secondary motions can be neglected. Therefore, the second experiment will investigate the flow and heat transfer in a rectangular duct with an aspect ratio of 10. Due to the design of the KASOLA facility, ducts with larger aspect ratios cannot be installed. Therefore, the second experiment is integrated in a new loop, being operated with the eutectic Gallium- Indium-Tin alloy. Due to its low melting point, < 11°C, the expenditures are reasonable low. The schematic arrangement of the experimental set-up is depicted in Figure 17. The main components of the experimental loop are: the rectangular test section, a flow meter, a heat exchanger and a pump.

The flow meter is commercial inductive flow meter. The heat exchanger is of the shell-and-tube counter current flow type with InGaSn in the tube and water in the shell. This heat exchanger is connected to a commercial chiller unit The pump is an electromagnetic one basically consisting of a motor with two discs placed next to the flow channel. The two discs are equipped with permanent magnets. The motor rotates the discs and by means of the Lorentz force the electrically conducting InGaSn is accelerated. Furthermore, the loop contains of a storage tank and several valves, pressure taps, thermo-couples, etc. for operation and control of the experimental loop.

The realization of the test section itself is shown in Figure 18. The test section is basically a rectangular channel with an aspect ratio (width-2W to depth-H) and internal flow dimensions of 250 mm x 25 mm. The test section is 2000 mm long and over the whole length, 85 thermo-couples are installed at five different levels (17 per level). At each level, the thermo-couples are arranged in measurement crosses to obtain temperature data over the whole width and depth. The levels of these crosses are indicated in Figure 18 by means of red lines They are located at the positions x/H = 25, 50, 75, 100 and 125 composed of a set of sensors located at the fluid wall interface, at y/H=0.25, 0.5 and 0.75 and z/2W= 0.0, 0.5 and equidistant spaced in transverse.

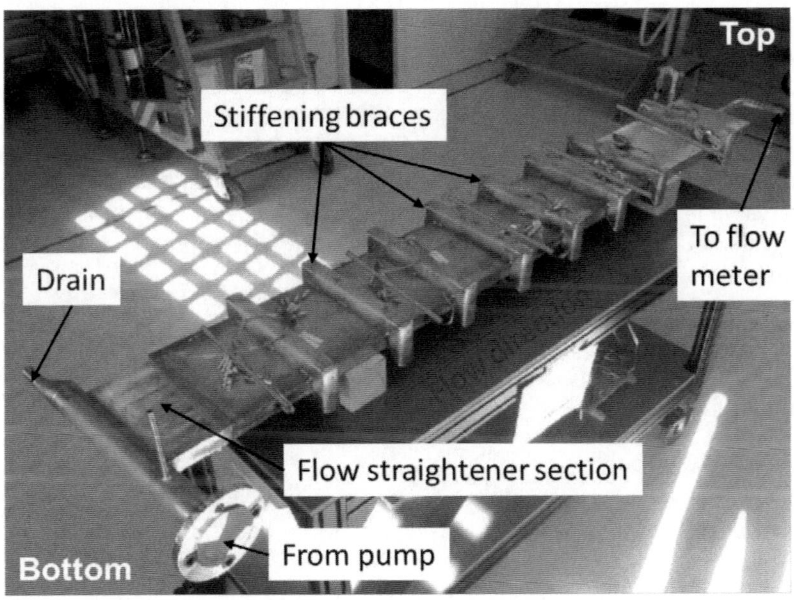

Figure 18. Rectangular test section of the experimental loop.

To investigate the heat transfer behavior under transitional behavior (forced convection to mixed convection to free convection) a heating plate is used, which is attached to the flow channel on one outside of the wall of the rectangular channel. The heating plate covers the entire width 2W of the flow channel, and is equipped with standard resistance heating wires, capable to provide up to 400 W per installed meter. With a total length of 8200 mm the heater can produce up to 3280 W, which corresponds to a mean heat flux of q= 26 kW/m^2 applied to the channel wall over a length of x/H= 20. The entire section is thermally insulated by by rockwool with a heat conductivity of 38 mW/(m·K) to limit the thermal losses. The test section is supplied fluid flow by means of a MHD pump with a pressure difference of Δp= 2 bar, providing a maximum flow of 1 l/s. By this set-up a Reynolds number of up to Re = 15000 and Richardson number of up to Ri= 0.4 can be achieved covering a large range of technical interest.

For the heat removal, a counter current shell and tube design heat exchanger is installed at the top of the loop. The inner pipe is the downward pipe of the loop piping. The shell side envelops this pipe. On the shell side, water is cooling the heated Indium-Gallium-Tin alloy. This heat exchanger is connected to a commercial cooler system.

6. Summary

This article addresses the vital engineering challenge to reliably predict the heat removal from a heated wall by liquid metals in different flow regimes, as they occur naturally during start-up/shut-down of pumps, normal operational transients etc. Accurate temperature and flow predictions using conventional numerical tools necessitate modified turbulence models to take the peculiarities of liquid metals into account. In this context, a new algebraic turbulent heat flux model for low molecular Prandtl number flows has been derived and carefully analysis in terms of the budget terms and algebraic truncation of the turbulent heat flux equations. In a first step this model is validated for turbulent channel flows and exhibits only marginal deviations to available DNS and LES data from the literature. In a second step the performance of the model is studied for a turbulent square duct flow at moderate Reynolds numbers. In the absence of DNS or LES data for liquid metal flows, an LES-simulation within forced and mixed convection regime has been set-up to derive a data set for mandatory to serve as model validation. The present algebraic model shows in this benchmark exercise a satisfactory agreement with the LES data, especially within the forced convection regime. Due to the high complexity of the equations, when the temperature acts as an active scalar contributing to the buoyancy terms, the RANS approach shows still an acceptable agreement with the LES results but deviations are increasing.

In further investigation the influence of the underlying thermal two-equation model will be evaluated. Additionally, LES for several different heating conditions and Prandtl numbers have to be executed to analyse this impact on the budget terms and to judge more completely on the capabilities of the model. Only then a holistic assessment of the present approach in comparison with other turbulent heat flux models from the literature can be made.

The ultimate proof of validity will be given only after a comprehensive code to experiment comparison. Therefore, two different experimental campaigns are under development to provide experimental data to obtain, among others, Nusselt numbers during different convection schemes and the transition among them.

Acknowledgments
The financial support of the Helmholtz Alliance "Liquid Metal Technologies – LIMTECH" is gratefully acknowledged.

References

[1] Pacio J 2015, Low Prandtl number thermal-hydraulics, pp. 647-730 in OECD, Handbook on Lead-bismuth Eutectic Alloy and Lead Properties, Materials Compatibility, Thermal-hydraulics and Technologies, NEA No. **7268**

[2] Wetzel T 2015, Instrumentation, pp. 731-839 in OECD, Handbook on Lead-bismuth Eutectic Alloy and Lead Properties, Materials Compatibility, Thermal-hydraulics and Technologies, NEA No. **7268**

[3] Onea A, Perez-Martin S, Jaeger W, Hering W and Stieglitz R 2017, Liquid metals as heat transfer fluids for science and technology pp. 305-376 in Minea A A (Ed.), Advances in new heat transfer fluids, CRC Press

[4] Abe H, Kawamura H and Matsuo Y 2004 *Int. J. Heat and Fluid Flow* **25** 404-19

IOP Conf. Series: Materials Science and Engineering **228** (2017) 012011 doi:10.1088/1757-899X/228/1/012011

[5] Shikazono N and Kasagi N 1996 *Int. J. Heat Mass Transfer* **39** 2977-87
[6] Carteciano L N 1996 *Wissenschaftliche Berichte FZKA* **5775**
[7] Baumann T 2012, *Ph.D Thesis Karlsruhe Instituite of Technology (KIT)*
[8] Dehoux F 2012 *Ph.D thesis SIMMEA Université de Poitiers*
[9] Manceau R and Hanjalić K 2002 *Phys. Fluids* **14** 744-54
[10] Dol H S, Hanjalić K and Kenjereš S 1997 *Int. J. Heat Fluid Flow* **18** 4-14
[11] Rodi W 1976 *Zeitschrift für angewandte Mathematik und Mechanik* **65** 219-21
[12] Lai Y G and So R M C 1990 *Int. J. Heat Mass Transfer* **33** 1429-40
[13] So R M C and Sommer T P 1994 *Trans. ASME.* **116** 844-54
[14] Duponcheel M, Bricteux L, Manconi M, Winckelmans G and Bartosiewicz Y 2014 *Int. J. Heat Mass Transfer.* **75** 470-82
[15] Kawahara G, Ayukawa K, Ochi J, Ono F and Kamada F 2000 *Trans. JSME B* **66** 95-102
[16] Pinelli A, Uhlmann M, Sekimoto A and Kawahara G 2010 *J. Fluid Mech.* **644** 107-22
[17] Grötzbach G 2011 *Nucl. Eng. Des.* **241** 4370-90
[18] Sommer T P, So R M C and Zhang H S 1994 *J. Heat Transfer* **116** 855-63
[19] Manservisi S and Menghini F 2014 *Nucl. Eng. Des.* **273** 251-70
[20] Kempe T, Fröhlich J 2012 *J. Comput. Phys.* **231** 3663–3684
[21] Niemann M, Fröhlich J 2016 *Int. J. Heat Mass Transfer* 101 1237–1250
[22] Ma L-D, Li Z-Y, Tao W-Q 2007 *Heat Mass Transfer* **44** 229–250.
[23] Jaeger W, Hering W, Lux M and Portes F 2015 *Proc. Int. Conf. on Nuclear Engineering*, Chiba, Japan, May 17-21
[24] Jaeger W, Trimborn F, Hering W, Pritz B and Gabi M 2016 *Proc. Int. Congr. on Advances in Nuclear Power Plants,* San Francisco, April 17-20
[25] Jaeger W, Hering W, Stieglitz R Schum T, Frohnapfel B, Niemann M and Fröhlich J 2017 *IOP Conf. Ser.: Mater. Sci. Eng.* (this volume)

IOP Conf. Series: Materials Science and Engineering **228** (2017) 012012 doi:10.1088/1757-899X/228/1/012012

Liquid metals for solar power systems

J Flesch[1], K Niedermeier[1], A Fritsch[2], D Musaeva[3], L Marocco[1,4], R Uhlig[2], E Baake[3], R Buck[2] and T Wetzel[1]

[1] Karlsruhe Institute of Technology (KIT), Institute for Nuclear and Energy Technologies, Hermann-von-Helmholz Platz 1, 76344 Eggenstein-Leopoldshafen, Germany
[2] German Aerospace Center (DLR), Institute for Solar Research, Pfaffenwaldring 38-40, 70569 Stuttgart, Germany
[3] Leibniz Universität Hannover (LUH), Institute of Electrotechnology, Wilhelm-Busch-Str. 4, 30167 Hannover, Germany
[4] Politecnico di Milano, Department of Energy, via Lambruschini 4, 20156 Milan, Italy

E-mail: thomas.wetzel@kit.edu

Abstract. The use of liquid metals in solar power systems is not new. The receiver tests with liquid sodium in the 1980s at the Plataforma Solar de Almería (PSA) already proved the feasibility of liquid metals as heat transfer fluid. Despite the high efficiency achieved with that receiver, further investigation of liquid metals in solar power systems was stopped due to a sodium spray fire. Recently, the topic has become interesting again and the gained experience during the last 30 years of liquid metals handling is applied to the concentrated solar power community. In this paper, recent activities of the Helmholtz Alliance LIMTECH concerning liquid metals for solar power systems are presented. In addition to the components and system simulations also the experimental setup and results are included.

1. Liquid metals as heat transfer fluids in solar thermal electricity (STE) generation

Concentrating solar thermal power plants are capable of electricity generation that is dispatchable and renewable. The future competitiveness of this technology depends on the reduction of the cost of electricity generation.

State-of-the-art solar thermal power plants use thermal oils, water/steam, air or molten salts as heat transfer fluids (HTF). Those fluids should ideally be chemically stable at high temperatures, have a high boiling temperature, a low melting point, low cost, enable safe operation and have good corrosion properties.

Although water and air are highly available and cheap fluids, they are not suitable for directly storing thermal energy at high temperatures. Oils can typically not operate at temperatures that would allow for high efficiency in the thermal-to-electric power conversion process. Molten 'solar salt' (mass specific composition: 60 % $NaNO_3$ and 40 % KNO_3) is cheap and can be used as direct storage medium due to its high heat capacity. For these reasons, it is the state-of-the art HTF for STE plants with thermal storage. Solar salt melts at 220 °C. Under strong solar irradiation, it can be operated up to an outlet bulk temperature of 565 °C, but not higher in order to avoid film temperatures above 600 °C where it chemically decomposes.

Advanced heat transfer fluids should have a low melting and a high boiling temperature resulting then in an increased operating temperature range. This reduces expensive auxiliary heating of the piping in non-irradiated sections and enables higher outlet temperatures for advanced power conversion cycles. These may lead to reduced costs in electricity generation ([1], [2]).

Pure liquid metals and their alloys have high thermal conductivities and thus show high potential for large heat transfer rates.

Pacio and Wetzel [3] performed a screening of liquid metal candidates as heat transfer fluids in STE systems. Based on a review study of liquid metals in solar power systems, they found that the candidates with the highest potential are sodium, lead-bismuth eutectic alloy (LBE) and tin. For example, Kesselring et al. [4] published on practical experiences with a sodium cooled receiver and Falcone [5] on thermal energy storage with sodium. More recent publications by Singer et al. [6] and Boerema et al. [7] on sodium and Kotzé et al. [8] on sodium-potassium assessed the thermodynamic possibilities of liquid metals in STE systems, however lacked the state-of-the-art of the technology associated.

Sodium and LBE have vast operational experience and engineering handbooks exist ([9] and [10]). They have a lower melting point than salt and a higher upper temperature limit: sodium boils at 883°C and LBE at 1670°C. Sodium, however, violently reacts with water and therefore requires special safety considerations in loop operation and LBE shows significant corrosion with structural steels unless the dissolved oxygen content is controlled. Due to their low heat capacities compared to salt, direct thermal storage will not be economically feasible in two-tank systems and special storage design is required. Possible indirect storage options using sodium are discussed in section 2.2 of this paper.

Pacio and Singer [11] assessed and summarized the thermo-hydraulic properties of sodium and LBE compared to solar salt with special attention to the thermal efficiency of the receiver and the conversion cycle. By using fluids capable of achieving high heat transfer rates in an economically feasible way the concentration of the incident light can be increased. A reduced receiver aperture could then generate the same power output at reduced losses, being radiation and convection losses proportional to the surface area of the receiver. They found a possible increase of receiver efficiency and outlet temperature in liquid metal cooled receivers. They showed a 20 % higher thermal efficiency for flux densities of 2 MW/m² and fluid outlet temperatures of 900 °C for liquid metal-cooled receivers and conversion cycles compared to molten salt reference case. Such high flux density may be possible due to the potential of obtaining very large convective heat transfer coefficients with liquid metals, especially with sodium.

The thermo-hydraulic analysis of the potential of receiver operation at higher temperature and increased cooling rate was therefore promising.

Pacio et al. [12] proposed basic concepts for solar plants with liquid metal temperatures of up to 700 °C that should, from a thermodynamic point of view, show high advantages over molten salt.

Singer et al. [13] assessed the cost reduction potential for a plant with a supercritical steam cycle, which could be combined with a solar power plant if the HTF outlet temperature is 635 °C, a too high value to be reached with solar salt. They found no economic advantage from the increased temperature and steam pressure when salt (the development of which is an open task as of today) was used, due to the additional costs for the power block.

While at that state positive thermo-hydraulic perspectives existed, uncertainty remained with regard to the costs of a system operated with liquid metals as heat transfer fluid. Also the technological readiness was unclear. Therefore, the different groups in LIMTECH Project B2 have addressed these issues both numerically and experimentally. The different aspects are described in the following sections.

2. Simulations

In order to assess liquid metals as possible HTF for STE plants, numerical simulations have been performed to analyze the heat transfer under asymmetrical heating (section 2.1) and to determine the best thermal storage configuration when using sodium as HTF (section 2.2). Furthermore, a detailed

IOP Conf. Series: Materials Science and Engineering **228** (2017) 012012 doi:10.1088/1757-899X/228/1/012012

techno-economic study has been conducted to obtain a fair cost comparison between solar salt, LBE and sodium in a real plant condition (section 2.3) and finally, the potential for an electromagnetic pump for the liquid metal circulation has been evaluated (section 2.4).

2.1 Heat transfer in a non-uniformly heated single tube

The central receiver is typically arranged as a series of parallel tubes in which the cooling medium flows and that are irradiated by the concentrated sunlight on their outer surface. The latter condition results in a circumferentially and longitudinally non-uniform heat flux, resulting in high thermal stresses in the tube walls. Therefore, for properly designing the receiver, knowledge of the maximum wall temperature values is required. For this purpose, the convective heat transfer coefficient should be determined. Several Nusselt number correlations are available in literature, also for liquid metals, but all of them are valid in principle for a uniform heat flux on the tube's wall. This creates doubts about the applicability of them to the present configuration, where the tube is irradiated from only one side. Therefore, computational fluid dynamics has been used to analyze the conjugate heat transfer in the receiver tube of a solar thermal tower operated with a liquid metal. A representative longitudinally and circumferentially varying heat flux, shown in **Error! Reference source not found.** has been applied on only half of the tube´s surface. Appropriate values for engineering applications of the solid-to-fluid thermal conductivity, wall thickness ratio, Péclet number and diameter-to-length ratio have been considered.

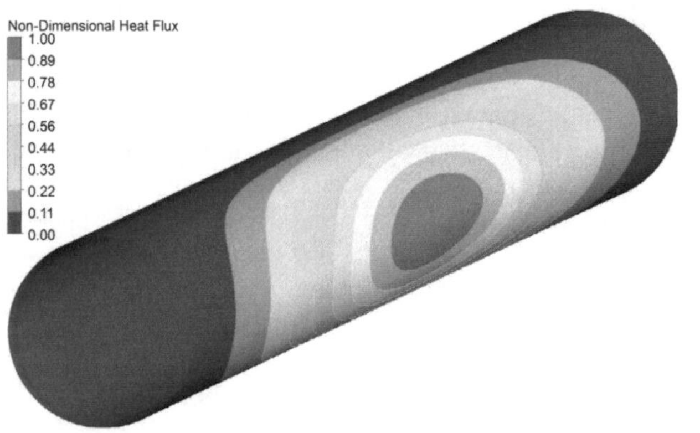

Figure 1. Non-dimensional heat flux distribution on the outer tube's wall. Reproduced from [14] with permission. Copyright © 2016 Elsevier Masson SAS. All rights reserved.

Due to the very low Prandtl number, the heat transfer mechanism of liquid metals differs from that of medium-to-high Prandtl number fluids like air, water or oils. As a consequence, the Reynolds analogy, which results in a constant and almost unitary turbulent Prandtl number, does not apply to liquid metals. In the simulations, the Boussinesq assumption is used for the turbulent heat fluxes, which then result proportional to the mean temperature gradient through the turbulent thermal diffusivity. This is calculated as the ratio of the turbulent viscosity to the turbulent Prandtl number. While the first derives from the solution of an equation for the turbulent kinetic energy and one for its dissipation rate, two different approaches have been used to evaluate the turbulent Prandtl number. On the one hand using a locally varying correlation and on the other hand by solving two additional transport equations, namely one for the temperature variance and one for its dissipation rate. This last approach is computationally more expensive and can give rise to stability issues but it also allows to consider the dissimilarities between the thermal and dynamical turbulence fields. For a thorough description of the equation solved, the models used and the numerical setup the interested reader is referred to Ref. [14] and references therein.

Detailed results of the inner, outer, fluid bulk temperature, turbulent Prandtl number and Nusselt number are also reported in Ref. [14]. In what follows only some results, particularly relevant for the receiver design, are discussed.

The circumferentially averaged Nusselt numbers do not vary significantly by using either a proper correlation or two additional transport equations to evaluate the turbulent Prandtl number. This suggests that, at least for this type of attached boundary layer-like flow, the former simpler approach can be used for the simulations.

When comparing the circumferentially averaged Nusselt numbers along the tube axis with those obtained with a correlation developed for a uniformly applied heat flux, marked discrepancies are evident, highlighting the inadequacy of the available correlations for this type of configuration.

The same big differences do not appear when comparing for different Péclet numbers the circumferentially and longitudinally averaged Nusselt numbers with those obtained with a correlation for uniform heat flux over the tube's perimeter. Indeed, in this case and for the Péclet numbers here considered, the computed Nusselt values are within ±10 % of the correlation. The reason for this is the averaging procedure along the longitudinal direction that contributes in canceling out the local differences. It should be remarked that for both cases the error between simulation and correlation increases with increasing Péclet number.

Two recommendations, useful when numerically simulating a solar receiver tube with liquid metal flowing inside, can be derived from this analysis:

A correlation depending on the local turbulent viscosity, ν_t, can be used for evaluating the turbulent Prandtl number, Pr_t:

$$Pr_t = 0.85 + \frac{0.7}{Pr \cdot \frac{\nu_t}{\nu}} \quad (1)$$

When performing simplified one-dimensional simulations, it is important to use a circumferentially averaged but axially varying Nusselt number derived from a detailed CFD simulation, in order to obtain reliable wall temperatures, which are crucial for the design of the receiver.

2.2 Thermal energy storage for a STE plant with sodium as HTF

The STE technology has the advantage of being able to store thermal energy for later use directly, efficiently and cost-effectively. Currently, the levelized cost of electricity (LCOE) of a solar thermal power plant with a 9 hour storage is 40 % lower than the corresponding photovoltaic plant including a battery [15].

Operational solar power plants like Crescent Dunes [16] and Gemasolar [17] use a two-tank configuration including a "cold" tank to store the molten salt before heating it up in the thermal receiver and a "hot" tank for the hot liquid leaving the receiver.

This configuration has also been used in the IEA-SSPS sodium facility in Almeria in the 1980s with a storage capacity of 5 MWh_{th} using 60 tons of sodium to provide the power block with 1 MW_{th} for 5 hours [18]. Furthermore, Pomeroy [19] suggested a packed bed storage with iron spheres and indirect storage with molten salt as alternatives to a direct two-tank configuration. To the authors' best knowledge, no further studies relating to thermal energy storage for a liquid metal solar power plant have been carried out since.

Therefore, a thorough evaluation of the potential storage options for a STE plant with sodium as HTF were performed. Suitable thermal energy storage systems were selected by applying the following five criteria on thermal energy storage systems available in the literature: storage medium cost, storage density, cycling behavior, maturity level and suitability for sodium. Three different kinds of thermal energy storage types were considered: sensible (direct and indirect), latent and thermochemical. This assessment showed that one promising storage system for sodium is a direct thermocline system with filler material, mainly due to low cost and relatively high storage density.

By including a filler material in the storage tank (porosity $\varepsilon = 0.25$), the storage medium costs can be reduced to less than one third (from 27.2 €/kWh_{th} to 7.5 €/kWh_{th}) and the volumetric storage density can be more than doubled compared to a sodium-only system (from 277.8 MJ/m^3 to 725.6 MJ/m^3). Physical properties and cost data of the materials for this calculation can be found in Ref. [20].

For a comparison with molten salt and a detailed evaluation of all the considered storage options the interested reader is also referred to Ref. [20].

The selected thermocline storage system with filler material is explained in the following. A principle scheme of an STE plant including the storage system is shown in Figure 2. The hot and the cold fluid are separated by a thermocline layer due to different densities of the hot and the cold fluid.

If thermal energy is in excess, hot fluid enters from the top, "pushes" the thermocline towards the bottom and cold fluid exits the storage tank (charging). However, if thermal energy is needed, e.g. after sunset, cold fluid enters from the bottom, pushes the thermocline towards the top and hot fluid exits at the top of the tank (discharging). A part of the fluid is replaced by solid filler material. Therefore, during charging, heat is transferred to the filler material and, during discharging, the heat is returned from the hot filler material to the cold fluid.

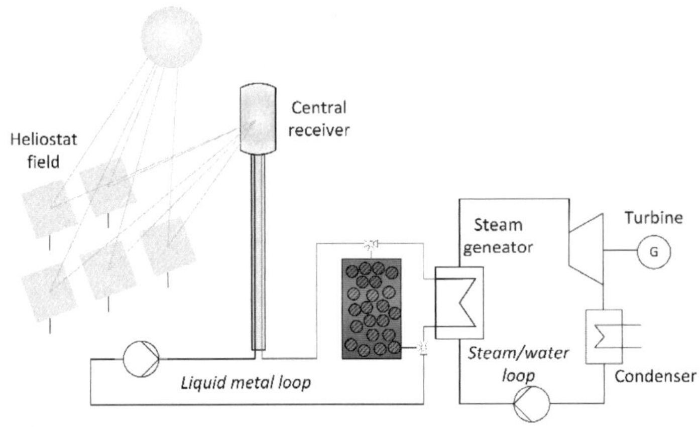

Figure 2. Principle scheme of a liquid metal-based solar tower plant with a single-tank storage system.

Single-tank thermal energy storage systems with filler material have been investigated experimentally and theoretically before. Experimental studies on thermocline storage systems with filler material have firstly been performed in the STE plant Solar One in the 1980s with a storage capacity of 170 MWh_{th} [21]. The storage tank filled with rocks and sand with thermal oil as HTF proved the potential of single-tank storage systems. Pacheco et al. [22] published results of a pilot-scale experiment of 2.3 MWh_{th} in Sandia Laboratories with molten salt as HTF and a quartzite rock and sand mixture as filler. Recently, lab-scale experiments with molten salts, oil or air as HTF have been performed to re-demonstrate the feasibility of a thermocline storage system and to obtain new data for validation purposes ([23], [24]). Furthermore, the German Aerospace Centre (DLR) is currently building a pilot-scale facility for material testing and demonstration purposes [25].

To the authors' best knowledge, neither an experimental nor a systematic theoretical study have been conducted for liquid metals in a thermocline storage system with filler material yet. Therefore, a two-dimensional two-phase model will be developed to test the feasibility of liquid metal in a thermocline system with filler material.

2.3 Commercial plant configuration and LCOE analysis

The receiver tests with liquid sodium in the 1980s at the Plataforma Solar de Almería (PSA) already proved the feasibility of liquid metals as HTF in STE systems. The conducted tests showed high receiver efficiency over 90 % [26]. Operation at high flux and power levels even resulted in higher efficiencies. From a thermodynamic point of view, the use of liquid metals is therefore beneficial. But in practical applications like power plants the electricity production costs are more important than the efficiency. This section contains a techno-economic analysis of such a power plant which allows a precise assessment of the liquid metal technology. The first step in such an analysis is to identify promising configurations. A concept study with STE systems using liquid metals as HTF and/or storage material was carried out [27] with innovative systems including component / material selections and temperature ranges.

Solar salt was the choice of HTF in the Solar Two plant (1995) [28], in Gemasolar (2011) [17], in Crescent Dunes (2015) [16] and is the choice for most of the announced big solar thermal projects: e. g. Atacama-1, Chile with P_{el} = 110 MW; Redstone, South Africa with P_{el} = 100 MW; Copiapo, Chile with P_{el} = 260 MW; Supcon, China with P_{el} = 50 MW [16]. Therefore, the reference system in this study is based on such a molten salt system (see Figure 3 left). All liquid metal concepts are designed and calculated with the same tools and compared to this reference system. In all cases, external receivers with 360°-field are used. The reference case for the comparison is a large scale power plant (P_{el} = 125 MW) at the location Postmasburg, South Africa.

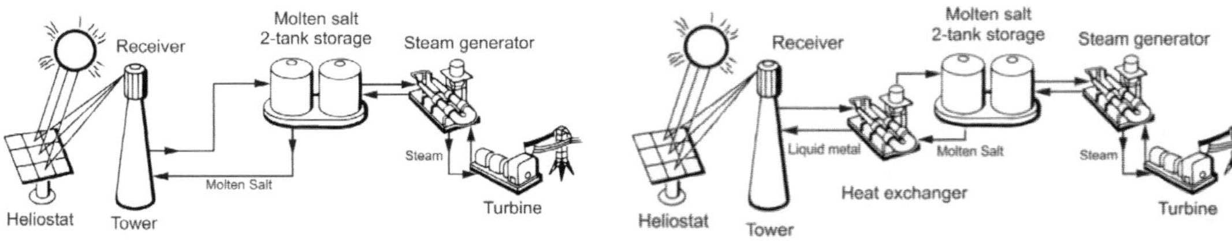

Figure 3. Left: Flow schematic of a molten salt central receiver system, right: Flow schematic of a liquid metal/molten salt binary central receiver system [5].

The salt temperature in the loop ranges between 290 – 565 °C. The software HFLCAL [29] is used for the field layout and the tower height.

As discussed in section 1, liquid metals inherit the possibility to increase the upper temperature limit for higher system efficiencies. In this case, a high temperature storage would be necessary, which is up to now not available or too expensive [27]. Additionally, the comparability between two concepts decreases the more different they are. For this reason all systems analyzed in this paper have the same power block and the same thermal storage system. The only differences are the receiver system with the HTF and the associated heliostat field (see Figure 3 right). The presented results are therefore the lower limit of the potential benefit of liquid metals. Higher system temperatures and innovative storage concepts (see section 2.2) can further fulfil the potential of liquid metals.

The presented work consists of a comparison between two different liquid metal systems (one with sodium and the other with LBE) and a reference system with solar salt as HTF.

A major part of the parasitic losses in the solar salt system (see Figure 3 left) is due to the high pressure head of the pump which works in an open loop, because the 2-tank storage is at atmospheric pressure. Typically, the salt is pumped up the tower and then throttled at the tower base before it enters the hot storage tank. The liquid metal receiver and the intermediate heat exchanger could be designed as a close loop system in which the work required to circulate the liquid metal must only overcome the friction losses of the piping. In this case, however, the intermediate heat exchanger has to be placed at the ground and riser and downcomer are filled with the liquid metal. If sodium is used as HTF, the static pressure at the ground would be approximately half of that with solar salt. For LBE, however, the static pressure at the ground would be over five times higher (approx. 1 bar/m). For commercial plants, such a high static pressure would require very thick tubes for riser and downcomer. For this reason, the intermediate heat exchanger in the LBE system is located just below the receiver, followed by an open salt loop. Only the sodium system is designed as a close loop, resulting in lower pumping losses. Additionally, only EM pumps (see section 2.4) are considered for sodium, while mechanical pumps are foreseen for solar salt and LBE.

All receiver designs presented in this paper are calculated and optimized with the ASTRID code [30], which was developed within the project B2 of the LIMTECH alliance. An external interface to a ray tracing software calculates the heat flux distribution on the absorber tubes. The detailed FEM model also considers radiation exchange between surfaces and forced convection due to influence of wind. Reducing the absorber area results in higher receiver efficiency and lower receiver cost. But at the same time, this leads to higher spillage losses and therefore to a lower optical efficiency of the heliostat field, which increases the required number of heliostats and therefore the cost. In order to find the optimal trade-off between receiver size and optical efficiency a detailed analysis of field layout, ray tracing and receiver arrangement based on annual system performance is necessary. The solar salt receiver results in a mean heat flux density of 0.51 MW/m^2 with aiming strategy to reduce the peak flux to approx. 1.0 MW/m². For liquid metal receivers no aim point strategy is necessary, resulting in a mean heat flux density of 1.06 MW/m² and a peak value of 2.9 MW/m².

All components are optimized in terms of cost and efficiency to improve the complete system. The calculated efficiency characteristic of the optimized system including heliostat field, receiver, storage and power block together with the hourly direct normal irradiance (DNI) and weather data is used to predict the annual energy yield of the plant and finally calculate the LCOE.

Both the sodium and the LBE system show lower LCOE. Compared to the reference system with solar salt, the LBE system turns out in 6 % and the sodium system in up to 16 % lower LCOE. The LCOE reduction is composed by two parts: Firstly, the higher annual efficiency of the liquid metal systems and on the other hand the lower investment cost. Figure 4 shows on the left the annual energy turnover for all systems. The overall system efficiency (net electricity divided by the solar input) is about 2 %-points higher for the liquid metal systems. The second part of the LCOE reduction with liquid metal systems is due to the lower investment cost (see Figure 4 on the right). The higher system efficiency of the liquid metal systems leads to slightly smaller and cheaper heliostat fields. However, the main contribution to the lower LCOE is due to the size reduction of the high flux liquid metal receiver, which results in significantly lower receiver cost.

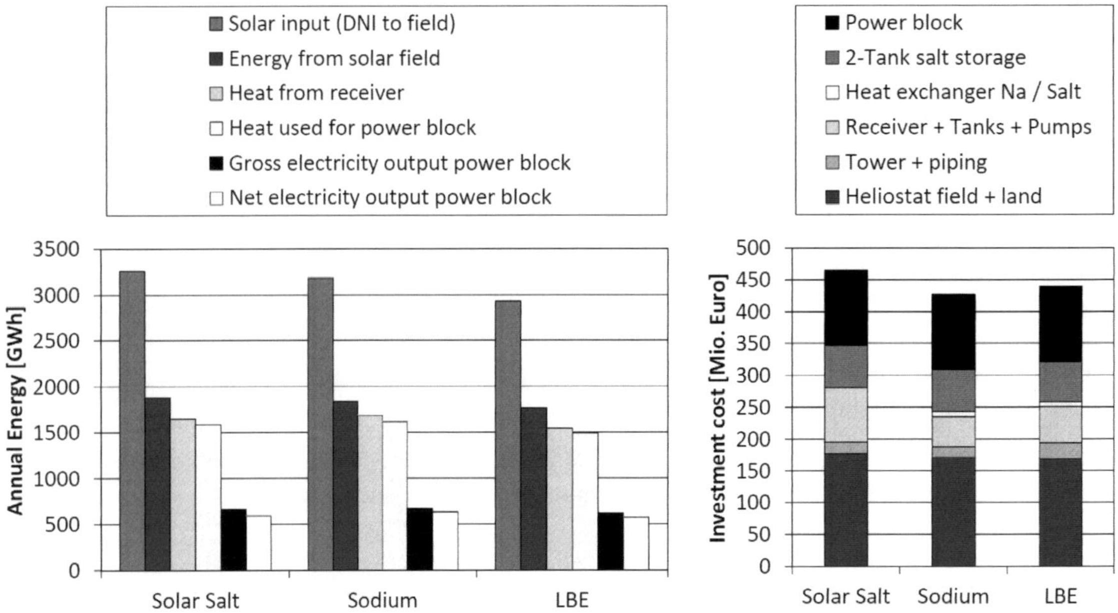

Figure 4. Comparison between the reference system with solar salt, the sodium system and the LBE system, left: annual energy turnover, right: investment cost of main components.

2.4 Electromagnetic pumps for solar power plants

The results of the previously made analysis have shown an advantage of liquid metals used as a heat transfer fluid. Nervertheless, considering the height of the tower of a concentrated solar power system, the question of circulating with high-density fluids such as LBE as a liquid metal has arisen. An expedient way to operate the liquid metal flow and provide a required pressure head in the system is the use of an electromagnetic (EM) field ([31], [32], [33]). An inductive EM pump was chosen as an option to fulfill these requirements in an STE plant operating with sodium designed within the LIMTECH B2 Project. Following advantages of induction pumping were taken into account: 1) high efficiency at high flow rates (2 - 8 % for heavy metals, 10 – 40 % for alkali metals); 2) absence of moving parts; 3) ability to work at high temperatures ([34], [35]).

A numerical model of an annular linear induction pump (ALIP) for sodium was derived at Leibniz University of Hannover (LUH) to examine its abilities to operate with planned solar plant's parameters. A numerical model of an ALIP was built in order to investigate the abilities of a travelling magnetic field (TMF) (operating frequency is from 5 to 40 Hz) for sodium transporting.

Within the project, a design of an STE plant was developed [27]. The hydraulic parameters of the solar plant coolant loop, which should have to be provided by the EM pump, are presented in the Table 1.

Table 1. Pumping requirements for two STE plants with different capacity and the properties of heat transfer medium.

Parameter	Value
Thermal power receiver (MW)	700
Heat transfer fluid	Sodium
Operating temperature (°C)	290
Electrical conductivity (S/m)	$1{,}3 \cdot 10^{-7}$
Dynamic viscosity (Pa·s)	$2{,}5 \cdot 10^{-4}$
Density (kg/m^3)	840
Mass flow (t/h)	120
Volume flow (m^3/h)	8500

There are two most usual types of induction pumps – Flat Linear Induction Pump (FLIP) and Annular Linear Induction Pump (ALIP). In FLIP the windings are placed on either side of the channel, or in one side for one sided FLIP. A disadvantage of such a geometry is a high side-effect loss, which appears due to the current distribution along the width of the channel. In ALIP windings surround the annular pipe with a ferromagnetic core in the center of the annulus. Induced currents flow in circular paths thus neglecting side effect loss. The annular geometry of the pipe also permits the use of higher hydraulic pressure, thus the annular type of the pump was chosen for the simulation [36].

Table 2. Values of the basic parameters of the studied ALIP (Figure 5).

Parameter	Value
Operating frequency of TMF, (Hz)	5 – 40
Voltage (V)	1400
Current (A)	1000
Number of windings (for each pole)	9
Distance between the coil and the duct (mm)	3
Inner diameter of the outer pipe (mm)	1000
Outer pipe wall thickness (mm)	75
Duct end length (mm)	200
Duct wall thickness (mm)	10
Winding height (mm)	220
Winding thickness (mm)	27

The developed model of ALIP is presented in Figure 5. A 2D symmetric formulation was used to obtain electromagnetic and hydrodynamic solutions. The pump consisted of the windings (1), which generated magnetic field B (2), and were connected in poly-phase for creation a travelling magnetic field. The angle between the phases was 60°. The windings were positioned on both sides of the duct (3). A half of the pipe (4), where the direction of the fluid flow is opposite to the melt in the pump, was simulated to consider its possible influence on the pumps efficiency. Table 2 provides information about the dimensions of the pump's components and applied EM parameters.

In the pump currents within the liquid metal are induced by means of alternating magnetic field *B*.

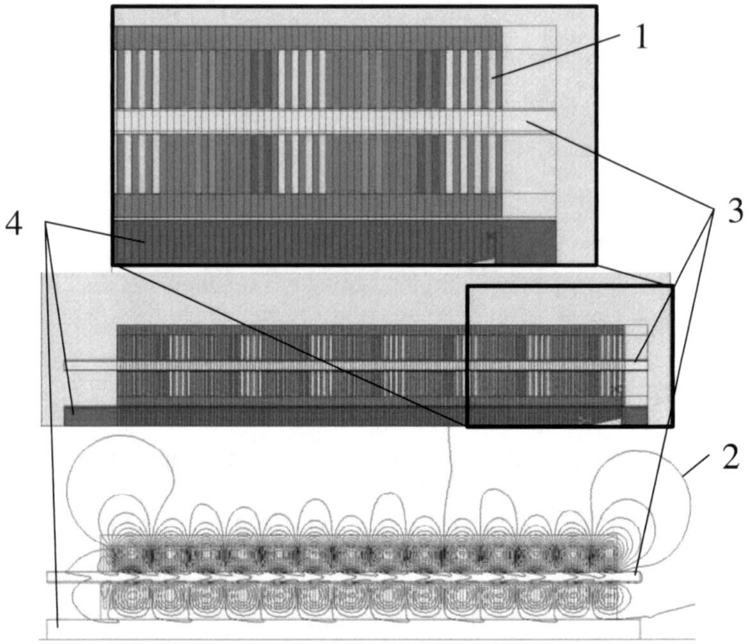

Figure 5. The numerical model of ALIP used in simulation of EM process. Here (1) – windings (poles), (2) – resulting magnetic field, (3) – duct, (4) – half of the pipe with the flow in opposing pumping direction.

The numerical simulation cycle included harmonic EM analysis and hydrodynamic solution and was performed with the ANSYS software packages (Mechanical APDL and Fluent, respectively).

The EM analysis provides the driving forces that are used as source terms in the hydrodynamic simulation of the flow field. For the latter, the k-ω SST turbulence model is used. The obtained fluid velocity field is then used again in the EM model to update the source terms. This iterative process ends when the pressure on the duct inlet and outlet stabilizes.

As shown in Figure 6, although the efficiency of the pump decreases with increased volume flow, the pressure head is still high enough to provide a required circulation even for large STE plants (pressure of 1 bar for 700 MW).

The numerical simulation has shown that ALIP is suitable for the needs of concentrating solar power plant, due to the wide range of developing mass flow (from 5 to 250 t/h) with high enough pressure head (from 4 to 16 m), ability to work at high temperatures (at 290 °C for sodium) and efficient operation (71 %).

3. Experiments – The SOMMER test facility

At KIT's liquid metal laboratory KALLA a small demonstrating system is designed, constructed and will be operated. The demonstration system incorporates a thermal receiver, a pump for liquid metal circulation in a closed liquid metal loop, a heater, a cooler, auxiliary heating systems, measurement and control. The equipment is sized to make the later addition of a thermal storage system possible. The cooler is used as a general heat sink. Except for the heater, which enhances the limited solar power input into the system, the components are representative for a solar power plant.

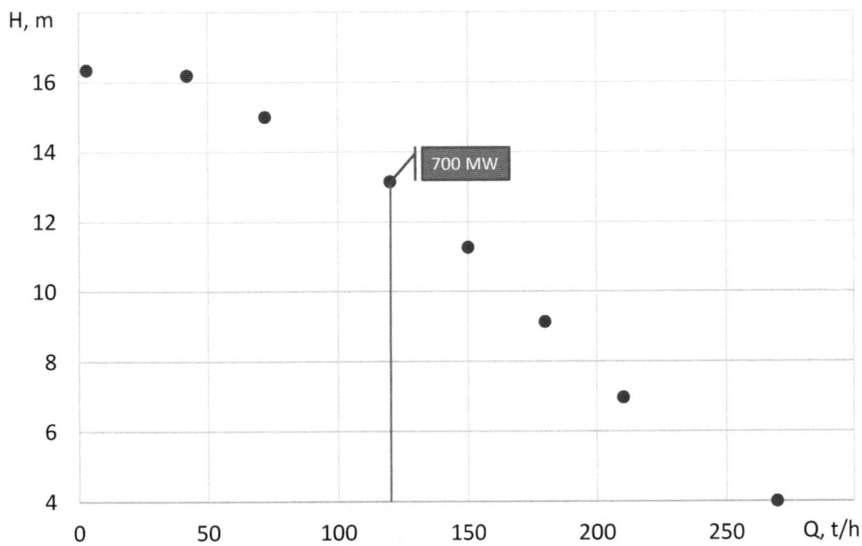

Figure 6. Maximal pressure head (H), which can be provided by the designed ALIP dependent on the mass flow (Q). The vertical line marks the mass flow required in the loop of the solar plant.

3.1 Selection of LBE as the heat transfer fluid for the demonstration loop

Among tin, LBE and sodium, the best suited liquid metal for use in a demonstration loop had to be selected. Tin was eliminated due to its very corrosive interaction with high-nickel alloys, as concluded in Ref. [3], and sodium was eliminated due to the effort for operational safety, such as the welding of tube connections instead of flanging. This effort may be legitimate in a commercial facility. For an experimental facility, however, where the system may frequently be adjusted and improved this effort was estimated to be inacceptable. LBE inherently provides the benefit of a lower operational risk and has been tested in the past as coolant of rod bundles in nuclear power plants.

Even though there are past working experiences of STE plants using sodium as HTF, to the authors' best knowledge LBE has only been considered as thermal storage fluid, as discussed by Kim et al. [37]. Therefore, the planned experiments will be helpful to evaluate the possibility to use also LBE as HTF in a solar plant.

3.2 Goals of the demonstration loop

The following goals are aimed for during the operation of the LBE-loop:

- The temperature at the outlet of the thermal receiver at full load shall be above 600 °C
- The receiver shall be operated at a solar flux of 1 MW/m^2 in average and a peak flux higher than that by a factor of two to three
- The operation shall be demonstrated at fluctuating supply of solar power

The high outlet temperature is intended to demonstrate the capability of liquid metals to achieve operating temperatures higher than the state-of-the-art molten salt. A high solar flux illustrates the high cooling capability of liquid metals and the fact that high liquid metal temperatures are not a threat to the fluid as is the case with molten salt. A fluctuating power supply makes active flow control a requirement for in-field operation. The thermal power rating for the receiver shall be 10 kW.

3.3 Thermal energy storage

The thermal energy storage options are investigated at KALLA separately from the demonstration loop for the thermal receiver (see section 2.2).

3.4 The light source

As light source for the receiver the sun is used for the SOMMER (Solar furnace with a Molten MEtal-cooled Receiver) project. The receiver is located in the focal point of a solar furnace. That is an arrangement of two mirrors, one of which is flat and has the capability of tracking the sun (heliostat), constantly reflecting the sun rays onto a second, parabolic mirror. The latter mirror concentrates the light in a focal point. The heliostat mirror in the SOMMER facility has an aperture area of 32 m² and the parabolic mirror of 16 m². The DNI is measured by a Hukseflux DR-02 Pyrheliometer mounted on an EKO STR-21G Sun Tracker. Both are installed approximately 10 meters away from to the heliostat mirror's center of rotation. At the facility's location in Eggenstein-Leopoldshafen peak DNI values of up to 950 W/m² have been measured.

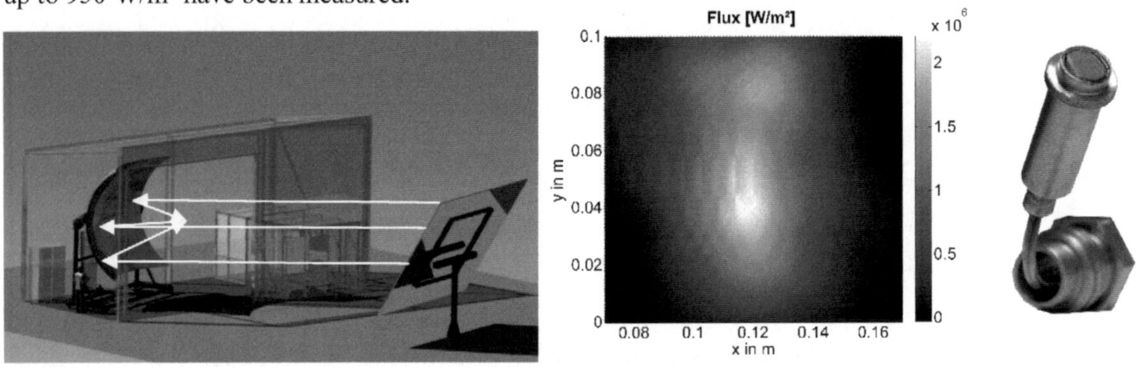

Figure 7. Left: CAD drawing of the solar furnace arrangement, center: measured flux distribution in the focal plane at a DNI value of 770 W/m² and at horizontal position of the shutter blinds, right: close-up of a heat flux micro sensor used to obtain the image in the center.

The heliostat is made of 1 mm thin glass mirror using silver as reflective layer manufactured by the company ToughTrough. The glass is mounted on a supporting sandwich structure that is connected to the hydraulic tracking mechanism. Silver coated mirrors have a specular reflectance of more than 93 % at a view angle of 45° [38]. The heliostat is slightly concentrating due to canted facets with a focal length of 500 m. The parabolic mirror is a Forbes Solar dish with Flabeg mirror facets. Taking the two-fold reflection in a solar furnace into account on a good day a power of the sunlight of 13 kW is delivered by the solar furnace.

The solar power level can be adjusted with a shutter (stock horizontal venetian blinds with PC control interface) which is installed at the entrance of the laboratory hall between heliostat and parabolic mirror.

In the event of a power outage or other irregular conditions the light source has to be shut off immediately. This is accomplished with an opaque curtain that drops into the light passage driven by gravity when triggered. Also, the heliostat is connected to an auxiliary power supply allowing it to move into stow position even during a power outage.

For the receiver's performance evaluation it is required to know precisely how much energy is delivered to the receiver. Also the distribution of the flux and the position of the peak intensity on the receiver is of interest. For this purpose a device was developed using a heat flux micro sensor that is being traversed in a plane parallel to the receiver across the focal point. The sensor moves on a circular path, however, this motion is superimposed with a linear motion component. As a result the flux distribution is being scanned on 'spiral' paths and an image with Cartesian coordinates can be generated from the angular and linear position values. Such resulting image is shown in Figure 7 in the center after post-processing steps including interpolation. For a measurement, the parabolic mirror is driven away from the receiver. The measurement then happens exactly in the distance from the mirror

IOP Conf. Series: Materials Science and Engineering **228** (2017) 012012 doi:10.1088/1757-899X/228/1/012012

where previously the receiver had been positioned. For that purpose, the parabolic mirror is installed on rails and can be driven by a linear motor along a distance of 0.5 m.

3.5 The solar receiver

The receiver is designed so that its performance in operation can be used to draw conclusions for a future receiver with thermal power one or two orders of magnitude higher. The design Reynolds and Nusselt number of the flow in the receiver's tubes were kept constant with respect to a 42 MW reference receiver under full load. The tube diameter was reduced to 50 % of that of the reference in order to obtain a model of a larger section of one of the reference receiver tubes. The tube length of the reference receiver is 2.7 m so the scaled tube should be 1.35 m in length. In order to fit such a tube into the available aperture area it is divided into separate short sections which are passed by the fluid flow in sequential manner. In order to connect the tube sections the receiver is designed as a spiral as shown in Figure 8, with a straight tube section after each bend long enough to generate a developed velocity and temperature profile prior to entering the irradiated zone. With its 11 windings, the accumulated heated tube length in the receiver is 1 m, representing 73 % of one of the reference's tubes. At full load the fluid temperature increase is 70 K in the model receiver. The receiver in SOMMER is designed for a thermal power of 10 kW at a flux density of 1 MW/m^2. The resulting approximately square aperture area is therefore 0.01 m^2.

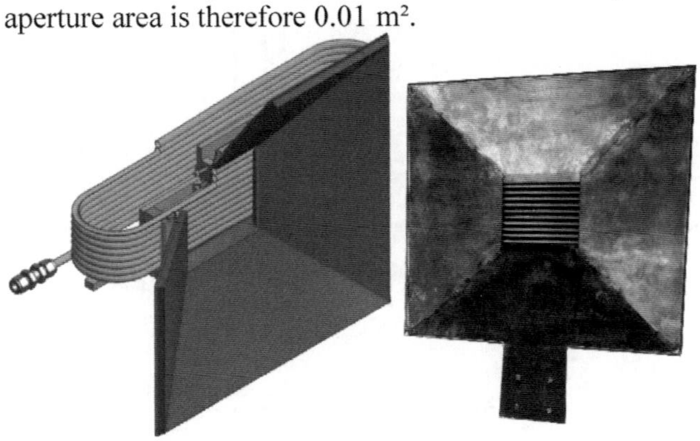

Figure 8. Left: CAD model of the receiver, right: nearly completed part.

The design decision of using a spiral, however, is a trade-off. The flux distribution along the liquid flow is much less homogeneous than in a reference tube.

Due to the scaling, the model receiver tube requires a smaller mass flow rate at a higher pressure drop than the reference tube. Thus, 0.1 kg/s at a pressure drop of 6 bar along the spiral are the calculated process parameters.

The receiver tube has an inner diameter of 9 mm and a wall thickness of 0.5 mm. For the tests at temperatures above 550 °C the austenitic steel 1.4828 is used as tube material while at lower temperatures 1.4571 is used. In both cases, welded tubes are used. The tube is coated with a 40 µm thick layer of Pyromark 2500 absorptive coating, a paint for which data for its spectral optical properties exist in literature ([39], [40]).

The tube sections and surroundings outside of the absorbing zone are protected by a large copper plate. In case of a tracking error of the heliostat, resulting in an erroneous position of the focal point, the heat is absorbed by the copper and redistributed over the entire volume. This provides enough protection and time to shut down the light source. The receiver design allows for quick replacement of the spiral.

3.6 Design of the hydraulic loop

In order to achieve the hydraulic parameters needed by the receiver a gear pump was selected. It is a WITTE CHEM10,2-2 mechanical gear pump, which delivers 0.1 L/s at 1000 turns/min. It has a hermetically sealed magnetic coupling and the pump head is submerged in the fluid inside a pump tank as indicated in the flow diagram in Figure 9. The pump can operate up to a temperature of

380 °C. The gears have a chromium-nitrogen coating for increased hardness and a reduced friction coefficient.

The loop is constructed so that after an emergency or during down time all fluid can be collected in a sump tank. Then maintenance work can be performed. The sealing of the pump allows for evacuation of the loop prior to filling it again with liquid lead-bismuth. This way no gas enclosures remain after the filling process that could otherwise influence the flow rate and pressure measurements. Filling is accomplished by pushing the liquid up in the loop through an over pressure in the gas cover of the sump tank.

The flow rate is obtained with a differential pressure measurement around a Venturi nozzle. Due to the high temperature of the fluid the pressure is measured at the end of extended, thermally not insulated stagnant branches departing from the main line so that a sufficiently low temperature is obtained.

The tube material used in the loop is 1.4571 stainless steel.

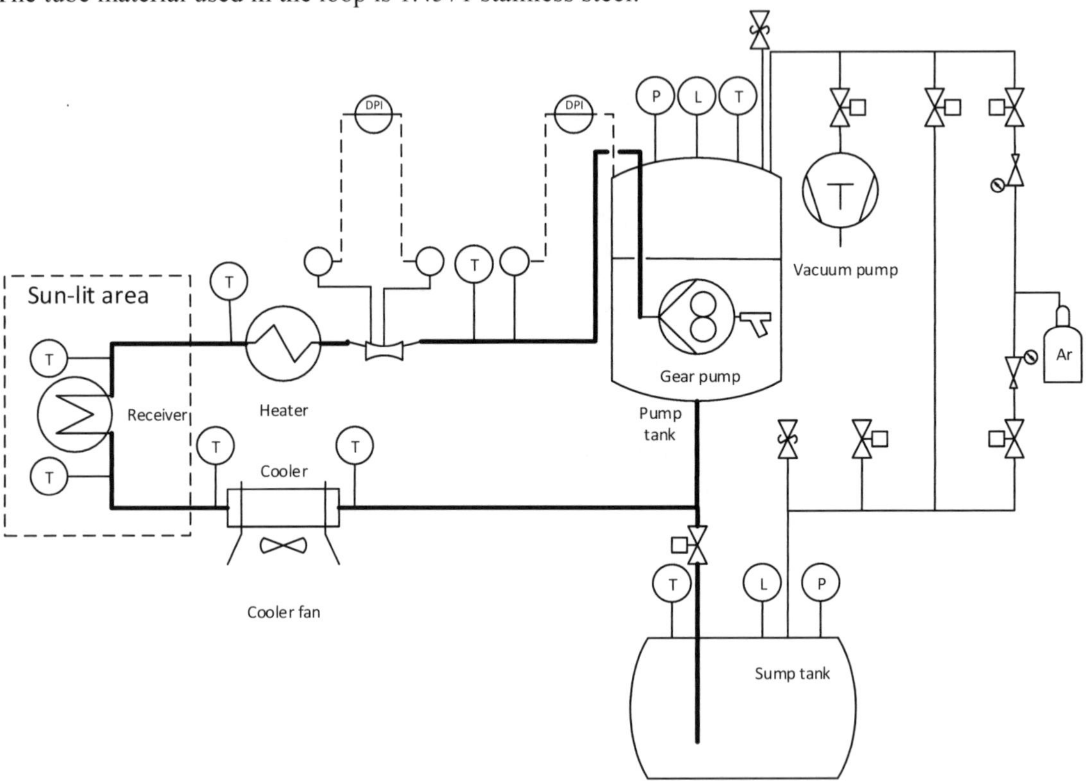

Figure 9. Process and instrumentation diagram of the SOMMER loop (simplified).

Prior to entering the receiver the fluid is passing an electric heater with a maximum power of 30 kW. As mentioned, the LBE flow temperature is raised in the receiver by only 70 K. In order to achieve the goal of an outlet bulk temperature of the receiver of 600 °C the fluid has to be heated from 380 °C at the pump outlet to 530 °C before entering the receiver. The receiver is followed by an air cooler in order to cool the fluid down to 380 °C before entering the pump again. The heater is only required in this demonstration setting and because of the upper temperature limit of the pump. Also the solar power achievable by the existing solar furnace is limited to 10 kW. In larger-scale solar power loops centrifugal pumps can be used which could operate at higher temperatures if required. However, in such a loop more likely the HTF will be cooled down to the lowest possible temperature before returning to the pump.

The oxygen content of the LBE fluid is set by its equilibrium oxygen solubility at the temperature in the storage tank, where sufficiently oxygen resides to saturate the liquid. The temperature during storage is selected such that an equilibrium oxygen content is obtained, that generates a safe operation within the entire temperature range. Thus, neither significant precipitation of solid oxides nor removal of protective oxide layers from the inner tube walls should occur.

3.7 Loop control

The incident solar power will fluctuate during operation depending on the weather conditions. The passage of a cloud will lead to a very quick drop in input power. Partial cloud passages will at least reduce the power. The goal of the loop control is to keep the outlet temperature of the fluid at the receiver constant in spite of these transients. This is done by adjusting the flow rate accordingly. In addition to the temperature value at the receiver's outlet the change in the DNI value can be used to initiate an adjustment of the pump speed. The heater's power input can be adjusted, too, in order to increase the magnitude of the change in power input.

4. Conclusions

This paper highlights the achievements within the LIMTECH project "Liquid metals for solar power systems", both numerical and experimental.

A numerical study on the turbulent heat transfer in a liquid metal flow suggests to evaluate the turbulent Prandtl number locally with a correlation depending on turbulent viscosity. Furthermore, it points out the importance of using an axially varying Nusselt number, which derived from a detailed CFD simulation.

As thermal energy storage system for an STE system with sodium as HTF, a direct thermocline system with filler material is proposed due to low cost and high storage density.

A thermo-economic analysis of the overall system results in lower LCOE for both sodium (16 % lower LCOE) and LBE (6 % lower LCOE) compared to the reference case with solar salt.

A detailed numerical study shows that an electro-magnetic pump can operate with high efficiencies over a wide range of mass flow rates at sufficiently high working temperatures.

Finally, the demonstration loop SOMMER using LBE as heat transfer fluid is presented with its solar furnace and the liquid metal loop including the thermal receiver. The SOMMER facility will be operated at a comparable solar flux as in a large scale solar tower plant with fluid temperatures above 600 °C at full load.

All in all, this paper shows the broadness of this topic and confirms the necessity of further experimental and numerical studies with liquid metal as heat transfer fluid as well as the importance of having different work groups working together on this topic.

Acknowledgments

The authors wish to acknowledge the support of the Helmholtz Association in the framework of the Helmholtz-Alliance LIMTECH (Liquid Metal Technologies) which funded this work.

References

[1] Moore R, Vernon M, Ho C, Siegel N and Kolb G 2010 *Design Considerations for Concentrating Solar Power Tower Systems Employing Molten Salt* (Sandia National Laboratories)
[2] Kolb G J 2011 *An Evaluation of Possible Next-Generation High-Temperature Molten-Salt Power Towers* (Sandia National Laboratories)
[3] Pacio J and Wetzel T 2013 *Sol. Energy* **93** 11–22
[4] Kesselring P 1986 *Iea/Ssps Solar Thermal Power Plants: Central Receiver System* (Springer-Verlag)
[5] Falcone P K 1986 *A handbook for solar central receiver design* (Sandia National laboratories)
[6] Singer C, Buck R, Pitz-Paal R and Mueller-Steinhagen H 2010 *J. Sol. Energy Eng.* **132** 041010
[7] Boerema N, Morrison G, Taylor R and Rosengarten G 2012 *Sol. Energy* **86** 2293–2305
[8] Kotzé J P, Erens P J and Von Backstroem T W 2012 *Proceeding of SolarPaces 2012, 11- 14 September, Marrakech, Morocco.*
[9] Foust O J 1976 *Sodium-NaK engineering handbook* (New York: Gordon & Breach)

[10] OECD-NEA 2015 *Handbook on Lead-bismuth Eutectic Alloy and Lead Properties, Materials Compatibility, Thermal-hydraulics and Technologies* (OECD/NEA Nuclear Science Committee Working Party on Scientific Issues of the Fuel Cycle Working Group on Lead-bismuth Eutectic)
[11] Pacio J, Singer C, Wetzel T and Uhlig R 2013 *Appl. Therm. Eng.* **60** 295–302
[12] Pacio J, Fritsch A, Singer C and Uhlig R 2014 *Energy Procedia* **49** 647–655
[13] Singer C, Giuliano S and Buck R 2014 *Energy Procedia* **49** 1553-1562
[14] Marocco L, Cammi G, Flesch J and Wetzel T 2016 *Int. J. Therm. Sci.* **105** 22–35
[15] Feldman D, Margolis R and Denholm P 2016 *Exploring the Potential Competitiveness of Utility-Scale Photovoltaics plus Batteries with Concentrating Solar Power, 2015-2030* (National Renewable Energy Laboratory (NREL))
[16] NREL Concentrating Solar Power Projects Database, National Renewable Energy Laboratory Web page: https://www.nrel.gov/csp/solarpaces/ last visited: Jan. 25, 2017
[17] Burgaleta 2014 *Gemasolar, the first tower thermosolar commercial plant with molten salt storage* (Torresol Energy)
[18] Casal F G 1987 *Solar thermal power plants : achievements and lessons learned exemplified by the SSPS Project in Almeria/Spain* ed P Kesselring and C-J Winter (Springer)
[19] Pomeroy B D 1979 *Sol. Energy* **23** 513–515
[20] Niedermeier K, Flesch J, Marocco L and Wetzel T 2016 *Appl. Therm. Eng.* **107** 386–397
[21] Faas S E 1986 *10 MWe Solar Thermal Central Receiver Pilot Plant: Thermal Storage Subsystem Evaluation - Final Report* (Albuquerque, New Mexico: Sandia National Laboratories)
[22] Pacheco J E, Showalter S K and Kolb W J 2002 *J. Sol. Energy Eng.* **124** 153-159
[23] Bruch A, Fourmigue J F and Couturier R 2014 *Sol. Energy* **105** 116–125
[24] Yin H, Ding J and Yang X 2014 *Appl. Therm. Eng.* **62** 293–301
[25] Breidenbach N, Martin C, Jockenhoefer H and Bauer T 2016 *Energy Procedia* **99** 120–129
[26] Schiel W J C and Geyer M A 1988 *Sol. Energy* **41** 255–265
[27] Fritsch A, Flesch J, Geza V, Singer C, Uhlig R and Hoffschmidt B 2015 *Energy Procedia* **69** 644–653
[28] Pacheco J E 2002 *Final Test and Evaluation Results from the Solar Two Project* (Albuquerque, New Mexico 87185 and Livermore, California 94550: Sandia National Laboratories)
[29] Schwarzboezl P, Buck R, Sugarmen C, Ring A, Crespo M J M, Altwegg P and Enrile J 2006 *Sol. Energy* **80** 1231–1240
[30] Frantz C. Fritsch A. Uhlig R. 2016 *Solar-Paces International Conference*
[31] Jaross R and Barnes A 1958 Design and operation of a 10.000 gpm dc electromagnetic sodium pump and 250000 ampere homopolar generator *Second U.N. International Conference on the Peaceful Uses of Atomic Energy, Tech. Rep.*
[32] Baker R and Tesser M 1987 *Handbook of electromagnetic pump technology* (New York: Elsevier Science Publishing Co. Inc)
[33] Mackay R 2004 *The Practical Pumping Handbook* (Elsevier)
[34] Verkamp J and Rhudy R 1966 *Electromagnetic alkali metal pump research program* (NASA)
[35] Blake L 1957 *Proceedings of the IEE – Part A: Power Engineering* vol 104 pp 49–67
[36] Sharma P, Sivakumar L S, Prasad R R, Saxena D K, Kumar V A S, Nashine B K, Noushad I B, Rajan K K and Kalyanasundaram P 2011 *Energy Procedia* **7** 622–629
[37] Kim J-S, Dawson A, Wilson R, Venkatesan K and Stein W 2015 *Proceedings of the ASME 2015 9th International Conference on Energy Sustainability*
[38] Good P, Cooper T, Querci M, Wiik N, Ambrosetti G and Steinfeld A 2016 *Sol. Energ Mat Sol C* **144** 509-522
[39] Ho C K, Mahoney A R, Ambrosini A, Bencomo M, Hall A and Lambert T N 2012 *Proceedings of the ASME 2012 6th International Conference on Energy Sustainability* (ASME International)
[40] Coventry J and Burge P 2017 *AIP Conference Proceedings* **1850**, 030012.

IOP Conf. Series: Materials Science and Engineering **228** (2017) 012013 doi:10.1088/1757-899X/228/1/012013

Liquid metal batteries - materials selection and fluid dynamics

T Weier[1], A Bund[2], W El-Mofid[2], G M Horstmann[1], C-C Lalau[2], S Landgraf[1], M Nimtz[1], M Starace[1], F Stefani[1] and N Weber[1]

[1] Institute of Fluid Dynamics, Helmholtz-Zentrum Dresden - Rossendorf, Bautzner Landstr. 400, 01328 Dresden, Germany

[2] Electrochemistry and Electroplating Group, Technische Universität Ilmenau, Gustav-Kirchhoff-Str. 6, 98693 Ilmenau, Germany

E-mail: t.weier@hzdr.de

Abstract. Liquid metal batteries are possible candidates for massive and economically feasible large-scale stationary storage and as such could be key components of future energy systems based mainly or exclusively on intermittent renewable electricity sources. The completely liquid interior of liquid metal batteries and the high current densities give rise to a multitude of fluid flow phenomena that will primarily influence the operation of future large cells, but might be important for today's smaller cells as well. The paper at hand starts with a discussion of the relative merits of using molten salts or ionic liquids as electrolytes for liquid metal cells and touches the choice of electrode materials. This excursus into electrochemistry is followed by an overview of investigations on magnetohydrodynamic instabilities in liquid metal batteries, namely the Tayler instability and electromagnetically excited gravity waves. A section on electro-vortex flows complements the discussion of flow phenomena. Focus of the flow related investigations lies on the integrity of the electrolyte layer and related critical parameters.

1. Introduction

The increasing penetration of renewables into the energy system poses a number of challenges. One major aspect is that electricity from sun and wind is available only intermittently and with varying timescales. This is in stark contrast to traditional energy systems where electricity generation is driven by demand. Electricity grids with a large share of intermittent input call for storage if electricity demand shall be satisfied independently of daytime and weather conditions [1]. At least equally important, grid stability may suffer from insufficient storage if generation and demand cannot be balanced [2]. Obviously, cost is a determining factor in selecting storage options [3] especially if the massive capacity needed by the economy of a whole country is considered.

Liquid metal batteries (LMBs) might be part of a solution to the gap in massive and economically feasible large-scale stationary storage [4]. LMBs consist of three liquid layers segregated on the basis of their differences in density (ρ), see Fig. 1. An alkaline or earth alkaline metal floats on top, a heavy metal or half metal fills the bottom space and a molten salt immiscible with both metals is sandwiched in between. Obviously, the operating temperature of an LMB has to exceed the melting temperatures of all cell ingredients and therefore LMBs belong to the high temperature systems. Hence, reactions are comparably fast which implies

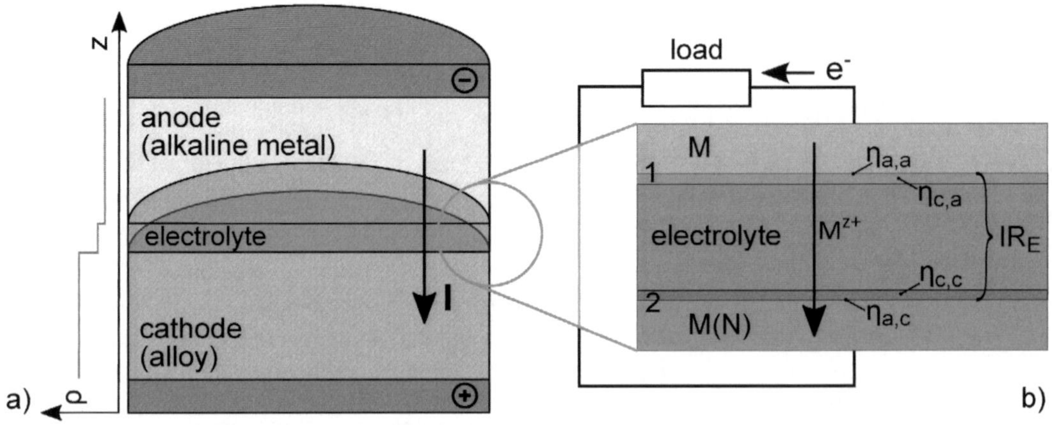

Figure 1. Schematic view of an isothermal liquid metal cell just after starting discharge (a) and sketch of the discharge process including activation and concentration polarizations (b).

rapid charge transfer but also increased corrosion rates. Likewise, diffusion is much faster than in solids and the absence of solid structures eliminates many mechanisms otherwise causing cell degradation. Fast transport and rapid kinetics can result in extreme current densities of up to 130 kA/m^2 [5] allowing for the simple cell construction sketched in Fig. 1a without compromising on power density. This simple cell construction, in turn, facilitates scale-up on the cell level.

Already in the 1960, LMBs were intensively investigated (see, e.g., [6]), mainly as components of energy conversion systems named "thermally regenerative electrochemical systems" (TRES); an overview can be found in [7, 8]. However, using LMBs for off-peak stationary storage was envisioned as well [5]. Active material combinations used at that time were Na||Sn [9, 10] and K||Hg [11] for the reason of good thermal separability. Higher cell voltages could be achieved, e.g., with Na||Bi, Li||Bi, Li||Te, and Li||Se, [6] where in the latter two systems the bonds between Li and the cathode metal are too strong to allow for separation by distillation [12]. According to a recent review by Kim et al. [4] interest in LMBs waned in the 1970s because the low specific energy rendered LMBs unattractive for mobile applications.

Currently, we are observing a renaissance of LMBs sparked and led by the MIT group around Donald Sadoway. Primary targets are now a cost driven development and grid-scale energy storage. Active materials currently under investigation include Mg||Sb [13], Ca||Bi [14], Li||Sb-Pb [15], Li||Bi [16], and Na||Pb-Bi [17].

The completely liquid interior and the high current densities characterizing LMBs can cause a number of fluid flow phenomena that will be primarily important for the operation of future large cells. However, individual aspects, especially mixing in the cathodic compartment, are of immediate relevance for today's smaller cells as well. Following a section on electrochemistry, we will discuss three electromagnetically driven flow phenomena, namely the Tayler instability, electro-vortex flows, and interface instabilities.

2. Electrochemistry

Usually a liquid metal cell is a simple concentration cell and the electromotive force (EMF) or open circuit voltage (OCV) of the cell is solely given by the activity of the alkaline metal in the cathode alloy. This can easily be seen by considering the equations for the transfer reactions at the two interfaces (see Fig. 1b). For the anode/electrolyte interface (1) and the electrolyte/cathode interface (2) during discharge, we have

$$\mathrm{M} \rightarrow \mathrm{M}^{z+} + z\mathrm{e}^- \tag{1}$$
$$\mathrm{M}^{z+} + z\mathrm{e}^- \rightarrow \mathrm{M(N)} \tag{2}$$

with M denoting an alkali or earth-alkali metal (e.g., Li, Na, Mg) and N refering to the heavy or half metal of the cathode (e.g., Bi, Pb, Sb). z is the valency, i.e., 1 for alkali and 2 for earth alkali metals. The equilibrium potentials φ_0 of both half-cells are then

$$\varphi_0(1) = \varphi_{00} + \frac{RT}{zF} \ln \frac{a_{\mathrm{M}^{z+}}}{a_{\mathrm{M}}} \tag{3}$$
$$\varphi_0(2) = \varphi_{00} + \frac{RT}{zF} \ln \frac{a_{\mathrm{M}^{z+}}}{a_{\mathrm{M(N)}}} \tag{4}$$

with φ_{00} denoting the standard potential, R the universal gas constant, F the Faraday constant and a the activity of the metal in the pure (M), ionic (M^{z+}) and the alloyed (M(N)) state. From this, the cell's OCV determined by the difference of the two electrode potentials $\varphi_0(2)$ and $\varphi_0(1)$ results as

$$E_{\mathrm{OC}} = -\frac{RT}{zF} \ln a_{\mathrm{M(N)}} \tag{5}$$

since the standard potential of both half cells is identical and the activity of the pure anode metal is one by definition. The terminal voltage E available under current flow is the difference of the OCV and several terms describing voltage losses, i.e., polarizations (cf. [12] and Fig. 1b):

$$E = E_{\mathrm{OC}} - IR_E - \eta_{c,a} - \eta_{c,c} - \eta_{a,a} - \eta_{a,c} \tag{6}$$

with the ohmic losses in the electrolyte of resistance R_{E} due to the current I, the concentration polarizations at the anode $\eta_{c,a}$ and cathode $\eta_{c,c}$ and the corresponding activation potentials $\eta_{a,a}$ and $\eta_{a,c}$.

As mentioned in the introduction, LMBs have to operate at elevated temperatures in order to keep the active materials and the electrolyte in the liquid state. Conductive heat loss through the walls is proportional to the temperature gradient and can be understood as self-discharge. The higher the temperature, the faster the chemical reactions and the more serious material degradation phenomena become. This alone would be motivation enough to aim at lowering the operating temperature of the cells. Concerning flow measurements, temperatures of less than 230 °C are required for typical Ultrasound Doppler Velocimetry (UDV) transducers to be directly applicable to the cells. This requirement considerably restricts the available options and is not easily fulfilled. While the typical alkali metals used for the anodes (Na, Li) have low enough melting temperatures ($\mathrm{T_m}$) by themselves and suitable cathode alloys with $\mathrm{T_m}$ well below 200 °C are likewise available (e.g., Pb-Bi), the melting points of traditionally used molten salt systems exceed 350 °C and often lie beyond 450 °C. Lower melting salt systems typically contain components that are highly hygroscopic (e.g., NaOH) and/or are not stable in direct contact (e.g., $\mathrm{NaAlCl_4}$) with liquid alkali metals. While interesting developments in terms of low melting salts are underway [17, 18], room temperature ionic liquids (RTILs) pose an alternative option to molten salts. RTILs [19] have found increasing interest during the last decades due to their non-volatility, non-flammability, wide temperature range of operation and high conductivity. In addition, RTILs are already investigated as electrolytes for rechargeable batteries [20, 21]. Therefore, testing the applicability of RTILs as possible low temperature electrolytes in LMBs seemed a logical step to try. The remainder of the section is divided into a first part dealing with traditional molten salt based systems and a second one describing cells using RTILs.

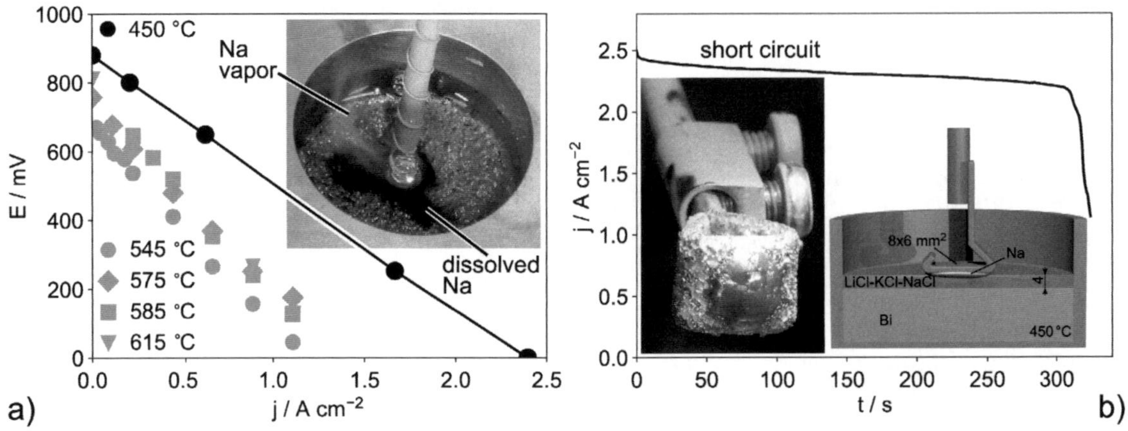

Figure 2. Na||Bi cell with foam anode (insets in b): terminal voltage vs. current density compared to measured values of [6] (gray symbols, a) and full discharge under short circuit (b). Na dissolution (black clouds in the molten salt) and evaporation (green-yellow vapor) shown in the inset in a).

2.1. Molten salt electrolytes

Na is the natural first choice for the anode material because of its abundance [22] and its low price. A combination with Bi gives a comparably high cell voltage (~0.9 V near full charge, see [6]) that exceeds that of several other pairings, e.g., with Pb, Sn, or Hg [23, 24].

Several open cells based on crucibles of 30 mm diameter containing Bi and the molten salt mixture were designed with the aim to provide maximum optical access. The negative pole was either formed by a Ta spiral holding a drop of liquid Na or an Inconel® alloy 625 foam wetted with Na, see the insets in Fig. 2a and Fig. 2b. An eutectic mixture of KCl-LiCl-NaCl (36-55-9 mol%, melting point around 346 °C [25]) served as electrolyte. Single cells were operated in an Ar-filled glove-box (c_{O_2} and c_{H_2O} < 0.1 ppm) at 450 °C. Heating for the cell was provided by a hot plate situated underneath the cell. This kind of cell setup entails an excess of Bi: typically about 100 g of Bi (0.48 M) were combined with ca. 40 mg of Na (1.7 mM) resulting in a relatively large and comparably stable cell voltage. Current densities (calculated with the projected area of the foam anode) shown in Fig. 2a approach 2.5 A cm^{-2} during short circuit (Fig. 2b) and are in the range of those measured by [6] for the Na||Bi system. As already observed by those authors, the $E(j)$-dependency is almost linear over the whole current density range. This suggests that all polarizations listed in Eq. (6) can be neglected compared to the ohmic losses. This is remarkable because unlike the salt mixture (NaCl-NaF-NaI, 31.6-15.2-53.2 mol%) used by Cairns et al. [6] that contains only Na^+-ions, here mixed cations with only 9 mol% Na^+ are present in the electrolyte. Especially under high current conditions, mass transfer limitations in the electrolyte could therefore be expected. Their absence signals good mixing in the electrolyte and indeed intense convection could be observed (with velocities of about 4 mm s^{-1} to 34 mm s^{-1} depending on the cleanness of the cathode/electrolyte interface). Flow visualization in a limited sense is enabled by the solubility of Na in its salts resulting in the black clouds that form in the electrolyte near the drop anode (Fig. 2a inset) and disperse rapidly in the electrolyte. The experimental setup enforces convection by heating from below and leaving the large free surface surrounding the Na-drop open to the environment of the glove-box. The electrolyte surface is cooled by convection, radiation, and evaporation of Na since its vapor pressure at the cell temperature is considerable. Na vapor forms the greenish-yellowish clouds seen in the inset of Fig. 2a.

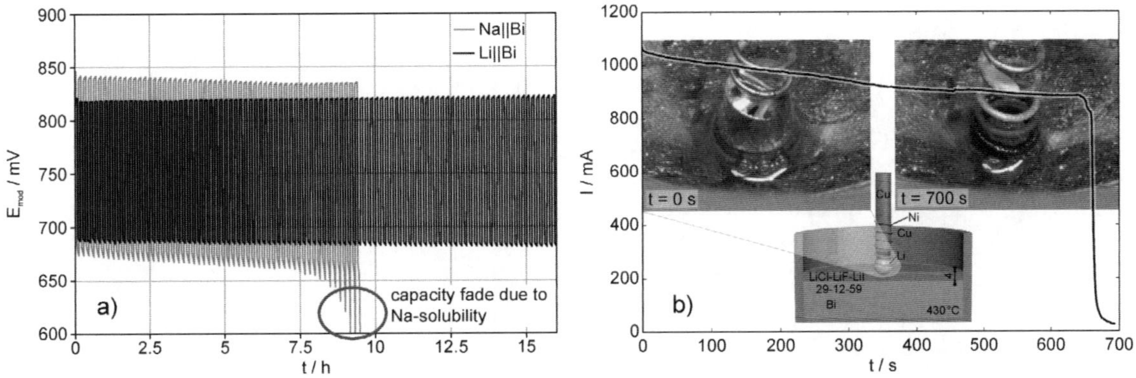

Figure 3. Cycling of a Na||Bi and a Li||Bi cell with 100 mA (a) and discharge of a Li||Bi cell under short ciruit (b). Insets b: drop anode at start and end of discharge and sketch of the Li||Bi cell. E_{mod} is the cell voltage shifted so that $E_{\text{OV}}(\text{Na}) = E_{\text{OV}}(\text{Li})$ for ease of comparison.

Na solubility in the molten salt electrolyte additionally leads to partly electronic conductivity thereby limiting Coulombic efficiency. Fig. 3a shows the cycling behavior of two open cells with an alkali metal drop anode. Na solubility leads to a steady decline of the open circuit voltage over time and eventually to a situation where the Na available in the anode is not sufficient for a full discharge anymore (~9 h). In contrast, Li||Bi shows a stable E_{OC} for the full time span. While Li itself is almost insoluble in its molten salts, the intermetallic phase Li_3Bi has some solubility in molten salts. The tinting of the electrolyte visible in Fig. 3b is possibly caused by dissolved Li_3Bi. According to Vogel et al. [26] Li_3Bi solutions have a reddish color. During short circuit discharge, current densities beyond 2 A/cm^2 were measured. The Ni spiral pinning the Li drop anode can be completely emptied during discharge. Recharging of the fully depleted spiral is possible as well.

While cells containing molten alkaline metals should generally be operated in the absence of oxygen and moisture, Li anodes require N_2 removal that is not necessary in case of Na. However, a well controlled Ar atmosphere is sufficient in both cases.

2.2. Ionic liquid electrolytes

The design and operation of a low temperature liquid metal battery based on an ionic liquid as electrolyte was reported for the first time by our group in [27].

The Na||Pb-Bi liquid metal cell operated at 160 °C has shown promising results, although system functionality was limited up to now to only four charge/discharge cycles at about 50% depth of discharge [27]. Molten Na metal-electrolyte interaction enhanced by the presence of ppm water traces in the RTIL are probably the reason for the limited performances. Further investigations are under way in order to understand the mechanism behind this interaction. Based on these findings we extended the investigations and focused on a more stable chemistry with improved characteristics. It is comprised of Li and Ga as the negative and positive electrodes, used in combination with an RTIL as an electrolyte, namely the 1 mol/l lithium bis(trifluoromethylsulfonyl)imide (Li[TFSI]) in 1-butyl-1-methylpyrrolidinium bis (trifluoromethylsulfonyl)imide ([BMP][TFSI]). The battery operates at 220 °C which is a relatively low temperature for an LMB and showed good electrochemical performance at low current densities [28]. Several cells of ca. 200 mAh theoretical capacity were assembled inside an argon atmosphere glove-box and galvanostatically cycled at different current densities ranging from 10 mA cm^{-2} (corresponding to a C/40 rate) to 32 mA cm^{-2} (corresponding to a C/12 rate). The batteries cycled at 10 mA cm^{-2} to a 50% depth of discharge were operated continuously for

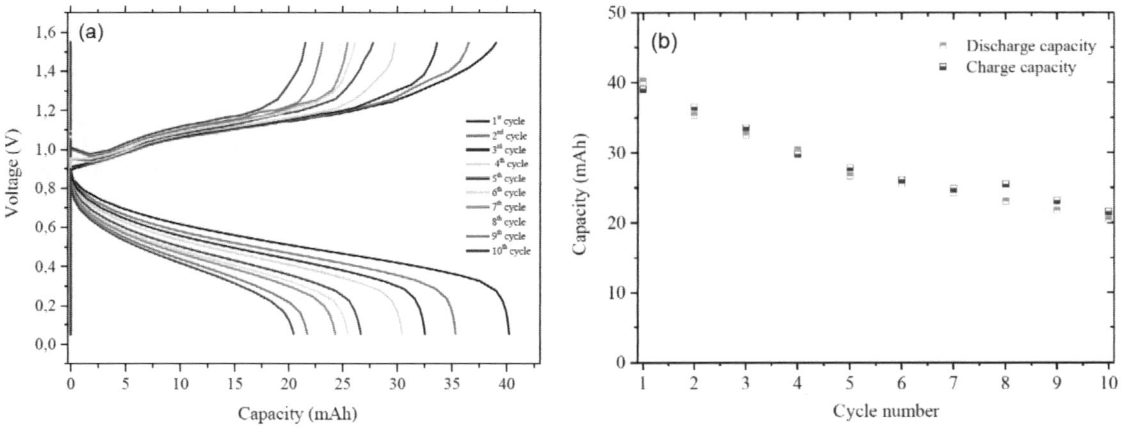

Figure 4. Electrochemical performances of a Li||Ga cell cycled at $\pm 4\,\mathrm{mA\,cm^{-2}}$ and operated at 220 °C: voltage curves as a function of capacity for the ten first cycles (full discharge, a) and evolution of the capacity with cycling (b).

about 600 hours, i.e. over 14 charge-discharge cycles with stable cycling performances. The near-constant voltage plateau observed in the galvanostatic voltage-capacity profile and the absence of a sharp voltage drop showed that during the operation of the battery no intermetallic compounds are formed, in agreement with the data from the Li-Ga binary phase diagram. During the tests the cells exhibited a round-trip Coulombic efficiency of 97% with an average value of voltage efficiency of 66% resulting in an overall energy efficiency of 65% [28].

In the case of higher charge/discharge current densities ($16\,\mathrm{mA\,cm^{-2}}$ and $32\,\mathrm{mA\,cm^{-2}}$), the battery exhibits higher voltage drop and rapid discharge capacity fade. X-Ray Photoelectron Spectroscopy (XPS) reveals that during battery operation the RTIL electrolyte decomposes already after the first discharge cycle and the reduction products accumulate in the Ga-cathode. Next to the metallic Ga in the bulk of the cathode, two Li-Ga intermetallic species are also detected: Ga_7Li_2 and $Ga_{14}Li_3$. Electrochemical impedance spectroscopy indicated mass transport limitations of the Li cations in the Ga positive electrode [28]. Further, the cycling behavior of a Li||Ga cell with ca. 40 mAh theoretical capacity is shown in Fig. 4. The cell was assembled inside the glove-box under a dry and high purity argon atmosphere, then transferred outside the glove-box and heated to 220 °C by means of a tube furnace. A continuous and similar evolution of the potential with time can be perceived. The shape of the charge-discharge curves shows then a good reversibility of the Li^+ intercalation-deintercalation process. However, a continuous capacity fade with each cycle is observed which is due to the depletion of Li cations in the RTIL electrolyte and in the Ga bulk by the formation of intermetallic compounds. Further investigations are needed in order to stabilize the tested low temperature systems and enhance the cyclability and mass transport within the cells. The operating temperatures of 160 °C and 220 °C allow the use of commercially available UDV sensors, which is indispensable for collecting operating experience with UDVs in the multilayer systems inside model LMB cells.

3. Fluid Dynamics

As already mentioned in the introduction, fluid dynamics of LMBs is quite involved and includes a multitude of fluid flow phenomena like Rayleigh-Bénard convection driven by the Joule heating of the electrolyte layer, Marangoni convection due to surface-tension differences resulting from concentration and temperature gradients, solutal convection in the cathode, etc. The remaining sections will deal mainly with magnetohydrodynamic (MHD) aspects that will become important for LMBs with large cell sizes. We start with the Tayler instability, discuss electro-vortex flows, and conclude with a paragraph on interfacial instabilities.

3.1. Tayler instability

The Tayler instability (TI) is a current driven kink-type instability akin to the instability of the z-pinch in plasma physics. A cylindrical conductor with a circular cross section bearing an axially directed homogeneous current density J_0 produces a magnetic field B_0 with a purely azimuthal component B_φ. The Lorentz force resulting from J_0 and B_0 is an inwardly directed pinch force that is prone to an axisymmetric "sausage" or a non-axisymmetric "kink" instability mode. For liquid metals that possess a considerable kinematic viscosity ν when compared to plasmas and have a finite electrical conductivity σ, the onset of the TI can be described by a critical value of the Hartmann number

$$\mathrm{Ha} = B_0 R \sqrt{\frac{\sigma}{\rho\nu}} = I\frac{\mu_0}{2\pi}\sqrt{\frac{\sigma}{\rho\nu}} \tag{7}$$

where $B_0 = B_\varphi(R)$ is the azimuthal magnetic field at the cylinder radius R, ρ the liquid's density, I the total current, and μ_0 the vacuum permeability, respectively. TI will set in as a flow with a dominant azimuthal mode $m = 1$ if Ha exceeds a value of about 20 as was predicted by Rüdiger et al. [29] and was experimentally confirmed only recently by Seilmayer et al. [30]. The characteristic velocity of the TIs saturated state scales with Ha^2. According to Eq. (7), the critical Ha can directly be transferred to a critical current depending only on the material properties of the liquid metal considered. Doing so (cf., Tab. A.1 in [31]) it becomes apparent that - without considering possibly different aspect ratios - typical anode materials (K, Li, Na, Mg) with critical currents between 1 and 2 kA are more prone to the TI than the denser and less conductive metals used for the cathodes. That means that TI is less likely to contribute to cathode mixing but should be considered as a possible culprit endangering the integrity of the electrolyte layer. Note that it is indeed the total current and not the current density that determines criticality in case of the TI.

As can be seen from the stability criterion derived by Vandakurov [32] and Tayler [33]

$$\frac{\partial(rB_\varphi^2(r))}{\partial r} > 0, \tag{8}$$

the kink-type instability depends on the radial (r) dependence of B_φ. In turn, Eq. (8) opens the possibility to stabilize against the TI by changing $B_\varphi(r)$ as suggested by Stefani et al. [34]. This type of stabilization can be achieved by applying a bore around the cylinder axis changing the cross section from a cylindrical one to an annular gap. This simple measure has proven to be effective in numerical [31] as well as experimental [30] investigations. Increases in critical current are however moderate and the cell shape changes in a way that decreases the overall energy density. Using a small isolated inner bore to hold a massive wire feeding back the cell current along the axis can be shown to stabilize the cell against TI for any current [34] in the inviscid and non-resistive case. Numerical investigations [31] indicate that about one third of the cell current might suffice for stabilization if the beneficial effect of viscous and resistive damping is taken into account.

In plasma physics it has long been recognized that current-driven instabilities can be stabilized by applying an axial magnetic field. The Kruskal-Shafranov condition describes basically a ratio of the axial to azimuthal field above which kink-type instabilities cease to occur. Consequently, adding an axial field to a cylindrical LMB by means of a Helmholtz-like coil can stabilize against the TI as well. Even a horizontal magnetic field has a similar but weaker effect [31]. Stabilization by different magnetic field arrangements was compared by Stefani et al. [35]. Seemingly an axial

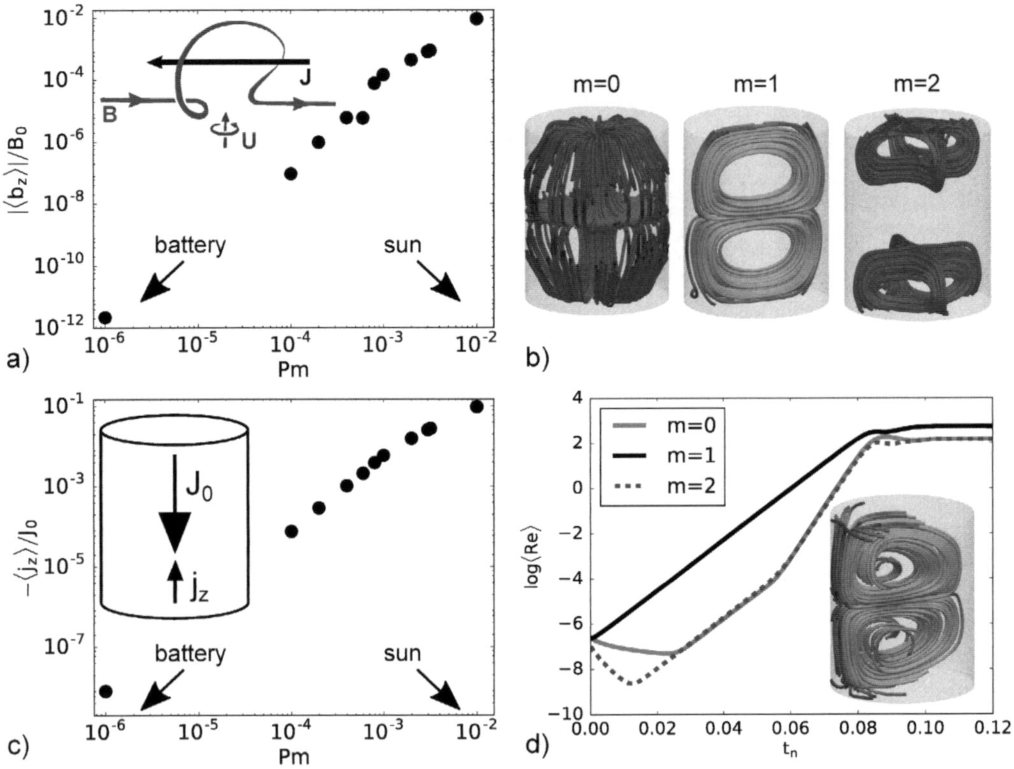

Figure 5. Saturation mechanisms vs. magnetic Prandtl number (a, c) and hydrodynamic saturation for low Pm (b, d). ⟨Re⟩ denotes a dimensionless mean velocity corresponding to the given mode and t_n a time scale normalized with the viscous time. Streamlines of the three first modes (m=1, 2, 3); (b) and of the total flow corresponding to the saturated TI (d, inset).

magnetic field performs best in terms of the ratio of stabilizing to cell current. However, one should keep in mind that axial fields are able to amplify the interfacial instabilities discussed later in section 3.3. LMBs with a capacity limited to a few hours will have a low height to width ratio. Under such a condition, the preferred wave length of the TI cannot develop in the shallow geometry and the critical current is thereby drastically increased (cf. Fig. 13 of [36]).

It is however inspirational to compare these technical stabilization measures with saturation mechanisms [37] found at different magnetic Prandtl numbers (Pm=$\nu\mu_0\sigma$) as is done in Fig. 5. For astrophysical bodies like the sun, magnetic Prandtl numbers are typically high and in the order of 10^{-2}. To access the modification of the electromagnetic fields by the flow we can resort to mean-field MHD [38] and the α effect introduced by Krause and Rädler for helical turbulence. The α effect drives an electromotive force parallel to the prevailing large scale magnetic field $\bar{\mathbf{B}}$. In a similar way turbulence leads to an increase in resistivity by the β effect. The mean electromotive force can than be written as $\varepsilon = \alpha\bar{\mathbf{B}} - \beta\nabla\times\bar{\mathbf{B}}$. α essentially produces a mean axial magnetic field $\langle b_z\rangle$ while β induces a mean axial current $\langle j_z\rangle$ opposing the applied one J_0. For high Pm $\langle b_z\rangle$ and $\langle j_z\rangle$ are large enough to interfere with B_0 and J_0 (Fig. 5, a, c) in such a way that TI goes into saturation. A similar mechanism lies behind the TI suppression with superposed azimuthal or axial magnetic fields. However, for low Pm the situation changes completely. Both, $\langle b_z\rangle$ as well as $\langle j_z\rangle$, are much too small to interact with the main fields. Instead, the saturation mechanism is a purely hydrodynamic one (Fig. 5, b, d): the quadratic

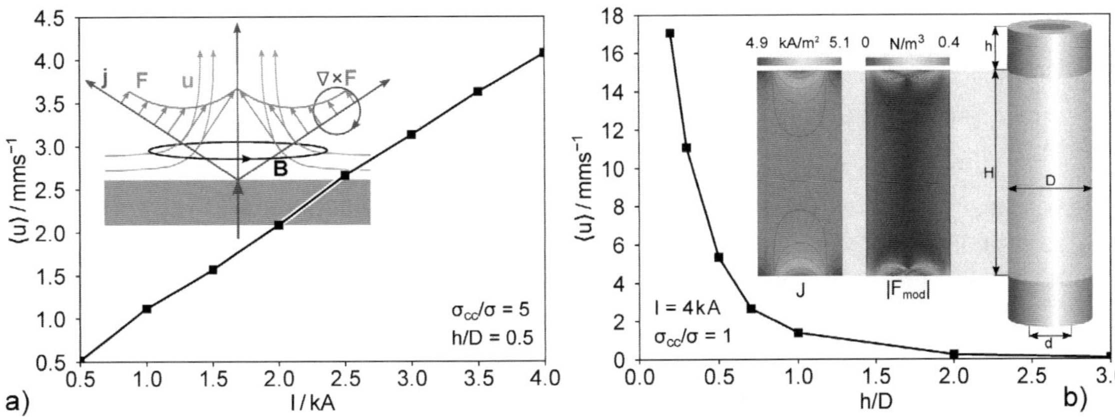

Figure 6. Sketch of the driving mechanisms of electro-vortex flows after [39] (inset in a), distribution of current density and residual force density in a cell (inset in b), mean velocity vs. total current (a), and mean velocity vs. current collector height/diameter (b). $D = 1\,\mathrm{m}$, $d/D = 0.5$, $H/D = 2.4$, material properties of Na at $580\,^{\circ}\mathrm{C}$ (see [40]).

combination of the $m = 1$ velocity perturbations produces $m = 0$ and $m = 2$ velocity components which suppress the further growth of the TI.

3.2. Electro-vortex flows

The Tayler instability (TI) discussed in the previous section can be understood as a generic case of a current driven instability under the ideal condition of a perfectly uniform current. In this sense it constitutes sort of an upper bound for a current-carrying fluid to remain at rest. Non-uniform current distributions are more typical for real settings. They give rise to rotational Lorentz force distributions and will thereby also generate electro-vortex flows (EVFs). Inhomogeneities in the current density distribution will, e.g., occur if the positive and/or negative current collector at the cell has a finite thickness, as is always the case. An extreme case discussed by Shercliff [39] and sketched in the inset of Fig. 6a is that of a point current source embedded in an insulating flat plate. When the current enters the conducting half space above the plate, it spreads radially outwards. The resulting Lorentz force distribution has - in contrast to the canonical base state of the TI - a non-vanishing rotational part. This rotational part is present for currents of any strength and drives a flow that cannot be balanced by pressure but only by friction. For the inviscid case, a singularity results as was shown by Shercliff [39, 41]. EVFs are relevant for a number of industrial processes, e.g., electroslag remelting [42], weld pools [43], and aluminium refining [44]. For a comprehensive overview of EVFs refer to Bojarevičs et al. [41].

Here we look at a much milder case than that considered by Shercliff [39] with comparably gentle gradients of the electric field, see the left contour plot in Fig. 6b. The setup used for the computations is sketched in the inset of Fig. 6b. A cylinder of diameter D and height H filled with a liquid metal of conductivity σ is bounded by current collectors of the same diameter, height h, and conductivity σ_{cc}. Over the circular area with diameter d a constant potential is applied at the outer surfaces of the current collectors. The numerical solution shows in good approximation a linear dependence of the mean velocity $\langle u \rangle$ on the total applied current I (Fig. 6a). This finding is somewhat in line with the observations of Vlasyuk [45] who found that for the EVF in a cylinder the relationship between total current and axial flow velocity is quadratic for $S < 10^3$ and linear for $S > 10^5$, where $S = \mu_0 I^2 4^{-1} \pi^{-2} \rho^{-1} \nu^{-2}$ is a dimensionless parameter [46] characterizing the ratio of the electromagnetic force to the fluids resistance due to viscosity and density [43]. The lowest current shown in Fig. 6a corresponds to $S \sim 10^8$. The mean velocities are in the order of magnitude that is expected for confined EVFs [47].

Since upper and lower current collector are symmetric in the setup of Fig. 6b, so are the electric field and the residual Lorentz force F_{mod} distributions, where F_{mod} is the difference between the Lorentz force distribution resulting from a perfectly uniform current and the actual one. Consequently, a symmetric flow comprising two toroidal vortices develops shortly after switching on the current. However, this flow situation is not stable under the given parameters and is eventually replaced by an irregular flow akin to thermally driven convection. The smaller the current collectors height, the more pronounced are the electric field gradients and the stronger the rotational part of the Lorentz force. The mean velocity increases steeply with decreasing h/D, see Fig. 6b. If the current is above the critical one for TI, a current collector's height to diameter ratio $h/D \gtrsim 3$ is necessary to ensure that the EVF is weak enough to allow the TI to develop for the current densities used here. Already at $h/D = 2$ the base state is so different from that of a fluid at rest that TI does not occur. Instead the initial EVF prevails and saturates into a completely different flow pattern [40].

Strong enough EVFs are indeed able to disrupt the electrolyte layer of LMBs [35]. If the characteristic flow velocity is known, a proper criterion to asses the danger for the electrolyte layer's integrity is the Richardson number

$$\text{Ri} = \frac{g\Delta\rho h_0}{\rho_m u^2} \tag{9}$$

where $\Delta\rho$ is the density difference between metal and electrolyte, h_0 the initial electrolyte thickness, ρ_m the metal density, and u a characteristic velocity of the metal. Ri describes the ratio of the potential energy of stratification E_{pot} to the kinetic energy of the flow E_{kin}. Its critical value is in a good approximation 0.5 corresponding to an equality of potential and kinetic energy [35]. Similar reasoning can be applied to the action of the TI as was done by Herreman et al. [48]. Kelley and Sadoway [49] studied experimentally the interplay of heat driven convection and EVFs in a molten Bi-Pb alloy simulating the cathode of an LMB.

3.3. Interfacial instabilities

By far the largest electrical resistance of the cell is due to the electrolyte layer. Their electrical conductivity is in most cases four orders of magnitudes less than that of the electrode metals. So any small displacement of the electrolyte layer influences the current distribution in the whole cell. The situation is related to that found in aluminium reduction cells (ARCs) and indeed the mechanisms of wave excitation are similar. In Fig. 7a a cylindrical LMB subject to a vertical magnetic field B_z is sketched. Such a field component is typically generated by the feed lines connecting single cells to the current supply or load and is well known from large installations of ARCs. This B_z gives rise to an instability mechanism explained for ARCs by Sele [50]. If one or both of the electrolyte/electrode interfaces are inclined from the horizontal position, the formerly homogeneous current density will redistribute to minimize the total resistance. Due to its low conductivity, current passes nearly vertically through the electrolyte. Horizontally directed currents I_h will appear mainly in the metal layers. These horizontal currents together with the vertical magnetic field B_z generate Lorentz forces pointing in azimuthal direction. The forces give rise to motions that culminate in rotating waves with the help of constant redirections by the cell walls. In ARCs the rotating wave is generated in the bottom layer consisting of pure aluminum. In LMBs Lorentz forces with opposing directions are produced in the anode (top) and in the cathode (bottom) layers. However, for typical material combinations mainly the anode/electrolyte interface will develop waves because of the much lower density difference and therefore weaker restoring gravity force compared to the electrolyte/cathode interface. A typical wave shape is shown in Fig. 7b. The wave's surface is colored with the vertical current density

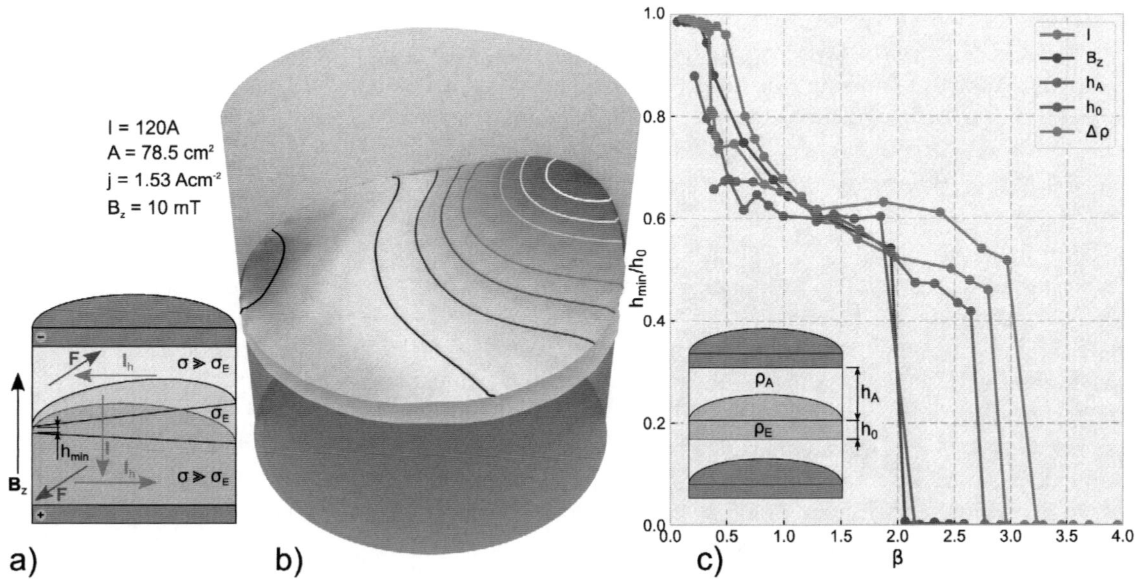

Figure 7. Electromagnetic excitation of gravity waves in LMBs. Mechanism (a, see text for explanation); shape of a rotating wave (b) colored by vertical current density (red: high, blue: low) with contour lines of the electrolyte height; and dimensionless minimal electrolyte height vs. Sele's parameter β (c).

(red: high, blue: low). As expected, the vertical current density is largest where the distance $h_{\min}$ between the electrodes is smallest. Overall, the vertical current density distribution is similar to the level curves shown by solid contours.

The wave's onset as well as the electrolyte layer rupture can be described in good approximation using Seles criterion expressed by the parameter β

$$\beta = \frac{I \cdot B_z}{h_0 h_{\mathrm{A}}(\rho_{\mathrm{E}} - \rho_{\mathrm{A}})g} \tag{10}$$

with h_0 denoting the initial height of the electrolyte, h_{A} the height of the anode metal layer, and ρ_{E} and ρ_{A} the densities of electrolyte and anode metal, respectively. If β, the ratio of Lorentz force to buoyancy, exceeds a certain critical value, sloshing will set in. Please note that cells with circular or quadratic cross sections are always unstable against sloshing, if a vertical magnetic field is present and the damping influence of several dissipation mechanisms is neglected, see Bojarevičs and Romerio [51]. A proper measure against sloshing in ARCs is to use rectangular cross sections with short and long sides.

The numerics used to study sloshing in LMBs [52,53] included ohmic and viscous dissipation, therefore there is a critical value of β beyond which sloshing sets in for the circular cell considered here. As can be seen from Fig. 7c this value is around $\beta_{\mathrm{cr}} \approx 0.35$ for the considered setup. Zikanov [54] used a mechanical analog related to that proposed by Davidson and Lindsay [55] and arrived at a somewhat different formulation.

The ordinate in Fig. 7c displays the minimum electrolyte height $h_{\min}$ normalized by its initial value h_0. Different curves belong to variations of different single parameters noted in the legend. For β above but near β_{cr} a wave develops but approaches a saturated state were $h_{\min}/h_0$ reaches a stationary finite value. Electrolyte rupture happens only later for $\beta_{\mathrm{cr}} \approx 2$. It is obvious from Fig. 7c that β is still a reasonable parameter choice but far from perfect to describe electrolyte layer rupture. This discrepancy might have various reasons, one of them is that for thin electrolyte layers waves of the upper and lower interface start to interact as was shown by Weber et al. [53]. In addition, the transition from an intact electrolyte layer to a ruptured one is rather abrupt changing from $h_{\min}/h_0 \approx 0.4$ to $h_{\min}/h_0 = 0$ in one β-increment. A better description of the electrolyte layer rupture needs further investigation of the processes involved and is underway.

4. Conclusions

Liquid metal batteries are promising candidates for future economical large-scale electricity storage as required for energy systems dominated by volatile sources. While present-day cells are still relatively small, scalability on the cell level is one of the appealing aspects of the concept. Larger cells will show a multitude of flow phenomena that should be properly understood in order to profit from beneficial aspects like mixing and to prevent potentially dangerous features like electrolyte layer ruptures. Molten salt based cells are generally high temperature devices complicating measurements. Using room temperature ionic liquids is a promising way towards low temperature cells, even more so if electrical conductivities of the electrolytes could be increased and degradation phenomena prevented.

Two magnetohydrodynamic instability mechanisms were discussed, the Tayler instability and electromagnetically excited surface waves. The latter are more dangerous for the integrity of the electrolyte layer than the former if a sufficiently strong vertical magnetic field exists. Electro-vortex flows will most likely be present in real life liquid metal batteries since perfectly uniform current distributions are seldomly met in large installations. They should be taken into account as possibly endangering the electrolyte layer's integrity but are good candidates to promote gentle mixing in the cathode as well.

Due to space limitations we had to omit own ongoing work on thermally driven convection but the interested reader may refer to Zikanov and Shen [56, 57] and Köllner et al. [58] for information on this aspect of LMB fluid dynamics.

Acknowledgments

This work was supported by Helmholtz-Gemeinschaft Deutscher Forschungszentren (HGF) in frame of the Helmholtz Alliance Liquid metal technologies (LIMTECH). Computations were performed on the Bull HPC-Cluster "Taurus" at the Center for Information Services and High Performance Computing (ZIH) at TU Dresden and on the cluster "Hydra" at Helmholtz-Zentrum Dresden - Rossendorf. Fruitful discussions with Pascal Beckstein, Valdis Bojarevičs, Vladimir Galindo, Wietze Herreman, Douglas Kelley, Caroline Nore, Takanari Ouchi, Donald Sadoway, and Oleg Zikanov on several aspects of liquid metal batteries and metal pad roll instabilities are gratefully acknowledged.

References

[1] Wittingham M S 2012 *Proc. IEEE* **100** 1518–1534
[2] Nardelli P H J, Rubido N, Wang C, Baptista M S, Pomalaza-Raez C, Cardieri P and Latva-aho M 2014 *Eur. Phys. J. Spec. Top.* **223** 2423–2437
[3] Spatocco B L and Sadoway D R 2015 *Advances in Electrochemical Science and Engineering: Electrochemical Engineering Across Scales: From Molecules to Processes* vol 15 ed Alikre R, Bartlett P and Lipkowsi J (Wiley-VCH) chap 7
[4] Kim H, Boysen D, Newhouse JM Spatocco B, Chung B, Burke P, Bradwell D, Jiang K, Tomaszowska A, Wang K, Wei W, Ortiz L, Barriga S, Poizeau S and Sadoway D 2013 *Chem. Rev.* **113** 2075–2099
[5] Cairns E and Shimotake H 1969 *Science* **164** 1347–1355
[6] Cairns E, Crouthamel C, Fischer A, Foster M, Hesson J, Johnson C, Shimotake H and Tevebaugh A 1967 Galvanic cells with fused-salt electrolytes Tech. Rep. ANL-7316 Argonne National Laboratory
[7] Chum H and Osteryoung R 1980 Review of thermally regenerative electrochemical systems volume 1: Synopsis and executive summary Tech. Rep. SERI/TR-332-416 Solar Energy Research Institute

[8] Chum H and Osteryoung R 1981 Review of thermally regenerative electrochemical systems volume 2 Tech. Rep. SERI/TR-332-416 Solar Energy Research Institute
[9] Weaver R, Smith S and Willmann N 1962 *J. Electrochem. Soc.* **109** 653–657
[10] Agruss B 1963 *J. Electrochem. Soc.* **110** 1097–1103
[11] Agruss B 1963 *Proc. 17th Annual Power Sources Conf.* (Fort Monmouth, NJ) pp 100–103
[12] Swinkels D 1971 *Advances in Molten Salt Chemistry* vol 1 ed Braunstein J, Mamantov G and Smith G (New York: Plenum Press) chap 4, pp 165–223
[13] Bradwell D, Kim H, Sirk A and Sadoway D 2012 *J. Am. Chem. Soc.* **134** 1895–1897
[14] Kim H, Boysen D, Ouchi T and Sadoway D 2013 *J. Power Sources* **241** 239–248
[15] Wang K, Jiang K, Chung B, Ouchi T, Burke P J, Boysen D A Bradwell D J, Kim H, Muecke U and Sadoway D R 2014 *Nature* **514** 348–350
[16] Ning X, Phadke S, Chung B, Yin H, Burke P and Sadoway D R 2015 *J. Power Sources* **275** 370–376
[17] Spatocco B L, Ouchi T, Lambotte G, Burke P J and Sadoway D R 2015 *J. Electrochem. Soc.* **162** A2729–A2736
[18] Ashour R F, Yin H, Ouchi T, Kelley D H and Sadoway D R 2017 *J. Electrochem. Soc.* **164** A535–A537
[19] Armand M, Endres F, MacFarlane D R, Ohno H and Scrosati B 2009 *Nat. Mater.* **8** 621–629
[20] Sakaebe H, Matsumoto H and Tatsumi K 2005 *J. Power Sources* **146** 693–697
[21] Ejigu A, Greatorex-Davies P A and Walsh D A 2015 *Electrochem. Commun.* **54** 55–59
[22] Yaroshevsky A A 2006 *Geochem. Int.* **44** 48–55
[23] Bartlett H E, Neethling A J and Crowther P 1970 *J. Chem. Thermodynamics* **2** 583–590
[24] Itoh M and Kozuka Z 1991 *J. Mater. Sci.* **26** 5221–5228
[25] Janz G J, Allen C B, Downey Jr J R and Tomkins R P T 1978 Physical properties data compilations relevant to energy storage I. molten salts: Eutectic data Tech. Rep. NSRDS-NBS 61, Part 1 National Bureau Of Standards Washington, D.C.
[26] Vogel R C, Levenson M and Masten F R 1964 Chemical engineering division semiannual report Tech. Rep. ANL-6900 Argonne National Laboratory
[27] Lalau C C, Ispas A, Weier T and Bund A 2015 *J. Electrochem. Plating Techn.* 4808
[28] Lalau C C, Dimitrova A, Himmerlich M, Ispas A, Weier T, Krischok S and Bund A 2016 *J. Electrochem. Soc.* **163** A2488–A2493
[29] Rüdiger G, Schultz M and Gellert M 2011 *Astron. Nachr.* **10** 1–7
[30] Seilmayer M, Stefani F, Gundrum T, Weier T, Gerbeth G, Gellert M and Rüdiger G 2012 *Phys. Rev. Lett.* **108** 244501
[31] Weber N, Galindo V, Stefani F and Weier T 2014 *J. Power Sources* **265** 166–173
[32] Vandakurov Y V 1972 *Astron. Zh.* **49** 324–333
[33] Tayler R J 1973 *Mon. Not. R. astr. Soc.* **161** 365–380
[34] Stefani F, Weier T, Gundrum T and Gerbeth G 2011 *Energ. Convers. Manage.* **52** 2982–2986
[35] Stefani F, Galindo V, Kasprzyk C, Landgraf S, Seilmayer M, Starace M, Weber N and Weier T 2016 *IOP Conf. Ser.: Mater. Sci. Eng.* **143** 012024
[36] Weber N, Galindo V, Stefani F, Weier T and Wondrak T 2013 *New J. Phys.* **15** 043034
[37] Weber N, Galindo V, Stefani F and Weier T 2015 *New. J. Phys.* **17** 113013
[38] Krause F and Rädler K H 1980 *Mean-field magnetohydrodynamics and dynamo theory* (Berlin: Akademie-Verlag)
[39] Shercliff J A 1970 *J. Fluid Mech.* **40** 241–250
[40] Weber N, Galindo V, Priede J, Stefani F and Weier T 2015 *Phys. Fluids* **27** 014103
[41] Bojarevičs V, Freibergs Y, Shilova E I and Shcherbinin E V 1989 *Electrically Induced Vortical Flows* (Dordrecht: Kluwer)
[42] Jardy A, Ablitzer D and Wadier J 1991 *Metall. Trans. B* **22B** 111–120
[43] Ajayi O O, Sozou C and Pickering W 1984 *J. Fluid Mech.* **148** 285–300
[44] Frary F C 1925 *Transactions of the American Electrochemical Society* **47** 275–286
[45] Vlasyuk V K 1988 *Magnetohydrodynamics* **23** 434–439
[46] Butsenieks I É, Peterson D E, Sharamkin V I and Shcherbinin É V 1976 *Magnetohydrodynamics* **12** 70–74
[47] Davidson P A 2001 *An Introduction to Magnetohydrodynamics* (Cambridge University Press)
[48] Herreman W, Nore C, Cappanera L and Guermond J L 2015 *J. Fluid Mech.* **771** 79–114
[49] Kelley D H and Sadoway D R 2014 *Phys. Fluids* **26** 057102
[50] Sele T 1977 *Met. Trans. B* **8B** 613–618
[51] Bojarevičs V and Romerio M V 1994 *Eur. J. Mech. B/Fluids* **13** 33–56
[52] Weber N, Beckstein P, Galindo V, Herreman W, Nore C, Stefani F and Weier T 2017 *Magnetohydrodynamics* **53** 129–140
[53] Weber N, Beckstein P, Herreman W, Horstmann G M, Nore C, Stefani F and Weier T 2017 *Phys. Fluids* **29**

054101
[54] Zikanov O 2015 *Phys. Rev. E* **92** 063021
[55] Davidson P A and Lindsay R I 1998 *J. Fluid Mech.* **362** 273–295
[56] Shen Y and Zikanov O 2016 *Theor. Comput. Fluid Dyn.* **30** 275–294
[57] Zikanov O and Shen Y 2016 *10th PAMIR Int. Conf. - Fundamental and Applied MHD* (Cagliary, Italy: DIEE, University of Cagliary) pp 522–526
[58] Köllner T, Boeck T and Schumacher J 2017 *Phys. Rev. E* **95** 053114

IOP Conf. Series: Materials Science and Engineering **228** (2017) 012014 doi:10.1088/1757-899X/228/1/012014

Design and construction of the ATEFA facility for experimental investigations of AMTEC test modules

N Diez de los Rios Ramos[1,4], W Hering[1], A Weisenburger[2], M Stüber[3], A Onea[1], M Lux[1], S Ulrich[3] and R Stieglitz[1]

[1]Institute for Neutron Physics and Reactor Technology (INR), Karlsruhe Institute of Technology (KIT), Hermann-von-Helmholtz-Platz 1, 76344 Eggenstein-Leopoldshafen, Germany
[2]Institute for Pulsed Power and Microwave Technology (IHM), Karlsruhe Institute of Technology (KIT),
[3]Institute for Applied Materials – Applied Materials Physics (IAM-AWP), Karlsruhe Institute of Technology (KIT)
[4]nerea.diez@kit.edu

Abstract. The Alkali Metal Thermal-to-Electric Converter (AMTEC) is an electrochemical cell that requires a high temperature heat source to generate electricity. At KIT the AMTEC technology is being investigated focusing on the use of concentrating solar energy as heat source. First a review on AMTEC technology is given. Further, the design and realization phases of the AMTEC Test Facility (ATEFA) and AMTEC test cell are presented, including the data acquisition and control system and two key technology developments: a ceramic to metal joint for high temperatures (800 – 1000 °C) and the magnetron sputtering of cathode layers on the ceramic electrolyte. The sheet resistance of several electrode samples has been analyzed using the 4-point probe technique and the microstructure of the cathode layer has been examined using the scanning electron microscopy (SEM).

1. Introduction

Liquid metals are possible candidates for expanding the temperature application range in concentrating solar power (CSP) plants and thus leading to an increased efficiency of central tower solar systems. The same objective can be supported by the Alkali Metal Thermal-to-Electric Converter (AMTEC), which operates with sodium in two main modes: liquid- or vapor-anode modes depending on the applied temperature range (up to 800 – 1000 °C).

At Karlsruhe Institute of Technology (KIT) the AMTEC technology is being investigated focusing on the use of concentrating solar energy as heat source of the electrochemical converter. The concept proposed by Hering et al. [1] uses liquid metals such as sodium for a hybrid CSP plant combined with the AMTEC technology. The system should increase the overall efficiency of the power plant compared to conventional CSP systems. As first step towards the present concept, several liquid sodium facilities have been designed and constructed at the Institute for Neutron Physics and Reactor Technology (INR) – KIT [2]; in particular the AMTEC Test Facility (ATEFA) serves for research on single AMTEC test cell characteristics and performance evaluation, as well as process optimization in short term tests. A SOLTEC (SOdium Loop to TEst materials and Corrosion) facility has also been designed at INR-KIT for long term tests on power degradation issues [3].

IOP Conf. Series: Materials Science and Engineering **228** (2017) 012014 doi:10.1088/1757-899X/228/1/012014

Taking into consideration the main developments reported in the literature on AMTEC cells and the still existing challenges, a basic single cell design was proposed based on the work of Heinzel et al. [4]. The aim of the present work was to develop the layout and engineering solutions to construct the AMTEC Test Facility (ATEFA) and AMTEC test cell using new available technologies within the Helmholtz AMTEC Center (HAC).

2. Alkali Metal Thermal to Electric Converter (AMTEC)

AMTEC, originally called Sodium-Heat-Engine, arose in the sixties at the Ford Motor Company as a spinoff of the sodium-sulfur-battery and was further investigated at the Jet Propulsion Laboratory (JPL) [5] focusing on electrical power generation mainly for spacecraft operations. Already in the '90s AMTEC was a promising power conversion device for space missions thanks to its fast development.

In Germany, the group of Heinzel at the Forschungszentrum Karlsruhe (presently KIT) worked in cooperation with the Asea Brown Boveri (ABB) in several AMTEC experimental studies for terrestrial applications, especially for CSP applications. Elaborated tests were made from 1991 until 1993 to analyze different electrodes, current collectors and current leads [6].

Several authors observed that the conductivity of the electrolyte decreased over time, influencing significantly the power output. Lodhi et al. [7] found that during the first 7000 h of operation almost the entire power loss was attributable to the ionic resistance of the ceramic electrolyte. After that, other components like electrodes began to be relevant in the power degradation mechanisms too. At the end of 12000 h, the ionic resistance of the electrolyte represented 58 % of the total cell resistance.

Contrary to the disclosed results, Hunt and Rasmussen [8] reported in 2006 the long term constant performance of a small, multi-tube AMTEC over 5 years of continuous operation. This result shows the first AMTEC cell operated over several years without power degradation. Although no explanation is given by the authors to these positive results, one possible reason is the operation at relative low temperature and very low current density of < 1.4 A/cm^2.

2.1. *Operating principle*

AMTEC is a concentration cell that uses an alkali metal such as sodium and a high temperature heat source to generate DC electricity. The key component is the ß''-alumina solid electrolyte (BASE) that is able to conduct sodium ions while acting as a barrier for electrons. The high temperatures (600 – 1000 °C) in the cell are needed for an improved ion transport in the ceramic electrolyte. The BASE separates the cell into the anode side, where sodium is heated up and pressurized to 0.1 – 0.2 MPa, and the cathode side that contains sodium vapor at much lower pressure (~ 100 Pa) and temperature (< 400 °C).

The pressure difference applied across the electrolyte generates an electrochemical disequilibrium. As a result the system tends to balance it by reducing sodium atoms at the anode surface and further diffusion of sodium ions through the electrolyte to the cathode surface. When connecting electrically both sides of the cell, electrons set free in the sodium reduction travel through an external load from the anode to the cathode generating thereby electricity. After reaching the cathode surface, electrons recombine with sodium ions to give sodium vapor. The redox equation for AMTEC cells is based on the sodium ionization:

$$Na \leftrightarrow Na^{+} + e^{-} \quad (1)$$

Since the oxidation and reduction products in AMTEC are the same, AMTEC cells can be considered a "zero" net fuel consumption technology if the obtained sodium at the cathode is condensed and directed back to the anode for instance by means of an electromagnetic pump or a wick that uses capillary forces.

If no electrical charge exchange between electrodes is allowed, an open circuit voltage (OCV) arises in the system at electrochemical equilibrium. The theoretical OCV for an AMTEC cell is represented by the Nernst equation (2) in terms of the sodium vapor pressure at the anode and cathode, p_a^{oc}, p_c^{oc}.

IOP Conf. Series: Materials Science and Engineering **228** (2017) 012014 doi:10.1088/1757-899X/228/1/012014

$$V^{oc} = \frac{RT_B}{F} \ln\left[p_a^{oc}/p_c^{oc}\right] \tag{2}$$

where T_B represents the temperature of the BASE, R the gas constant and F the Faraday constant. The values obtained with the Nernst equation are considered to be the maximum for the ideal case. Typical measured values are $V^{oc} = 0.6 - 1.2$ V.

The two main issues related to AMTEC devices that require further investigations are the lower real efficiency (~ 20 % at > 800 °C) [9] compared to the theoretical value (~ 45 % at > 800 °C) [6] and power degradation occurring in long time operation due to changes in BASE and electrodes.

2.2. Polarization and cell losses

The isothermal cycle of AMTEC approximates the Carnot cycle and therefore the theoretical efficiency of AMTEC can reach high values of about ~ 45 %. This also means that the temperature difference between BASE and condenser and therefore the heat losses play an important role in a coaxial configuration of the cell (see Figure 5). Heat radiation losses can be reduced for instance by adding radiation shields or by using the condensed molten sodium film on the condenser surface as radiation shield.

Figure 1 shows typical characteristic curve of a coaxial AMTEC cell from an electrochemical point of view and reflects the impact of ohmic cell losses and polarization losses in the electrodes. Polarization losses in the anode are negligible thanks to the good wetting at high temperatures and the good electrical conductivity of liquid sodium. The curve has a rapid initial fall; the main source of voltage drop in this region is the activation polarization losses that appear due to the slow reaction kinetics on the surface of the electrodes. After the activation polarization region, the voltage continues decreasing with a linear slope, characteristic of the region where the increase of ohmic losses predominates. At high current densities the voltage drop increases due to the concentration polarization losses; in this region the diffusion rate of sodium vapor is considerably lower than the reaction kinetics, which leads to an increase of the sodium concentration at the surface of the cathode. Most of the reported AMTEC cells in the literature do not exhibit the concentration polarization region owing to a relative good sodium vapor transport; the linear voltage-drop region prevails even at current densities as high as 2 A/cm² and voltages of ~0.2 V.

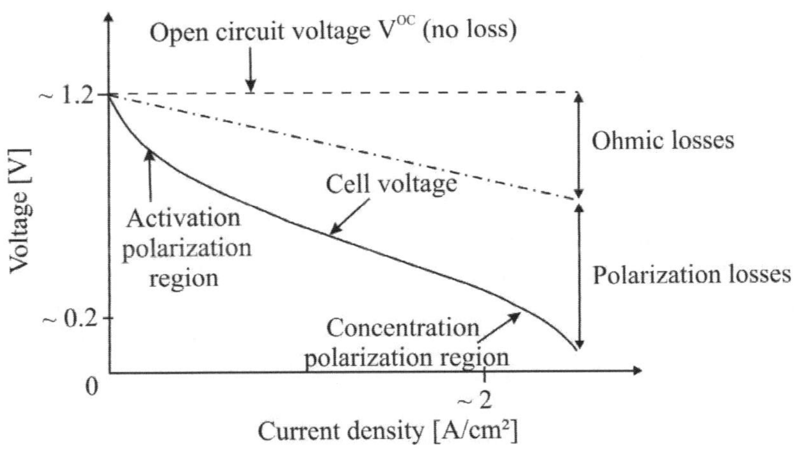

Figure 1. Representation of a typical AMTEC characteristic curve (solid line) and the electrochemical loses for a coaxial configuration of the cell.

The internal resistance of the cell is composed by the ionic resistance of the BASE (B), the electrical resistance of the electrode (e), the current collector (ic), the current leads (l) and the contact resistance (con):

$$R_{cell} = R_B + R_e + R_{ic} + R_l + R_{con} \tag{3}$$

Taking all these losses into consideration, the effective electromotive force of the cell V^{cc} can be calculated as the open-circuit voltage V^{oc} minus the voltage loss originated at anode ξ_a and cathode ξ_c and the ohmic losses ξ_r generated by the internal resistance of the cell:

$$V^{cc} = V^{oc} - (\xi_a - \xi_c) - \xi_r \tag{4}$$

Liquid-anode AMTEC cells have negligible polarization losses in the anode $\xi_a \sim 0$ thanks to the good wetting of liquid sodium at high temperatures. After an elaborated derivation of the Butler-Volmer equation and assuming that interfacial and bulk sodium and sodium-ion concentrations are equal, and that the electrochemical transfer coefficient is ½, the current density at the cathode interface J can be calculated as follows [10]:

$$J = J_0^{ex} \frac{p_B\, e^{-0.5 V^{cc} f} - p_c^{oc}\, e^{0.5 V^{cc} f}}{p_B + K_f J_0^{ex}\, e^{0.5 V^{cc} f}} \tag{5}$$

where $f = F / RT_B$, J_0^{ex} is the exchange current density of the cell at an equilibrium potential obtained with saturated sodium vapor in contact with the electrode, p_B, p_c^{oc} represent respectively the saturated vapor pressure of sodium at T_B and the pressure of sodium vapor at the cathode open circuit conditions, and K_f is the factor of linear pressure increment in the cathode. Clearing the cell potential V^{cc} from equation (5) and combining it with equation (4), the cell potential can be written as [10]:

$$V^{cc} = \frac{2}{f} \ln\left\{ \frac{p_B}{2(J K_f + p_c^{oc})} \left[\frac{-J}{J_0^{ex}} + \left(\left(\frac{J}{J_0^{ex}} \right)^2 + \frac{4(J K_f + p_c^{oc})}{p_B} \right)^{1/2} \right] \right\} - J R_{cell} \tag{6}$$

which reduces to the Nernst equation for J = 0. The right part of the equation (6) corresponds to the ohmic losses ξ_r and the left part of the equation (6) is the expression of the rest of the terms from equation (4): $V^{oc} - (\xi_a - \xi_c)$.

2.3. *Technical implementation*

AMTEC devices have five main components that are relevant for the performance of the cell: BASE electrolyte, electrodes, current collector, the working fluid and the configuration of the cell. The latter will not be addressed in this work, however it should be considered from a thermodynamic point of view for optimized AMTEC cells.

2.3.1. ß''-alumina solid electrolyte (BASE). The BASE electrolyte emerged in the field of sodium batteries and has a complex crystal structure (see figure 2). It consist of spinel-like blocks of Al^{3+} and O^{2-} ions, built with octahedral and tetrahedral prisms, and connected to each other through Al-O-Al bonds. The large oxygen ions create "open" planes between spinel blocks, known as conduction planes, through which the conduction takes place by migration of small sodium cations vacancies. The conductivity is limited to these planes only, since the movement along the perpendicular direction is notably difficult.

There are two main BASE polycrystalline structures that have been used in AMTEC cells: ß- and ß''-alumina. They have a different stacking arrangement of the spinel blocks and conduction planes: ß-alumina has two conduction planes that act simultaneously as mirror planes and ß"-alumina has three conduction planes however these are not mirror planes. The stacking arrangement of the ß''-structure is repeated every three spinel-type blocks, while ß-alumina has a two block structure. This arrangement is responsible for the higher ion conductivity of ß''-alumina. This is the reason why materials with the ß''-structure have been widely used in AMTEC cells rather than ß-structure. Polycrystalline ß''-alumina is a non-stoichiometric ceramic and contains an excess of Na^+ ions. The ß''-structure is stable only in the presence of ions like Li^+ and Mg^{2+}. The substitution of Al^{3+} with these stabilizers (mono- or divalent ions) balances the excessive positive charges of Na^+ and increases the conductivity of the ceramic. Single crystal structures have also been proposed in the literature, due to their higher ion conductivity compared to polycrystalline structures; however the high costs connected with synthesizing single phases make them not competitive against polycrystalline ceramics.

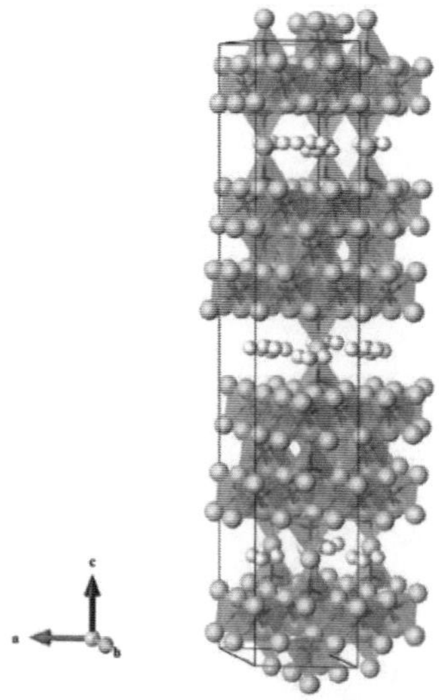

Figure 2. Crystal structure of Na-ß''-alumina. Spinel-like blocks of Al-O are separated by conduction planes, where the Na^+ and O^{2-} ions are located [11].

2.3.2. Electrode. To ensure a high current density and high efficiency, the cathode must satisfy the following properties: i) high electrical conductivity, ii) good adhesion to BASE's surface, iii) similar coefficient of thermal expansion to that of BASE, iv) high corrosion resistance to sodium vapor, v) large triple phase boundary (TPB), vi) low resistance to transport of sodium vapor, and vii) stability at high temperatures (slow grain growth and material migration). Refractory and ceramic electrode materials have been commonly used in AMTEC devices due to their suitable properties [12]; nevertheless their long-term stability at high temperatures needs to be further investigated. Table 1 classifies some of the most relevant literature results on AMTEC cells attending to electrode materials. T_{Na} represents the temperature of sodium in the anode, P_e is the power output of the cell, t_e is the thickness of the electrode and A_e the electrode area. According to the present table, materials with the highest power output seem to be Pt/W, Rh/W and TiC. However an accurate comparison requires considering identical conditions of temperature, electrode surface, current collector structure, operating time, vapor pressure, etc. Therefore further experimental tests should be performed in this field to confirm this statement. In this work TiN and TiC electrodes have been selected for coating development. Despite the lower power density of TiN compared to TiC, it is one of the electrode materials with largest documentation reported in AMTEC literature.

The morphological aspects of the cathode are directly connected with the polarization losses in AMTEC. Table 2 shows the favored grain size, porosity and thickness tendencies to decrease polarization losses. A certain porosity is needed in the cathode for a good sodium vapor transport and thus low concentration polarization; however a too large porosity can affect negatively the activation and ohmic losses. The TPB length would decrease and simultaneously the sheet resistance of the electrode will increase. Therefore an optimum morphology needs to be obtained experimentally for different materials. Detailed experimental data analysis can be found in the literature [13].

Two main factors were identified to describe the properties of different electrodes: the temperature-independent charge-exchange coefficient B_{ex} [$AK^{1/2}/Pa\ m^2$], determined experimentally, represents the ability of the electrode to recombine sodium atoms, and the electrode pressure loss factor G [-] stands for the transport process of vapor sodium in the cathode. An ideal perfect cathode should have $B_{ex} = \infty$

IOP Conf. Series: Materials Science and Engineering **228** (2017) 012014 doi:10.1088/1757-899X/228/1/012014

and $G = 0$, meaning that the length of the TPB is infinite long and the resistance to vapor transport is negligible.

Table 1. Literature review on electrode materials in AMTEC

Electrode	T_{Na} [°C]	P_e [W/cm²]	t_e [µm]	A_e [cm²]	Operation [hours]	Vacuum [Pa]	Ref.
Mo	800	0.5	1	0.5	Few	-	14
	800	0.217	1	30	Few	-	17
	927	0.62	~ 1	5	5	-	18
		0.46	~ 1		90	-	
TiN	687	0.25	5	7	Few	7×10^{-4} (without Na)	16
	746	0.315	-	60	-	-	17
	750	0.075	1	30	Few	-	18
		0.16	5		Few	-	
	850	0.47	15…50	6	Few	-	19
TiN/TiO2	850	~ 0.38	15…50	6	Few	-	
TiC	800	0.11	3…5	0.5…1	-	130	13
		0.4	3…5			0.07	
		0.75	5			0.5	
TiB_2	800	0.24	-	30	10	-	18
W	927	0.35	50	-	-	-	20
Pt/W	827	0.25	-	5	475	1×10^{-5} (without Na)	21
	887	0.8	0.7…1	1…5	160	$7 - 8 \times 10^{4}$	22
Rh/W	817	~ 0.53	1.2	< 10	150	-	5
		0.5	1.2		500		
	857	0.8	1.5…2.5	5	160	7×10^{-4} (without Na)	16
NbN	800	0.12	2…3	-	-	130	13
		0.36	6			0.5	
LaB_6-B	800	0.54	-	-	-	40	23

Table 2. Optimum cathode-morphology factors to reduce polarization losses. Contrary effect of parameters.

	Grain Size	Porosity	Thickness
Activation polarization	Small	Small	-
Ohmic polarization	Large	Small	Large
Diffusion polarization	-	Large	Small

Current collector. A rather small number of studies are reported on the current collector morphology in the literature. Though, it has been shown to have a great impact on the power output of the cell. Knödler et al. [15] showed experimentally that large electrodes exhibit lower power densities than smaller ones. This fact was explained by a poor contact of the current collector and a non-uniform current distribution. To reduce this effect Knödler et al. [15] tested different current collector structures and confirmed the great impact on the power density. They increased power output between 50 – 100 % by improving the current collector structure. This great increase of the power output can be explained by an improved active electrode area. A poor or no contact between current collector and electrode surface in some regions favours the deviation of the current to those other regions where the contact is good due to the lower electric resistance. Therefore the active electrode surface is reduced.

Reducing the thickness of the electrode layer would decrease the losses due to sodium vapor transport; however the sheet resistance and thus the total resistance of the cell would increase too. Therefore an ideal current collector should have an infinite number of contact points with the electrode to compensate the high electric sheet resistance of the electrode. Moreover the current collector should have gradual structure (increasing thickness of the wires and decreasing number of wires until reaching the current lead), made of a single piece in order to eliminate internal contact resistances and simultaneously to have a good porosity to allow a free transport of sodium vapor.

Due to the cylindrical geometry of the ceramic electrolyte, obtaining a good mechanical (and therefore electrical) contact to the electrode surface becomes a challenging issue. Here small tolerances in geometry and tightness between electrode and current collector influence significantly the final contact resistance, which predominates among the electric losses. An experimental study of different current collector structures was carried out, which confirmed large significance of the contact resistance in a cylindrical AMTEC cell. Tests were performed under standard ambient conditions (25 °C, 1 atm, air) and assuming a copper rod as an ideal electrode. Among the tested current collectors, the knitted mesh used to contact directly the electrode seems to be promising thanks to the malleability of the structure that is given independently of the material properties. These results need to be still validated for AMTEC conditions at high temperature and sodium vapor environment.

2.3.3. Alkali metals for AMTEC. Sodium and potassium can be considered the best candidates as working fluid in AMTEC cells due to their favorable physical properties. There is an extensive experience in AMTEC working with sodium as opposed to potassium, which has not been yet widely spread as working fluid especially ought to the difficulties in manufacturing K-BASE electrolyte with common techniques (very high vapor pressure of K_2O at very high sintering temperatures) and to the rather limited database on the use of potassium cells. Nevertheless the interest in K-AMTEC cells is growing since 2000 thanks to the vapor phase method developed by Virkar et al. [24].

The most notable advantages of sodium compared to potassium are: i) the higher surface tension in fluid state supports the capillary process (used to form a radiation shield with liquid sodium), ii) the lower molecular weight that helps also the capillary process and decreases the pressure losses on the cathode side and iii) the ionic resistivity of the BASE is about three times lower than that of the K-BASE at ~ 1100 °C. Potassium has also several advantages: i) the higher vapor pressure (120 kPa at 1050K versus 35 kPa for Na) allows a higher anode vapor-pressure and a lower hot temperature, ii) smaller latent heat of vaporization, which reduces the thermal heat input and lowers source temperature in vapor-anode cells, and iii) lower specific heat capacity of liquid potassium.

In this work sodium was selected based on the following reasons: i) the advantages of sodium properties in AMTEC mentioned above, ii) the excellent properties of sodium as heat transfer fluid for CSP, and iii) the background knowledge available in Na-AMTEC at the institute.

3. Experimental setup.

The main target of the work was the development and construction of the ATEFA and the AMTEC test cell [25, 26]. A detailed description of the facility and test cell is given in this chapter.

3.1. ATEFA

ATEFA is a liquid sodium facility developed for research on AMTEC technology. It was built to operate at a maximum temperature of 1000 °C and a maximum pressure of 1.7 bar during rather short term experiments, limited to a total sodium content of 3 liters. Figure 3 shows the piping and instrumentation diagram of the sodium loop in ATEFA. Sodium is driven by pressurized argon from a first tank (NA-ST-01) into the AMTEC cell; where it is heated up to 800 – 1000 °C inside the ceramic. After the diffusion through the electrolyte, sodium vapor is cooled down on the condenser surface (air cooled loop) and collected in a secondary tank (NA-ST-02). Purification of sodium takes place at the cold trap located in the KASOLA facility at INR-KIT [27].

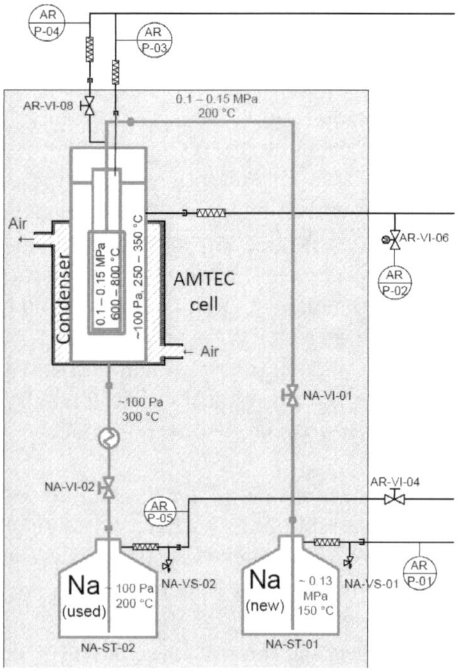

Figure 3. Piping and instrumentation diagramm of the sodium loop in ATEFA.

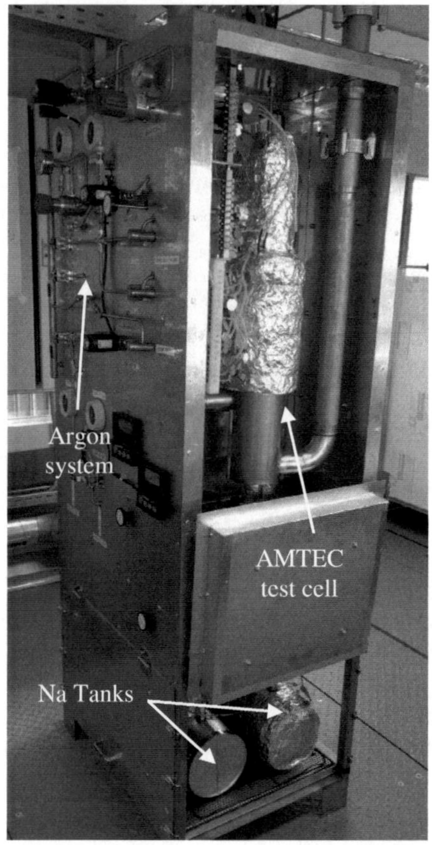

Figure 4. ATEFA with opened containment. Argon system is located outside the containment and sodium system inside the containment.

ATEFA is characterized by a compact construction, which is enclosed in a containment of 1.9 m height and 0.8 m width (see figure 4). The sodium system that includes the AMTEC test cell and the two sodium tanks is insulated through double metallic walls and thus physically separated from the argon system, which is located outside the containment. The argon system serves to control the mass flow rate and pressure of sodium in cell and tanks. All manual valves and displays of the sensors are located on a panel for an easy handling.

During operation, sodium is maintained molten at 150 - 200 °C in tanks and piping thanks to a heat tracing system. The highest temperature of 1000 °C is reached only inside the ceramic electrolyte and is therefore isolated from the rest of the system. Different structural materials were chosen according to the working conditions to which they will be subjected. For the low temperature region, stainless steel EN 1.4571 was selected due to its suitability in corrosive mediums. In the high temperature region of the cell, where corrosion effects are considerably more pronounced, nickel based alloys (EN 2.4663) and ceramics were considered.

Among the safety considered measures in the design and construction phases are a low sodium content of ~ 0.3 liters in the facility (without tanks), a low maximum pressure in the sodium system, the high temperature region restricted only to the ceramic electrolyte and the construction of a physical and thermal barrier between the sodium system and the operators thanks to the metallic containment (see figure 4). At any malfunction or in case of a leakage operational safety is provided through: i) a constant supervision of temperature and pressure in the system, ii) a rapid draining of sodium into the tanks (~10 seconds), iii) the possibility of floating the metallic containment with argon, iv) a ventilation system installed in the laboratory coupled to an oxygen sensor that will activate the suction of aerosols and argon excess out of the laboratory, v) an emergency shutdown of the facility and as ultimate measure, vi) the collection of leaked sodium in a metallic sump located on the bottom of the facility containment that will provide a fast freezing.

3.2. AMTEC test cell

The structure of the cell was designed to allow a simple assembly and disassembly of all parts and defines a favorable handling of each component of the cell (see figures 5 – 7). The central part of the AMTEC test cell is the BASE, which has a tube form (D_{out} = 32 mm) and is closed at the lower end. The cell is divided by the BASE in two main regions: the inner side of the tube is filled up with liquid sodium anode and the outer surface of the tube is coated with a thin porous and electric conductive cathode.

The ceramic electrolyte is held by a niobium transition part welded to the metallic structure of the cell. The cell will be heated up through an electrical resistance. A current lead is submerged in molten sodium and collects the electrons from the anode. Three porous cathodes of ~ 45 cm² were placed on the outer surface of the BASE by reactive d.c. magnetron sputtering. The outer cell containment represents the condenser and is cooled by air.

Titanium carbide (TiC) has been selected in this work as cathode material due to the similar coefficient of thermal expansion with that of the BASE and the good results reported in the literature [8]. Current collectors were used to reduce the sheet resistance of the thin cathodes and spread the electrons homogeneously over the cathode surface. They consist of a nickel foam fastened with nickel porous sheet around the BASE tube.

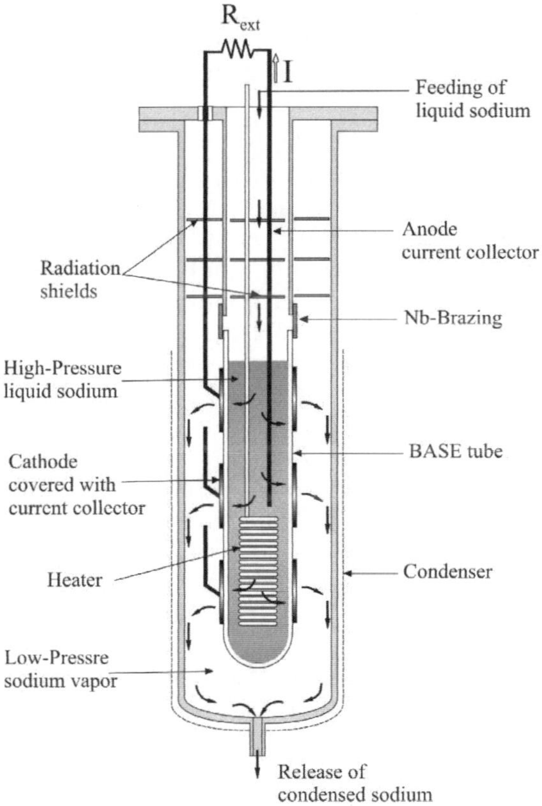

Figure 5. Schematic representation of AMTEC test cell components.

Figure 6. AMTEC test cell with integrated cooling system.

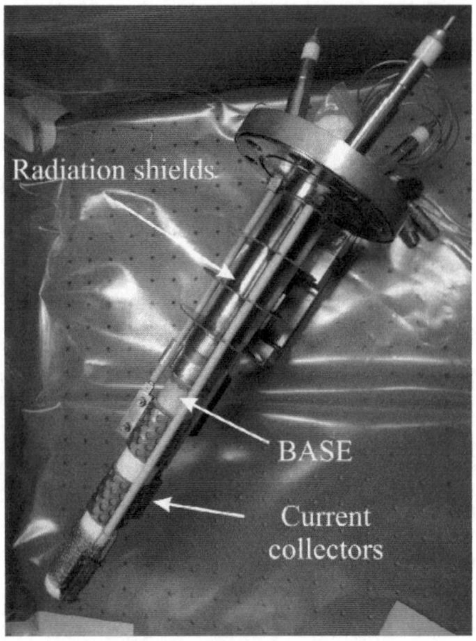

Figure 7. Central flange of the AMTEC test cell with the ceramic electrolyte as main component.

Electrically isolated feedthroughs are provided to lead the current in and out of the cell. Miniature feedthroughs have also been embedded for the thermocouples and heater at the upper flange. These were electrically isolated. Thermocouples were also placed at the condenser and were introduced through the outer cell containment. In the presented prototype BASE and condenser are located coaxially. To reduce radiation losses shielding sheets were placed in both sides of the cell and the inner wall of the condenser was polished.

During the assembly phase of the cell, special attention was paid to the development of a high-temperature ceramic to metal brazing and the coating of the BASE electrolyte. These two issues are presented in chapter 4. Also a thermal and stress evaluation of the assembly was performed using the ANSYS V14.5 software. The main goals were: i) the verification of the cell design for the AMTEC operating conditions and ii) identification of pressure and temperature limits of the structure. Results have been reported in [12] and show that no material failure should occur in the test cell, since all safety factors determined are sufficiently large above unity. The upper brazing between the niobium transition part and the metallic tube was determined to be the weakest component in the cell, due to the difference in the thermal expansion of the materials.

3.3. Data acquisition and control system

All the instrumentation of ATEFA is located in the test cell, sodium system and the immediately attached argon system. For a better evaluation of the temperature distribution in the cell, six different cross section planes were chosen to measure temperature difference in the radial direction, between anode and condenser. In the sodium system several thermocouples were placed to control the heat tracing system of the facility. Four absolute-pressure transmitters and a vacuum transmitter were located in the experimental facility in the argon system.

The cell and each of the tanks and were supplied with level sensors. The miniature level sensor developed for the cell was tested in the glovebox in hot sodium at 250 °C and showed a rapid reaction to sodium level changes and good wetting conditions even at moderate oxide layered free surface.. Moreover tests showed the validity of this sensor even with a moderate sodium oxide layer. Two scales were located below the sodium tanks to measure changes in the weight of the tanks, therefore the transport of sodium between tanks can be supervised. To measure the characteristic curve of the cell an electronic load is used, which adapts the electric circuit for different currents.

ATEFA is equipped with several heaters distributed all over the sodium system. The distribution was established depending on temperature range and also by design, which allows separable cell, valves and tanks. Different heating circuits regulate the power of the heaters in the system depending on the temperature regime and the heat capacity of each part. The total power needed for the heating system of the facility at steady state conditions is about 1.5 kW. The automatic control of the supplied heat power was done in the PC using a PID controller. For the monitoring of the data acquisition and control system a program was developed using LabView software and connected to the TwinCAT3 system Manager of the Beckhoff modules via Automation Device Specification (ADS) communication. In addition, the LabView software allows simulating the operational steps in ATEFA, which helped to optimize the control system.

Figure 8 shows the main states of the facility during operation and the respective transition processes. The facility starts at cold conditions T_{Na}<100 °C (first two states "system on/off") then it is cleaned from air and moisture and heated up to T_{Na} ~ 200 °C ("system clean"). The filling of the cell is first done at moderate temperatures T_{Na} ~ 200 °C ("cell full") and finally pressure and temperature are adjusted to the AMTEC conditions up to 800 - 1000 °C ("cell ready"). Finally the cell is operating after closing the electric circuit ("cell operating"). The draining of the facility is facilitated through pressurized argon and is supervised through temperature and pressure measurements in the facility. During maintenance and before filling up the cell/facility, the AMTEC test cell and sodium tanks can be physically disconnected from the facility for examination or refill after shutting down the system. Several subroutines have been programmed to be executed independently from the state or transition, for instance the constant automatic monitoring of temperature and pressure in the sodium system.

In case of a sodium leakage "emergency shutdown" will be initiated; automatically the complete heating system and the electric circuit of the cell will be switched off; and the outlet valves will start the floating process of the sodium containment with argon, so that any possible sodium fire can be extinguished. Simultaneously the facility will be depressurized.

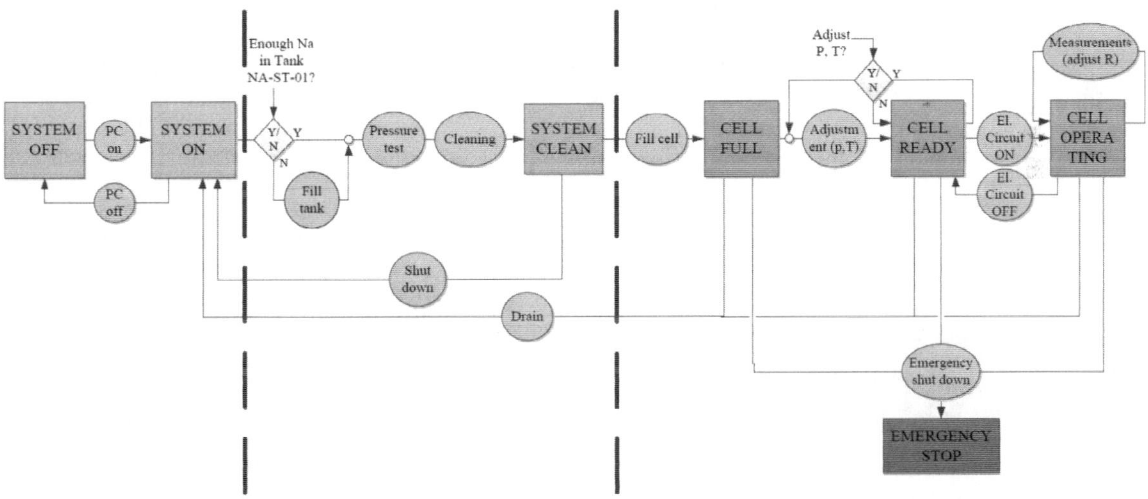

Figure 8. Process diagram of ATEFA. The boxes symbolize the stable states of the facility during operation and the green circles represent the transition processes between the states.

4. Key technology developments

The ceramic to metal joint and the magnetron sputtering of cathode layers on the BASE belong to the main key technology developments in this work.

4.1. Ceramic to metal joint

Active brazing is a one-step process where a specifically designed brazing filler metal is placed directly between two main materials and then heated in vacuum or in an inert atmosphere to avoid oxidation. The brazing filler material contains a reactive element to guarantee a good wetting by chemical reaction with the ceramic. Optimum brazing filler metals must have a low viscosity to ensure a good capillary action and provide a good distribution over the brazing area, a low surface tension to provide a good wetting of the joint surfaces, a reduced surface tension and a low volatilization of alloying elements of the brazing filler metals at brazing temperatures. Further, for a good quality of the brazing an adequate surface preparation; a proper joint design and clearance; and a proper brazing time and temperature are fundamental for a good quality braze.

TiCuNi brazing filler metal was tested by Williams et al. [5] and Knödler et al. [15]. In both cases the brazing was not recommended for high temperature (> 800 °C) AMTEC cells, due to the volatility and high solubility of Cu in Na. In the nineties at KIT (former Forschungszentrum Karlsruhe) Cu was substituted with Nb. Nb helped to reduce the thermal stresses due its thermal expansion coefficient similar to that of BASE. Among the joints tested in literature [5, 15] it was observed that Ti reacted strongly with the BASE. However the processes occurring in the brazing are still not completely understood; therefore further analysis and technology development is needed.

The brazing presented in this work has been developed at the Institute for Pulsed Power and Microwave Technology (IHM) [26]. It consists of a Ni-based brazing filler metal that includes a certain amount of Nb and Ti. The Nb particles act as filler particles in the wide gap and Ti particles were used as active material in the brazing for a good interaction with the ceramic. Tests revealed a strong influence of the gap size on strength and quality of the brazing. Optimization efforts were performed by decreasing significantly the tolerances of the ceramic. A helium leak tightness of 10^{-5} mbar l/s was achieved at the best ceramic-Nb brazing. Optimizations on the brazing are foreseen in ATEFA under AMTEC conditions, where the effects of the high temperature and the corrosion in sodium environment are considered.

4.2. Magnetron sputtering of cathode layers

The magnetron sputtered cathode layer on the BASE component is being developed at the Institute for Applied Materials – Applied Materials Physics (IAM-AWP). Titanium nitride and titanium carbide samples have been obtained using reactive d.c. magnetron sputtering [26]. A picture of a TiC coated BASE ceramic is presented in Figure 9.

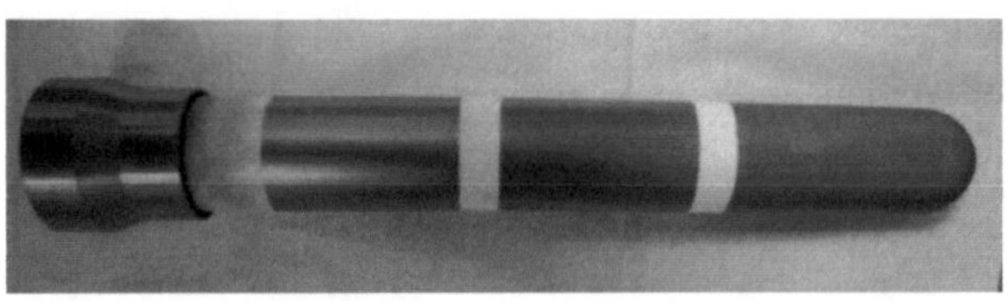

Figure 9. BASE ceramic coated with TiC sputtered layer and connected to the niobium transition tube through a Ni-based braze filler material.

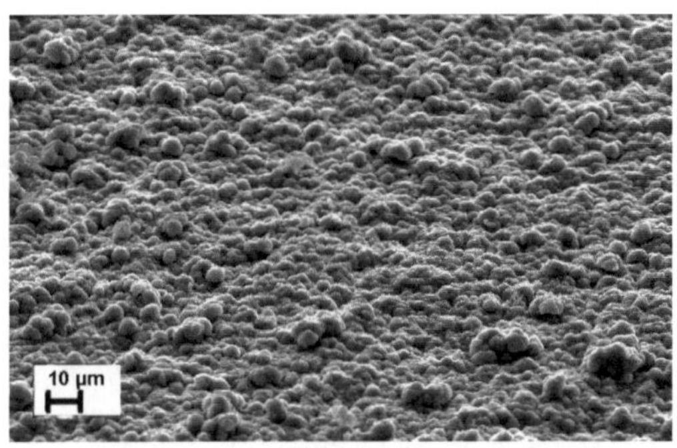

Figure 10. SEM top-view image of the TiC sputtering layer on BASE electrolyte.

Several of the magnetron sputtered electrode layers were analysed using the 4 point probe technique to measure the electrical sheet resistance of the electrode surface by using the correction factor described in [28]. A comparison of one of the TiN coated samples and two of the TiC coated samples developed in this work and the literature can be found in Table 3. As can be observed, results obtained for the TiN coating show a developed sputtering layer with a significant low resistance. The sheet resistance of TiC layers could be reduced by a factor of 10 (TiC-2 compared to TiC-1) mainly by an increase of the layer thickness, which was one of the optimized parameters. The homogeneity of the layer thickness distribution seems to play an essential role considering the large surface roughness of the ceramic.

Despite the larger electrical resistivity of bulk TiC compared to the rest of the analysed materials, it shows very good results in thin sputtered layer configuration. A direct comparison of the sheet resistances between materials should be done taking into consideration the layer thickness and the porosity; unfortunately parameters like porosity are not easy to measure and are therefore rare to be found in the literature.

Some TiC coated BASE samples were analysed using the scanning electron microscope (SEM). Figure 10 shows exemplarily a top-view image of a coated electrode. Some parts of the TiC layer had been detached after several hours/days, therefore the microscopic structure was not perfectly homogeneous. Possible causes of the moderate adhesion may lie on: i) the surface quality of the BASE, ii) pretreatment before the coating (cleaning, brazing, etc.) and/or iii) coating process. The high chemical reactivity of BASE to absorb for instance moisture makes the handling process difficult. Further optimization and analysis is therefore needed.

Table 3. Comparison of measured sheet resistance of AMTEC electrodes for different materials.

Sputter-Layer	ρ_{bulk} (20 °C) [10^{-8} Ωm]	ρ_{sheet} [Ω/□]	t_e [μm]	Ref.
Mo	5.5	11…16	-	16
WPt	W 5.5 Pt 10.3	18.8…98.1	0.7…1	22
WRh	W 5.5 Rh ~5	40.9…92.1	0.7…1	22
NbN	Nb 15	20…100	2…3	29
TiN-1 TiN-2	20	13.8…19.1 17.7…4.1	1.1…1.3 0.9…2.0	30
TiN	20	0.18…0.70	-	This work
TiC-1 TiC-2	60	132…351 21…24	TiC-1 < TiC-2 TiC-2 ≤ 5	

5. Conclusions

The aim of the present work was to present the layout and engineering solutions, the construction of ATEFA and the AMTEC test cell using new technologies available in KIT. The research was focused on AMTEC systems intended to be applied in concentrating solar power plants. The main components of the cell have been addressed and analysed: BASE, electrodes, current collectors and working fluid. In addition technology developments achieved in the field of ceramic-to-metal brazing and electrode coating of BASE ceramics were addressed, including the analysis of the electrode samples.

Acknowledgments
The present work was performed at the Helmholtz AMTEC Center (HAC) initiated within KIT, integrated in the Helmholtz Alliance LIMTECH (Liquid Metal Technology) and financially supported by the Helmholtz Energy Materials Characterization Platform (HEMCP).

References
[1] Hering W, Stieglitz R and Wetzel T 2012 *EPJ Web of Conf.* 33:03003.1–7
[2] Hering W, Diez de los Rios Ramos N, Onea A and Stieglitz R 2016 *Proc.10th PAMIR Int. Conf. (Cagliari)*
[3] Onea A, Hering W, Reiser J, Weisenburger A, Diez de los Rios Ramos N, Lux M, Ziegler R, Baumgärtner S and Stieglitz R 2017 *IOP Conf. Ser.: Mater. Sci. Eng.*
[4] Heinzel V, Huber F, Peppler W and Will H 1991 *J. Mat. Science For.* **76** 261–4
[5] Williams R M, Ryan M A, Jeffries-Nakamura B, Underwood M L, O'Connor D and Kikkert S 1991 *Proc. 8th Symp. on Space Nuclear Power Systems* 464–71
[6] Heinzel V, Huber F, Peppler W and H. Will 1991 *J. Key Eng. Mater.* **59–60** 381–394
[7] Lodhi M A K, Vijayaraghavan P and Daloglu A 2001 *J. Power Sources* **96** 343–51
[8] Hunt T K and Rasmussen J R 2006 *AIP Conf. Proc.* **813** 567–72
[9] Hunt T K 1986 *Proc. Worksh. on Solar Appl.of En. Conv. Tech.* 65-77.
[10] Williams R M, Loveland M E, Jeffries-Nakamura B, Underwood M L, Bankston C P, Leduc H and Kummer J T 1990 *J. Electr. Soc.* **137** 1709-15
[11] Koganei K, Oyama T, Inada M, Enomoto N and Hayshi K 2014 *Solid State Ionics* **267** 22–6
[12] Onea A, Diez de los Rios Ramos N, Hering W, Palacios J L and Stieglitz R 2015 *Magnetohydrodynamics* **51** 249–61
[13] Kato A, Nakata H and Tsuchida K 1993 *Proc. 28th IECEC* **1** 1.819–22
[14] Knödler R, Reiss K and Westhoven B 1992 *J. Mat. Science Lett.* **11** 343–5
[15] Knödler R, Bossmann H P, Kranzmann A and Harbach F 1992 *J. Electr. Soc.* **139** 3029–34
[16] Underwood M L, Williams R M, Jeffries-Nakamura B, Bankston C P and Cole T 1988 *Proc. 23rd IECEC* **1** 227–33
[17] Hunt T 1990 *Proc. 25th IECEC* 420–5
[18] Fang Q and Knödler R 1992 *J. Mat. Sci.* **27** 6725–9
[19] Fletcher R W and Schwank 2003 J W *AIP Conf. Proc.* **654:1** 722–9
[20] Wheeler B L, Williams R M, Jeffries-Nakamura B, Lamb J L, Loveland M E, Bankston C P, and Cole T 1988 *J. App. Electrochemistry* **18** 410–6
[21] Ryan M A, Jeffries-Nakamura B, Williams R M, Underwood M L, O'Connor D and Kikkert S 1991 *Proc. 26th IECEC* **5** 463–8
[22] Williams R M , Jeffries-Nakamura B, Underwood M L, Wheeler B L, Loveland M E, Kikkert S J, Lamb J L, Cole T, Kummer J T and Bankston C P 1989 *J. Electr. Soc.* **136** 893–4
[23] Tsuchida K, Nagata T, Nakata H and Kato A 1998 *J. Mat. Sci.* **33** 755–62
[24] Virkar A V, Jue J F and Fung K Z 2000 *Alkali-metal-ß- and ß''-alumina and gallate polycrystalline ceramics and fabrication by a vapor phase method* Patent (United States) 6,117,807
[25] Diez de los Rios Ramos N, Onea A, Scherrer S, Weisenburger A and Hering W 2015 *Proc. 5th Int. Youth Conf. Energy (Pisa)*
[26] Diez de los Rios Ramos N, Weisenburger A, Onea A, Hering W, Stüber M, Ulrich S and Stieglitz R 2016 *Proc. 10th PAMIR Int. Conf. (Cagliari)*
[27] Onea A, Hering W, Lux M and Stieglitz R 2017 *Int. J. Heat Mass Transfer* **113** 984
[28] Smits F M 1958 *Bell System Tech. J.* **37:3** 711–8
[29] Kato A, Tsuchida K, Hashimoto T, Nakata H and Nagata T 1992 *Proc. 27th IECEC* **3** 3.1–6
[30] Asakami O, Tsuchida K and Kato A. 1990 *J. Mat. Sci. Lett.* **9** 892–4

IOP Conf. Series: Materials Science and Engineering **228** (2017) 012015 doi:10.1088/1757-899X/228/1/012015

Development of high temperature liquid metal test facilities for qualification of materials and investigations of thermoelectrical modules

A Onea[1,4], W Hering[1], J Reiser[2], A Weisenburger[3], N Diez de los Rios Ramos[1], M Lux[1], R Ziegler[2], S Baumgärtner[2] and R Stieglitz[1]

[1]Institute for Neutron Physics and Reactor Technology, Karlsruhe Institute of Technology (KIT), Hermann-von-Helmholtz-Platz 1, 76344 Eggenstein-Leopoldshafen, Germany
[2]Institute for Applied Materials (KIT-IAM),
[3]Institute for Pulsed Power and Microwave Technology (KIT-IHM),
[4]Alexandru.Onea@kit.edu

Abstract. Three classes of experimental liquid metal facilities have been completed during the LIMTECH project aiming the qualification of materials, investigation of thermoelectrical modules, investigation of sodium transitional regimes and fundamental thermo-dynamical flows in concentrating solar power (CSP) relevant geometries. ATEFA facility is dedicated to basic science investigation focussed on the alkali metal thermal-to-electric converter (AMTEC) technology. Three SOLTEC facilities are aimed to be used in different laboratories for long term material investigation sodium environment up to a 1000 K temperature and for long term tests of AMTEC modules. The medium scale integral facility KASOLA is planned as the backbone for CSP development and demonstration.

1. Introduction

Liquid metals have very good thermodynamic properties (large thermal conductivity, large temperature range as liquids), which makes them efficient heat transfer fluids (HTF) for a wide range of heat transfer processes [1, 2]. They are considered to extend the application range of concentrating solar power to temperature up to 1000 °C, the temperature with the maximum efficiency of central tower systems. With that objective, the operation range exceeds single phase liquid metal application of sodium and two phase flows are to be considered. The same is true for the Alkali Metal Thermal-to-Electric Convertor technology, which can operate in two modes focused on the fluid conditions on anode and cathode side: vapor/liquid and vapor/vapor. The AMTEC technology has been proposed as topping cycle for sodium operated CSP facilities [3].

AMTEC cells are thermoelectric devices that directly convert heat in the range 600-1000 °C into DC electricity at total efficiency of ~ 30 % [4, 5]. The technology relies on β''-alumina solid electrolyte (BASE) ceramics to allow the transport of Na ions, while blocking the flow of electrons. Three issues have to be solved: 1. materials to operate at these high temperatures on a long time basis 2. electric isolations and feedthroughs 3. processes to combine different materials so that their noteworthy properties for high temperature applications can be used. Since the proof of concept has been achieved several years ago, the next step was initiated to profit from modern manufacturing developments and to build a new AMTEC prototype.

The present paper aims to give an overview of the liquid metal test facilities erected at KIT within the frame of the LIMTECH project. Financial support for the construction of the facilities has been also provided by the Helmholtz Energy Materials Characterization Platform (HEMCP).

2. New concentration solar power concept

Among all liquid metals, sodium has been identified as probably the most appropriate heat transfer fluid for CSP plants [1, 2]. Compared to other liquid metals and to conventional HTFs such as molten salts, sodium has several favourable properties that make it a very promising heat transfer fluid. It has a large thermal conductivity (68.2 W K^{-1} m^{-1} at 500 °C) that allows a fast conductive thermal transport. Its temperature operating range (98 – 883 °C at atmospheric pressure) as liquid is significantly higher than for conventional HTFs and the low melting temperature ensures lower costs for trace heating. Further, the demands on the pumps are low, since sodium has a low density (834 kg m^{-3} at 500 °C) and a viscosity in the range of water. The main disadvantages of sodium are related to its reactive character with water and air. These issues can be tackled by appropriate safety measures and careful maintenance and procedures in order to completely suppress the sodium-water contact. Such an approach is based on knowledge base in handling sodium as well as on the KIT expertise.

In the above mentioned context, a new concept for a CSP plant that uses sodium as HTF instead of conventional molten salt has been proposed in [3]. Depending on the chosen material for the pipelines of the receiver, the proposed plant can be operated up to temperatures of ~900 °C at the level of the receiver. The fluctuations of the solar energy are compensated by a thermal energy storage device, which can be appropriately dimensioned to allow a continuous operation on a 24/7 basis. The basis loop of the plant operates below ~550 °C, therefore conventional steels such as AISI 316Ti can be employed. The peaks in the solar energy occurring in the range 600 – 900 °C are planned to be used by a cluster of thermoelectrical converters that can convert directly the heat into DC electricity.

3. Experimental capabilities established within Helmholtz AMTEC Center (HAC)

Such a concept raises several issues that need appropriate and dedicated investigation. Namely, several major tasks were identified:

- Experimental investigation of fundamental thermo-dynamical aspects of sodium flows in relevant geometries (e.g. flows in a CSP receiver, flows in channels with sudden expansion).
- Generation of an experimental database for sodium flows to qualify and improve the anisotropic heat flux and momentum models in simulation codes.
- Investigation and qualification of appropriate materials for high temperature applications and pipelines able to endure such a high sodium temperature level under steady-state and transient conditions (e.g. creep fatigue investigations).
- Stress-corrosion cracking analysis as well as creep fatigue evaluation of the materials to be performed in hot sodium environment on a long time basis and under isothermal and temperature gradient conditions.
- Investigation of corrosion/erosion rate of high temperature steels in hot sodium environment under steady-state and transient conditions.
- Material qualification for thermoelectric converters in hot sodium environment.
- Short-term and long-term investigations of thermoelectric devices to assess the performance behavior under different operating conditions, as well as on a long time basis.

For such a wide range of experimental investigations, several experimental facilities have been developed as summarized in table 1.

3.1. ATEFA facility

The AMTEC TEst FAcility (ATEFA) [6, 7] is intended for short-term tests of single AMTEC cells and targets the investigation of the cell performances, the coating materials and of the ceramic-metal brazing. Temperatures up to 1000 °C can be achieved locally in the ceramic at a maximal overpressure of 0.5 bar. Sodium is directed by pressurized argon from a tank into the test cell, where it is heated up at the required test temperature. After diffusion through the ceramic, the sodium is condensed and collected in a second tank. Taking into consideration the low amount of sodium of only ~3 liters, the facility can operate for approx. 36 h until the tanks can be exchanged. The argon cover gas controls the system pressure and the sodium flow rate and is divided in a supply branch and a pressure branch. The first branch controls the sodium flow towards the cell, while the second one controls the pressure in the cell.

Table 1. Overview of the liquid metal experimental facilities erected at KIT.

Facility	Description and experimental tasks	Fluid/ Amount	Facility size	Operating conditions (Max. temperature, max. gauge pressure)
ATEFA	Small scale facility for short term AMTEC single cell tests	Na (~ 3 liter)	0.8 ×1.0×2.0 m^3	< 1000 °C, < ~0.5 bar
SOLTEC-1	Small scale facility for material qualification (low cycle fatigue; long term tests)	Na (~ 14 liter)	1.2×1.6×1.9 m^3	< 720 °C, < ~3.5 bar
SOLTEC-2	Small scale facility for investigation of steel corrosion in hot sodium (long time tests)	Na (~ 14 liter)	1.2×1.6×1.9 m^3	< 720 °C, < ~3.5 bar
SOLTEC-3	Small scale facility for long time tests of modular AMTEC cells	Na (~ 14 liter)	1.2×1.6×1.9 m^3	< 950 °C, < ~2.5 bar
KASOLA	Versatile large scale infrastructure	Na (7 tonnes)	Ø 7.6 m Height 12 m	< 550 °C, < ~3 bar

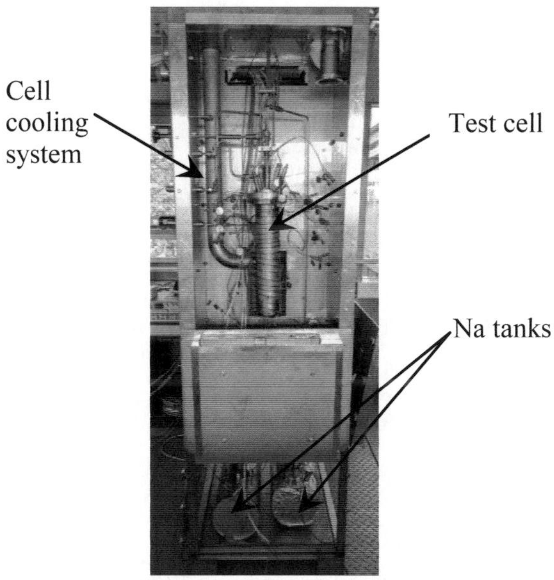

Figure 1. Frontal view of the ATEFA facility.

The sodium side consists of the sodium pipeline, the test cell, a sodium-air spiral heat exchanger and two sodium tanks. The entire sodium side is in-housed in a thermally insulated metallic frame separated from the argon system and the control system. Additional information about the test facility is reported in [6] and [7].

An overview and analysis of the candidate materials for the coating of the BASE (beta"-alumina solid electrolyte) ceramic is reported in [8]. Based on this analysis, the coatings performed up to now are molybdenum (Mo), titanium nitride (TiN) and titanium carbide (TiC) [6].

3.2. SOLTEC facilities

For the SOLTEC (SOdium Loop to TEst materials and Corrosion) facilities it was decided that three independent facilities will be developed, each having its own test section. The SOLTEC-1 and -2 are similar facilities for material investigations having a maximal operating temperature of 720 °C at maximal 3.5 bar overpressure. SOLTEC-3 loop is planned for long term tests of thermoelectric converters to be operated at maximal 950 °C up to maximal 2.5 bar overpressure. While the differences between SOLTEC-1 and SOLTEC-2 are limited to the test section and operating conditions, the SOLTEC-3 facility has a different configuration than the other two loops due to the different test conditions, see Figures 2 and 3.

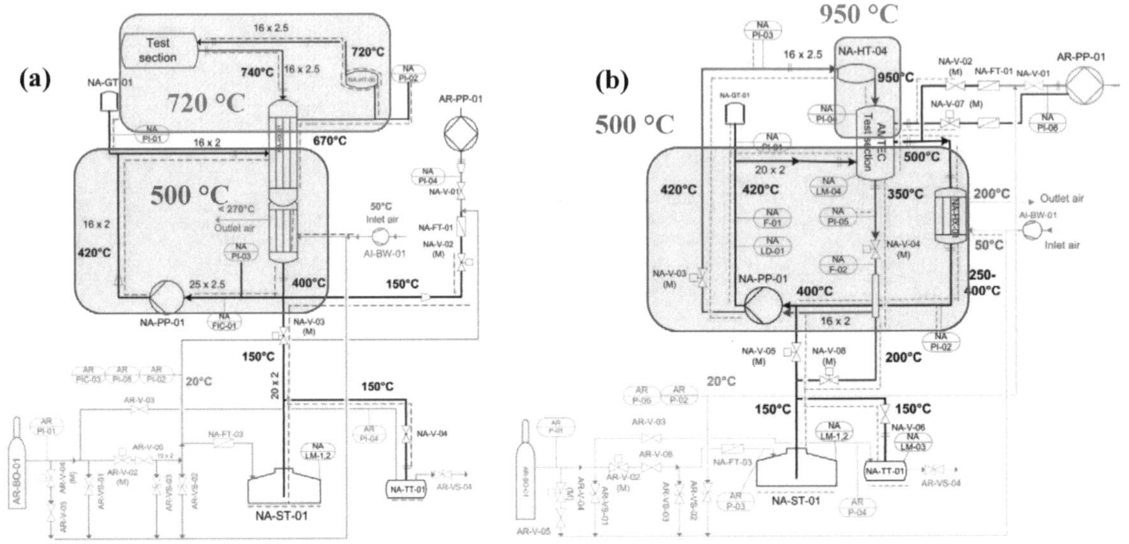

Figure 2. Piping and instrumentation diagrams of SOLTEC-1,-2 (a) and SOLTEC-3 (b).

For all facilities, a very compact construction is considered, as presented in table 1. For SOLTEC-1 and -2, the test sections are placed outside of the facility, while SOLTEC-3 integrates the test section within the framework. All three facilities have a high temperature side, where the test section is located. The main components of the loops are located in the low temperature side, where temperatures of up to ~500 °C can be reached, as highlighted in the piping and instrumentation diagrams in Figure 2. Due to the high temperature level nickel-based superalloys are used for the high temperature side, while for the low temperature side stainless steel AISI 316Ti is used. At the maximal sodium mass flow rate planned, i.e. 300 kg/h, the sodium velocity can reach 0.31 m/s in the low temperature side and 1.1 m/s in the high temperature side. The maximal velocity occurs in the test sample in SOLTEC-1, where velocities up to 4.8 m/s can be reached. For SOLTEC-1 and -2 the total pressure loss in the loop at maximal flow rate is estimated to be about 1.2 bar. All facilities have the same pump type, namely a 3 kW permanent magnet pump able to operate up to 450 °C. Argon is used as cover gas to protect sodium against oxidation and to facilitate the filling and drainage of the facility. All facilities have their own vacuum pump used to extract air and moisture to avoid their potential reaction with sodium. All major components (valves, pumps) have been placed in the low temperature side of the loop, where the thermal stresses are lower.

For SOLTEC-1 and -2 a combined 27 kW Na-Na heat recuperator and 7.5 kW Na-air heat exchanger has been placed at the interface between the high temperature side and the low temperature side. This combined heat exchanger provides a compact and efficient solution to recuperate most of the sodium heat after the test section. For the SOLTEC-3 loop only a Na-air heat exchanger has been considered. Due to the expected temperature level, the Na-Na heat recuperator will be manufactured from Inconel steel, while standard stainless steel AISI 316Ti is used for the Na-air heat exchanger.

The safety concept is based on the experience gained in the design of other sodium loops [9, 10] in KIT and relies on several constructive and operational measures. To ensure a safe and fast drainage of the sodium from the hot side in less than 30 s, the sodium inventory in the loop is low (restricted for SOLTEC-1 and -2 to about 12 liters) and all relevant sodium valves have a normally open configuration. In case of a leakage or test sample rupture occurring in the test section, several measures were foreseen to automatically detect the situation and sodium will be immediately released in the storage tank. No water based systems are allowed in the loop and possible fire scenarios can occur only due to the contact of hot sodium to air, e.g. in case of cracks occurring in the pipelines. In this case, the leak sensor placed on the pipelines will detect the sodium leaving the crack and initiate the drainage. Before leaving the thermal insulation most of the sodium can be collected inside the outer hull. Therefore, only a small amount of sodium can access the encapsulated atmosphere of the metallic housing of the facility. In the event of cracks occurring in the storage tank, sodium can drain in a collection tray integrated in the bottom part of the framework of the housing. In this case, the associated fire and smoke will be limited within the insulated metallic housing. As an operational safety measure, the loops will be operated at low overpressure.

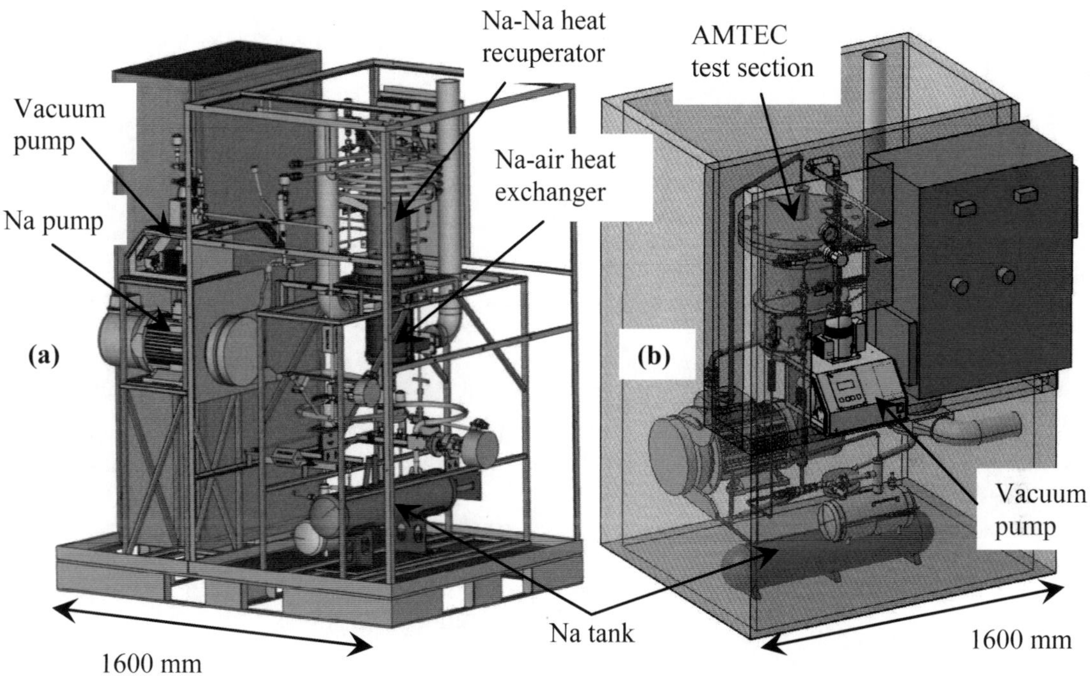

Figure 3. 3D-models of SOLTEC-1 and -2 (a) and SOLTEC-3 (b).

The main operating modes of the facilities are filling, normal operation, drainage and emergency. The argon side of the loop will be used to drive the sodium from the 4 liter transport tank (NA-TT-01) into the 14 liter storage tank (NA-ST-01), see Figure 2. Prior to the filling of the loop with sodium, the facility will be evacuated using a vacuum pump (AR-PP-01). The evacuation will be performed with the sodium side heated at about 150-200 °C, so that besides air, moisture rests can be also extracted. Once the loop has been evacuated, the sodium will be driven by pressurized argon in the loop, which is moderately heated at about 200 °C. As a difference to other sodium loops, the present facility does not contain a dedicated expansion tank. Such a tank should be placed in the highest location of the facility, i.e. in the high temperature side. In this zone no components are desired, in order to avoid the constructive complex measures associated to the connections that have to be welded in this high temperature side. Instead, the storage tank is used as an expansion tank in the following way: once the loop has been filled with sodium, it will be heated up to the operational temperature required. Since the connection between the loop and the storage tank will be kept open, the sodium will expand in the storage tank, where the pressure can be controlled by adjusting the argon pressure (green pipelines in Figure 2). Once the loop has reached its desired operating temperature, the valve situated at the interface between the loop and the storage tank (NA-V-04 in Figure 2a and NA-V-05 in Figure 2b) can be closed. In this way the loop is constructively simplified and the operational safety is significantly increased. A small gas bubble trap is considered in the low temperature side.

The in situ low cycle fatigue investigation tests will be performed in a Zwick/Roell Z100 universal traction facility that has been installed in the rotunda building that houses the sodium loop KASOLA facility at KIT, as displayed in Figure 4. In the traction facility traction and compressive forces up to 50 kN can be measured at a cycle frequency of max. 1000 Hz. A Maytec vacuum oven has been installed in the facility, where test samples can be heated up and tested at temperatures above 1000 °C in vacuum down to about 10^{-5} mbar. Test samples up to about 10 cm in length and 6 cm in diameter can be tested. Among the materials planned to be investigated are conventional steels, such as AISI 316Ti, 1.4988 and 1.4970. Furthermore, new materials for high temperature applications, such as advanced PM2000, innovative W-Cu laminates and ceramic-metal joints for advanced thermoelectrical modules are planned for investigations.

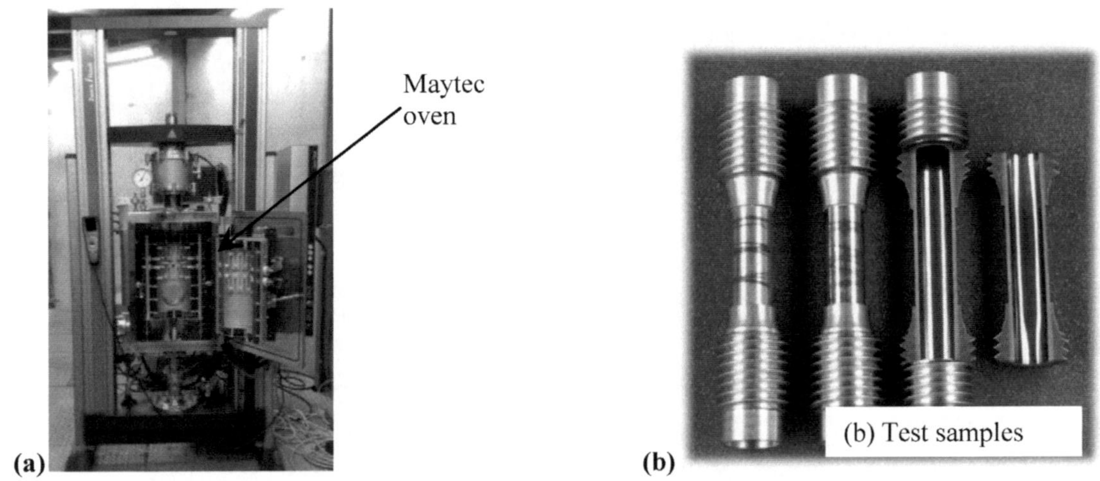

Figure 4. (a) Zwick/Roell universal traction facility for low cycle fatigue tests (SOLTEC-1 loop); (b) W-Cu samples for high temperature applications.

Prototypic samples of tungsten-copper laminates planned to be investigated in SOLTEC-1 in flowing hot sodium are displayed in Figure 4b. The aim is to employ them in high temperature receivers for CSP, as proposed in [11]. Tungsten has the highest melting temperature of all metals in pure form and is an excellent absorber material, therefore it is one of the best candidates for high temperature applications. However, tungsten has low fracture toughness at room temperature and high brittle-to-ductile transition temperature. Therefore, it cannot be used directly as structural material. As reported in [12], the tungsten-copper laminates have the potential to by-pass these drawbacks by cold working the tungsten, significantly lowering the temperature of the brittle-to-ductile transition below ~125 °C. Hence, pipes made of tungsten laminates can extend the operational range of chrome steels and nickel-based superalloys, to meet the requirements of future CSP plants.

Although some experimental data are available regarding corrosion of austenitic materials in hot sodium [13], there is a lack of experimental data regarding the effect of temperature excursions on structural materials at such high temperatures as planned in SOLTEC.

While data are available for temperatures up to 550 °C [14, 15], for SOLTEC-2 one of the tasks is to generate reliable long term data based on a set of experiments in sodium at temperatures above

650 °C. The second task addresses thermal cycling of receiver tubes that in real operation are caused by day/night and cloud shadowing. For the investigations of steel corrosion in sodium environment at constant temperature and thermal cycling, the test section contains a U-type pipe made of materials to be investigated, which will be connected to the upper part of the SOLTEC-2 facility. During the experiments hot Na leaving the loop with up to 720 °C will flow through the tube sample and will be heated by an inductive heater. The inductive heater cooled by a thermal oil to minimize any risk is placed in a vacuum vessel (see Figure 5a). The oxygen and moisture content in the vacuum chamber will be permanently surveyed and if required adjusted. A pressure sensor controls any pressure increase, which can result from a leakage of the vacuum vessel or by a rupture of the U-type tube specimen, and gives signal for fast draining of the loop and the shutdown of all heaters.

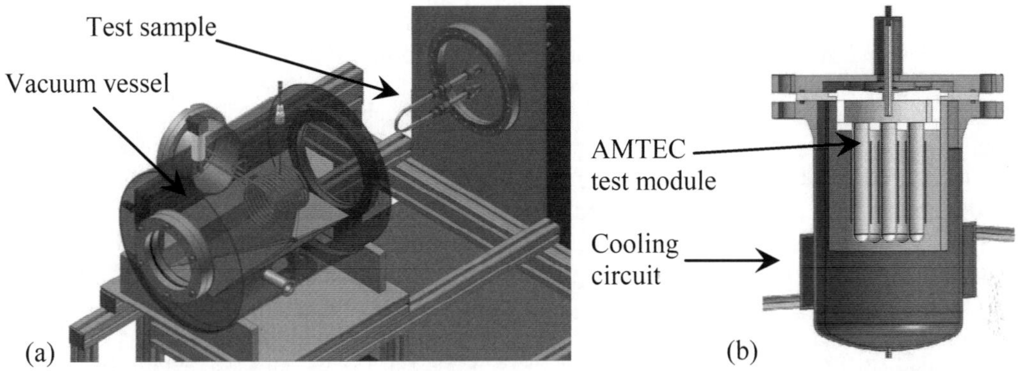

Figure 5. (a) High temperature test chamber for investigations of new steels in SOLTEC-2 loop (b) Universal test section for investigations of different AMTEC modules in SOLTEC-3 loop.

Among the materials planned for investigation are austenitic steels with variable chrome content, nickel-based superalloys, Inconel-based superalloys and W-Cu laminate pipes previously mentioned. The probes will be analysed by scanning electron microscope (SEM) method and the material surface will be characterized using a profilometer.

Long term investigations of AMTEC modules in the range of weeks are planned in SOLTEC-3. The loop has a universal test section (see Figure 5b) that can incorporate modules of different configuration with a limited number of parts to be replaced at every module test. Temperatures up to 950 °C in the module can be sustained, due to the use of ceramic parts. For SOLTEC-3, the facility consists of two independent loops, the test loop being characterized by high temperature and low mass flow rate, while the cooling loop has a lower temperature ranging between 250 – 450 °C and larger mass flow rate. The cooling loop is used for the cooling of the AMTEC test chamber.

The design of the facility and the numerical calculations for the layout of the components and of the facilities have been performed by KIT. The loops are presently in the final stage of construction by the firms SAAS GmbH and Sowec GmbH & Co. KG, Germany. The first firm developed the instrumentation and control technology, while the second firm performed the mechanical and thermo-dynamical calculations of the heat exchangers, optimized the 3D-model and manufactured the main components. The set-into-operation phase is planned for the mid 2017.

3.3. Integral facility KASOLA

The KArlsruhe SOdium LAboratory experimental research facility is a versatile sodium loop, able to embed test sections up to 6 m height. It contains a base loop with a purification system that incorporates a cold trap with integrated heat recuperator [16], the test ports for experiments and the 7m^3 sodium storage tank situated in a neighboring dedicated building. Argon is used as cover gas to protect sodium from oxidation, to maintain the pressure balance in the loop and to facilitate the filling and drainage of the loop. The entire facility is located in a three floor steel containment, approximately

12 m in height. The maximum operational temperature foreseen is 550 °C. A 75 kW magneto hydrodynamic annular linear induction pump provides a maximum liquid sodium mass-flow rate of app. 150 m^3/h within the base loop. Almost all main components of the base loop are gathered in the machine room, while the test section is located above the MHD pump.
The scientific program of the experiments foreseen in the KASOLA facility is reported in [17], while additional information about the facility can be found in [7]. The numerical investigations of the sodium flows in KASOLA performed with TRACE are reported in [18-20].

To assess the applicability of a sodium operated thermal energy storage (TES) tank, a prototype sodium TES is foreseen to be investigated in KASOLA. The thermocline tank considered has a total volume of about 500 liters and a thermal energy storage capacity of 50 kWh. It contains a floatable plate that separates sodium in cold and hot regions, avoiding therefore heat losses by conduction and convection. The proposed design can be appropriately dimensioned to meet the requirements set in [3], in order to allow overnight plant operation and cooling capabilities for the AMTEC cluster.

4. Summary

The present paper gives an overview of the experimental facilities erected at KIT in the frame of the Helmholtz Alliance on Liquid Metal Technology LIMTECH. The construction of the facilities was financially supported by the Helmholtz Energy Materials Characterization Platform.

The ATEFA facility is used for basic short-term investigations of thermoelectrical converters. The SOLTEC loops are planned to be used for long term investigations and qualification of innovative materials and steels in hot sodium environment for high temperature applications as well tests of innovative thermoelectrical converters. Among the materials planned for investigations are W-Cu compounds, advanced PM2000, austenitic steels and nickel-based superalloys. The research is motivated by the need of material development for high temperature applications and by the lack of experimental data for in-situ experiments of creep fatigue in hot sodium. The versatile medium size KASOLA facility will integrate prototypical components for future CSP plants to test a scaled CSP demonstrator.

Acknowledgment

The present work was performed in the frame of the Helmholtz AMTEC Center (HAC) at KIT. The authors acknowledge the financial support received from the Helmholtz Alliance on Liquid Metal Technology (LIMTECH) and the Helmholtz Energy Material Characterization Platform (HEMCP).

References

[1] Heinzel A et al. 2017 *Energy Technol.* **5** 1
[2] Onea A, Perez-Martin S, Jäger W, Hering W and Stieglitz R 2017 Liquid metals as heat transfer fluids for science and technology *Advances in new heat transfer fluids*, ed A. Minea CRC Press, Taylor & Francis Group chapter 12 pp 305 - 376
[3] Hering W, Stieglitz R and Wetzel T 2012 *EPJ Web of Conferences* 33:03003.1–03003.7.
[4] Tanaka K 2001 *Heat Transfer – Asian Research* **30** 234
[5] Tournier J-M and El-Genk M 2003, *Space Techn. Appl. Int. Forum – STAIF 2003*
[6] Diez d.l. Rios Ramos N, Weisenburger A, Onea A, Hering W, Stüber M, Ulrich S and Stieglitz R 2016 *Proc. 10th PAMIR Int. Conf.*
[7] Diez d. l. Rios Ramos N, Hering W, Weisenburger A, Stüber M, Onea A, Ulrich S and Stieglitz R, 2017 *IOP Conf. Ser.: Mater. Sci. Eng.*
[8] Onea A, Diez N, Hering W, Palacios J and Stieglitz R 2016 *Magnetohydrodynamics* **51** 249
[9] Onea A, Hering W, Homann C, Jianu A, Lux M, Scherrer S and Stieglitz R 2013 *Proc. 15th Int. Topical Meeting on Nucl. Reactor Thermal-Hydraulics NURETH-15* (Pisa, Italy)
[10] Diez N, Onea A, Scherrer S, Weisenburger A and Hering W 2015 *Proc. 5th Int. Youth Conf. Energy* (Pisa, Italy)

[11] Reiser J et al. 2015 *Advanced Eng. Mat.* **17** 491
[12] Reiser J, Hoffmann J, Jäntsch U, Klimenkov M, Bonk S, Bonnekoh C, Rieth M, Hoffmann A and Mrotzek T 2016 *Int. J. Refractory Metals and Hard Mat.* **54** 351
[13] Borgstedt H, Frees G and Jesper H 1989 *Werkstoffe und Korrosion* **40** 525
[14] Weisenburger A et al. 2011 *J. Nucl. Mat.* **415** 260
[15] Weisenburger A, Jianu A, An W, Fetzer R, Del Giacco M, Heinzel A, Müller G, Markov V and Kasthanov A 2012 *J. Nucl. Mat.* **431** 77
[16] Onea A, Hering W, Lux M and Stieglitz R 2017 *Int. J. Heat Mass Transfer* **113** 984
[17] Hering W, Stieglitz R, Jianu A, Martin L, Onea A, Scherrer S and Homann C 2013 *Proc. Fast Reactors Related Fuel Cycles: Safe Techn. Sust. Scenarios FR 13* (Paris, France)
[18] Jäger W, Hering W, Lux M and Portes F 2015 *Proc. Int. Conf. on Nuclear Engineering* (Chiba)
[19] Jäger W, Trimborn F, Hering W, Pritz B and Gabi M 2016 *Proc. Int. Congr. On Advances in Nuclear Power Plants* (San Francisco)
[20] Jäger W, Hering W, Stieglitz R Schum T, Frohnapfel B, Niemann M and Fröhlich J 2017 Thermohydraulic flow in a sudden expansion, *IOP Conf. Ser.: Mater. Sci. Eng.*

IOP Conf. Series: Materials Science and Engineering **228** (2017) 012016 doi:10.1088/1757-899X/228/1/012016

Carbon dioxide free production of hydrogen

L Stoppel[1], T Fehling[2], T Geißler[1], E Baake[2], T Wetzel[1]

[1] Karlsruhe Institute of Technology, IKET, Karlsruhe Liquid metal Laboratory (KALLA), Hermann-von-Helmholtz-Platz 1, 76344 Eggenstein-Leopoldshafen, German

[2] Leibniz University of Hannover, Institute of Electrotechnology, Wilhelm-Busch-Strasse 4, 30167 Hannover, Germany

Corresponding author e-mail: thomas.wetzel@kit.edu

Abstract. The present report summarizes the theoretical modelling and experimental investigation results of the study on the direct thermal methane cracking. This work is a part of the LIMTECH-Project (Liquid Metal Technologies) funded of Helmholtz Alliance and was carried out from 2012 to 2017. The Project-part B5 "CO_2-free production of hydrogen" focused on experimental testing and particularly on modelling the novel methane cracking method based on liquid metal technology. The new method uses a bubble column reactor, filled with liquid metal, where both the chemical reaction of methane decomposition and the separation of gas fraction from solid carbon occur. Such reactor system was designed and built in the liquid metal laboratory (KALLA) at KIT. The influences of liquid metal temperature distribution in reactor and feed gas flow rate on methane conversion ratio were investigated experimentally at the temperature range from 930°C to 1175 °C and methane flow rate at the reactor inlet from 50 to 200 mL_n/min. In parallel with experimental investigations, a thermochemical model, giving insight in the influence of the above mentioned parameters has been developed at KIT and a CFD model was developed at LUH to get an overview about the bubble dynamics in the reaction system. The influence of different bubble sizes and shapes, multi-inlet coalescence effects as well as the potential of electromagnetic stirring have been investigated.

1. Introduction

The carbon dioxide emission (CO_2) from human activities influences the earth's ecosystem significantly. The worldwide economic growth and development plays a key role in the upward trend in CO_2 emission. As a result, there are many ongoing research works on development of efficient technologies that will be able to reduce the CO_2 emission. One of the technologies, whose usage led to producing a considerable amount of carbon dioxide, is the steam methane reforming (SMR). Depending on the feedstock, this technology provides production of 9 to 12 tons of CO_2 as a by-product per one ton of produced hydrogen. In the future, current SMR process might be replaced by other environment-friendly hydrogen production technologies based on carbon free sources. However, from the economic point of view, it is not to expect that in the near future hydrogen will be completely produced from renewable and sustainable resources. The proposed direct methane cracking method can in relatively short time provide an attractive industrial solution that leads to the reducing of global carbon dioxide emission while using fossil feedstock, i.e. natural gas.

The chemical reaction of the direct methane cracking can be described by an overall endothermic reaction:

$$CH_4(g) \rightarrow C(s) + 2H_2(g) \qquad \Delta H_R^0 = +74.85 \frac{kJ}{mol} \tag{1}$$

The reaction products are gaseous hydrogen and solid elemental carbon. The energy requirement for this chemical reaction is moderate, but a considerable shift of the thermodynamic equilibrium in direction of the reaction products is only achievable beyond 800°C (1 atm). At 1200°C, the theoretical methane conversion efficiency is about 95%.

The direct thermal decomposition of methane is not a new idea, see [2-4] for further reference. In the past, a lot of research works were dedicated to the investigation and realizing this process experimentally. The main problem in the practical implementation was the very short operating time of the developed systems. Indeed, the produced carbon is deposited in the reaction zone and this causes the blockage of the reactor. In comparison to previous approaches, the main advantage of the suggested method is the continuous separation and transport of the produced solid carbon from the reaction zone. It occurs owing to a difference of density between liquid metal and carbon and it is supported by the rising gas bubbles. After reaching the liquid metal surface, the carbon powder can be removed from the system with available conventional methods used in the industry (skimming, gas floatation, etc.).

Due to the opaqueness of the liquid metal, visualization of bubble dynamics in the reactor is hardly possible. Numerical simulations have therefore been done in order to investigate dependencies of bubble residence times, bubble shapes, potential carbon sediment locations etc. on operation parameters like gas flow rate. To identify potential optimizations, reactor geometry, inlet types and the heating system have been varied. Especially in this article the aspect of multi-inlets as well as the effect of electromagnetic stirring with regards to increase of hydrogen yield is put in the focus.

2. Basic process description and experimental set-up

The simplified process flow diagram is shown in Figure 1. The methane cracking reaction occurs in a vertical bubble column reactor filled with liquid metal. Natural gas or pure methane can be used as the feed gas. It is injected at the bottom of the column reactor. The gas bubbles formed at the injector rise to the top of the reactor. Thereby, methane is heated up to liquid metal temperature level and split into gaseous hydrogen and solid carbon.

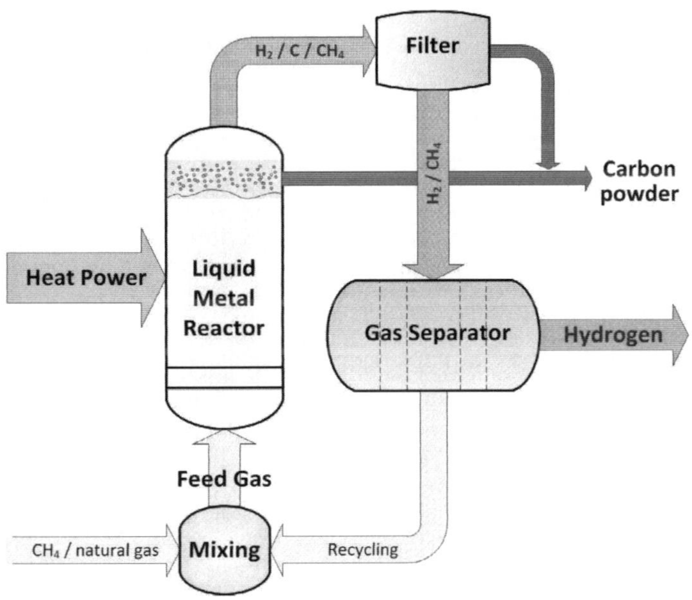

Figure 1. Process flow diagram [1].

The separation of produced hydrogen from the methane-hydrogen mixture takes place in a separator. The unreacted methane and other possible gaseous by-products can be mixed with the feed gas via mixing unit and moved again to the reactor inlet. Most of the solid carbon formed in the cracking reaction collects on the liquid metal free surface and can be removed from the reactor relatively easy. The filter installed within the outlet gas line holds back residual, gas-borne carbon particles.

The construction of experimental test facility HELiS (Hydrogen production Experiment in Liquid Sn) has been part of this project and has been operated at the KArlsruhe Liquid metal LAboratory (KALLA) of KIT. In particular, the facility consists of a gas supply system, an experimental port for the liquid metal reactor, an outlet gas analysis system and an instrumentation and control system. The gas supply system is designed for feeding the working gas in required mixing proportion (methane/argon or nitrogen) into the chemical reactor. It consists of a manifold of standard cylinders with commercial high purity argon at pressure of 200 atm, pipelines, pressure gauges, and gas flow sensors. The use of liquid tin (Sn) as the working fluid in the bubble column reactor has been based upon a careful selection process along several criteria like inert behaviour towards the reaction gases. Major advantages are of course the high thermal conductivity, high density compared to carbon, non-toxicity and long-term chemical stability. The strong corrosion attack of liquid tin at high temperatures towards metallic, particularly steel container materials poses special challenges, which will be detailed in the following.

During the study, two types of the reactor design have been developed and tested: stainless steel reactor and combined quartz glass-steel reactor designs. In general, stainless steel is the favoured construction material due to the great design flexibility and the availability of connectors and flanges even for high temperature applications. However, in contact with molten tin, stainless steel alloying elements are dissolved increasingly with increasing temperature. In the experimental setup, 316 and 1.4549 steels were used as reactor materials. In both cases, experiments showed that materials are corroded quite fast. The lifetime of these reactors was insufficient. In comparison with stainless steel, the quartz glass is chemically stable and showed no corrosion impact towards tin within the planned temperature range. The main idea of this reactor design concept is to couple the advantages of the chemically stable glass material and the mechanical stability of stainless steel. The non-corrosive quartz glass tube getting in contact with liquid metal is inserted into the steel tube. This design proved reliable during the experimental campaigns [2, 3, 4, 5].

In detail, the chemical reactor used in the experiments is a vertical quartz glass tube with inner diameter of 40.6 mm and length of 1268 mm, which is inserted into a stainless steel tube with inner diameter of 49.25 mm and length of 1150 mm (Fig. 2). A single orifice (Ø 0.5 mm), at the bottom of the quartz glass tube was used as gas injector. The combined reactor was placed inside of an 8 kW electric furnace with a maximum operating temperature of 1200 °C. The upper part of the reactor, so called gas zone, is ending above the furnace and was insulated separately. The temperature inside of the reactor was continuously measured during the operating time by applying an alumina tube with type K thermocouples at 10 positions along the whole reactor length. The temperature measured at a level of 600 mm in the middle of the reaction zone was used as reference temperature. Glass fragments or cylindrical rings made of quartz glass have been used to form packed beds. The space porosity of the packed beds was varied from 76% to 84%. The separate supply tank placed above the rector was used for melting of tin prior to filling up the reactor. The gaseous reaction products were filtered via sinter metal elements made of stainless steel with pore size 0.5 μm. The analysis of product gas was realized by using a gas chromatograph (GC). During the experiments, the mole-fraction was continuously measured for methane, hydrogen, nitrogen, ethane, ethylene and acetylene at various operating conditions.

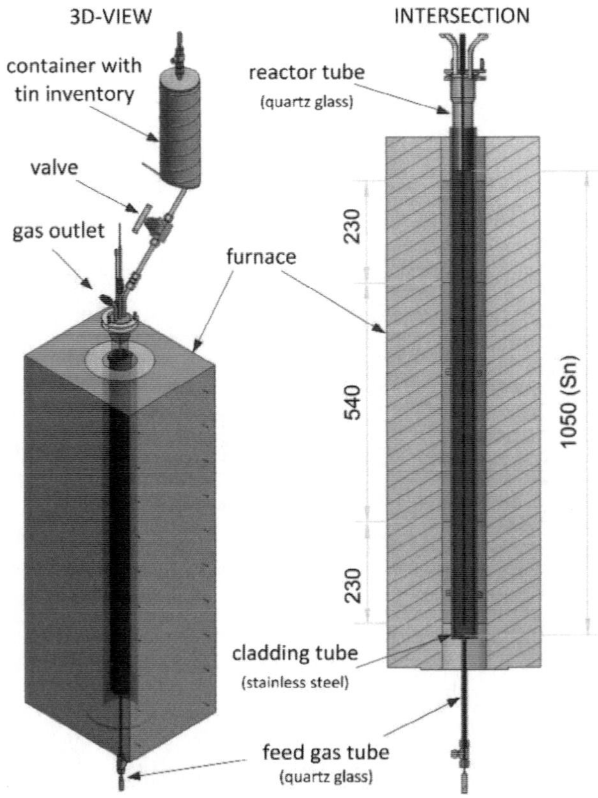

Figure 2. Experimental set-up.

3. Thermo-chemical modelling and experimental results

The following section describes the thermo-chemical modelling and the experimental results and extends our previous work [3].

3.1. Thermo-chemical modelling

To estimate the influence of different parameters on the process as well as for future process description and hydrogen yield predictions, a thermo-chemical (TC) model was developed. The model is based on three coupled one dimensional partial differential equations with spherical coordinates, solved using MATLAB® parabolic PDE solver. The energy equation (3), the species equation (4) and a pressure equation (5) are solved inside of a bubble traveling through the liquid metal reactor. The bubble is assumed as a spherical body with a constant radius and a constant residence time in the reactor depending on the inventory material and height. As the bubble radius is kept constant, volume expansion due to the pressure drop and the chemical reaction along the liquid metal reactor length is not included in the model. The initial bubble radius in equation (2) is calculated by Tate´s law, which bases on a force balance between buoyancy and surface tension [6] using temperature dependent surface tension and density from Alchagirov et al. [7], [8].

$$r_{B,0} = \left(\frac{3 d_{orf} \sigma_0}{2 \rho_0 g} \right)^{1/3} \tag{2}$$

where d_{orf} is the diameter of the orifice, σ_o and ρ_o are the surface tension and density respectively.

In the thermo-chemical model, all gas properties are functions of the present local temperature, pressure and gas composition. The source terms in equation (3) and (4) imply temperature dependent reaction rates, following Arrhenius law and assuming a first order chemical reaction.

Energy equation

$$\rho_G c_{p,G}\frac{\partial T}{\partial t}=\frac{1}{r^2}\frac{\partial}{\partial r}\left(r^2\lambda_G\frac{\partial T}{\partial r}\right)-\frac{k_0\Delta H_R p\, y_i}{R_g T}\exp\left(-\frac{E_a}{R_g T}\right) \tag{3}$$

where ρ_G is the gas density, $c_{p,G}$ is the specific heat capacity of the gas, T is the temperature, r is the bubble radius, λ_G is the heat conductivity of the gas, k_0 is the pre-exponential factor, ΔH_R is the reaction enthalpy, p is the absolute pressure, R_g is the universal gas constant, y_i is the mole fraction, E_a is the activation energy.

Species equation

$$\frac{p}{T}\frac{\partial y_i}{\partial t}=\frac{1}{r^2}\frac{\partial}{\partial r}\left(r^2\frac{pD_{i,j}}{T}\frac{\partial y_i}{\partial r}\right)-\frac{k_0 y_i p}{T}\exp\left(-\frac{E_a}{R_g T}\right), \tag{4}$$

where D is the diffusion coefficient.

Pressure equation

$$\frac{\mathrm{d}p}{\mathrm{d}t}=-\frac{(p_0-p_1)}{\tau_G}, \tag{5}$$

where τ_G is the residence time.

Assuming linear pressure drop from the bottom to the top of the reactor, whereby the local liquid metal hydrostatic pressure in the column equals the pressure inside of the bubble. The hydrostatic pressure is calculated using the temperature dependent tin density equation from Alchagirov et al. [8].

The feed gas conditions at the bottom of the reactor are used as the initial conditions in equation (3) – (5).

$$T=T_0; \quad y_i=0; \quad p=p_0 \tag{6}$$

Symmetry boundary conditions are set at the center of the bubble for the bubble temperature, the species fraction and the pressure (equation (10)). At the bubble interface, symmetry boundary conditions are set for the species fraction and the pressure (equation (11)). As the bubble is rising in the liquid metal due to density differences, and assuming the reference coordinate system fixed to the travelling bubble, the liquid metal is flowing around the bubble. Assuming further, that gas convection inside of the bubble can be neglected and the bubble behaves like a solid sphere, the following Nusselt correlation for forced convection over a sphere from Melissari et. al [9] can be used to model heat transfer from the tin to the bubble. The correlation is valid in a wide range of Pr and Re numbers, including the present Pr number for tin (Pr = 0.0031) at the lowest investigated temperature.

$$Nu=2+0.47Re^{0.5}Pr^{0.36}, \tag{7}$$

with

$$Re=\frac{u_B d_B}{\nu_{LM}}, \tag{8}$$

where u_B is the bubble velocity, d_B is the bubble diameter, ν_{LM} is the kinematic viscosity of the liquid metal.

Even at low Re numbers, resulting from small bubble rise velocities and diameters, the Nu number always leads to a value of 2 or higher. Evaluating the heat transfer limitation, the Bi number in equation (9) reveals a minimum value of 520 for the highest investigated temperature

$$Bi = \frac{Nu\lambda_{Sn}}{\lambda_{CH_4}}, \tag{9}$$

where λ_{Sn} is the heat conductivity of tin and λ_{CH_4} is the heat conductivity of methane.

Consequently the heat transfer limitation is on the gas phase side and the interfacial bubble temperature is set to the local liquid metal temperature and changes with time, and thus with the liquid metal height.

Inside of the bubble ($at\ r = 0$), the conditions are defined as follows

$$\frac{\partial T}{\partial r} = 0; \quad \frac{\partial y_i}{\partial r} = 0; \quad \frac{\mathrm{d}p}{\mathrm{d}r} = 0 \tag{10}$$

On the gas liquid interface of the bubble ($at\ r = r_B$), the conditions are defined as follows

$$T = T_{LM}(t); \quad \frac{\partial y_i}{\partial r} = 0; \quad \frac{\mathrm{d}p}{\mathrm{d}r} = 0 \tag{11}$$

To estimate the performance of the process using the measured gas mole fractions from the experiments, the hydrogen yield, the methane conversion and the ratio between the produced hydrogen and the reacted methane are investigated. As the product gas analysis in the GC only considers the components in the gas phase, without taking into account the produced solid carbon, basic calculations to clarify the difference between the measured mole fractions in the GC and the mole fractions of the components, considering carbon production, are necessary. For pure methane as feed gas, assuming the behaviour of an ideal gas, the initial total molar flow rate is given by

$$\dot{N}_{total,0} = \dot{N}_{CH_4,0} = \frac{p_0 \dot{V}_{CH_4,0}}{R_g T_0}, \tag{12}$$

where $\dot{N}_{CH_4,0}$ is an initial molar flow rate of methane, $\dot{V}_{CH_4,0}$ is an initial volume flow rate of methane. With the definition of the methane conversion

$$X_{CH_4,1} = \frac{\dot{N}_{CH_4,0} - \dot{N}_{CH_4,1}}{\dot{N}_{CH_4,0}} \tag{13}$$

the molar flow rates of the different product gas components, considering no intermediates or by-products, are as follows

$$\dot{N}_{H_2,1} = 2X_{CH_4,1}\dot{N}_{total,0} \tag{14}$$

$$\dot{N}_{CH_4,1} = \dot{N}_{total,0}\left(1 - X_{CH_4,1}\right) \tag{15}$$

$$\dot{N}_{C,1} = X_{CH_4,1}\dot{N}_{total,0} \tag{16}$$

Using Equations (14) - (16), the total molar flow rate at the outlet of the reactor is calculated by

$$\dot{N}_{total,1} = \dot{N}_{total,0}\left(1 + 2X_{CH_4,1}\right) \tag{17}$$

To estimate the output mole fraction of each component the following term is valid

$$y_{i,1} = \frac{\dot{N}_{i,1}}{\dot{N}_{total,1}} \tag{18}$$

Usually, given by mass conservation, the sum of all components present in the chemical reaction leave the reactor. In this process, the carbon stays inside of reactor, which reduces the total mass and consequently the total molar flow of the product gas. This leads to a new total molar flow entering the GC, changing the mole fraction of each component in the product gas, without taking the carbon mole fraction into account. With the new GC molar flow rate, the calculation of the methane conversion, using GC mole fraction data and Equations (14) - (18), result in

$$X_{CH_4} = \frac{1 - y_{CH_4,GC}}{1 + y_{CH_4,GC}}, \tag{19}$$

whereby the hydrogen yield is given by

$$Y_{H_2,CH_4} = \frac{\dot{N}_{H_2,1}}{2\dot{N}_{CH_4,0}} = \frac{y_{H_2,GC}\left(1 + X_{CH_4}\right)}{2} \tag{20}$$

Besides the hydrogen yield, the ratio between the produced hydrogen and the reacted methane is mandatory to estimate possible deviations due to the formation of intermediates

$$\frac{\dot{N}_{H_2,prod}}{2\dot{N}_{CH_4,reac}} = \frac{y_{H_2,GC}\left(1 + X_{CH_4}\right)}{1 - y_{CH_4,GC}\left(1 + X_{CH_4}\right)} \tag{21}$$

This ratio leads to a value of 2, in case of all the methane in the feed gas converts to hydrogen and carbon without the formation of intermediates.

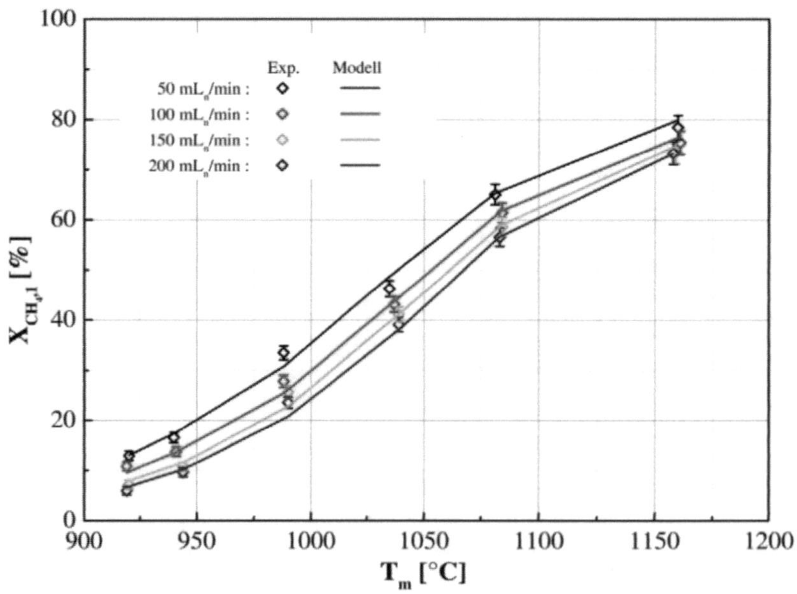

Figure 3. Measured and calculated methane conversion as a function of the average temperature [2].

3.2. Comparison of calculation and measurements

As mentioned above, to quantify the chemical process of the direct methane cracking in liquid metal, several experimental campaigns were performed at the liquid metal laboratory (KALLA). The hydrogen yield measured by the gas chromatograph was used as an indicative parameter to estimate the potential of the process. At the experiments, special focus was given on the influence of the feed gas volume flow rate at different liquid metal temperatures and on the reactor clogging due to carbon formation.

The figure 3 shows the experimental results of methane conversion compared to calculated data as a function of the average liquid metal temperature in reactor at different feed methane volume flow rates. The model prediction showed very good agreement with the experiments.

Especially the liquid metal temperature has a significant influence on the resulting methane conversion, whereby the dependency of the applied methane volume flow rate is quite moderate. The influence of the gas residence time in the liquid metal reactor is only slightly visible by changing the methane feed volume flow rate as the packed bed itself enhances the residence time significantly compared to a reactor design without a packed bed. Nevertheless, applying a pure methane volume flow rate of 200 mL_n/min at 1175 °C results in an almost 4 times higher overall hydrogen standard volume flow rate at the outlet compared to a hydrogen output applying a 50 mL_n/min initial methane feed volume flow rate.

During the experiments, almost no intermediate reaction products occurred, apart from a maximum mole fraction of 0.5 mol-% ethane and 1.0 mol-% ethylene in the product gas could be detected, whereby no acetylene was measured. As a consequence, the hydrogen selectivity in this process is almost one, leading to a potential product portfolio of only hydrogen and solid carbon.

While disassembling the quartz glass reactor, a thin carbon layer with a thickness in the range of 10 μm was found between the inventory and the quartz glass reactor wall. In the upper part of the reactor, above the metal surface, a mixture of the packed bed and the produced carbon, appearing as a powder,

Figure 4. Carbon powder produced during the experiment [1].

was found (Figure 4). By means of the observed amount, most of the produced carbon accumulated in this area. A smaller amount, transported by the product gas, was collected in the filter elements between the off-gas tube and the GC.

IOP Conf. Series: Materials Science and Engineering **228** (2017) 012016 doi:10.1088/1757-899X/228/1/012016

4. Numerical investigation of fluid dynamic effects within the reactor

To get an idea about the bubble dynamics inside the reactor system of the HELiS facility and to evaluate potential optimization opportunities, additionally to the experiments numerical simulations have been done. According to an efficient production, the dynamics and residence times for different bubble diameters as well as optimization methods like electromagnetic stirring or multi-inlet usages were discovered. Simulations were done using the open source software package *OpenFOAM*.

4.1 Phase description models for numerical bubble simulation

To describe the physics of a two phase bubble flow and to get a numerical description of the dynamics, generally two different methods are available. The discrete phase models for particle tracking based on *Euler-Lagrange* description and *Euler-Euler* approaches with a Eulerian description based on an inertial observer formulation for both phases. The Euler-Euler-approaches can be also subdivided into Volume of Fluid, mixture and Euler methods. Prior simulations, especially magnetohydrodynamic modeling have been done using Volume of Fluid method. In the Volume of Fluid method, a volume fraction equation is solved to calculate volume-fraction-averaged parameter that is used to solve the momentum equation. A detailed explanation can be found in [10].

In actual simulations the Eulerian model, described in [11], is used. A solver for coupling this method also with electromagnetic calculations in the software *getDP* has been developed. One advantage is a more exact calculation for greater bubble volume fractions and greater Stokes numbers, especially at the surface and for different velocities $\vec{v}$ of the phases. In contrast to the Volume of Fluid model the Eulerian model solves a separated set of equations for every phase. The coupling between the phases is reached by pressure- and phase-exchange coefficients. The continuity equation, given by

$$\frac{\partial \alpha_q}{\partial t} + \vec{v} \cdot \nabla \alpha_q = \frac{1}{\rho_q}\left(\sum_{p=1}^{n} \dot{m}_{pq} - \dot{m}_{qp} \right) + S_q \tag{22}$$

is used to solve the volume fraction α_q in the momentum equation for every fluid phase:

$$\begin{aligned} &\frac{\partial}{\partial t}\left(\alpha_q \rho_q \vec{v}_q\right) + \nabla \cdot \left(\alpha_q \rho_q \vec{v}_q \vec{v}_q\right) = -\alpha_q \nabla p + \nabla \cdot \bar{\bar{\tau}}_q + \alpha_q \rho_q \vec{g} + \\ &\alpha_q \rho_q \left(\vec{F}_q + \vec{F}_{lift,q} + \vec{F}_{vm,q}\right) + \sum_{p=1}^{n}\left(K_{pq}\left(\vec{v}_p - \vec{v}_q\right) + \dot{m}_{pq} \vec{v}_{pq}\right) \end{aligned} \tag{23}$$

Here ρ_q describes the density, ∇p the pressure gradient and $\dot{m}_{pq}$ the mass transfer between phase p and q. The qth phase stress tensor $\bar{\bar{\tau}}_q$ is a function of phase fraction and velocity. External forces $\vec{F}_q$, lift forces $\vec{F}_{lift,q}$ and virtual mass forces $\vec{F}_{vm,q}$ can be calculated by correlations described in [12]. The phase exchange coefficient K_{pq} between the dispersed phase q (methane) and continuous phase p (tin) is a function of drag and will be calculated by a *Schiller-Naumann-Correlation* [13] in the following studies. For the description of energy transfer between the two phases, a separate enthalpy equation has to be solved for each phase:

$$\frac{\partial}{\partial t}\left(\alpha_q \rho_q h_q\right) + \nabla \cdot \left(\alpha_q \rho_q \vec{v}_q h_q\right) = \alpha_q \frac{\partial p_q}{\partial t} + \bar{\bar{\tau}}_q \div \nabla \vec{v}_q - \nabla \cdot \vec{q}_q + S_q + \sum_{p=1}^{n}\left(Q_{pq} + \dot{m}_{pq} h_{pq}\right) \tag{24}$$

The parameter h_q describes the enthalpy, S_q includes possible source terms. In addition, it includes the heat flux Q_{pq} that is a function of heat transfer coefficient between the phases. It is dependent on the Nusselt number. In the following studies the correlation from *Ranz and Marshall* [14] is used for its description. The fluid dynamics in the system methane and tin are turbulent. To describe the turbulent flow structures in the system a *mixture-k-epsilon-model* from Behzadi et. al [15] is used. The main idea of this model is based on the assumption that turbulence is dictated by the continuous phase, so turbulent kinetic energy and dissipation for each phase can be linked with constants.

4.2 Bubble size studies and slug flow dynamics

In the following studies simulation results of bubble flow in reactor using an Eulerian model will be described. In this section bubble residence times will be analyzed as a function of bubble diameters. As well as the system temperature the residence time is also coupled with the hydrogen yield and offers a potential for optimization. In these models, a tin temperature of 900°C is assumed. Varying the temperature would also influence the rates of ascents. Generally, it has been shown that carbon sediments occur also at walls near the bottom of the reactor, so a greater range from 1 mm bubbles up to bubble sizes near the reactor diameter were discovered (see figure 5.b-d.). The maximum possible bubble size due to reactor diameter is limited to 40 mm. The results of the diameter studies are shownin figure 6. According to experimental results with the system air/water, a spherical bubble shape occurs at smaller bubble diameters, which results in lower rates of ascent and respectively higher residence times. For the spherical shape, rates of ascents increase with bubble diameter until the point of 7 mm. For greater diameters, oscillations occur that results in lower bubble velocities. Further increasing the bubble diameter leads to bubble shape change due to greater Eötvös-number that increases the velocity again. For bubble diameters greater than 20 mm another effect develops. As it can be seen in figure 5.b, stronger wall effects occur, especially at the lower part of the reactor, these effects also lead to bubble dispersion. This is also a possible explanation for carbon sediments occurrence at the reactor walls near the lower part of the reactor and the described velocity reduction. After dispersion effects at the lower part of the reactor, coalescence occur after a short time, so a single bubble leaves the melt after the described residence time (figure 5.c).

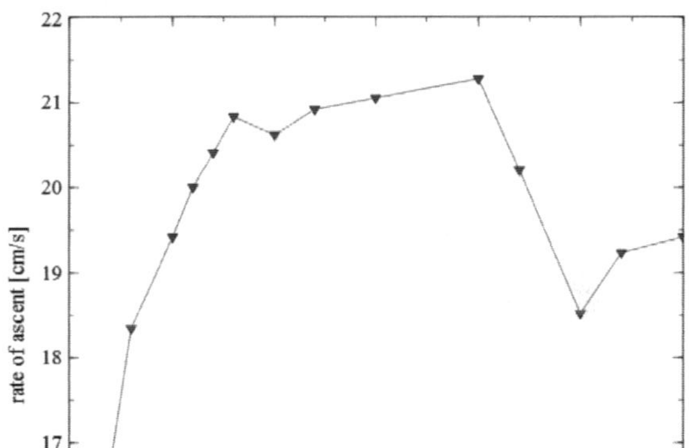

Figure 5. Volume fraction 10mm (a), 30mm (b, c) and relative velocities (d).

Figure 6. Rates of ascent for different bubble sizes.

For the simulation of constant volume flows used in experiment, the effect of generating greater bubbles and splitting can be seen, as well as the occurrence of wobbling with increasing volume flow, which also fits to other experiments. Another opportunity of comparison, the change of inclination angle of the reactor has been discovered. Increasing the angle, the residence time decreases down to a minimum at 45° and then goes up for greater angles, as it can be seen in experiments for air/water.

IOP Conf. Series: Materials Science and Engineering **228** (2017) 012016 doi:10.1088/1757-899X/228/1/012016

4.3 Magnetohydrodynamic flow control

To influence bubble residence times, one opportunity is given by the electromagnetic stirring and heating. This can be implemented by an inductor around the bubble column. To calculate the influence of an electromagnetic field on the bubble flow, additional relations, in detail described in [16], are necessary. The Lorentz force that is caused by the induced electromagnetic field of an inductor, has an influence on the velocity field of the melt. This influence and the coupling between hydrodynamic and electromagnetic effects can be described by the following equations:

$$\nabla^2 \vec{B} / (\mu_{mag}\sigma_{el}) + (\vec{B}\cdot\nabla)\vec{v} - (\vec{v}\cdot\nabla)\vec{B} = 0 \tag{25}$$

$$\vec{j} = \sigma_{el}\left(-\nabla\phi + [\vec{v}\times\vec{B}]\right) \tag{26}$$

$$\nabla\cdot\vec{v} = 0 \tag{27}$$

$$\partial\vec{v}/\partial t + (\vec{v}\cdot\nabla)\vec{v} = -\nabla p/\rho_l + \nu_l\nabla^2\vec{v} + \vec{f} \tag{28}$$

Here $\vec{B}$ describes the magnetic field induction, $\vec{j}$ is the current density, μ_{mag} the magnetic constant, σ_{el} the electrical conductivity of the liquid, φ the electrical potential. The vector $\vec{f}$ describes the volume forces that include gravity g and Lorentz force $\vec{f}_L = \vec{j} \; x \; \vec{B}$. In this case an additional force vector $\vec{F}$ appears in the momentum equation described in section 4.1.

The Lorentz forces generate whirl structures in the melt with velocity maxima near the reactor wall (see figure 4.4). As in many melting processes, electromagnetic stirring can be achieved, depending on the inductor frequency and the corresponding skin depth. To optimize the bubble dynamics and the resulting hydrogen yield, the idea is to use the turbulent shear force of the induced melt velocities to split bubbles into blebs. According to result in section 4.2., especially for bubble diameters around 7 mm, a diameter reduction results in higher residence times, besides the surface for heat transfer and reaction increases.

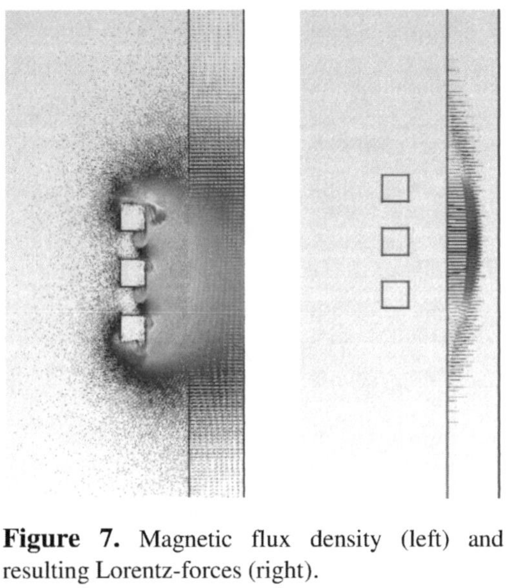

Figure 7. Magnetic flux density (left) and resulting Lorentz-forces (right).

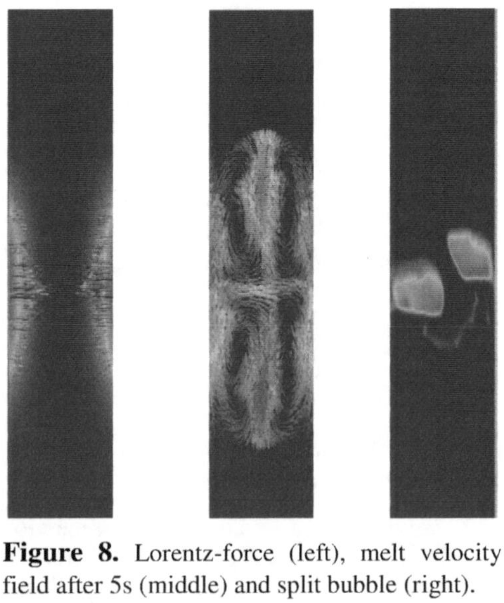

Figure 8. Lorentz-force (left), melt velocity field after 5s (middle) and split bubble (right).

In former studies a volume of fluid method and a k-ω-SST-turbulence model has been implemented in the commercial tool *Ansys*. To discover the influence of an electromagnetic field on the bubbles, a coil with three windings and a width of 10 mm each is implemented in the lower part of the reactor using frequencies of 1 kHz and currents of 1 kA. The field has been switched on before the bubbles were inserted to get the necessary whirl structures. Former simulations in Ansys Fluent have shown that bubbles with diameters bigger than 6 mm have been split up. For greater bubble diameters around 9 mm an increase of residence time has been shown. In actual investigations a solver for dynamical coupling the Eulerian method with electromagnetics has been developed in OpenFOAM. The electromagnetic field and the resulting Lorentz forces are calculated via the open source tool *getDP* (figure 7) and coupled with Eulerian method in *OpenFOAM* (figure 8). In actual studies inductor current and frequencies have been varied. A higher skin depth respectively a decreased the frequency, will increase the stirring effect due to modified distribution of the Lorentz forces. Bubble splitting can be better achieved when decreasing the skin depth by increasing the inductor frequency. But if the current is not changed, a higher power lever would also be needed. Depending on the relation of reactor diameter and skin depth, energy losses reduces with higher frequency which can be used for splitting the bubbles. Generally, it has to be noticed that the change of coupling between the Lorentz forces and the multiphase flow cannot be done with every time step. To identify such phenomena clearly also additional validations using neutron radiography are planned.

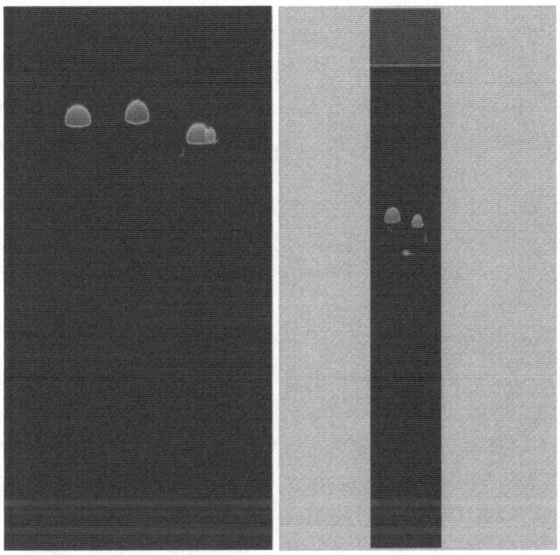

Figure 9. Volume fractions for 3 inlets for higher (left) and dR/dB with resulting coalescence (right).

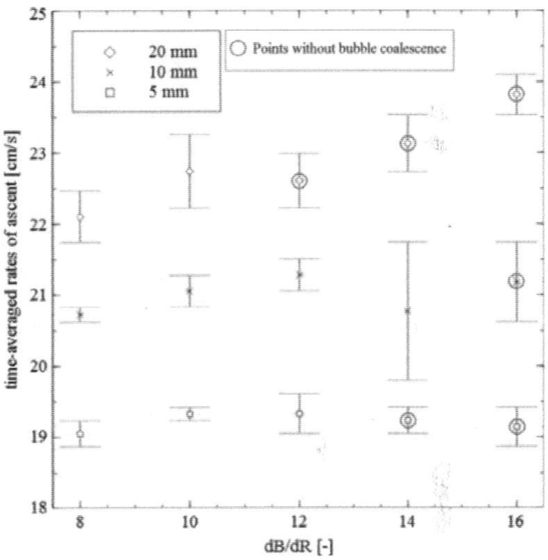

Figure 10. Rates of ascents and coalescence smaller points for 3 inlets.

4.4 Multi-inlet coalescence studies

For a possible scaling up to industrial scale it is interesting to use more than one inlet in the reactor to increase the efficiency. In previous studies, a porous sparger has been used as an alternative inlet. The problem, that is numerical described in [17], is a strong coalescence effects of the resulting bubbles above the porous media. To avoid this, the following passage will include different studies for the relation between the characteristic lengths of the flow problem. This has also been done with regards to a possible scaling up. In these studies, three inlets were discovered. For this case the initial distances between rising bubbles as well as reactor diameter were varied. Decreasing the reactor diameter and the initial distances of the 3 bubbles down to a critical point leads to coalescence (figure 9). Here the reactor volume (dark blue area in figure 9) is reduced. Detailed results are shown in figure 10. Here dR describes the characteristic length of the continuous phase (reactor diameter), while dB can be described as the characteristic length of the dispersed phase (initial bubble diameter). In the diagram the points without bubble coalescence are marked with blue rings. A range for the rates of ascents for the three bubbles is marked. It can be seen that especially for 10 mm bubbles a greater distance between the bubbles is necessary and for 20 mm also smaller distances avoid coalescence effects. Especially for 5 mm bubbles, an avoidance of coalescence leads to higher residence times. For 10 mm no strong influence can be seen while for 20 mm bubbles residence times even decrease without coalescence. Generally, especially the middle of the three bubbles is influenced by the other ones, and also wobbling effects appear. As seen in section 4.2, an increase of bubble size on smaller bubbles, e.g. by coalescence, has stronger influence on the rates of ascents.

In these studies, the distance between the bubbles are equally separated to the reactor, so the relations dR/dB can be recalculated to initial bubble distances to bubble diameter (aB/dR). So dR/dB = 12 equals aB/dR = 2, dR/dB = 14 equals aB/dR = 2,5 and dR/dB=16 equals aB/dR = 3. For the case dR/dB = 16 (aB/dR = 3) in every discovered cases no coalescence appeared. In other words, the initial bubble distance has to be 3 times higher than initial bubble size to avoid bubble coalescence in the discovered range. It has to be noticed, that the zigzag behavior of the bubbles is increased with greater volume flows, which can be seen in simulations and also experiments using neutron radiography. If it is not possible to generate single bubbles with approximately same diameter, especially at higher constant volume flow, greater distances between the inlets would necessary to avoid the described coalescence effects.

5. Conclusion

The main objective of the presented work was to better understand the influence of some major influencing parameters on the direct thermal methane cracking within a liquid metal bubble column reactor. The feasibility of the method based on liquid metal technology was successfully demonstrated and presented in [2-4].

The results of a mechanistic thermo-chemical 1D-Model have confirmed the strong influence of liquid metal temperature on the methane cracking process. The model predictions are in good agreement with the experimental results. The developed model can be applied to investigate the influence of operating parameters and gas residence time in order to improve the process efficiency.

In the numerical investigation, dynamic studies for different bubble diameters were done. The simulations of volume flows as used in experiments show slug flow dynamics at highest volume flows. Generated bubble sizes vary with volume flow, but can be dispersed while dwelling in the reactor. Generally increasing the volume flow results in stronger bubble wobbling and appearance of strong swirls for higher volume flows. Another aspect, the splitting of methane bubbles into blebs via electromagnetic stirring was also explored. It has been shown that for bubbles with a sufficient high diameter bubbles can be split. Especially for greater bubbles around 9 mm, this split also result in greater residence times, which potentially lead to higher hydrogen yield. Instead of splitting the bubbles, it is generally interesting to avoid their coalescence. Several multi-inlets with ranges of bubble residence times as well as coalescence points were numerically discovered with regards to a possible scaling up of the system. The points of coalescence vary with bubble diameters and the relation between bubble diameter and reactor diameter.

The next steps in technological development will be scaling-up to prototype level and further investigations in view of increasing methane conversion efficiency. The possibility of efficient continuous carbon removal from the reactor may be the key factor for the practical implementation of the developed technology.

Acknowledgement
This study was financially supported by the Helmholtz Society within the Helmholtz Alliance for Liquid Metal Technologies (LIMTECH).
The authors would like to sincerely thank Prof. Carlo Rubbia, Prof. Alberto Abànades, Dr. Delia Salmieri and Dr. Stefan Stückrad of the Institute of Advanced Sustainability Studies, Potsdam, who piloted and enabled a previous complementary project on the topic and supported this project in multiple ways.

References

[1] Heinzel A et al. 2017 *Energy Technology* **5** 1026 - 1036
[2] Geißler T 2017 *Methanpyrolyse in einem Flüssigmetall – Blasensäulenreaktor.* (Karlsruhe: Karlsruher Institute of Technology)
[3] Geißler T et al. 2015 *Hydrogen Energy* **40** 14134 - 14146
[4] Abánades A et al. 2016 *Hydrogen Energy* **41** 8159–8167
[5] Geißler T et al. 2016 *Chemical Engineering* **299** 192–200
[6] Zhang Y et al. 2003 *Colloids and Surfaces A: Physicochemical and Engineering Aspects* **223** 45–54
[7] Alchagirov B B et al. 2007 *Russian Journal of Physical Chemistry* **81** 1281-1284
[8] Alchagirov B B and Chochaeva A M 2000 *High Temperature* **38** 44-48
[9] Melissari B and Argyropoulos S A 2005 *Heat and Mass Transfer* **48** 4333-4341
[10] Hirt C W and Nichols B D 1981 *Journal of Computational Physics* **39** 201-225
[11] Drew D A 1983 *Ann. Review Fluid Mech.* **15** 261-291
[12] Drew D A and Lahey R T 1993 Particulate Two-Phase Flow, ed M C Roco (Butterworth-Heinemann, Boston) pp 509-566
[13] Schiller L and Naumann Z 1935 *VDI Zeitung* **77** 318-320
[14] Ranz W E and Marshall W R 1952 *Chem. Eng. Progr* **48** 141-146,
Ranz W E and Marshall W R 1952 *Chem. Eng. Progr* **48** 173-180
[15] Behzadi A et al. 2004 *Chemical Engineering Science* **59** 759-770
[16] Tucs A et al. 2013 *Proc. 11th Int. Conf. on Numerical Analysis and Applied Mathematics.* (Rhodes) p 21-27
[17] Fehling T et al. 2017 *International Journal of Applied Electromagnetics and Mechanics* **53** 111-120

IOP Conf. Series: Materials Science and Engineering **228** (2017) 012017 doi:10.1088/1757-899X/228/1/012017

Magnetic flow control in growth and casting of photovoltaic silicon: Numerical and experimental results

A Poklad[1], J Pal[2], V Galindo[2], I Grants[2], V Heinze[1], D Meier[1], O Pätzold[1], M Stelter[1], G Gerbeth[2]

[1] Institute of Nonferrous Metallurgy and Purest Materials, TU Bergakademie Freiberg, Leipziger Str. 34, 09599 Freiberg, Germany
[2] Institute of Fluid Dynamics, Helmholtz-Zentrum Dresden - Rossendorf, Bautzner Landstraße 400, 01328 Dresden, Germany

g.gerbeth@hzdr.de

Abstract. A novel, vertical Bridgman-type technique for growing multi-crystalline silicon ingots in an induction furnace is described. In contrast to conventional growth, a modified setup with a cone-shaped crucible and susceptor is used. A detailed numerical simulation of the setup is presented. It includes a global thermal simulation of the furnace and a local simulation of the melt, which aims at the influence of the melt flow on the temperature and concentration fields. Furthermore, seeded growth of cone-shaped Si ingots using either a monocrystalline seed or a seed layer formed by pieces of poly-Si is demonstrated and compared to growth without seeds. The influences of the seed material on the grain structure and the dislocation density of the ingots are discussed. The second part addresses model experiments for the Czochralski technique using the room temperature liquid metal GaInSn. The studies were focused on the influence of a rotating and a horizontally static magnetic field on the melt flow and the related heat transport in crucibles being heated from bottom and/or side, and cooled by a crystal model covering about 1/3 of the upper melt surface.

1. Introduction

Multi-crystalline silicon (mc-Si) is still an important material for the production of solar cells [1,2]. However, the efficiency of the cells is limited by structural defects in mc-Si wafers, such as grain boundaries and dislocations. An improvement of the structural quality of the material is therefore a great challenge for crystal growers. In case of vertical Bridgman-grown mc-Si ingots, so called seeded or seed-assisted growth methods are ascribed a promising potential to address this challenge. Several approaches for seeding are currently under investigation, such as the use of monocrystalline seed plates for a growth of quasi-mono (QM) ingots, firstly described in [3], and the use of a seed layer of polycrystalline Si pieces for a growth of high-performance multi (HPM) ingots [4].

Recently, a novel setup for a Bridgman-type growth of mc-Si ingots in an induction furnace was presented [5], which is characterized by the use of a SiO_2 crucible and graphite susceptor with cone-shaped bottom parts. In this paper, a numerical study on the heat and mass transport in this setup is presented, and the growth of conical ingots using the abovementioned seeding methods is demonstrated.

The Czochralski (Cz) technique is the most advanced crystal growth technique. About 95% of the worldwide electronic grade silicon production is based on it, and meanwhile also photovoltaic silicon wafer produced by the Cz method dominate the market compared to the ingot solidification of mc-Si.

IOP Conf. Series: Materials Science and Engineering **228** (2017) 012017 doi:10.1088/1757-899X/228/1/012017

For the stabilization and optimization of the Cz growth process externally applied magnetic fields were investigated during past decades [6,7]. This covers static (DC) magnetic fields and alternating (AC) ones such as rotating (RMF) or travelling (TMF) magnetic fields. It is typically attributed to DC fields that they slow-down the flow and damp related turbulent fluctuations, whereas AC fields allow an active generation of a variety of flow structures. A main benefit of AC fields consists in their typically much lower energy consumption compared to DC fields.

A cylindrical Rayleigh-Benard (RB) cell represents likely the simplest model of the Cz process. It consists of a heated bottom, a cooled upper surface and an adiabatic side wall. For this RB case it was shown that an RMF of a few Millitesla in strength may delay the RB instability and suppress turbulent temperature fluctuations significantly. The latter is due to a transition of the flow structure from a buoyancy driven large-amplitude low-frequency flow structure to a low-amplitude high-frequency flow field if the RMF driven flow dominates [8]. The reduction of the turbulent temperature fluctuations may enable starting the Cz process with a higher initial melt level compared to the standards of today. Besides the RMF, horizontal DC magnetic fields are in industrial use for the Cz process and are known to provide a well-controllable low level of oxygen in the final Si wafer. For the rotating RB case the basic flow field and its linear stability were numerically analysed in [9]. If the bulk flow is dominated by the DC magnetic field then the instability has the form of field aligned convection cells.

For both cases, the RMF and the horizontal DC magnetic field, we report in the following recent experimental results for a physical modelling of the melt flow and the related temperature field which is more close to the real Cz process compared to the RB case.

2. Seed-assisted growth of multi-crystalline silicon in a cone-shaped setup

2.1 Numerical modelling

Numerical modelling of the cone-shaped setup is focused on calculations of the melt flow, as well as the temperature and concentration fields in the melt. It bases on the following programs:

i) CrysMAS: The commercial software package CrysMAS is designed to solve global heat transport phenomena in high-temperature furnaces with complex, axisymmetric geometries [10]. It includes induction and resistance heating, nonlinear heat conduction and radiative heat transfer, which is treated with the method of view factors. In this paper, the program is used for a global thermal modelling of the induction furnace to figure out realistic thermal boundary conditions for a local simulation of the melt.

ii) OPERA 3D: OPERA 3D (© Cobham plc) is a commercial finite element software package for solving three-dimensional electromagnetic simulations on the basis of Maxwell's equations. The simulation tools are particularly suited for modelling the electromagnetic fields and related physical phenomena in complex experimental setups with an induction coil and electrically conducting components. In this paper, Opera is used to compute the induced current and Lorentz force in the induction furnace, which are needed for a simulation of the Joule heating effect and the electromagnetically driven flow, respectively.

iii) OpenFOAM: OpenFOAM is a free, open source finite volume software for computational fluid dynamics (CFD). It is used for a local simulation of the melt flow, as well as the temperature and concentration fields in the melt in a similar manner as it was done in [11].

Figure 1 shows the CrysMAS model of the central part of the induction furnace with a cone-shaped graphite setup consisting of a conical SiO_2 crucible and a graphite susceptor with a conical inner bottom face. The calculated temperature field results from a global thermal simulation with a certain power of the induction coil and T = 300°C at the outer walls of the furnace. From such calculations, the temperatures at the inner and outer crucible walls were taken to formulate realistic heat flux boundary conditions for a subsequent local modelling of physical phenomena in the melt. The corresponding heat flux profiles at the melt surface and the inner crucible walls are also given in figure 1.

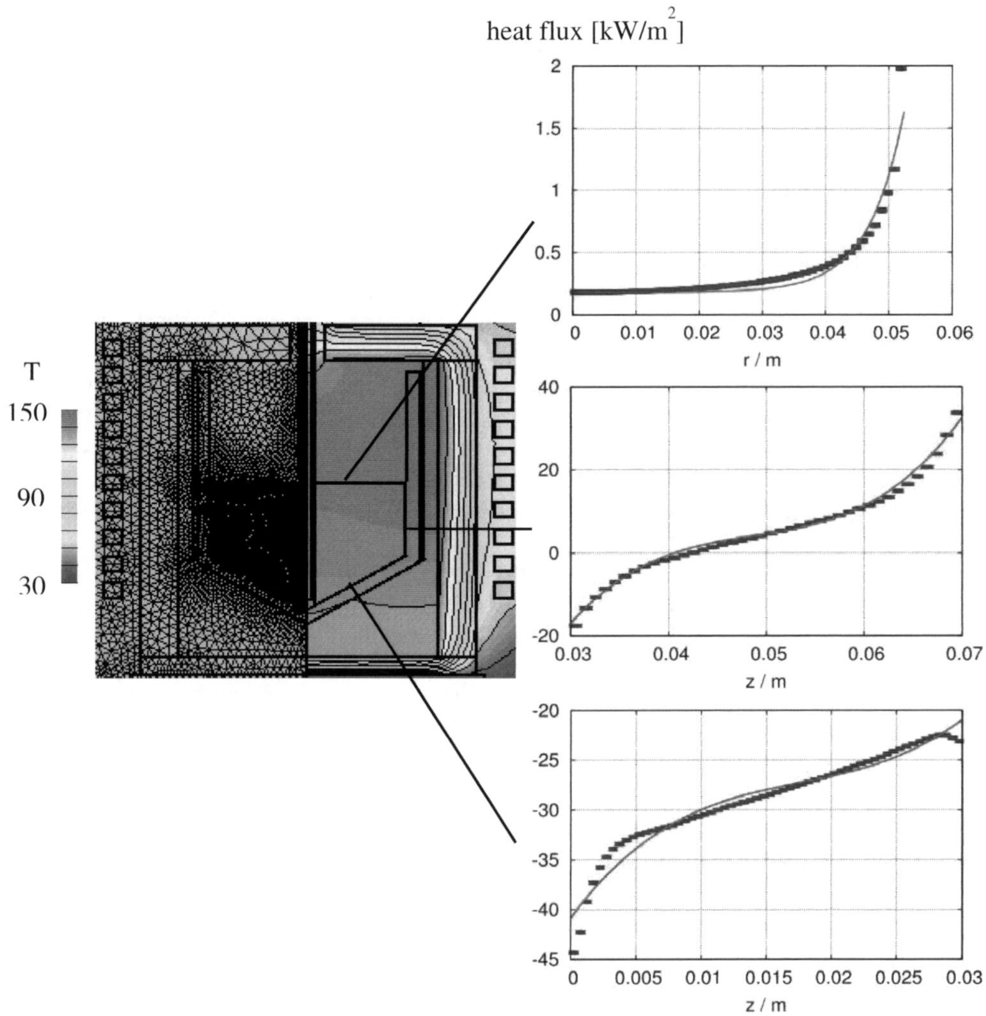

Figure 1. Left: Cross-sectional view of the conical growth setup including a temperature field resulting from a global thermal simulation of the induction furnace. Right: Heat flux profiles (dots – numerical values, lines – adjusted profiles) taken as boundary conditions for a local modelling of the melt (see figure 2).

In figure 2 results of a local modelling of the melt in the conical crucible are presented. Figure 2a shows the Lorentz force density $\boldsymbol{f}_L = < \boldsymbol{j} \times \boldsymbol{B} > / \rho$ induced by the induction coil. The melt flow and the temperature field are shown in figures 2b and c. They were computed by solving the coupled system of Navier-Stokes and energy equations in Boussinesq approximation including Lorentz force and Joule heating terms, which were transferred from the OPERA simulation of the induction furnace:

$$\frac{\partial \boldsymbol{u}}{\partial t} + (\boldsymbol{u} \cdot \nabla)\, \boldsymbol{u} = -\nabla p + \nu \nabla^2 \boldsymbol{u} - \beta\,(T - T_0)\, \boldsymbol{g} + \boldsymbol{f}_L \tag{1}$$

$$\rho c_p \left(\frac{\partial T}{\partial t} + (\boldsymbol{u} \cdot \nabla)\, T \right) = k\, \nabla^2 T + < j^2 > / \sigma \tag{2}$$

The brackets $< \cdots >$ in the expression for the Lorentz force density $\boldsymbol{f_L}$ and in equation (2) specify mean quantities averaged over one AC period. The boundary condition for the flow field at the solid container walls is the no-slip condition $\boldsymbol{u} = 0$. For the melt surface either $\boldsymbol{u} = \boldsymbol{0}$ or the conditions for a stress-free, non-deformable surface $u_n = 0$ and $\partial\, u_t \,/\, \partial\, z = 0$ are applied, depending on whether the melt flow is evaluated in an open or enclosed container. As mentioned above, the boundary conditions for the temperature field were taken from a global simulation with CrysMAS in form of given normal heat flux profiles $-\,k\,\boldsymbol{n} \cdot \nabla T$ at the edges of the melt (see figure 1). Here ν is the kinematic viscosity, β is the volume expansion coefficient, $\boldsymbol{g}$ is the gravity vector, ρ is the density, c_p is the specific heat capacity, k is the thermal conductivity and σ is the electrical conductivity.

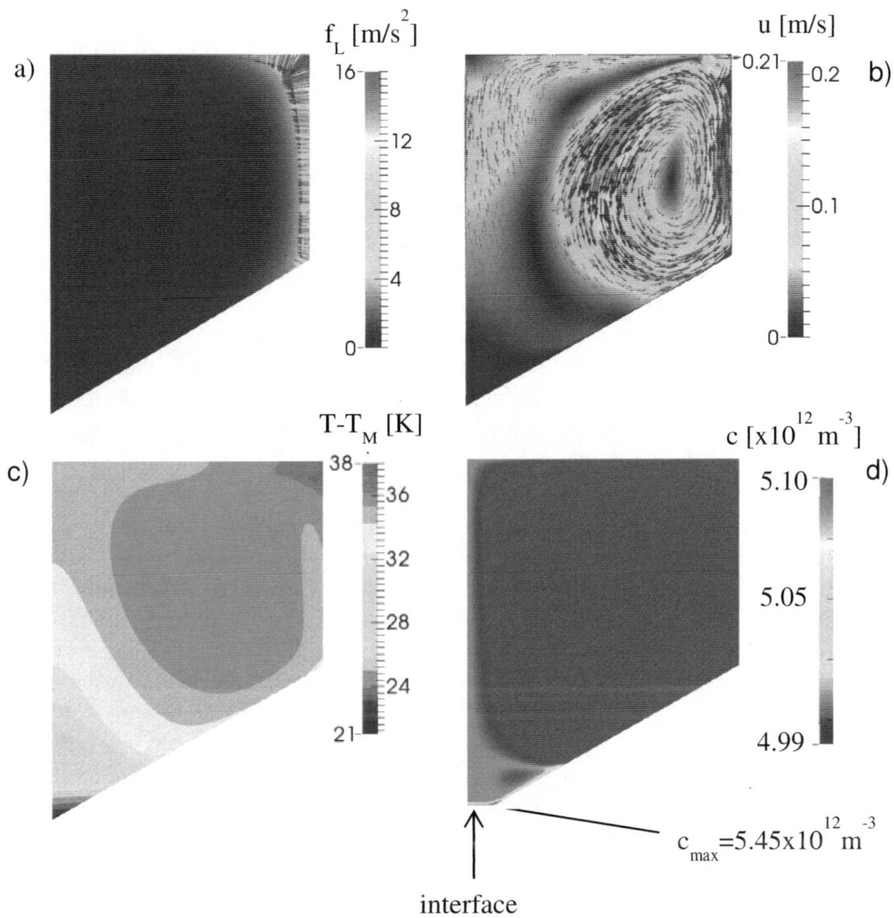

Figure 2. Local simulation of the melt in the induction furnace with the conical growth setup for a coil current and frequency of $I_{eff} = 260$ A and f = 6.3 kHz, respectively: (a) mass density related Lorentz force density, (b) melt flow, (c) temperature field relative to the melting temperature, and (d) carbon concentration for a growth rate of 20 mm/h, an initial concentration of $5x10^{12}$ m^{-3}, a diffusion coefficient of carbon in liquid silicon of $2x10^{-8}$ m^2/s, and a partition coefficient of carbon in silicon of 0.07.

The simulation of the concentration field (figure 2d) aims at the distribution of impurities in the silicon melt. For a graphite setup, which is characterized by the use of a graphite susceptor and graphite-based insulation material, carbon is the most important impurity [12,13]. Prior to solidification, the concentration of carbon in the melt is at about the solubility limit of $5x10^{-12}\,m^{-3}$. Therefore, the focus

IOP Conf. Series: Materials Science and Engineering **228** (2017) 012017 doi:10.1088/1757-899X/228/1/012017

of simulation was on the concentration at the beginning of the directional solidification process, when the position of the solid-liquid interface is in the vicinity of the cone tip. For the calculation of the carbon concentration, we solved the diffusion equation for the species transport:

$$\frac{\partial c}{\partial t} + (\boldsymbol{u} \cdot \nabla)\, c = D\, \nabla^2 c \tag{3}$$

At the interface, which is assumed to be an ideally flat surface intersecting the cone tip (see figure 2d), the segregation boundary condition $-D\, \partial c/\partial n = u_G c\, (1 - k_0)$ is used, where D is the diffusion coefficient, u_G is the growth rate and k_0 is the equilibrium partition coefficient.
The following conclusions can be drawn from the results shown in figure 2:

i) The melt flow is mainly driven by the Lorentz force of the induction coil (figure 2a). On the other hand, the buoyant flow is much less intense by more than one order of magnitude.
ii) The flow velocity in the cone region of the crucible is much lower than in the cylindrical part (figure 2b) leading to a limited convective heat and mass transfer in this region.
iii) There is a significant axial temperature gradient in the conical part of the crucible (figure 2c), which favors the control of the crystal-melt interface.
iv) The isotherms near the cone tip are almost flat (figure 2c). This means, the radial temperature gradient, which is associated with local thermal stresses, tends to zero.
v) The carbon concentration at the beginning of solidification reveals a significant supersaturation at the assumed interface near the cone tip (figure 2d). This is the consequence of the low solubility of carbon in solid silicon (partition coefficient 0.07) and the poorly mixed melt due to the low flow velocity near the cone tip. In a real growth experiment, this would result in the formation of SiC inclusions, as was indeed observed [5].

2.2 Growth experiments

Si ingots with a cone angle/height of about 120°/25 mm and diameter/height of the cylindrical part of 105/45 mm were produced with different seeding methods. High-performance multi (HPM) and quasi-mono (QM) crystals were grown using conventional, polycrystalline Si chunks as feedstock material. In the case of HPM growth, the crucible was entirely filled with poly-Si, whereas a single, monocrystalline Si cone was placed in the conical part of the crucible for the growth of QM ingots. In a growth experiment, the feedstock was first molten from above, until the position of the melt-crystal interface was in the conical part of the crucible. The remaining solid silicon formed the seeds for subsequent directional solidification processes. So the solidification of an ingot started either from a poly-Si seed layer or from a monocrystalline Si seed resulting in HPM or QM ingots, respectively.
Vertical and horizontal slices were cut from the grown crystals. The slices were polished and etched in KOH or Secco [14] solutions to reveal the grain structure and dislocation etch pits, respectively. The etch pit density (epd) was detected by optical microscopy.
The experimental results are shown in figure 3, where HPM and QM crystals are compared with a conical crystal grown without seed. They can be summarized and discussed as follows:

i) Seed-assisted growth of Si ingots in the conical setup works well. A stable seeding is possible in the conical part of the crucible, because of the significant axial temperature gradient.
ii) Solidification from a seed layer of poly-Si chunks results in an ingot with a fine grain structure and a low dislocation density, mainly below the detection limit of $10^4\,cm^{-2}$ (see figures 3c,d middle). Compared with conventionally grown mc-Si (figure 3d right), the dislocation density is significantly reduced. This is attributed to an increased ratio of random-type grain boundaries, which prevent the formation of extended dislocation clusters [4].
iii) The solidification from a cone-shaped Si monocrystal gives an ingot with a large monocrystalline area, which shows also a low dislocation density (see figures 3c,d left). This is in contrast to results on QM ingots grown in a cylindrical setup from a set of seed plates, where pronounced dislocation clusters originating from the joints between seed plates were

found [15]. Hence, with respect to the dislocation density of QM ingots, the use of a single Si seed would be preferable.

iv) Extended polycrystalline areas are detected near the edges of the QM ingot (see figures 3b,c left). Obviously, polycrystalline growth starts at the same time as the solidification process, because first grains appear in the vicinity of the melt – crystal – crucible triple point of the initial seeding interface. On solidification, the grains spread toward the central area of the ingot, indicating a concave shape of the interface and radially outward heat flux. These results show the challenges of the growth of QM ingots with the conical setup. There is a need for further improvements of the seeding process and the temperature field to reduce or even avoid polycrystalline growth.

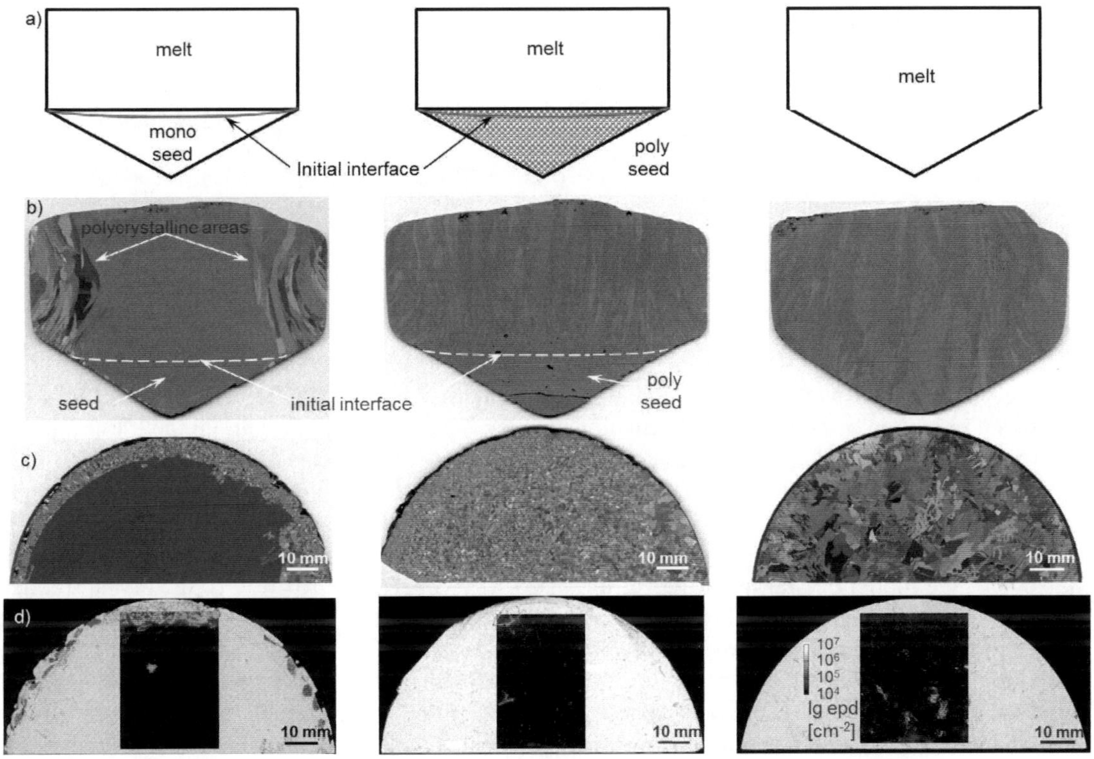

Figure 3. (a) Scheme of conical Si ingots grown with a monocrystalline seed (left), with a poly-seed layer (middle), and without seed (right). Grain structure of (b) vertical and (c) horizontal slices. (d) Etch pit density of horizontal slices.

3. Physical modelling of the Cz process

3.1 Experiments with an RMF

The modelling experiments for the Cz process followed an approach of increasing complexity. Starting with the generic RB case [8], a modified RB case was then considered with an upper cooled wall smaller than the melt cross-section modelling the solidified crystal [16]. As model melts always GaInSn or pure Ga were used as it is almost standard today [17,18]. Next, a set-up with a realistically shaped, double-walled glass crucible, flown through by a heating fluid, and a water cooled crystal model was installed, both differentially rotatable [19]. Eventually a surface cooling system was added in order to provide realistic heat transfer conditions at the upper free surface of the melt [18]. This model set-up reproduces in a reasonable approximation the realistic flow and heat transfer conditions in a silicon Cz crucible of limited size. Figure 4 shows two photos and a scheme of this experiment and its components. Main idea behind this type of physical modelling is that the model case allows detailed local measurements of the velocity field and the temperature distribution, which is an essential basis for code validations.

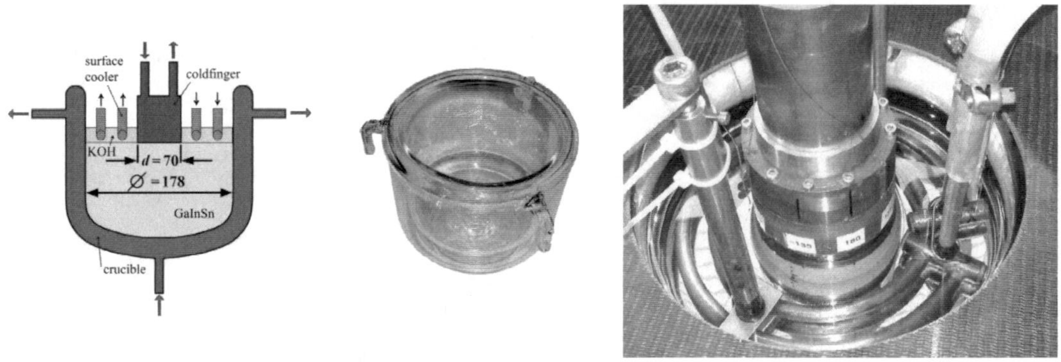

Figure 4. Scheme of the crucible, crystal model and surface cooler (left). Photo of the double-walled glass crucible (middle). Photo of the set-up with the GaInSn surface, the surface cooler and the rotatable water-cooled crystal model (right).

The use of an RMF was investigated mainly because it is considered to be an attractive tool for a significant reduction of temperature fluctuations, which would allow higher initial melt levels in the Cz process, thus increasing the yield of the growth process. Previous experiments for the RB case have shown such strong reductions of the turbulent temperature fluctuations very convincingly [8,16].

The results published in [18,19] examined the suppression of the temperature fluctuations close to the edge of the model crystal under the action of an RMF and/or the rotation of the crystal model. Stirring with an RMF as well as rotating the crystal model causes large-scale temperature oscillations since the single buoyant convection roll is driven around the crucible. Such strong oscillations are an adverse condition in Cz crystal growth. If the field strength of the RMF is increased the convection roll rotates azimuthally faster, and consequently the frequency of the oscillation increases, until the temperature fluctuations lose their periodicity. This transition from a large-scale low-frequency turbulent flow to a small-scale high-frequency turbulent flow was accompanied by a significant (ca. factor 30…50) reduction of temperature fluctuations in the RB and modified RB cases. In the more realistic Cz model as shown in figure 4, some reduction of temperature fluctuations was found as well but by far not as large as in the RB or modified RB cases. The interpretation is that the RMF is still able to break the large-scale thermal structures, but the heating from the side-wall in the Cz case gives rise to a hot jet in the direction of the crystal, which forms a region with a large temperature drop over a small distance. As result, only some reduction of temperature fluctuations is obtained at the rim of the crystal.

The investigations in [18,19] were limited to a fixed aspect ratio of $a = H/D = 0.59$ where H denotes the melt filling height and D the diameter of the crucible. This is, on one hand, slightly below the value at which in a related RB case a non-axisymmetric single roll appears but, on the other hand, distinctly higher than the filling levels used today in industrial Cz processes. It would be interesting to see whether the above described limited reduction of turbulent temperature fluctuations becomes more pronounced if the aspect ratio is further reduced.

3.2 Experiments with a horizontal DC magnetic field

As mentioned in the introduction, horizontal DC magnetic fields are in use at industrial Cz growth processes as they allow to control the oxygen concentration in a wide range. Their use is somehow surprising since they obviously destroy any axisymmetry, and the details of the flow field which finally gives rise to the oxygen control is not fully understood. Therefore, a series of model experiments has been performed addressing exactly such questions. The set-up under consideration (above named as the modified RB case) used a cylindrical cell as shown in figure 5. With the rotatable crystal model as in figure 4 it was placed in a horizontal DC magnet of a maximum flux density of 326 mT as shown in figure 5.

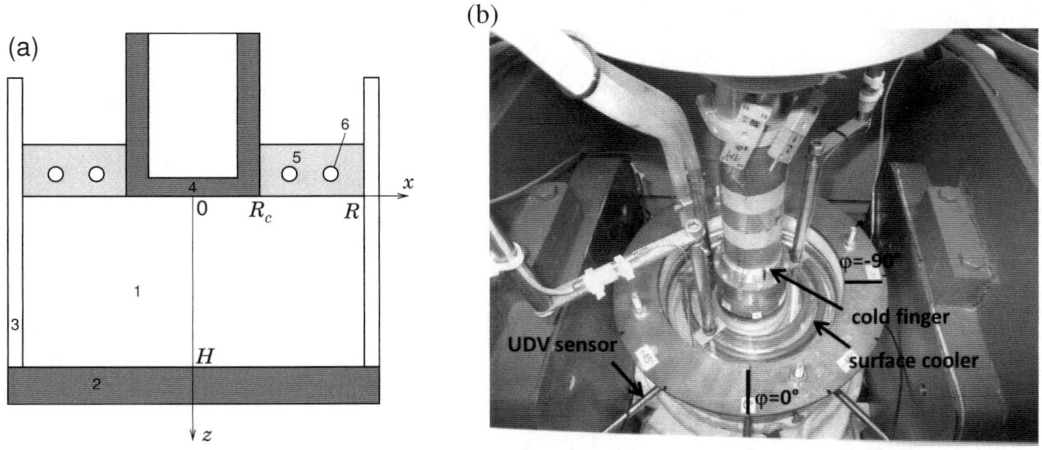

Figure 5. Scheme of the cylindrical cell (a) with 1 – liquid metal, 2 – copper bottom plate, 3 – glass wall, 4 – water cooled crystal model made of copper, 5 – surface coolant medium, 6 – cooler spiral. Top view photo of the experiment (b) with red pole faces of the horizontal DC field.

In [20] the heat transfer, the temperature azimuthal non-uniformity and the onset of oscillations were investigated in detail. It is observed that under certain conditions the integral heat flux may decrease with the magnetic field strength at the same time as the flow velocity increases. The horizontal magnetic field causes a significant azimuthal variation of the melt temperature near the crystal rim. This non-uniformity is explained by radial flow jets in the field perpendicular direction. The jets may be nearly suppressed by fine tuning of the crystal rotation rate resulting in an improvement of the temperature uniformity. The practical realization of such a fine tuning in a real Cz growth is, however, challenging and may be only possible under real-time monitoring of the melt surface temperature. The oscillatory onset in the Cz model with the horizontal DC magnetic field is much delayed in comparison to that in a rectangular RB cell. The difference may be attributed to an an "anchor" in in the form of a fixed cold spot that prevents the field perpendicular movement of rolls. The critical temperature difference increases approximately proportionally to the magnetic field strength in the considered parameter range. The onset conditions are little influenced by the crystal rotation although the basic flow varies significantly. The melt flow may be prone to hysteresis and structural instability leading to loss of symmetry and repeatability under certain conditions.

4. Summary

Numerical and experimental results on the growth of Si ingots in a cone-shaped setup were presented. The melt flow is found to be mainly driven by the Lorentz force of the induction coil. The temperature field is characterized by a significant axial gradient in the conical part of the melt.

With respect to impurities, such as carbon, the melt is poorly mixed near the cone tip. This leads to a local supersaturation of carbon on solidification, which can explain the formation of SiC inclusions observed in other studies. Seed-assisted growth to produce QM and HPM Si ingots is demonstrated. A stable seeding with the initial interface in the conical part of the crucible is possible, because of the high axial temperature gradient. The dislocation densities in QM and HPM ingots are found to be significantly lower than in conventional mc-Si grown without a seed.
For the Cz growth process a unique set of model experiments has been developed, which allows an almost complete measurement of local velocities and local temperatures in the GaInSn melt. Experiments with systematic parameter variations were performed focusing on the influence of either an RMF or a horizontal DC magnetic field on the melt flow and the related heat transfer. A closer comparison with numerical simulations would be highly desirable for future activities with the available Cz model experiments.

Acknowledgments
The financial support of the Helmholtz Alliance “Liquid Metal Technologies – LIMTECH” is gratefully acknowledged.

References
[1] Yu H J J, Popiolek N and Geoffron P 2014 *Prog. Photovolt. Res. Appl.* **24** 458
[2] 7th edition of the International Technology Roadmap for Photovoltaic (IPRPV) 2016 (http:{{www.itrpv.net)
[3] Stoddard N, Bei, W, Witting, I, Wagener M, Park Y, Rozgonyi G and Clark R 2008 *Sol. State Phenom.* **131-133** 1
[4] Yang Y M, Yu A, Hsu B, Yang A and Lan C W 2015 *Progress in Photovoltaics: Research and Applications* **23** 340
[5] Schmid E, Poklad A, Heinze V, Meier D, Pätzold O and Stelter M 2015 *J. Cryst. Growth* **416** 1
[6] Tomzig E, Virbulis J, von Ammon W, Gelfgat Y and Gorbunov L 2003, *Mat. Sci. Semicond. Process* **5** 347
[7] Vizman D 2015, *Handbook of Crystal Growth,* vol IIA, ed P. Rudolph (Amsterdam: Elsevier) p 909
[8] Grants I and Gerbeth G 2012 *Phys. Fluids* **24** 024103
[9] Grants I and Gerbeth G 2012 *J. Cryst. Growth* **358** 43
[10] Fainberg J, Vizman D, Friedrich J and Müller G 2007 *J. Cryst. Growth* **303** 124
[11] Galindo V, Niemietz K, Pätzold O and Gerbeth G 2012 *J. Cryst. Growth* **360** 30
[12] Raabe L, Pätzold O, Kupka I, Ehrig J, Würzner S and Stelter M 2011 *J. Cryst. Growth* **318** 234
[13] Funke C, Schmid E, Gärtner G, Reißenweber S, Fütterer W, Poklad A, Raabe L, Pätzold O and Stelter M 2014 *J. Cryst. Growth* **401** 732
[14] D´Secco Aragona F 1972 J. Electrochem. Soc.119 948
[15] Trempa M, Reimann C, Friedrich J, Müller G, Krause A, Sylla L and Richter T 2014 *J. Cryst. Growth* **405** 131
[16] Cramer A, Pal J and Gerbeth G 2013 *Eur. Phys. J. Special Topics* **220** 259
[17] Dadzis K, Niemietz K, Pätzold O, Wunderwald U and Friedrich J 2013 *J. Cryst. Growth* **372** 145
[18] Pal J, Cramer A, Grants I, Eckert S and Gerbeth G 2015 *J. Cryst. Growth* **432** 69
[19] Cramer A, Pal J and Gerbeth G 2014 *Flow Meas. Instrum.* **37** 99
[20] Grants I, Pal J and Gerbeth G 2017 *J. Cryst. Growth* **470** 58

Modeling electromagnetically driven free-surface flows motivated by the Ribbon Growth on Substrate (RGS) process

P Beckstein[1], **V Galindo**[1], **A Schönecker**[2] **and G Gerbeth**[1]
[1]Helmholtz-Zentrum Dresden-Rossendorf, Dresden, Germany
[2]RGS Development B. V., Netherlands

E-mail: p.beckstein@hzdr.de

Abstract. The Ribbon Growth on Substrate (RGS) technology is a crystallization technique that allows direct casting of silicon wafers and sheets of advanced metal-silicide compounds. With the potential of reaching high crystallization rates, it promises a very efficient approach for future photo-voltaic silicon wafer production compared to well-established processes in industry. However, a number of remaining problems, like process stability and controllability, need to be addressed for the RGS technology to eventually become a competitor in the near future. In this regard, it is very desirable to gain detailed insights into the characteristic process dynamics. To comply with this demand, we have developed a new numerical tool based on OpenFOAM (foam-extend), capable of simulating the free-surface dynamics of the melt flow under the influence of an applied alternating magnetic field. Our corresponding model thereby resolves the interaction of hydrodynamic and magnetodynamic effects in three-dimensional space. Although we currently focus on the RGS process, the modeling itself has been formulated in a more general form, which may be used for the investigation of similar problems, too. Here we provide a brief overview of these developments.

1. Introduction

Nowadays, photo-voltaic silicon wafers are mostly manufactured through a two-step process of growing and sawing large ingots. The ingots are either produced by directional solidification of multi-crystalline silicon or with the help of the Czochralski method in case of single crystals. For such a discontinuous procedure, unavoidable sawing losses are typically in the range of 40 to 50% of the feedstock. On the contrary, the Ribbon Growth on Substrate (RGS) technology, which was suggested and developed during the last decades [1, 2, 3], circumvents such a deficit in terms of energy and material almost completely.

Compared to the concept of growing and cutting ingots, the fundamental difference of the RGS process lies in a continuous and direct casting of silicon in near-net-shape. This is realized by means of feeding molten silicon into a casting frame without bottom, from where a solidified silicon foil is extracted side-wise on a sub-cooled moving substrate underneath. It results in both a close to perfect material yield and a low energy consumption due to the continuous nature of the processing. Concurrently, this also entails the crucial benefit of fully decoupled solidification and casting velocities.

Fig. 1 shows a schematic of the process principle. It shows one half of the core process assembly, where the foreground corresponds to a central sectional plane in the process direction.

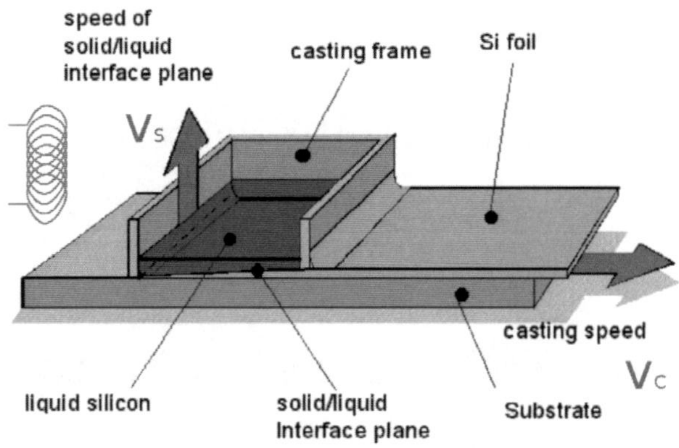

Figure 1. Scheme of the RGS process [4, 2]: one half with central cut in the process direction. A narrow slit region is situated between the casting frame and the substrate.

The silicon in the casting frame is melted by means of an induction heater with an alternating current (AC) of certain frequency. Within the RGS process there are still some effects that can negatively influence the properties and quality of the wafer. Two of them are, for example, the occurrence of flow instabilities and meniscus oscillations at the open slit, where the moving substrate enters and leaves the casting frame. The AC magnetic field, in first instance applied here for heating the silicon melt, is able to stabilize the melt at the slit since it behaves like a magnetic valve. It prevents leakage in the slit region and reduces oscillations at the extraction site of the silicon foil through electromagnetic forces. This Lorentz force field acts to counter the gravitational forces on the melt. A similar case of electromagnetic retention for liquid silicon was reported in [5]. The electromagnetic force field, however, typically gives rise to a dome-shaping of the melt in the casting frame, which represents a free surface problem complicating the numerical simulation of the system. There is only very limited literature about numerical simulations of the silicon melt flow for the RGS process (e.g., [6, 7]). Moreover, almost all related simulations were mainly focused on the crystallization process and solidification front shape of silicon ribbons. In contrast, we have performed numerical investigations in order to study the influence of the involved AC electromagnetic fields on the silicon melt during the RGS process.

2. Modeling

The casting region in the real RGS prototype machine [2] is complex. Especially the difference in scale between, e.g., the casting frame of width 156 mm in relation to the magnetic valve at the wafer exit side of less than 1 mm represents a challenge for numerical grid definition and computing capacity. Therefore, a simplified set of modeling parameters and casting environment geometry was used [8]. Within this study we have provided an overview of the magnetohydrodynamic (MHD) effects in the RGS process at its characteristic process parameters. It has been found that the Lorentz-force is dominating the melt-flow and that the deformation of the free-surface shape may not be neglected. A retention effect was demonstrated. The hydrostatic pressure at the bottom of the casting frame can be compensated by the magnetic pressure in the slit region. A parameter study revealed the total system inductivity as a function of the melt level, which might be usable for a fill-level sensing of the silicon melt. In a subsequent publication [9], we have conducted numerical simulations of the dome-shaping effect for the RGS process by means of a two-dimensional model, where we used COMSOL Multiphysics as tool for the magnetodynamic part. However, it turned out that the combination of OpenFOAM [10] and COMSOL Multiphysics is not suitable for large three-dimensional simulations. Conducting simulations of such (and similar) fully coupled MHD systems with a free surface is part of a broad field of active research to investigate industrial applications, where induction melting or levitation of liquid metals is involved. Recently, first three-dimensional simulations have been performed based on coupling ANSYS products [11], [12] and [13]. Coupling solvers or methods (e.g., Spectral, Finite Volume or Finite Element) may generally bring the need of interpolation. Thus, additional overhead is produced, which means increased computational costs.

To overcome all issues with computational overhead and external coupling of different numerical tools, we have concentrated on the development of a new solver concept in the finite volume framework of OpenFOAM (foam-extend) [14], where the magnetodynamics is modeled within the same three-dimensional finite-volume framework as the free-surface movement and the flow calculation. Based on this new solver, we were able to investigate the three-dimensional behavior of a free-surface flow under the influence of magnetic fields in the RGS process with the help of the k-ω-SST turbulence model [15]. The governing equations for the magnetic field and the flow dynamics will be presented in detail in the next section. Parts of our implementation in foam-extend were developed in close relation with [16], specially regarding the flexible and efficient application of Biot-Savart's law (c.f. chapter 3) and a multi-mesh concept to address problems with several coupled physical effects, which are each valid on different (possibly overlapping) domains.

In order to perform a more detailed Large-Eddy-Simulation (LES) of the RGS process, it is essential to be able to distribute the work load over several processors [17]. To achieve an acceptable load balance in combination with our multi-mesh technique, we have extended foam-extend to use the multi-constraint graph-partitioning feature of the software METIS [18]. With METIS, it is now possible to decompose mesh parts individually, such that each processor holds overlapping areas of all corresponding sub-meshes. No bulk communication is necessary for this to work.

On the way of realizing coupled simulations of the RGS process, we have developed a novel approach for the numerical solution of three-dimensional eddy-current problems using the finite volume framework of foam-extend [19]. To the best of our knowledge, this is the first time that such time-dependent (quasi-steady) electromagnetic problems are addressed with the help of OpenFOAM (foam-extend). Our implementation relies on a semi-coupled solution of the magnetic vector potential in a global region and the electric scalar potential in the conducting region. The method itself is memory efficient and due to the usage of block-coupled matrices numerically robust also for higher frequencies of the alternating inductor current. Even the modeling involves a non-conducting domain around the region of interest, the former is only used for the solution of the magnetic vector potential. By means of edge-based inductor models in combination with splitting the magnetic vector potential into an impressed and a reduced part, both the size of the non-conducting region and the computational costs of Biot-Savart's law can be minimized. In contrast to e.g. [20], our method is suitable also for very large simulations. Additionally, the basic concept of the so called Ghost-Fluid-Method (GFM) in analogy to [21] has been applied to account for discontinuities in the electrical conductivity of the modeled conductor. This development makes it possible to avoid several conducting sub-regions and allows for a simple setup. An extensive validation for all aspects of the numerical approach was accomplished and demonstrated in [19]. The method may be readily coupled to a variety of models of other physical phenomena within the library of foam-extend.

3. Governing equations

The magnetic fields are computed using the $\boldsymbol{A} - V$ – formulation of the quasi-static Maxwell equations in the magnetohydrodynamic approximation [22, 23]. It can be shown that due to the very small magnetic Reynolds number this approximation is valid for modeling the RGS process with sufficient quality. The magnetic fields can be described as follows with the applied Coulomb gauge $\nabla \cdot A = 0$:

$$\nabla^2 \boldsymbol{A} = \mu_0 \sigma (\partial_t \boldsymbol{A} + \nabla V) - \mu_0 \boldsymbol{j}_0 \tag{1}$$

$$\nabla \cdot (\sigma \nabla V) = -\partial_t \boldsymbol{A} \cdot \nabla \sigma \tag{2}$$

Here the magnetic vector potential is denoted by $\boldsymbol{A}$, the electric scalar potential by V, the time by t, the electric conductivity by σ and the vacuum permeability by μ_0. The total current density was divided into two terms $\boldsymbol{j} = \boldsymbol{j}_0 + \boldsymbol{j}_I$, where $\boldsymbol{j}_0$ is the contribution from the external sources and represent the effect of the excitation coils. The field $\boldsymbol{j}_I$ represents the induced current density and can be computed using the Ohm's law: $\boldsymbol{j}_I = -\sigma(\partial_t \boldsymbol{A} + \nabla V)$. To avoid meshing of the inductor it is possible to split the magnetic vector potential into an impressed and a reduced part [24], i.e. $\boldsymbol{A} = \boldsymbol{A}_0 + \boldsymbol{A}'$. We can easily compute the impressed magnetic vector potential in an algebraic way be means of Biot-Savart's law:

$$\boldsymbol{A}_0(\boldsymbol{x}) = \frac{\mu_0}{4\pi} \int_\Omega \frac{\boldsymbol{j}_0(\boldsymbol{r})}{|\boldsymbol{x} - \boldsymbol{r}|} \, d\Omega(\boldsymbol{r}) \tag{3}$$

where the integration has to be carried out only over the region of the excitation coil. This can be done by an exact integration of a parameterized geometry [19].

The induced current $\boldsymbol{j}_I$ in the electrically conducting region results in an electromagnetic force density (Lorentz force)

$$\boldsymbol{F} = \boldsymbol{j}_I \times (\nabla \times \boldsymbol{A}) \tag{4}$$

In most industrial applications, the source current $\boldsymbol{j}_0$ originates from an excitation coil with a time-harmonic alternating current at frequency f. If this is the case, a complex, quasi-steady formulation of equations 1 and 2 is more convenient, where all time derivatives are replaced with a complex-valued angular frequency ($\partial_t = i\omega$, $\omega = 2\pi f$). For a given frequency f, alternating magnetic fields appear in shape of a harmonically oscillating magnetic field $\boldsymbol{B} = \nabla \times \boldsymbol{A}$ and the induced current density $\boldsymbol{j}_I$. Both together cause a time dependent Lorentz-force $\boldsymbol{F} = \boldsymbol{j}_I \times (\nabla \times \boldsymbol{A})$. Due to the product of two harmonic functions, this force can be split in a steady part $\boldsymbol{F}_L =< \boldsymbol{F} >$ and a superposed part $\boldsymbol{F}'$ oscillating at twice the frequency. The constant part $\boldsymbol{F}_L$ is balanced mainly by friction forces, and the oscillating part is in equilibrium with inertial forces. The ratio of the corresponding velocities $U(< \boldsymbol{F}_L >)/U(\boldsymbol{F}')$ scales with $(\omega L^2)/\nu$. For liquid metals, this ratio is typically in the order of 10^6 to 10^8, which means that the oscillating part can be neglected. A more detailed explanation can be found in [25].

The flow dynamics in the electrically conducting melt is modeled by solving the incompressible Navier-Stokes equations

$$\rho(\partial_t \boldsymbol{u} + (\boldsymbol{u} \cdot \nabla)\boldsymbol{u}) = \nabla \cdot \tau + \boldsymbol{F}_L \; ; \quad \nabla \cdot \boldsymbol{u} = 0 \tag{5}$$

The equation contains the fluid velocity $\boldsymbol{u}$, the density ρ and the stress tensor $\tau = \eta(\nabla \boldsymbol{U} + (\nabla \boldsymbol{U})^T) - p'\boldsymbol{I}$ based on a modified diagonal fluid pressure $p' = p - \rho(\boldsymbol{g} \cdot \boldsymbol{x})$ with hydrostatic part according to the spatial coordinate $\boldsymbol{x}$ and the gravitational vector $\boldsymbol{g}$. The parameter η denotes the dynamic viscosity of the fluid.

The free-surface boundary is represented by means of a surface-tracking method on the basis of the Arbitrary Lagrangian-Eulerian (ALE) technique [26]. Boundary conditions for $\boldsymbol{u}$ and p can be derived from the force balance at the fluid interface. While the normal force balance results in a curvature and surface tension dependent Dirichlet-type condition for the pressure, the tangential force balance leads to a Neumann-type boundary condition for the gradient of the velocity. For more details about the conditions at the free surface, the reader is referred to [27] and [17].

4. Results

As mentioned before, an efficient solution algorithm with a solver named `eddyCurrentFoam` has been developed to solve electromagnetic (eddy-current) problems in OpenFOAM.

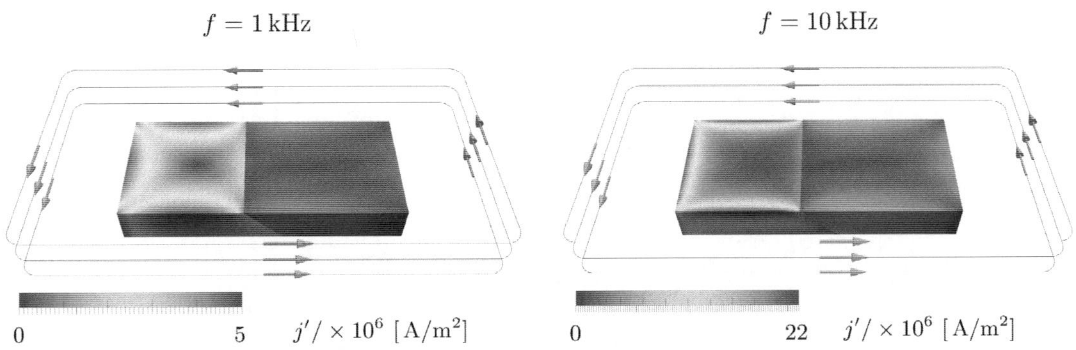

Figure 2. Results for the amplitude of the induced current density j' from simulations with jump discontinuity in the electrical conductivity ($\sigma_{left} = 12\,\sigma_{right}$) for two different frequencies.

For example both contour plots in Fig. 2 show numerical results of the induced current density from test simulations with a jump discontinuity in the electrical conductivity ($\sigma_{left} = 12\,\sigma_{right}$) at two different frequencies of 1 kHz and 10 kHz. For the case of the lower frequency, a mesh consisting of $\approx 1.8 \times 10^5$ cells was used and the calculation converged within ≈ 80 s to a normalized residual of 1×10^{-5}. The calculation for the higher value of the frequency is based on a mesh of $\approx 1.15 \times 10^6$ cells, while convergence was achieved after ≈ 840 s. The interface separating the two regions with different values of the electrical conductivity is inclined with an angle of 45 degrees with respect to the vertical direction. Details of the used setup can be found in [19]. With the application of our embedded discretization presented in [19], the flux continuity across the material boundary is preserved, while the jump in the tangential part of the current density is sharply resolved.

We have computed the induced Lorentz force density distribution in a circular cylinder caused by both a rotating (RMF) and a traveling magnetic field (TMF) [30]. In [19], the simulated results for the RMF case were validated with analytical expressions and numerical simulations with the commercial code Opera 3D [28].

Fig. 3 shows a comparison of results from a numerical simulations of the TMF Lorentz force density distribution: our implementation in foam-extend (solid) vs. results from the commercial software Cobham Opera 3D [28] (dashed) showing a very good agreement.The results from foam-extend are related to a computational grid of $\approx 7 \times 10^5$ cells and convergence at a residual of 1×10^{-8} was reached after ≈ 500 s with the help one processor. The right part of the figure includes additionally a comparison with the in the literature often used analytical expression $\boldsymbol{F}_L = \sigma\omega k_{\mathrm{TMF}} B_0^2\, r^2\, \boldsymbol{e}_z/8$ (dotted), where $k_{\mathrm{TMF}} = 2\pi/\lambda_{\mathrm{TMF}}$ denotes the TMF wave number [29, 30]. The TMF was realized by means of an axial array of multiple inductor coils supplied with alternating current and with different phase shifts. We have performed numerical simulations for a fluid region with radius R and height equal 30 mm, an AC angular frequency of $\omega = 2\pi 50$ Hz and an magnetic induction amplitude of $B_0 = 2$ mT in the center of the coil system. The electrical conductivity of the modeled liquid metal alloy Gallium-Indium-Tin (GaInSn) at room temperature was $\sigma = 3.289 \times 10^6$ S/m. The TMF wave length $\lambda_{\mathrm{TMF}} = 446,6$ mm and all the other parameters were chosen in agreement with [31].

The solution procedure for the fully coupled system of electrodynamics, flow motion and free surface deformation is separated according to physical effects. The coupling is addressed by means of iterative bulk data exchange and recurring updates of the mesh topology.

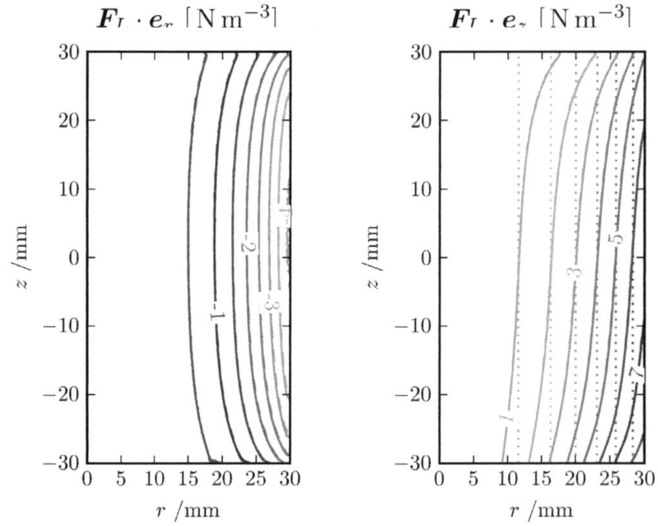

Figure 3. Comparison of results from a numerical simulation of the TMF Lorentz force density distribution: our implementation in foam-extend (solid) vs. results from the commercial software Cobham Opera 3D [28] (dashed, barely visible behind the solid lines due to the very good agreement between both methods). The dotted lines at the right part corresponds to an analytical expression linear in the radial coordinate [29, 30]

Fluid flow, surface-tracking and fluid mesh motion is numerically solved on the basis of a modified `interTrackFoam` solver algorithm which was derived from the originally published and implemented version by Tukovic and Jasak (2012) [27].

Two application examples are presented in the following as demonstration for the capabilities of our development. The first exemplary case is a simplified three-dimensional RGS process as examined and discussed in Beckstein et al. (2015) [9] and Beckstein et al. (2017) [15]. A typical snapshot of the instantaneous fluid velocity magnitude and time-averaged Lorentz-force density is illustrated in Fig. 4.

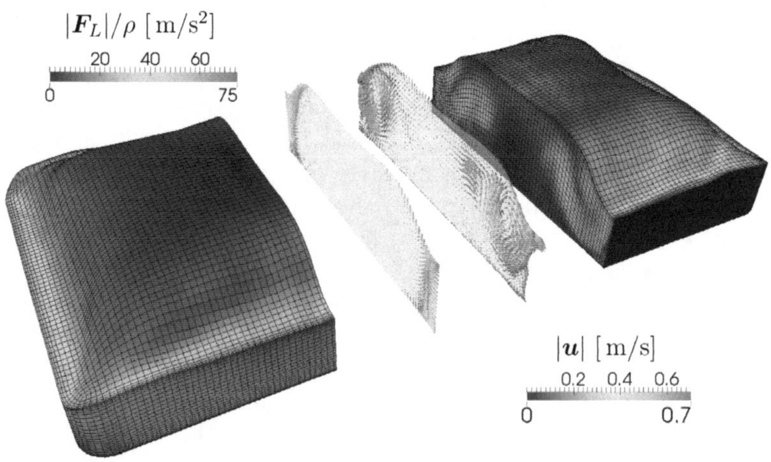

Figure 4. Deformed bulk melt shape of a three-dimensional RGS model (right part: velocity magnitude $|u|$, left part: Lorentz-Force density magnitude $|F_L|/\rho$)

IOP Conf. Series: Materials Science and Engineering **228** (2017) 012018 doi:10.1088/1757-899X/228/1/012018

The URANS-simulation presented in Fig. 4 was conducted with a mesh of $\approx 1.5 \times 10^6$ cells. Within one hour of calculation time, a simulation time of $\approx 0.05\,\mathrm{s}$ could be realized using a single CPU. A more detailed LES-simulation is already planned.

Another demonstration is given in Fig. 5. It shows the simulation of an electromagnetically levitated (EML) drop of silicon. The simulation and its parameters were chosen according to the experimental setup in [32]. In the related experiment, the levitated drop is held within a helical coil, which is modeled as five independent, toroidal loops. Due to the more complex coil geometry this simulation was realized using the reduced magnetic vector potential in combination with Biot-Savart's law as mentioned above. In this case, we have deployed LES-modeling with a corresponding spherical shaped mesh of $\approx 1 \times 10^6$ cells. The simulation runs revealed that within one CPU-hour of calculation time, $\approx 0.01\,\mathrm{s}$ of simulation time could be achieved.

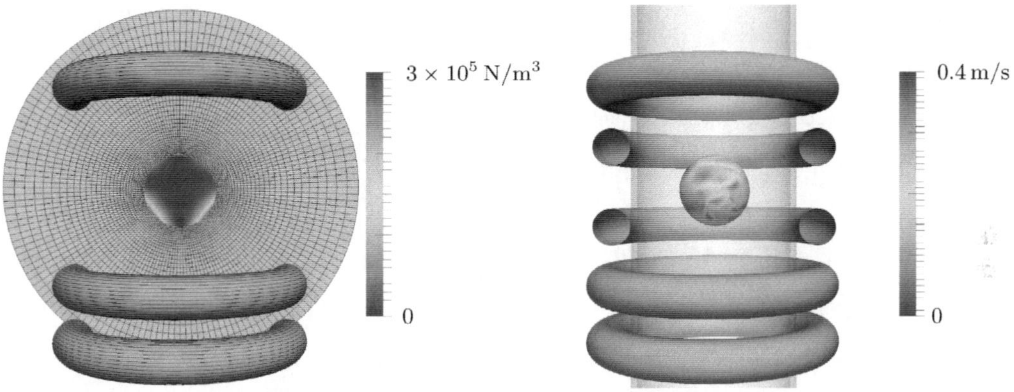

Figure 5. Numerical simulation of the induced electromagnetic force (left), the flow (right) and the shape deformation of a levitated liquid metal drop using the complete coupled code `interTrackEddyCurrentFoam` and the setup from [32].

5. Conclusions

We have presented a newly developed multi-physics method for the simulation of free-surface flows under the influence of alternating magnetic fields, which has been implemented in OpenFOAM (foam-extend). However, due to the complexity of the whole modeling we could only provide an overview of it in the scope of this paper. The key novelty is that electromagnetic and hydrodynamic effects are addressed in one single framework named `interTrackEddyCurrentFoam`. With our idea of using multiple overlapping meshes to capture individual physical effects, we have successfully created a versatile tool for a broad field of multi-physical applications, which is fully parallelized. As part of our work, a new solver for eddy-current problems called `eddyCurrentFoam` on the basis of the finite volume method was developed and validated. Additionally, the existing implementation of the surface-tracking algorithm of `interTrackFoam` has been improved and extended. In future work we would like address remaining problems like mesh-quality degradation of long simulation runs, time-step restrictions and general stability issues. This should also be accompanied with an extensive validation based on numerical and experimental data.

Concerning the RGS process, it was illustrated that a retention effect is correlated with the electromagnetically driven fluid flow. A parameter study revealed the total system inductivity as a function of the melt level, which might be usable for a fill-level sensing of the silicon melt in this process. Preliminary simulations of the flow and surface deformations including all relevant effects were conducted for the model geometry of the RGS process. A detailed insight into the flow and its surface dynamics is a key for controlling the RGS process by means of tailored magnetic fields which can be achieved by different coil geometries, power supply parameters as well as by a different electrical conductivity of the moving substrate. Such parameters studies and more detailed LES simulations with the help of our software may be used for further optimizing the RGS process.

Acknowledgments
This work was supported by Helmholtz-Gemeinschaft in frame of the LIMTECH Alliance. Fruitful discussions with Norbert Weber and Vuko Vukcevic are gratefully acknowledged.

References

[1] Lange H and Schwirtlich I A 1990 *Journal of Crystal Growth* **104** 108–112
[2] Schönecker A, Geerligs L and Müller A 2003 *Gettering and Defect Engineering in Semiconductor Technology X* (*Solid State Phenomena* vol 95) (Trans Tech Publications) pp 149–158
[3] Hahn G and Schönecker A 2004 *Journal of Physics - Condensed matter* **16** R1615–R1648
[4] Beckstein P, Galindo V and Gerbeth G 2016 *Proc. of the 10th PAMIR International Conference - Fundamental and Applied MHD* (Cagliari, Italy) pp 257–261
[5] Santara F, Delannoy Y and Autruffe A 2012 *Journal of Crystal Growth* **340** 41–46
[6] Jeong H M, Chunga H S and Lee T W 2010 *Journal of Crystal Growth* **312** 555–562
[7] Steinbach I, Apel M, Rettelbach T and Franke D 2002 *Solar energy materials and solar cells* **72** 59–68
[8] Beckstein P, Galindo V and Gerbeth G 2015 *Magnetohydrodynamics* **51** 385–396
[9] Beckstein P, Galindo V, Gerbeth G and Schönecker A 2015 *Proc. of the 8th international conference on electromagnetic processing of materials* EPM pp 167–170
[10] OpenCFD 2015 *OpenFOAM - The Open Source CFD Toolbox - User's Guide* OpenCFD Ltd. United Kingdom 3rd ed
[11] Spitans S, Jakovics A, Baake E and Nacke B 2011 *Magnetohydrodynamics* **47** 385–397
[12] Spitans S, Jakovics A, Baake E and Nacke B 2013 *Metallurgical and Materials Transactions B* **44** 593–605
[13] Spitans S, Baake E, Nacke B and Jakovics A 2014 *International Journal of Applied Electromagnetics and Mechanics* **44** 171–182
[14] Wikki Ltd. 2017 *The foam-extend project* 3rd ed URL `http://www.foam-extend.org`
[15] Beckstein P, Galindo V and Gerbeth G 2017 *International Journal of Applied Electromagnetics and Mechanics* **53** S43–S51
[16] Weber N, Beckstein P, Herremann W, Horstmann G, Nore C, Stefani F and Weier T 2017 *Physics of Fluids* **29** 054101
[17] Beckstein P, Galindo V and Gerbeth G 2017 *11th OpenFOAM Workshop* ed Nobrega J M and Jasak H (Springer) accepted
[18] Karypis G 2003 *Proceedings of the 2003 ACM/IEEE Conference on Supercomputing* SC '03 (New York, NY, USA: ACM) pp 56– ISBN 1-58113-695-1 URL `http://doi.acm.org/10.1145/1048935.1050206`
[19] Beckstein P, Galindo V and Vukcevic V 2017 *Journal of Computational Physics* **344** 623–646
[20] Djambazov G, Bojarevics V, Pericleous K and Croft N 2015 *Applied Mathematical Modelling* **39** 4733–4745 ISSN 0307904X
[21] Vukcevic V, Jasak H and Gatin I 2017 *Computers & Fluids* **153** 1–19
[22] Moreau R 1990 *Magnetohydrodynamics* (Springer) ISBN 978-0-7923-0937-6
[23] Binns K J, Lawrenson P J and Trowbridge C W 1992 *The Analytical and Numerical Solution of Electric and Magnetic Fields* (Wiley) ISBN 978-0-471-92460-9
[24] Biro O and Preis K 2000 *IEEE Transactions on Magnetics* **36** 3128–3130
[25] Davidson P A and Hunt J C R 1987 *Journal of Fluid Mechanics* **185** 67–106
[26] Hughes T J, Liu W K and Zimmermann T K 1981 *Computer Methods in Applied Mechanics and Engineering* **29** 329 – 349
[27] Tukovic Z and Jasak H 2012 *Computers & Fluids* **55** 70–84
[28] Cobham plc. 2016 *Opera-3D Design Software* URL `http://www.operafea.com`
[29] Grants I and Gerbeth G 2004 *Journal of Crystal Growth* **269** 630–638
[30] Grants I, Galindo V and Gerbeth G 2013 *The European Physical Journal Special Topics* **220** 215–225
[31] Pal J, Cramer A, Gundrum T and Gerbeth G 2009 *Flow Measurement and Instrumentation* **20** 241–251
[32] Gao L, Shi Z, Li D, Yang Y, Zhang G, McLean A and Chattopadhyay K 2016 *Metallurgical and Materials Transactions B* **47** 67–75

IOP Conf. Series: Materials Science and Engineering **228** (2017) 012019 doi:10.1088/1757-899X/228/1/012019

Experimental and Numerical Modeling of Fluid Flow Processes in Continuous Casting: Results from the LIMMCAST-Project

K Timmel[1], C Kratzsch[2], A Asad[2], D Schurmann[1], R Schwarze[2], S Eckert[1]

[1] Helmholtz-Zentrum Dresden-Rossendorf (HZDR), 01314 Dresden, Germany
[2] TU Bergakademie Freiberg (TUBAF), Institute of Mechanics and Fluid Dynamics, 09599 Freiberg, Germany

E-mail: s.eckert@hzdr.de

Abstract. The present paper reports about numerical simulations and model experiments concerned with the fluid flow in the continuous casting process of steel. This work was carried out in the LIMMCAST project in the framework of the Helmholtz alliance LIMTECH. A brief description of the LIMMCAST facilities used for the experimental modeling at HZDR is given here. Ultrasonic and inductive techniques and the X-ray radioscopy were employed for flow measurements or visualizations of two-phase flow regimes occurring in the submerged entry nozzle and the mold. Corresponding numerical simulations were performed at TUBAF taking into account the dimensions and properties of the model experiments. Numerical models were successfully validated using the experimental data base. The reasonable and in many cases excellent agreement of numerical with experimental data allows to extrapolate the models to real casting configurations. Exemplary results will be presented here showing the effect of electromagnetic brakes or electromagnetic stirrers on the flow in the mold or illustrating the properties of two-phase flows resulting from an Ar injection through the stopper rod.

1. Introduction

During the last decades, continuous casting has become the dominant process of steel casting, accounting for approximately 95 % of the annual world steel production [1]. Nevertheless, research is still being done with great effort to further develop and improve the process. It became obvious that the flow of the liquid steel in the tundish, the submerged entry nozzle (SEN) and the mold is one of the major issues, which is decisive for the steel quality at the end of the process [2]. In particular, the surface quality of the casted strand or the incorporation of inclusions is influenced by the structure and intensity of the mold flow. An efficient adjustment and control of the fluid flow and the related transport processes ensure a good quality of the product or facilitates a stable and efficient process. The steel flow is usually influenced by plant design, i.e. the geometry of the SEN, but this is fixed throughout the operation of the continuous caster. More flexible tools for flow control are electromagnetic fields, which are already in industrial use since about 30 years. There are basically two types of electromagnetic actuators in operation: so-called electromagnetic brakes (EMBr) generate a static magnetic fields for damping the turbulent flow and electromagnetic stirrers (EMS) use an alternating magnetic fields for stirring and pumping the liquid metal. However, the interaction between the turbulent steel flow and the diverse magnetic field appears to be rather complex and deserves further investigation.

Many numerical simulations have been conducted dealing with various aspects of the casting process, such as the submergence depth of the submerged entry nozzle [3], the design aspects of the nozzle ports [4], the particle wall interactions and nozzle clogging [5, 6] or the effect of electromagnetic fields [7, 8]. However, numerical models for multiphase flows and turbulent flows in large scales need to be validated by reliable experimental data. A comparison of the numerical results with plant trials require the availability of robust and suitable measuring techniques.

Flow measurements in liquid steel are very difficult. Because of the high temperature of about 1500 °C and the harsh environment at the casting machine, there are almost no measurement techniques available. Some rough information might be retrieved from observations of the free surface by dipping paddles or nail-boards into the melt [9–13], but flow measurements from deep inside the liquid steel in a real casting machine are still missing and there is no solution foreseeable for years to come. During the last decades water experiments are used for modeling the flow in the SEN and the mold. Respective investigations of the flow field were done by optical methods, e.g. [4, 14]. The modeling activities have to consider the significant differences in the material properties of water and liquid steel. While an appropriate scaling of the experimental model can achieve the similarity in terms of Reynolds and Froude number, the dramatic deviations in the surface tension and the thermal as well as the electrical conductivity render serious investigations of many flow problems actually impossible (like heat transfer, influence of magnetic fields, two-phase flows). This is illustrated by the mismatch in other dimensionless numbers, e.g. Morton, Hartmann or Prandtl number.

Model experiments in liquid metals provide an attractive possibility for investigations of fluid dynamic aspects in continuous casting. Three experimental facilities have been designed and assembled at HZDR in the framework of the LIMMCAST program [15]. The LIMMCAST-project within the LIMTECH alliance deals with the experimental and numerical modeling of the continuous casting of steel. Therefore, it covers both essential modeling strategies and allows a comparison of results and a validation of numerical models. This paper will give an overview about the experimental and numerical activities inside this project and present some recent results.

2. Modeling of the continuous casting process

2.1. The experimental models

The experimental facilities LIMMCAST and mini-LIMMCAST have been build up at HZDR for investigations of flow phenomena being relevant for the continuous casting process [15]. A third model setup has been realized for the visualization of two-phase flows in the submerged entry nozzle and the mold by X-ray radiography [16].

The design concept of all three experimental facilities contains the main components of a continuous caster, which are: the tundish, the submerged entry nozzle (SEN) and the mold. The facilities were planned and assembled as a flexible construction allowing for an easy modification of the setup at a reasonable effort and time. The transport of the liquid metal is achieved by induction pumps operating with permanent magnets. The liquid flow rate through the submerged entry nozzle is controlled by a stopper rod device. Experiments have been conducted using rectangular molds of different sizes and aspect ratios. Moreover, various geometrical configurations of submerged entry nozzles can be applied. A photograph of each setup is presented in figure 1 with indications of some major parts.

The dimensions of the model experiments are designed to meet the requirements with respect to the similarity. Main dimensionless numbers are the Reynolds number, Froude number, magnetic Reynolds number, Hartmann number or the magnetic interaction parameter. These

IOP Conf. Series: Materials Science and Engineering **228** (2017) 012019 doi:10.1088/1757-899X/228/1/012019

LIMMCAST mini-LIMMCAST X-LIMMCAST

Figure 1. Pictures of the three experimental setups

dimensionless numbers achieved at the experimental setups are well within the range of the literature values of real casters [17]. The maximum of the Hartmann number, which can be reached in the experiments with the actual existing electromagnets, is a little bit over 400 at mini-LIMMCAST [18] and around 1800 at LIMMCAST [15], whereas an exemplary reference from a real caster reaches a value of about 350. So, the electromagnets needn't to be operated at full power in the experiments to achieve similarity according to the Hartmann number in the given example. The simple geometric scaling of the model molds is in the range of 1:3 to 1:10. The cross sections of the prototype casters cover a wide region from thick to thin slab geometries. Basically, the experimental setups in this paper are modeling such slab casters. The rectangular cross sections of the molds have an aspect ratio of 1:4 to 1:6.7.

2.1.1. LIMMCAST
The LIMMCAST facility is the largest experimental facility available at HZDR for modeling the continuous steel casting process. The tin-bismuth-alloy ($Sn_{60}Bi_{40}$) is used as model liquid here. The physical properties of the alloy were measured and reported in [19]. The liquidus temperature of this alloy is 170 °C requiring operating temperatures in the range of 200 °C to 350 °C. All components and connecting pipes are made of stainless steel and are equipped with electric heaters and thermal insulation. In the first stage the facility is operated under isothermal conditions. Experiments with cooling and solidification are considered as a future option.

2.1.2. Mini-LIMMCAST
The mini-LIMMCAST setup is a small-scale model operating with the eutectic alloy GaInSn. That alloy is liquid at room temperature. The absence of electrical heaters eases the experimental effort and reduces potential interferences with sensitive measurement techniques significantly. The material properties can be found in [20]. The mold and the SEN are made of acrylic glass. The application of plastic materials enables a very flexible modification and a quick realization of complex geometries. The electrical wall conductivity plays an important role in experiments with an electromagnetic brake. Thin metal plates have been inserted at the inner mold wall for achieving electrically conducting boundary conditions in the mini-LIMMCAST mold.

2.1.3. X-LIMMCAST
The third experimental setup X-LIMMCAST was especially designed for the visualization of liquid metal / Argon two-phase flows by X-ray radiography. It is also made of acrylic glass and operated with GaInSn. The mold has a rectangular cross section with a maximum mold thickness

of 15 mm. This important restriction is caused by the high X-ray attenuation coefficient in the liquid metal. The damping of the X-ray becomes too strong at a larger thickness of the liquid metal domain and the number of photons is then no longer sufficient to provide a sufficient image contrast at sufficiently small exposure times. A scintillation screen located just behind the flow vessel to be investigated and a CCD-camera are used for the image acquisition. The observation window area can be shifted which allows for covering the area of gas injection at the stopper rod or the lower part of the SEN and the mold.

2.1.4. Velocity measurement techniques suitable for liquid metal modeling

The experiments at the liquid metal models of the LIMMCAST-family and the investigation of the related flow phenomena require an appropriate tool-box of measurement techniques. Flow measurements in liquid metals are a challenging task but nevertheless an inevitable condition for performing model experiments. The well-established optical methods in fluid dynamics cannot be applied due to the opaqueness of the metal melts. Other measuring principles relying on ultrasonic techniques or inductive methods can be used, however, ready-to-use commercial products for spatially and temporally resolved measurements in liquid metals are very rare. Therefore, new measurement techniques have been developed and qualified at HZDR for liquid metal flows. Ultrasonic techniques are a very attractive candidate for substituting the optical methods in the field of non-transparent liquids. In the 90's the Ultrasound-Doppler-Velocimetry (UDV) was established as a powerful tool for velocity measurements in fluid dynamics [21–23]. Meanwhile the equipment (sensors and instruments) were adapted to the specific operational conditions in liquid metal flows and the application range was extended towards higher temperatures, e.g. [24, 25]. This method is used as the main and reference method for the mold flow. A fully new approach for the reconstruction of three-dimensional flow pattern in liquid metals is the Contactless Inductive Flow Tomography (CIFT) [26, 27], The technique relies on the detection of flow-induced magnetic fields outside the liquid metal, which result from the interaction between the fluid flow and an externally applied excitation magnetic field. The capabilities of CIFT for measurements of the mold flow could be successfully demonstrated [28]. CIFT results were compared with UDV measurements and showed a very good agreement. Further research work has been done to improve the robustness of this method against disturbances. A detailed insight into this measurement principle and its first application to an continuous casting model experiment can be found in [28].

Another contactless velocimeter is the Local-Lorentz-Force-Velocimetry (LLFV). This technique is based on the principle of the Lorentz-Force flow meters [29]. The extension of the measurement devices towards local velocity measurement is a more recent development and had already been tested at a duct flow in a closed liquid metal loop [30]. The feasibility of LLFV measurements was tested at the mini-LIMMCAST facility.

2.2. Numerical model

The geometric dimensions for the calculation domain and corresponding material properties are taken from the experimental setups. The flow is assumed to be isothermal, incompressible, turbulent and under the influence of external magnetic fields.

2.2.1. Governing equations

The flow in the continuous casting process can be described by the conservation equations for mass and momentum:

$$\nabla \cdot \bar{\mathbf{u}} = 0 \tag{1}$$

$$\varrho \frac{\partial \bar{\mathbf{u}}}{\partial t} + \varrho \left(\bar{\mathbf{u}} \cdot \nabla \right) \bar{\mathbf{u}} = -\nabla \bar{p} + \nabla \cdot \left(\eta \nabla \bar{\mathbf{u}} \right) + \nabla \cdot \boldsymbol{\tau}^{\mathrm{mod}} + \mathbf{F}_{EM} + \mathbf{F}_{MP}. \tag{2}$$

Here, $\bar{p}$, $\bar{\mathbf{u}}$, $\mathbf{F}_{EM}$ and $\mathbf{F}_{MP}$ are pressure, velocity, electromagnetic forces and multiphase forces. The quantities are either Reynolds-averaged or filtered, depending on the turbulence modeling method used. ϱ and η denote the fluid density and the dynamic viscosity, respectively. The unknown stresses $\boldsymbol{\tau}^{\mathrm{mod}}$ are either Reynolds stresses in case of RANS-type turbulence modeling or subgrid stresses in case of LES. They have to be provided by suitable turbulence models.

2.2.2. Turbulence modeling
Two different approaches are considered to describe turbulence. The first turbulent closure is realized through the unsteady Reynolds-averaged Navier-Stokes (URANS) approach. By using this type of model, a typical RANS model keeps an unsteady term in its transport equations. This enables the ability to resolve large-scale in the inertial domain.

The second one are Scale Resolved Simulations (SRS). In this concept turbulent scales are resolved in some amount, while the rest is modeled. The classic approach is the Large Eddy Simulation (LES), where the larger eddy motions are directly calculated by the filtered equations. The drawback is that the amount of resolved eddies has to be very high since the so-called subgrid scale (SGS) models are tailored towards small scale physics. In this study, the σ-model fulfills all requirements [31, 32] for this complex fluid flow. The model is able to switch to a two-dimensional turbulence description. This is of importance since two-dimensional turbulence can occur in turbulent flows under strong magnetic fields [33, 34]. Moreover, the model uses a proper wall modeling with cubical decay of the SGS viscosity normal to the wall. It uses singular values of the resolved velocity gradient tensor and therefore incorporates some structure of the turbulence. Moreover, it is comparable in its behavior to a dynamic Smagorinsky SGS. In a recent study, this was found to be most suitable for MHD turbulence at low magnetic Reynolds numbers [35]. Nevertheless, the calculation time and the mesh size in LES can become incredibly high depending on the flow type. A way to overcome these problems is represented by hybrid turbulence models. The Delayed Detached Eddy Simulation (DDES) pursues the concept of blending over between a near-wall RANS-like modeling and an LES-like modeling in the core flow. This eliminates the need of a near wall refinement and reduces the calculation time. The DDES model here utilizes the Spalart-Allmaras model for the RANS task and the SGS part in the core region [36]. Another idea incorporates more features of the URANS modeling concept. The so-called Scale Adaptive Simulation (SAS), makes use of inherent flow instabilities to generate more unsteadiness as a traditional URANS can render. The reason is that these instabilities are not damped by turbulent viscosity. Hence, large turbulent eddies can be resolved. However, it is necessary to have a finer mesh in regions, where these instabilities arise. The method is often called the second generation of URANS, because of its close link to the URANS model [37]. The most common formulation is the combination of the SAS method [38] with the Shear Stress Transport (SST) variant of the k-ω RANS model [39]. Following the modeling equations, it can be observed that just one additional term is included in the specific turbulent dissipation rate equation that enables the SAS characteristics.

2.2.3. Electromagnetic forces
The electromagnetic forces $\mathbf{F}_{EM}$ in equation (2) can be generally described by the induction equation for a varying magnetic field $\mathbf{B}$. Since the length scale and the velocities in the experiment are small, the magnetic Reynolds number Re_m is smaller than unity. Hence, the

induced currents $\bar{\mathbf{j}}$ do not affect the imposed magnetic field $\mathbf{B}_{0,y}$, which allows a simplification called quasi-static approximation. All in all, it represents a one-way modeling of the electromagnetic forces [40]. The electromagnetic forces act as Lorentz forces in the Navier-Stokes equations and require only the solution of an additional Poisson equation for the electric potential $\bar{\psi}$:

$$\nabla^2 \bar{\psi} = \nabla \cdot (\bar{\mathbf{u}} \times \mathbf{B}_{0,y}) \tag{3}$$

$$\bar{\mathbf{j}} = \sigma \left(-\nabla \bar{\psi} + \bar{\mathbf{u}} \times \mathbf{B}_{0,y}\right) \tag{4}$$

$$\bar{\mathbf{f}}_L = \bar{\mathbf{j}} \times \mathbf{B}_{0,y}. \tag{5}$$

The subscript 0 in the description of the magnetic field denotes the static character whereas y denotes the alignment of the magnetic field with the y-direction in this study.

2.2.4. Multiphase modeling

Multiphase flow in the continuous casting mold appears in many ways. Here, a bubbly flow is considered, as it develops from argon gas injection at low and moderate gas flow rates. The resolution of the bubble interface is not expedient. Hence, a much cheaper technique is the point-particle approach within the Lagrangian frame. The bubble is seen as point mass which experiences different forces. The bubbles can move inside the Lagrangian domain, where the forces on the particle like drag, pressure, virtual mass [41], lift [42] and electromagnetic forces [43] are interpolated from the Eulerian domain. Integration of all the forces yields a bubble trajectory.

Since argon bubbles are subjected to break-up and coalescence, bubbles of different sizes can be observed. A recent study has measured these bubble size distribution in a liquid metal mold model [16]. This distribution was used for the simulations here.

The most important force acting on a moving bubble is the drag force. In a recent study [44] the authors found a large influence of bubble shape (by means of $C_{D,Dij}$) and swarm effects (by means of $C_{D,Rog}$) on the quality of the results using tailored drag models by [45, 46]. The drag model was then extended towards MHD effects (by means of $C_{D,Jin}$) using findings of a recent study [47]:

$$C_{D,Rog} = C_{D,Jin} + \left(\frac{22}{\mathrm{Eo} + 0.4}\right)(1 - \alpha_c)\, C_{D,Jin} \tag{6}$$

$$\text{with} \quad C_{D,Jin} = \begin{cases} C_{D,Dij}\left(1 + 1.5\mathrm{N} + 7.06\mathrm{N}^2\right); & \text{if} \quad 0 \le \mathrm{N} < 0.245 \\ 1.8\, C_{D,Dij}; & \text{if} \quad 0.245 \le \mathrm{N} < 0.65. \end{cases} \tag{7}$$

Here, α_c denotes the phase fraction of the continuous phase, Eo denotes the Eötvös number and N denotes the magnetic interaction parameter.

2.2.5. Numerical method and computational cost

The coupled Navier-Stokes MHD equations (Equation (1)-(5)) were discretized on a hexa-dominant unstructured mesh using the finite volume method within the open-source CFD library OpenFOAM. The Lorentz force is added as an explicit source in the momentum equations. After the flow development, the flow fields are collected for 20 s for the mini-LIMMCAST geometry and 80 s for the upscaled mold geometry.

The different turbulence modeling methods demanding for different mesh resolutions. In a previous study, the authors have shown that the SAS needs at least 1.5 million cells in the case of the mini-LIMMCAST geometry [48]. This is achieved by an overall cell size of 1,5 mm and a refined region of 0,75 mm cell size. During the paper, results of the SAS approach on a fine mesh are also shown, which consists of 3.5 million cells. The upscaled mold geometry consists of 5 million cells, where the majority of cells is located in the first 1,5 m of the strand. The simulation of the SAS on the medium size mesh of the mini-LIMMCAST takes about 6 days on 32 cores of 2,9 GHz Intel Xeon X5670 processors in the high-performance computation cluster at the TU Bergakademie Freiberg, while the fine mesh requires 25 days on 64 cores. The LES of the upscaled mold geometry takes about 30 days on 128 cores.

3. A short review on LIMMCAST results

The following sections will briefly summarize the research activities conducted by the LIMMCAST project within the LIMTECH alliance during the last five years.

3.1. Effect of electromagnetic fields on the mold flow

The impact of a static magnetic field on the mold flow was already a prominent topic in previous studies of the LIMMCAST program, e.g. [18, 49]. It was shown that the electrical boundary conditions at the mold wall have dramatic influence on the properties of the mold flow [17]. Insulating boundary condition can lead to dramatic flow oscillations whereas the flow becomes steady in case of conducting boundary conditions. Regardless of the electrical boundary conditions at the mold walls, a reverse flow close to the jet is triggered by the static magnetic field and a strong upward flow can be observed at the narrow mold wall above the jet. Moreover, a tendency to form two-dimensional flow structures along magnetic field direction becomes apparent. The insulating case showed some more special features, like an jet oscillation and the transition from a standard double roll flow in the mold to multiple rolls. In the lower mold one can now find a single roll over the whole mold width, which can also change its rotation direction over time. Generally, the symmetry was broken by the EMBr under insulating boundary conditions. These findings were also obtained by subsequent numerical simulations [7, 49–51]. The flow structures are sketched in figure 2. The standard double roll flow pattern under reference condition without magnetic field is shown in the left mold half. The effects of the EMBr especially at insulating boundary conditions is sketched in the right mold half.

The strategy for applying electromagnetic actuators can vary in the magnetic field design or the positioning of the magnetic coils at the mold. Therefore, experimental studies about the effect of variations of the EMBr position at the mold were performed for different nozzle designs [17, 52]. The same problem was investigated again by numerical simulations [53].

Another study addresses the action of a rotary electromagnetic stirrer (EMS) at the mini-LIMMCAST facility in a round mold [54]. The round mold in the experiments was a 1:3 scale model of an actual caster. The flow measurements were obtained by means of the UDV method. Three different situations were investigated in this work: the jet from the nozzle without EMS, a purely EMS-driven flow and a superposition of both. The EMS driven mold flow revealed the structure of a primary swirling flow and a secondary flow in the meridional plane consisting of two toroidal vortices. The secondary flow is significantly weaker as the primary flow, but, it is responsible for a redistribution of angular momentum in the melt. The combination of the jet flow with the stirrer showed a dramatic increase of free surface velocities with the creation of vortices and strong surface deflections near the nozzle [54]. The measurement results were used for the validation of numerical models in corresponding simulations [55] which are supposed to provide reliable predictions of the flow in the real caster.

3.2. Mold surface

In the industrial process of continuous casting the mold surface is of particular interest because it is the only access for visual observation by the operator or the positioning of crude testing

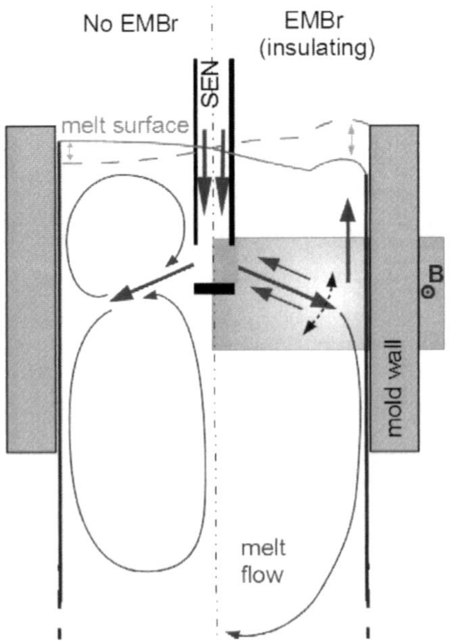

Figure 2. Sketch of the single phase mold flow observed at mini-LIMMCAST in the mid plane under reference condition (left mold half) and under the influence of an static magnetic field at insulating mold walls (right mold half).

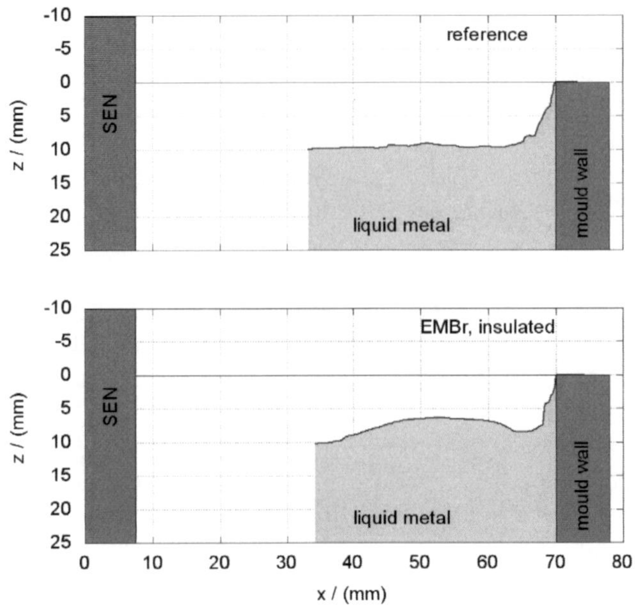

Figure 3. Mold level next to the narrow mold wall detected by a laser scanner for the reference condition without any electromagnetic actuator (top) and the applied EMBr with insulating walls (bottom).

probes. Moreover, the stability of the interface between the melt and the slag layer is regarded as a crucial issue with regard to achieving defect-free casting products [2]. Disadvantageous flow conditions can lead to entrainment of impurities, surface freezing, hook formation and other problems. This explains the desire for as low as possible surface disturbances and a sufficient heat transfer from the melt to the slag layer which ensures a sufficient melting of the flux powder for the lubrication of the strand. Therefore, some melt convection should exist just beneath the melt surface, but, the velocities should not exceed a certain limit to avoid the entrainment of impurities, slag or flux.

The melt surface was also examined in the mini-LIMMCAST experiments. Here, a slag layer does not exist, but the free surface of the liquid metal is covered by a certain oxide layer. In particular, the surface level in the mold and the tundish can be detected by an ultrasonic distance sensor. The sensor tracks the temporal behavior of the melt level at one location, e.g. [15]. A comparison of such results with an analytical assessment by the extended Bernoulli equation (including friction losses and accelerations) showed a good agreement [17]. As a second possibility, a video camera was applied for the visualization of the melt surface [17]. The recorded images revealed a specific frequency of the free surface in a slab configuration without electromagnetic actuators. This natural frequency is related to a surface wave with a wavelength of twice of the mold width [17]. Another approach uses line lasers in combination with video imaging for an improved quantitative evaluation of the surface behavior. A laser scanner provides time series of the surface profile along the central line at the mold surface [56]. For instance, the observation of the mold surface at the mini-LIMMCAST setup also showed the detrimental influence of an ill-posed EMBr at insulating boundary conditions. In this case, the melt surface becomes obviously unstable and shows intensive sloshing and bulging. The differences in the surface behavior are also sketched in figure 2 and an example of the surface profiles measured by the line laser is depicted figure 3. Such strong level oscillations were not observed in case of electrically conducting mold walls.

3.3. Argon injection

The effect of Argon gas injection has been studied at the X-LIMMCAST facility by X-ray radiography [16]. Two observation windows were chosen covering the injection point of argon gas at the tip of the stopper rod and the jet region in the mold, respectively. The monitoring of the two-phase flow in the injection region revealed the occurrence of large void zones in the upper part of the SEN just below the stopper rod. An exemplary picture from the injection point including the clearly visible void zones can be seen in the left of figure 4. The majority of the bubbles detaching from the stopper rod were captured by the void zones. The detachment of relatively large bubbles were observed at the end of the void zones. Smaller bubbles are formed by strong shearing in the bottom part of the SEN just before the nozzle ports [16]. The rather flat geometry of the X-LIMMCAST mold (see section 2.1.3) restricts the flow to an almost two-dimensional pattern. The liquid metal jets emerging from the SEN divide the mold flow in an upper and lower part. Because of the high inertia of the liquid metal jet the small bubbles are dragged into the lower recirculation zone in the mold. A direct ascending motion of the bubbles is blocked by the jet. The increasing gas concentration in the lower part of the mold promotes coalescence resulting in larger bubbles. The mean bubble size grows until the related buoyancy force becomes strong enough to overcome the jet barrier [16]. An sketch of the two-phase flow phenomena observed at X-LIMMCAST is depicted in the middle of figure 4. The right picture in figure 4 presents a snapshot of the two-phase flow in the mold. The two X-ray images in figure 4 reveals also the challenging task of identifying the fast moving bubbles.

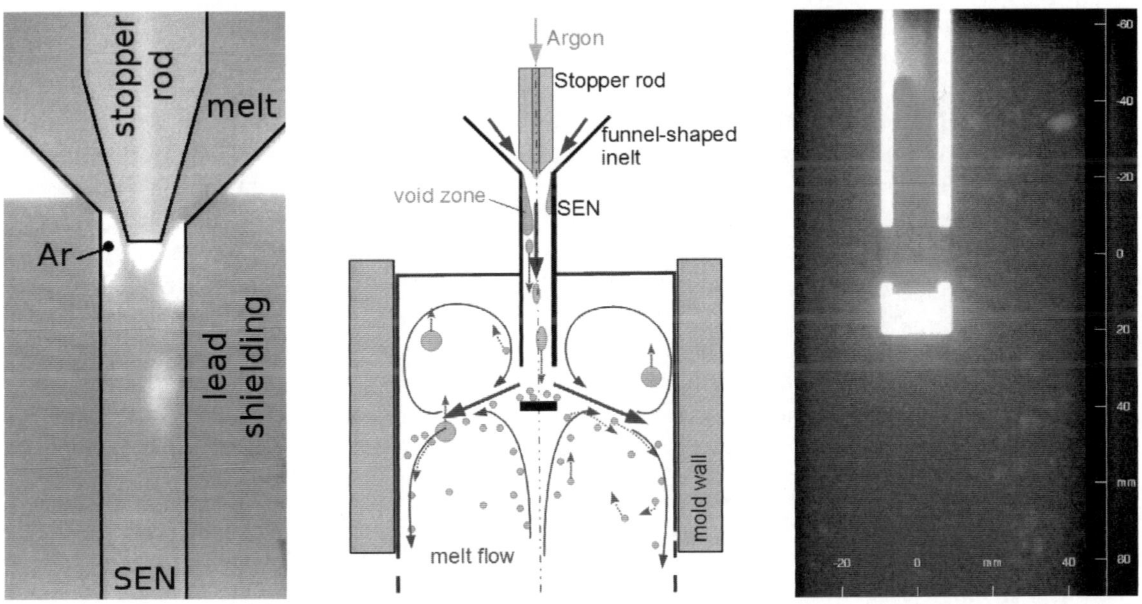

Figure 4. Sketch of two-phase flow in the SEN and the mold as it was observed at X-LIMMCAST (middle) and exemplary X-ray images of the argon injection (left) and the two-phase mold flow (right), both with an gas flow rate of 1,67 cm^3/s and an melt flow rate of about 110 cm^3/s.

IOP Conf. Series: Materials Science and Engineering **228** (2017) 012019 doi:10.1088/1757-899X/228/1/012019

3.4. Application and test of new measurement techniques

The development of the CIFT method for applications in the field of continuous casting has been reviewed in [57]. Further progress has been made to improve robustness and applicability of the method. A measuring configuration was designed and realized for the LIMMCAST facility. Related reconstructions of the flow field can be found in [58]. The detection of the weak induced magnetic field created by the melt flow is a challenging measurement task, especially in the presence of other disturbing external magnetic fields. A feasibility study about the applicability of the CIFT technique at a mold equipped with an EMBr was carried out for the mini-LIMMCAST setup [59]. It became obvious that the acquisition of the CIFT signals in presence of a much stronger EMBr can be successfully performed by specific induction coils [58, 60]. Another task is the application of gradiometric magnetic field sensors instead of flux gate sensors or single induction coils. Measurements by gradiometric coils are more robust against electromagnetic disturbances. First test measurements at mini-LIMMCAST are reported in [61].

The Local-Lorentz-Force-Velocimetry (LLFV) is dedicated for a localized and contactless measurement of the melt velocity. The applied LLFV sensor is able to measure the induced force as well as the torque in all three dimensions, resulting in six measurable parameters. Both, force and torque, can be used to determine the local velocity in all three directions. Test measurements were carried out at the mini-LIMMCAST setup [62]. These experiments revealed that the torque is less sensitive to noise than the force measurement [62]. The comparison of LLFV data with corresponding UDV measurements obtained in the mold center plane revealed some qualitative differences. These deviations can be associated with the restriction of the LLFV measuring volume to regions close to the mold wall and three-dimensional flow effects in the flat mold. A better agreement between LLFV and UDV data were obtained for UDV measurements conducted close to the mold wall [62].

3.5. Validation and comparison of different turbulence models

Since turbulence modeling concepts are based on canonical flow problems, the models are not universally applicable to every flow problem. Therefore, the first study was intended to find the most suited RANS turbulence model for mold flows [63]. The study came to the conclusion that the k-ω SST model is in favor of all the other models considered. Moreover, second order discretization was found to be crucial for reasonable results. The first observation with the OpenFOAM MHD numerical model has identified an active flow oscillation once an magnetic field is applied [48]. The study has also shown that conventional URANS modeling lacks of the MHD turbulence aspect which has either to be included in the model or tackled by models which allow more unsteadiness of the flow. The latter concept was pursued in subsequent studies [64, 65]. First, the SAS concept by Menter et al. [66] was utilized to the MHD flow problem. The model exhibits dependence on the mesh resolution of the refined mesh region. Nevertheless, the flow shows more fluctuation and unsteadiness (see Fig. 5 c)).

The already found oscillation is rendered very well and additionally turbulent scales can be resolved which then underlie Lorentz forces. Therefore, MHD turbulence can be described to some amount. The vertical component of the velocity along a line from the surface to the bottom of the mold is shown in Fig. 5 a) and b). In case without an EMBr, the profiles show a significant peak of the jet and a recirculation with upward directed velocities in the lower mold. The numerical and experimental data of the temporal average are in very good agreement. In contrast to that, the profiles of the MHD case show strong differences between the used modeling methods and a left-right asymmetry as well. The reason is the oscillation that is not perfectly periodic since a small shift in the magnetic field has drastic effects. In addition the measurement period only captures a few cycles of the oscillation. Among the turbulence modeling methods, the LES gives the most reasonable results. The upper recirculation and the jet progression matches quite well, despite differences in the jet angle. Moreover, deviations

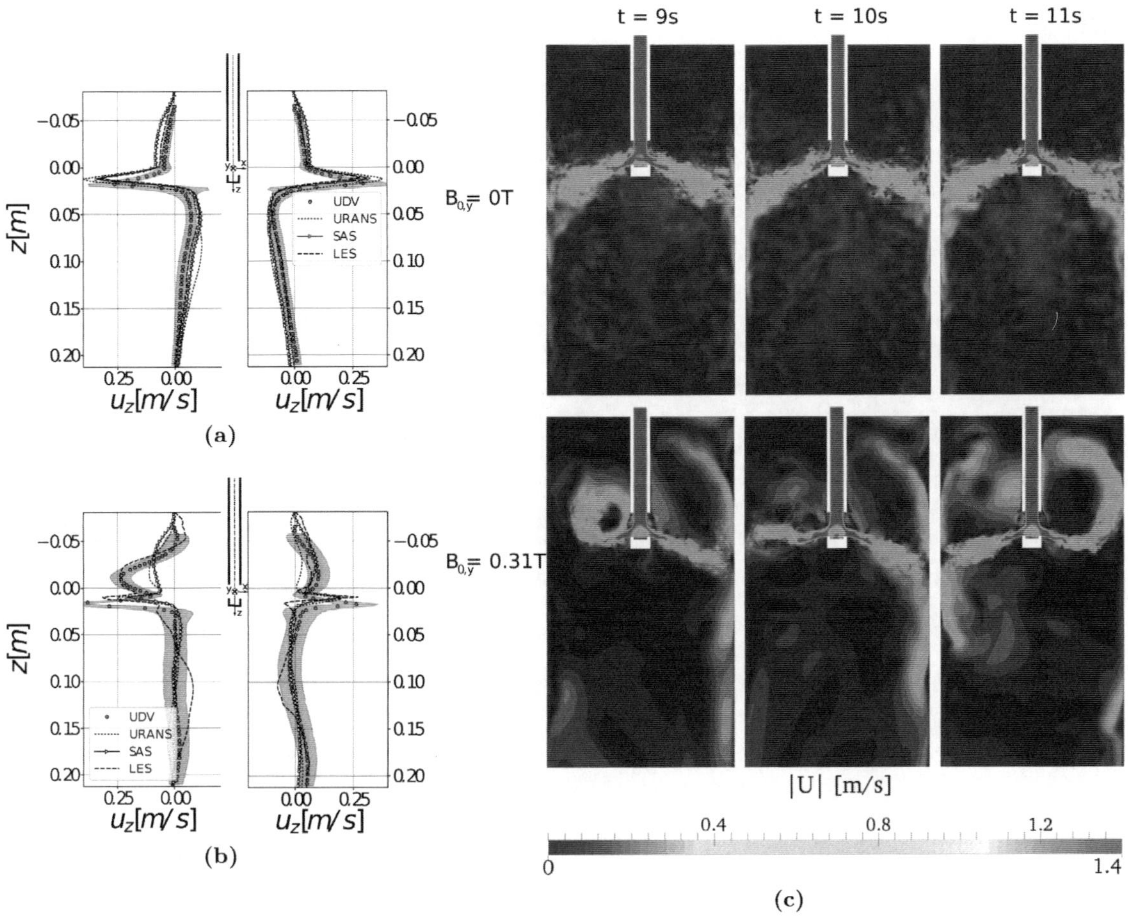

Figure 5. Left: Profiles of the time-averaged vertical velocity $\bar{u}_z$ in the midplane at $x = 20\,\mathrm{mm}$. (a) $B_{0,y} = 0T$; (b) $B_{0,y} = 0.31T$, insulating wall. UDV data filled with standard deviation. (c): Snapshots of the velocity magnitude $|\bar{\mathbf{u}}|$ distribution in the midplane for a SAS on a fine mesh. $B_{0,y} = 0T$ (top); $B_{0,y} = 0.31T$ (bottom).

from the experimental profile can be found in the lower mold because of modification of the outlets in the case of an LES.

Another study incorporates the DDES modeling concept by Spalart et al. [36] to the comparison [65]. Once again, the SRS models seem to be in better accordance with experimental results of the average flow as well as in instantaneous flow features. Apart from the turbulence modeling, the dynamics of the free surface level in mold flows was investigated in another study [67] by means of thin slab mold model at TU Delft [68]. As a next step, a scale-up of the mold model was done using different EMBr concepts (see section 4.3).

With regard to argon injection at the mold apparatus, a validation study was performed to identify dependencies of the results on bubble drag models and bubble inlet conditions in an air-water system [44]. Especially the scarcely discussed inlet conditions have shown their strong influence on the results. In addition drag models which incorporate swarm effects of the bubble ensemble have shown the best accordance with experimental results. The current investigations used these results to extend the numerical MHD model with the multiphase part of bubbly flow in a full scale mold model (see section 4.3).

4. Recent Results

4.1. Electromagnetic stirring at the SEN

Recent studies were concerned with the application of rotary electromagnetic stirring at the continuous casting setup. The CIFT method was used to investigate the effect of electromagnetic stirring at the nozzle on the flow in a rectangular mold [69]. Another experimental study about the influence of magnetic stirring in the submerged entry nozzle (SEN) for continuous casting of round blooms has recently been performed at the mini-LIMMCAST facility. The experimental setup consists of a round mold made of acrylic glass (PMMA) with an inner diameter of 80 mm. A rotating magnetic field (RMF) is applied to the SEN by rotating permanent magnets. Velocity measurements have been carried out using the Ultrasound Doppler Velocimetry (UDV). Both an horizontal and vertical arrangement of ultrasonic sensors was employed where the transducers were mounted at several positions along the side wall of the mold or were dipped directly into the liquid metal through the free surface. The points of origin of the coordinate axes x and y are located on axis of the cylindrical mold and pointing radial outwards, while the origin of the z-axis is placed on the top edge of the mold. The free surface is located at $z = 25$ mm, the SEN outlet at $z = 65$ mm.

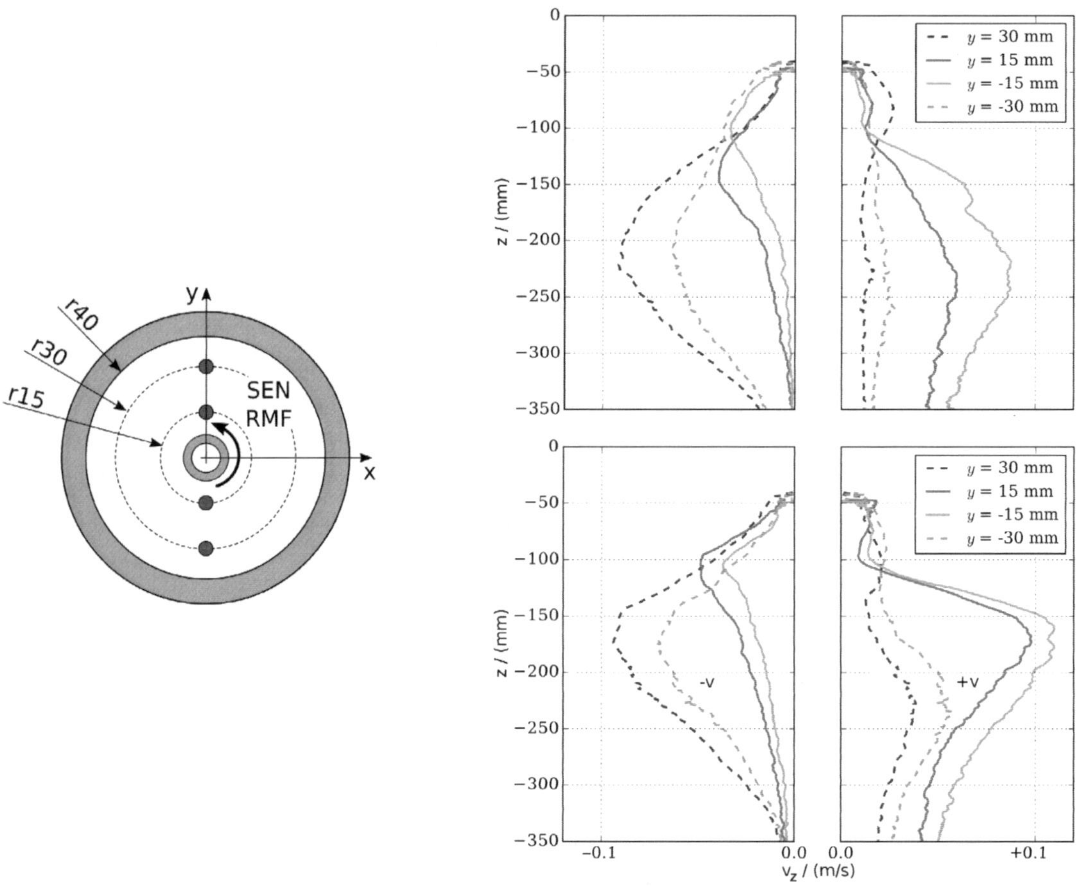

Figure 6. Sketch of the measurement configuration in the top view on the mold (left) and time average of the vertical velocity component split into negative and positive parts of four ultrasonic sensors in the vertical center-plane (at x = 0) with no RMF (right, top) and with RMF of $f = 45$ Hz at the SEN (right, bottom).

The distribution of the vertical velocity in the mold was measured in this setup up to a depth of more than 300 mm from the free surface. Because of the highly dynamic conditions, the measured velocities are split into positive and negative components before the time average is computed to avoid averaging to zero. In figure 6 (left) it can be seen that close to the center of the mold, at $y = \pm 15$ mm, downward (positive) velocities are predominant while closer to the outside of the mold, at $y = \pm 30$ mm, upward (negative) velocities can be observed. The application of a rotating magnetic field in the SEN moves the area of highest velocities closer to the free surface (figure 6 right).

A detailed discussion of the phenomena observed in our experiments is currently in preparation and will be published in the near future.

4.2. EMBr with conducting walls at LIMMCAST

Static magnetic fields are used for an efficient control of the jet flow and the suppression of surface fluctuations at continuous casting of steel [2]. The effect of a static magnetic field on the mold flow has been investigated experimentally at the LIMMCAST facility. The model mold made of stainless steel has a cross section of 400 mm times 100 mm. The SEN has an inner diameter of 35 mm and the pipe axis of the SEN determines the vertical z-axis with downward orientation. The top edge of the nozzle port marks the horizontal x-axis, which is oriented towards the narrow mold face. The location of the coordinates is also illustrated in figure 8. The melt level is controlled by two electrodes detecting the melt level and the adjustable pump rotation speed. The mean melt level in the mold was adjusted at a position of about −90 mm. The first ultrasonic sensors is located at a distance of 8 mm from the narrow mold wall. Neighboring ultrasonic sensors have a distance of 37 mm between each other. So, the horizontal positions in the coordinate system correspond to x =115 mm, 152 mm and 189 mm. The tips of the waveguides are positioned at a height of 13 mm above the upper port edge.

The mold flow without applied magnetic field was calculated by numerical simulations. The velocity magnitude in the center-plane is illustrated in figure 7. The figure depicts clearly the jet flow emerging from the nozzle ports. Further, the flow shows the typical double roll pattern, where the jet splits up in an up- and downward stream after impinging at the narrow mold wall.

These results of the numerical simulations are compared with corresponding measurements obtained by UDV. The ultrasonic measurement configuration is sketched in figure 8. Local flow measurements were carried out by means of the Ultrasonic-Doppler-Velocimetry applying specific waveguide sensors for high temperatures. Figure 9 presents the time-averaged vertical velocity measured by the waveguide-sensors at three different x-positions. The outermost sensor close to the narrow wall detected a strong ascending flow, which can be attributed to the upwards directed branch of the upper convection roll. The innermost sensor recorded the smallest velocities as it is located near the center of the roll pattern. The downward flow appearing at distances of about 100 mm can be related to the jet. The UDV results (solid lines) are compared to the outcome of the numerical simulation (dashed lines) and show a quite well agreement.

Next, the static magnetic field was applied to the flow. The top edge of the EMBrs pole shoes are located at a height of $z = -43$ mm. The pole shoes have a height of 200 mm and covered the nozzle ports completely. Figure 10 contains time-averaged profiles of the vertical velocity determined at the sensor position close to the narrow side wall for different values of the magnetic field strength, but otherwise for the same casting geometry like in figure 7 to 9. It becomes obvious that the upwards flow of the upper recirculation zone (above the jet and close to the narrow mold wall) is amplified by the magnetic field. Furthermore, the flow appears to be almost steady. This finding coincides very well with the data measured at the smaller mini-LIMMCAST setup for conducting walls [18], which indicates a similarity with respect to the electrical wall conductance ratio.

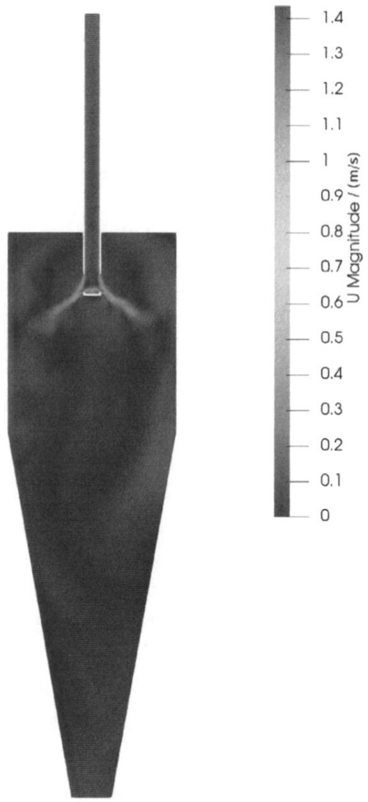

Figure 7. Snapshot of the velocity magnitude in the midplane of the mold from a numerical simulation under reference condition (without EMBr).

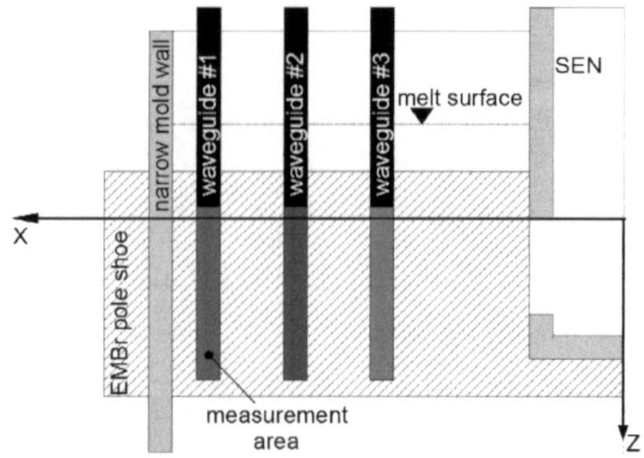

Figure 8. Sketch (non-scale) of the measurement configuration with positions of the ultrasonic sensor, the EMBr and the melt level. The measurement positions in x are: #1 = 189 mm, #2 = 152 mm and #3 = 115 mm.

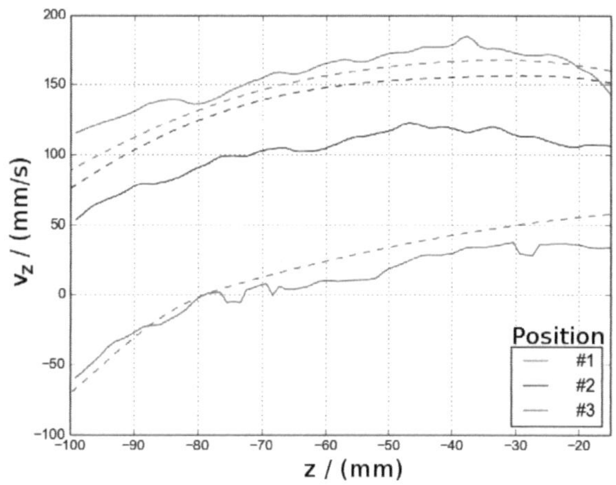

Figure 9. Comparison between numerical simulation (dashed lines) and UDV-measurement (solid lines) on the LIMMCAST setup for the reference condition without electromagnetic brake at three measurement positions. Line colors and position numbers correspond to those of figure 8.

4.3. Upscaling to a real casting geometry

Based on the successful validation of the numerical model using the experimental data from the mini-LIMMCAST facility, the numerical simulations were extended to melt flows in a continuous casting mold in a full-scale geometry as being relevant for real casting machines. For that purpose a rectangular mold of 1500 mm width and 250 mm thickness was chosen. The reduction of the liquid cross section with increasing distance from the flow surface due to progressing solidification from the side walls is considered in the model. Other parameters included in the model are a strand length of 5000 mm, a casting velocity u_c of 1,5 m/min and an SEN inlet velocity $\mathbf{u}_{in}$ of 1,91 m/s. The specific geometry of the SEN was taken from [70]. At the top surface, a free slip boundary condition was applied. Numerical calculations were performed for the case

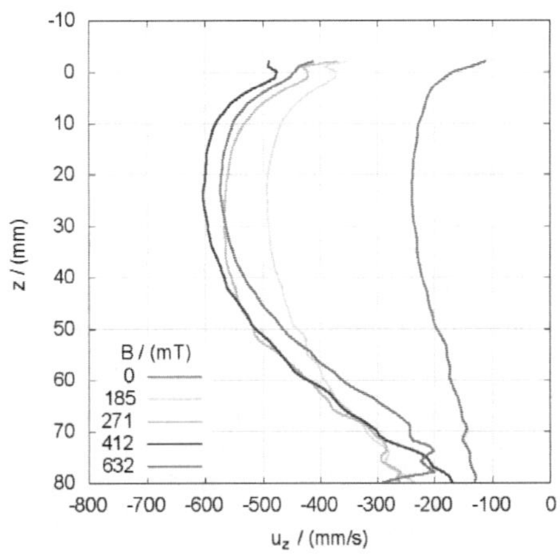

Figure 10. Mean vertical velocity close to the narrow mold wall (at $x = 189$ mm) for different strengths of the electromagnetic brake.

without magnetic field and for two different types of electromagnetic brakes. The first EMBr configuration concerns a ruler brake whose pole shoes cover the entire width of the mold. The other magnetic system represents the so-called Flow Control Mold (FCM) EMBr with two static magnetic fields, which are situated below the jets and on the level height of the free surface, respectively. Electrically insulating boundary conditions are assumed for the mold walls.

Fig. 11 shows LES results of the mean flow in the midplane of the mold. A double roll pattern can be identified in case without EMBr. The jets show a rather flat exit angle and significant spreading. Moreover, a slight asymmetry between the left and the right jet can be observed. The reason might be a self-sustaining long-wave oscillation having a period of about 8 seconds. The limited calculation time allows for capturing only a few oscillation cycles within the simulation. The surface velocities are moderate and in a range which is seen as tolerable for avoiding adverse phenomena like slag entrainment [71].

The application of a level EMBr to this flow leads to an upwards directed bending of the jets. The flow pattern reveals a strong asymmetry and distinct oscillations of the jets. These features appear to be similar as those found by the experiments in the mini-LIMMCAST facility. Compared to the mini-LIMMCAST experiments a smaller SEN submergence depth was chosen for the real-scale configuration. This might be the reason that the characteristic oscillations are not found in the upper recirculation rolls. The significant bending of the jets cause large flow velocities at the free surface. Such a situation bears the risk to enable shear layer instabilities at the slag interface which could cause the entrainment of slag droplets into the melt [71].

A completely different flow pattern is shown for the case of the FCM EMBr. The jets discharge from the nozzle with a flat exit angle and show a low spreading. The flow pattern appears to be almost symmetric and stable. The upper vortices are very weak and the lower recirculation is shifted downwards by the magnetic field. The surface velocity is drastically reduced which might create the risk of meniscus freezing.

The corresponding turbulent kinetic energy (TKE) is shown in Fig. 12. For the case without magnetic field it becomes obvious that the TKE is very high in the jet region. The TKE is produced in the jet shear layers, from there it is transported into the zone of the upper recirculation rolls. As a result, the surface TKE is rather high as well. At the top surface maximum values of the TKE are found in the narrow flow regions at both sides of the SEN. In the case with a Level EMBr the TKE in the jet is lower since the magnetic field dissipates

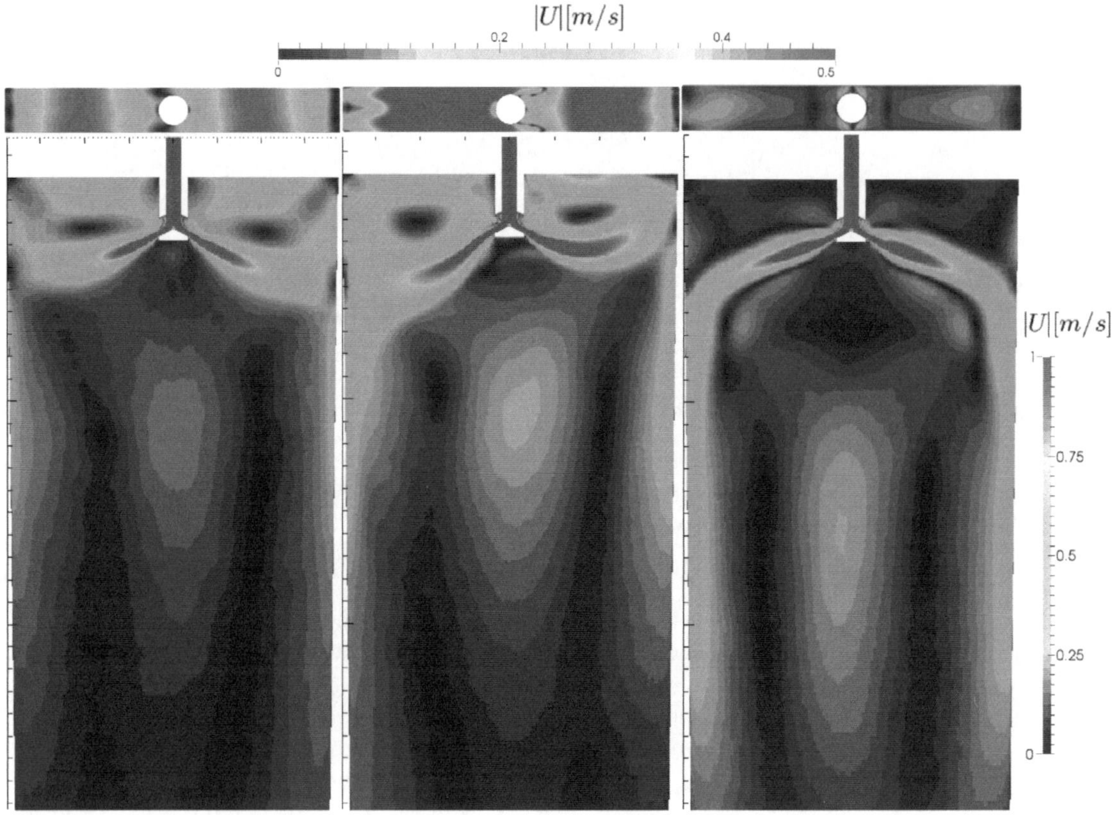

Figure 11. Distribution of the mean velocity in the midplane and top surface of the full scale mold. No EMBr case (left); Level EMBr case (middle); FCM EMBr case (right).

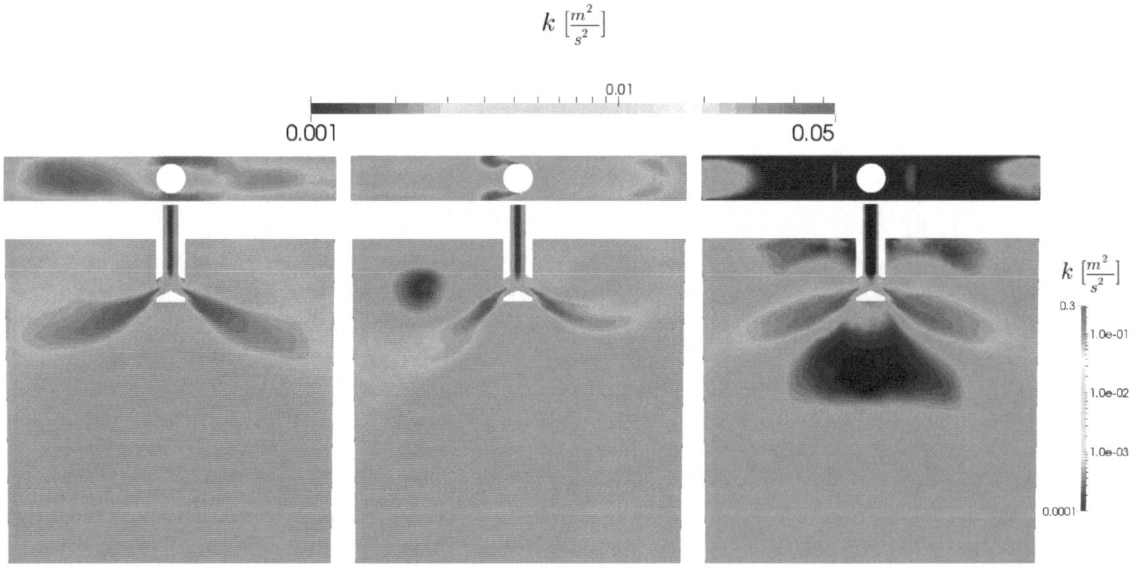

Figure 12. Distribution of the mean total turbulent kinetic energy (TKE) in the midplane and the top surface for the full scale mold. No EMBr case (left); Level EMBr case (middle); FCM EMBr case (right).

turbulent by Joule heating. This effect reduces the TKE at the free surface, too. However, the high velocities at the surface and the left-right asymmetry cause remarkable flow detachment at the SEN. As a consequence, the development of von Kármán vortices is facilitated which are detrimental to the steel quality [71]. For the FCM EMBr the contour plot of the TKE distribution reveals a broad jet, but, the striking feature are the very low TKE values occurring at the top surface around the SEN and in the region below the SEN. Due to the low surface turbulence, an entry of slag into the liquid metal appears to be quite unlikely.

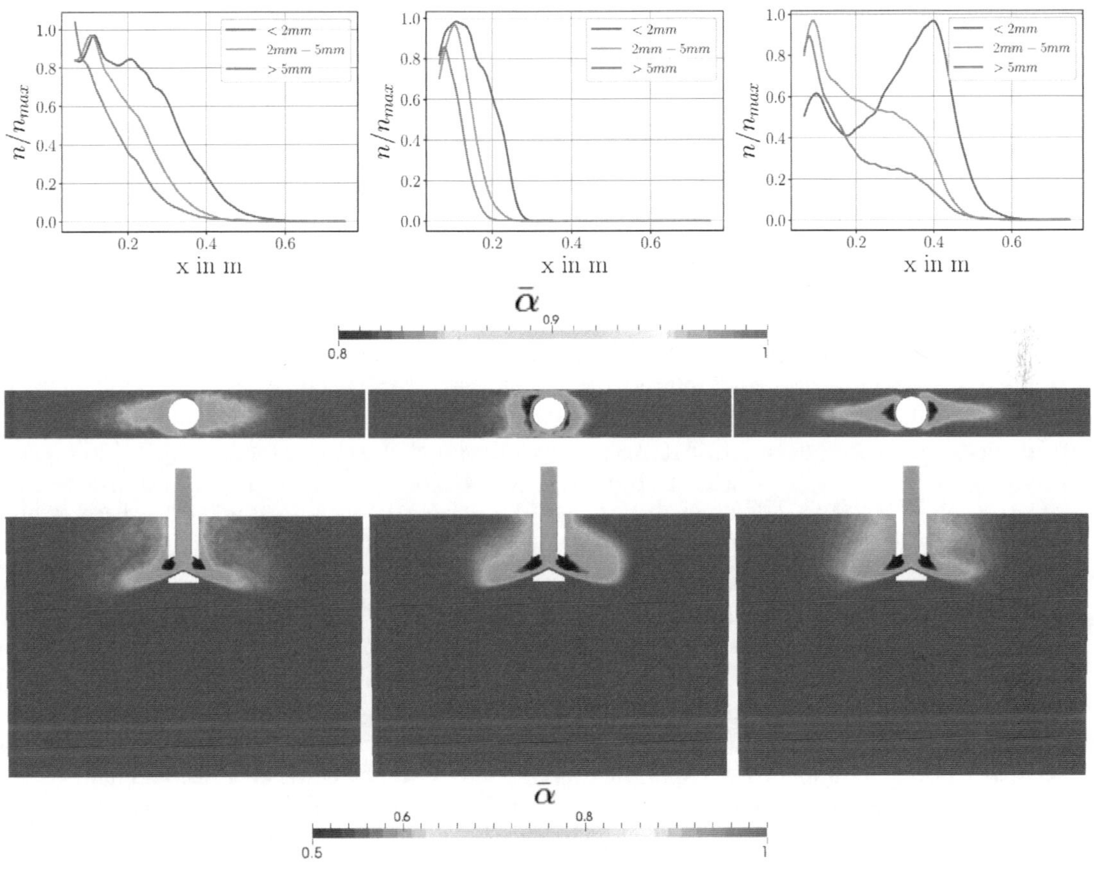

Figure 13. Top: Frequency of distribution of bubble size classes along right side of the top surface in the x-direction. Middle: Distribution of the mean liquid fraction $\bar{\alpha}$ at the top surface. Bottom: Distribution of the mean liquid fraction $\bar{\alpha}$ in the midplane. No EMBr case (left); Level EMBr case (middle); FCM EMBr case (right). All cases considered an argon flow rate of 8 % with respect to the liquid flow rate.

Another series of URANS calculations for two-phase flows in the mold was conducted using the k-ω SST model for the same three configurations as considered in the single-phase simulations: without magnetic field, ruler EMBr and FCM EMBr. For that purpose we assumed the injection of Argon bubbles at the SEN inlet. The URANS concept was chosen here because of expected uncertainties in the turbulent dispersion and the bubble trajectory obtained by SAS calculations. Fig. 13 shows the liquid volume fraction at the midplane of the mold and the free surface as well as the bubble size distribution along the top surface for an argon flow rate of 8 % with reference to the melt flow rate. Since larger bubbles are supposed to ascend quicker, their point of impact at the free surface is found to be closer to the SEN in all three cases. Smaller bubbles are more affected by dispersion due to the liquid flow and can be detected at larger distances from the SEN. It can be seen that the dispersion of argon bubbles along the x-direction is obstructed by a Level EMBr while such a damping effect on the horizontal bubble transport is not observed for the FCM EMBr. The distribution of the small bubbles in the FCM setup shows two peaks, one close to the SEN and the other almost in the center of the distance between SEN an narrow mold face. The second peak at a larger distance to the nozzle might be explained by the existence of quasi-two-dimensional vortices which are aligned with the magnetic field direction. Such vortices promote an effective bubble transport towards the side wall. The magnetic field also affects the transport of larger bubbles the distribution of which also shows higher bubble numbers at greater distances from the nozzle.

5. Summary & Conclusions

This paper presents a summary of experimental and numerical results produced during the LIMMCAST-project within the Helmholtz LIMTECH alliance. The project aims at the investigation of fluid flow aspects in the continuous casting process of steel. Special attention was paid to the influence of DC and AC magnetic fields on the flow in the mold. Furthermore, the issue of multiphase flows in the SEN and the mold due to Argon-injection at the stopper rod was addressed.

Three experimental facilities with liquid metals are operated at HZDR for modeling various essential aspects of fluid flow and related transport processes. Such an experimental program requires the availability of suitable measurement techniques for liquid metal flows. The measurements carried out at LIMMCAST rely on the ultrasonic Doppler velocimetry and inductive methods as the Contactless Inductive Flow Tomography (CIFT) and the Lorentz Force Velocimetry (LFV). In this respect, a fruitful cooperation was established with the Young Investigators Group of the LIMTECH alliance.

Several numerical schemes and different turbulence models were applied and verified. Finally, it can be concluded that the LES model provides the best results for the specific configurations considered here. This assessment is based on a comparison with the experimental data from the LIMMCAST experiments.

The action of static electromagnetic fields on the mold flow is one of the primary fields of interest in the project. The main finding is the non-trivial action of electromagnetic actuators on the mold flow. The electrical boundary conditions at the mold wall showed a dramatic influence on the temporal behavior of the melt flow under the influence of diverse electromagnetic actuators. For instance, the application of a DC magnetic field under insulating boundary conditions can promote the generation of biased flows characterized by large scale oscillations and instabilities of the jet. Different types of EMBrs usually employ steady horizontal magnetic fields normal to the wide side of the mold. Under such conditions the mold flow tends to become quasi-two-dimensional which becomes apparent by the formation of strong recirculation zones. Other experimental research activities are related to the application of AC magnetic fields for electromagnetic stirring in the mold and SEN.

The coordinated and complementary use of numerical simulations and model experiments allow the acquisition of essential insights into the flow phenomena occurring in the continuous casting of steel. The model experiments provide valuable measurement data which can be used for the validation of numerical models.

Despite the progress in modeling and understanding of the mold flow in continuous casting, there remain still a lot of work in experimental and numerical simulations as well as the enhancement of measurement techniques. There are a lot of different kind of electromagnetic actuators in industrial operation and up to now this project selected just two of them as an exemplary. Further, the experiments about the two-phase flow due to Argon injection were just a small beginning and even the combination of Argon injection with electromagnetic actuators is interesting topic but also connected to challenging measurement and numerical modeling tasks. The CIFT measurement technique has made huge progresses in the last years, but there is still some more work and development necessary to push it towards a potential application at a real caster. The simulations in this study considered insulating walls only. In the real process, the solidified shell allows the induced currents to enter the shell and therefore influence the Lorentz force distribution in the mold. Future studies should include this feature. Moreover, the model is missing a free surface treatment, which should be added by combining the current model with a Volume-of-Fluid (VoF) approach. By means of this, important questions regarding slag emulsification can be addressed more thoroughly.

Acknowledgments

The authors acknowledge the financial support from the German Helmholtz Association in the framework of the Helmholtz Alliance LIMTECH.

References

[1] worldsteel 2016 *Steel Statistical Yearbook 2016* (World Steel Association)

[2] Thomas B G 2003 *The Making, Shaping and Treating of Steel* (AISE Steel Foundation) chap 14 11th ed ISBN 9780930767044

[3] Jin K, Vanka S P and Thomas B G 2017 *Metall. Mater. Trans. B* **48** 162–178

[4] Calderon-Ramos I and Morales R D 2016 *Metall. Mater. Trans. B* **47** 1866–1881

[5] Mohammadi-Ghaleni M, Zaeem M A, Smith J D and O'Malley R 2016 *Metall. Mater. Trans. B* **47** 3056–3065

[6] Mohammadi-Ghaleni M, Zaeem M A, Smith J D and O'Malley R 2016 *Metall. Mater. Trans. B* **47** 3384–3393

[7] Singh R, Thomas B G and Vanka S P 2013 *Metall. Mater. Trans. B* **44** 1201–1221

[8] Sarkar S, Singh V, Ajmani S K, Ranjan R and Rajasekar K 2016 *ISIJ Int.* **56** 2181–2190

[9] Cukierski K and Thomas B G 2008 *Metall. Mater. Trans. B* **39** 94–107

[10] Ji C, Li J, Tang H and Yang S 2013 *Steel Res. Int.* **84** 259–268

[11] Gardin P, Galpin J M, Régnier M C and Radot J P 1996 *Magnetohydrodyn.* **32** 189–195

[12] Iguchi M, Kosaka H, Hayashi A, Terauchi Y, Mizukami H, Hanao M, Kawamoto M and Kawabata H 1999 *Metall. Mater. Trans. B* **30** 53–59

[13] Thomas B G and Zhang L 2001 *ISIJ Int.* **41** 1181–1193

[14] Salazar-Campoy M M, Morales R D, Nájera-Bastida A, Cedillo-Hernández V and Delgado-Pureco J C 2017 *Metall. Mater. Trans. B* **48** 1376–1389

[15] Timmel K, Eckert S, Gerbeth G, Stefani F and Wondrak T 2010 *ISIJ Int.* **50** 1134–1141

[16] Timmel K, Shevchenko N, Röder M, Anderhuber M, Gardin P, Eckert S and Gerbeth G 2015 *Metall. Mater. Trans. B* **46** 700–710

[17] Timmel K 2015 *Experimentelle Untersuchung zur Strömungsbeeinflussung mittels elektromagnetischer Bremsen beim kontinuierlichen Strangguss von Stahl* (*Freiberger Forschungshefte* vol A913) (Verlag der TU Bergakademie Freiberg) ISBN 9783860125076

[18] Timmel K, Eckert S and Gerbeth G 2011 *Metall. Mater. Trans. B* **42** 68–80

[19] Plevachuk Y 2010 *Int. J. Mater. Res.* **101** 839 –844

[20] Plevachuk Y, Sklyarchuk V, Eckert S, Gerbeth G and Novakovic R 2014 *J. Chem. Eng. Data* **59** 757–763

[21] Takeda Y 1986 *Int. J. Heat Fluid Flow* **7** 313–318

[22] Takeda Y 1991 *Nucl. Eng. Des.* **126** 277–284

[23] Takeda Y 1995 *Exp. Therm Fluid Sci.* **10** 444–453

[24] Eckert S and Gerbeth G 2002 *Exp. Fluids* **32** 542–546

[25] Eckert S, Gerbeth G and Melnikov V 2003 *Exp. Fluids* **35** 381–388

[26] Stefani F and Gerbeth G 2000 *Meas. Sci. Technol.* **11** 758–765

[27] Stefani F, Gundrum T and Gerbeth G 2004 *Phys. Rev. E* **70** 056306

[28] Wondrak T, Galindo V, Gerbeth G, Gundrum T, Stefani F and Timmel K 2010 *Meas. Sci. Technol.* **21** 045402

[29] Thess A, Votyakov E, Knaepen B and Zikanov O 2007 *New J. Phys.* **9** 299

[30] Heinicke C 2013 *Exp. Fluids* **54** 1560

[31] Toda H B, Cabrit O, Balarac G, Bose S, Lee J, Choi H and Nicoud F 2010 *Stanford Univ. Center for Turb. Res.: Proc. of the Summer Program 2010* pp 193–202

[32] Nicoud F, Toda H B, Cabrit O, Bose S and Lee J 2011 *Phys. Fluids* **23** 085106

[33] Sommeria J and Moreau R 1982 *J. Fluid Mech.* **118** 507–518

[34] Knaepen B and Moreau R 2008 *Annu. Rev. Fluid Mech.* **40** 25–45

[35] Amar K and Chandy A J 2016 *Flow, Turbul. Combust.* **96** 693–716

[36] Spalart P R, Deck S, Shur M, Squires K, Strelets M K and Travin A 2006 *Theor. Comput. Fluid Dyn.* **20** 181–195

[37] Fröhlich J and von Terzi D 2008 *Prog. Aerosp. Sci.* **44** 349–377

[38] Menter F, Kuntz M and Bender R 2003 *41st Aerospace Sciences Meeting and Exhibit*

[39] Menter F 1994 *AIAA J.* **32** 1598–1605

[40] Davidson P 2001 *An introduction to magnetohydrodynamics* (Cambridge university press) ISBN 9780521794879

[41] Jakobsen H A, Sannæs B H, Grevskott S and Svendsen H F 1997 *Ind. Eng. Chem. Res.* **36** 4052–4074

[42] Tomiyama A, Tamai H, Zun I and Hosokawa S 2002 *Chem. Eng. Sci.* **57** 1849–1858

[43] Leenov D and Kolin A 1954 *J. Chem. Phys.* **22** 683–688

[44] Asad A, Kratzsch C and Schwarze R 2016 *Eng. Appl. Comp. Fluid Dyn.* **11** 127–141

[45] Dijkhuizen W, Roghair I, Annaland M V S and Kuipers J 2010 *Chem. Eng. Sci.* **65** 1415–1426

[46] Roghair I, Sint Annaland M and Kuipers H J 2013 *AIChE J.* **59** 1791–1800

[47] Jin K, Kumar P, Vanka S P and Thomas B G 2016 *Phys. Fluids* **28** 093301

[48] Kratzsch C and Schwarze R 2014 *8th European Continuous Casting Conference* ISBN 978-3-200-03664-2

[49] Miao X, Timmel K, Lucas D, Ren Z, Eckert S and Gerbeth G 2012 *Metall. Mater. Trans. B* **43** 954–972

[50] Chaudhary R, Thomas B G and Vanka S P 2012 *Metall. Mater. Trans. B* **43** 532–553

[51] Miao X 2014 *Numerical study of a continuous casting process with electromagnetic brake* Ph.D. thesis TU Bergakademie Freiberg

[52] Willers B, Timmel K, Eckert S and Gerbeth G 2013 *5th Int. Conf. on Mod. and Sim. of Met. Proc. in Steelmaking* ed Society C M

[53] Thomas B, Singh R, SP V, Timmel K, Eckert S and Gerbeth G 2015 *J. Manuf. Sci. Prod.* **15** 93–104

[54] Willers B, Barna M, Reiter J and Eckert S 2017 *ISIJ Int.* **57** 468–477

[55] Barna M, Javurek M, Willers B, Eckert S and Reiter J 2016 *IOP Conf. Ser.: Mater. Sci. Eng.* **143** 012027

[56] Timmel K, Willers B, Wondrak T, Röder M, Shevchenko N, Gerbeth G and Eckert S 2016 *2016 EPD Congress* (John Wiley & Sons, Inc.) chap 3, pp 19–26 ISBN 9781119274742

[57] Ratajczak M, Gundrum T, Stefani F and Wondrak T 2014 *J. Sens.* **2014** 739161

[58] Ratajczak Mand Wondrak T, Timmel K, Stefani F and Eckert S 2015 *J. Manuf. Sci. Prod.* **15** 41–48

[59] Ratajczak M, Wondrak T, Stefani F and Eckert S 2015 *Magnetohydrodyn.* **51** 461–471

[60] Stefani F, Eckert S, Ratajczak M, Timmel K and Wondrak T 2016 *IOP Conf. Ser.: Mater. Sci. Eng.* **143** 012023

[61] Ratajczak M, Wondrak T and Stefani F 2016 *Philos. Trans. R. Soc., A* **374** 20150330

[62] Hernández D, Schleichert J, Karcher C, Fröhlich T, Wondrak T and Timmel K 2016 *Meas. Sci. Technol.* **27** 065302

[63] Kratzsch C, Timmel K and Eckert Sand Schwarze R 2015 *Steel Res. Int.* **86** 400–410

[64] Kratzsch C, Asad A and Schwarze R 2015 *J. Manuf. Sci. Prod.* **15** 49–57

[65] Kratzsch C, Asad A and Schwarze R 2016 *IOP Conf. Ser.: Mater. Sci. Eng.* **143** 012025

[66] Menter F and Egorov Y 2005 *AIAA J.* **1095** 1095

[67] Asad A, Kratzsch C and Schwarze R 2016 *Steel Res. Int.* **87** 181–190

[68] Kalter R, Tummers M, Kenjereš S, Righolt B and Kleijn C 2013 *Int. J. Heat Fluid Flow* **44** 365–374

[69] Wondrak T, Eckert S, Galindo V, Gerbeth G, Stefani F, Timmel K, Peyton A J, Yin W and Riaz S 2012 *Ironmaking & Steelmaking* **39** 1–9

[70] Singh R, Thomas B G and Vanka S P 2014 *Metall. Mater. Trans. B* **45** 1098–1115

[71] Hibbeler L C and Thomas B G 2013 *Iron Steel Technol.* **10** 121–136

IOP Conf. Series: Materials Science and Engineering **228** (2017) 012020 doi:10.1088/1757-899X/228/1/012020

Casting technology for ODS steels – dispersion of nanoparticles in liquid metals

M Sarma[1], **I Grants**[1,2], **I Kaldre**[2], **A Bojarevics**[2], **G Gerbeth**[1]

[1] Institute of Fluid Dynamics, Helmholtz-Zentrum Dresden - Rossendorf, Bautzner Landstraße 400, 01328 Dresden, Germany
[2] Institute of Physics, University of Latvia, Salaspils, Latvia

g.gerbeth@hzdr.de

Abstract. Dispersion of particles to produce metal matrix nanocomposites (MMNC) can be achieved by means of ultrasonic vibration of the melt using ultrasound transducers. However, a direct transfer of this method to produce steel composites is not feasible because of the much higher working temperature. Therefore, an inductive technology for contactless treatment by acoustic cavitation was developed. This report describes the samples produced to assess the feasibility of the proposed method for nano-particle separation in steel. Stainless steel samples with inclusions of TiB_2, TiO_2, Y_2O_3, CeO_2, Al_2O_3 and TiN have been created and analyzed. Additional experiments have been performed using light metals with an increased value of the steady magnetic field using a superconducting magnet with a field strength of up to 5 T.

1. Introduction

Nonmetallic nano-sized particles in a metal matrix can lead to improved mechanical, physical and chemical properties. One of the most important nanocomposites is the oxide dispersion strengthened (ODS) steel which has been proven of great value in nuclear fusion and fission systems [1,2] because of higher tolerance against radiation, higher operational temperature and creep resistance [3] than standard steels. Another important material group is high modulus steels (HMS) which aim in reducing the weight while increasing the stiffness of the materials [4,5].

Significant effort has been made in developing ways to produce MMNC, but until today they are mostly made in an expensive way by using powder metallurgy, ball milling or infiltration technique. A casting route in which the particles are stirred-in would be highly desirable. In general, a homogeneous dispersion just by stirring can be obtained for micro-sized particles, but not for nano-sized particles which quickly tend to agglomerate due to strong interfacial and van der Waals forces. Problems can also arise from weak particle wetting which supports the creation of particle clusters. A rather inexpensive solution is ultrasonic treatment of the melt – cavitation created by ultrasound could break particle clusters, clean the particle surface and support the dispersion [6,7,8]. Suslick et al. [9] showed that the collapse of bubbles can create "hot spots" with effective temperatures of 5000 K, pressures of 1000 atm, and heating and cooling rates above 10^{10} K/s.

Different methods which incorporate cavitation have been proposed and tested with various aluminum [10,11] and magnesium [12,13] alloys. Due to the much higher melting temperature of steel direct introduction of the ultrasonic probe into the melt is not possible. The interface between the probe and the metal is exposed to harsh working conditions which lead to fast degradation of the probe as well as contamination of the melt. Therefore, a contactless ultrasonic treatment of the melt is needed which can be achieved by electromagnetic vibration – a method first proposed by Vives [14].

A technique that could be used to create steel nanocomposites by a casting route was proposed by Grants et al. [15]. There a steady 0.5 T strong axial magnetic field is applied to a liquid metal zone heated by electromagnetic induction. Superimposition of both fields creates an alternating radial magnetic body force in a molten sample which results in an acoustic pressure field that generates cavitation. In a first step the experiments were limited to light metals. The current study describes the first admixing of different ceramic nanoparticles in stainless steel by using this contactless magnetic sonication of the melt. Additional experiments have been performed with an increased value of the steady magnetic field using a superconducting magnet of up to 5 T for SiC particles in an Al-6%wt.Mg alloy and in Sn [16].

2. Cavitation and nanoparticle dispersion in liquid metals

2.1 Experimental approach

The experimental set-ups can be seen in figure 1. A tablet containing a mixture of matrix metal and nanoparticles (1) is placed between two cooled steel supports (2) that limit the axial oscillations of the molten part. The melting is done in an argon atmosphere the pressure of which is varied between 5-500 mbar in an enclosed quartz tube (5) and water cooled copper caps (3). For additional safety a support from a refractory material (6) is used. An alternating current (AC) induction coil (7) is used for melting the metal and a direct current (DC) magnetic field is directed by the magnet poles (9). Copper caps together with the copper shielding (8) protect the electromagnet poles from the alternating magnetic field. At Institute of Physics of University of Latvia a similar set-up has been installed using for the DC field a superconducting magnet of up to 4 T with a 300mm bore diameter (see figure 1 (c,d)).

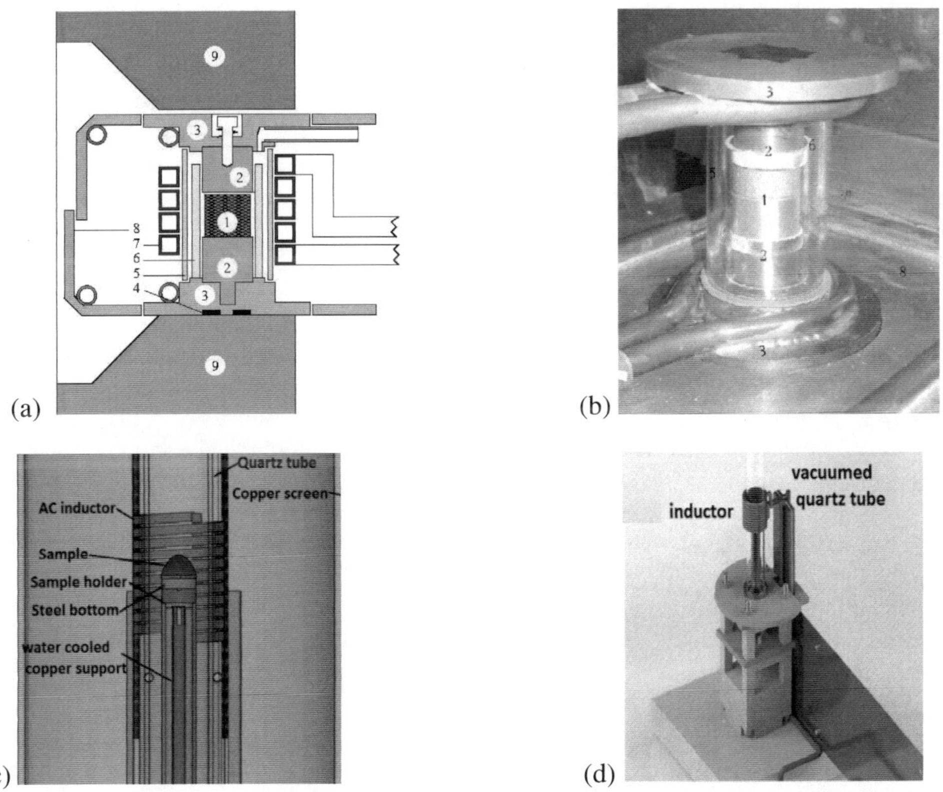

Figure 1. Scheme (a) and photo (b) of experiment in the DC field up to 0.5 T; central part scheme (c) and illustration (d) for the experiment in the superconducting magnet.

The combined action of the AC and DC magnetic fields gives rise to pressure oscillations in the sample. The amplitude p_A of those oscillations is, at least for large AC frequencies, simply given by [15]

$$p_A = \frac{B_{AC} B_{DC}}{\mu_0} \tag{1}$$

where μ_0 is the vacuum magnetic permeability. In the above described experiments this pressure amplitude reached values of up to 50 kPa in the experiments with a DC field up to 0.5 T, whereas amplitudes up to about 400 kPa have been reached in the experiments with the superconducting magnet. However, the occurrence of cavitation is not characterized a by definite threshold of this pressure amplitude but depends heavily on the availability of cavitation nuclei in the melt. Vives [14] estimated $p_A > 16$ kPa for the occurrence of cavitation, whereas Abramov [6] estimated a broad range of 150…800 kPa for it. Hence, for the present experiments not a continuous, but an intermittent observation of cavitation was to be expected.

The cavitation signal is measured by using four piezo elements (4), see figure 1(a). The height of the gap between the poles is 82 mm, the electromagnet provides up to 0.5 T in the melt zone. The maximum inductor r.m.s. current is I_{AC} = 1300 A which corresponds to B_{AC} = 0.130 T and the frequency of the AC source is around 14 kHz. Stainless steel powder (Cr16-18, Ni10-14, Mo2-3, C0.03, 5-20 μm) was mixed with TiB_2 (particle size 2-6 μm), TiO_2 (p.s. 0.01-0.03 μm), Y_2O_3 (p.s. 0.03-0.05 μm), CeO_2 (p.s. < 5 μm), Al_2O_3 (p.s. 0.02 μm) and TiN (p.s. 0.08 μm) with an approximate weight proportion of 2 % and afterwards the mixture was pressed in tablet form. The tablets were created by using a force of around 44 kN. After the melting is done a cylindrical sample with approximate dimensions of h = 45 mm, d = 20 mm is obtained which then is cut for examination.

For the experiments in the superconducting magnet the sample was prepared by compressing 50 nm SiC particles with coarse Al-6%wt.Mg alloy or Sn powder and then subjected to a superimposed 10 kHz AC field with 0.12 T amplitude and to a DC field of up to 4 T. The AC coil is placed inside the superconducting magnet, which is protected from the AC stray field by a 2 mm thick water cooled copper jacket. After melting of the sample the interaction is maintained for 10 min and then gradually reduced to solidify the sample.

2.2 Results

It is almost impossible to trace individual cavitating bubbles. Instead the noise emitted by numerous collapsing bubbles was recorded by using piezo elements. Cavitation theory suggests that signs of transient cavitation are sub-harmonics of the drive frequency $f_n = f_0/n$, n = 2, 3… as well as ultra-harmonics $mf_0 \pm f_n$, m = 1, 2… emitted as sound by the collapsing bubbles [17]. The recorded sound spectra in case of steel with different particles can be seen in figure 2. The graphs show the above mentioned uncertainty of cavitation – with initial as well as working conditions in all cases being more or less the same, the power and character of the observed cavitation differs. The case with Al_2O_3 (a) shows a very weak sign of the phenomenon as a signal of half frequency which starts around t = 1000 s, disappearing in the middle of the experiment and returning before the end. The case with TiB_2 (b) has higher signal strength, however, the cavitation is not steady and has frequent interruptions. With TiO_2 (c) it was possible to achieve cavitation only in the final period of the experiment, around t = 2300 s, while the sample with Y_2O_3 (e) experiences a rather strong and continuous influence of cavitation. Two experimental trials should be highlighted, namely CeO_2 (d) and TiN (f). The increase in pressure in the melting cell leads to the appearance of a wide spectrum noise (marked by a dashed arrow). This change might be explained by cavitation going from stable to unstable oscillating bubble fields [18]. While in theory the ambient pressure should dampen the cavitation and first experiments in [15] were done in an evacuated cell in order to promote cavitation, it was found here that an increase in pressure promotes the occurrence of such unstable oscillating bubble fields. Likely, gas diffusion in the melt increases the bubble sizes. Apparently, this increase has outweighed the decrease of negative pressure. That may explain the observed stronger cavitation signal under elevated ambient pressure.

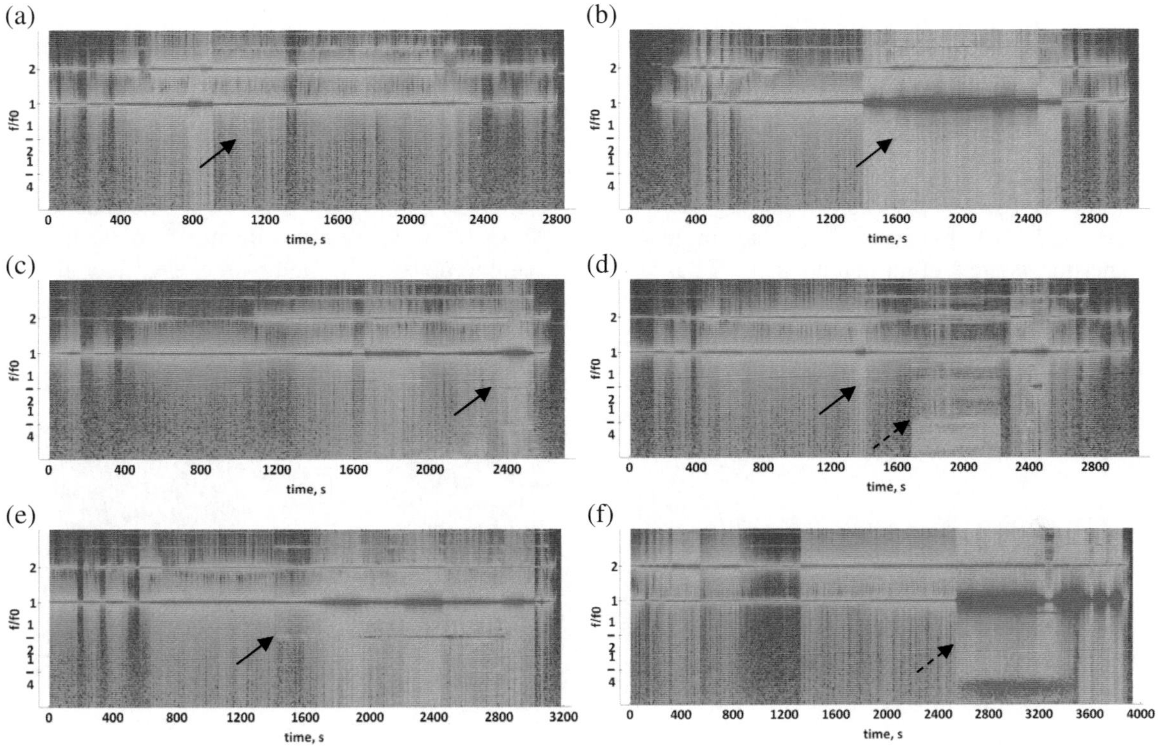

Figure 2. Temporal dependency of the sound power spectrum when cavitation is applied to 6 different samples – with Al_2O_3 (a), TiB_2 (b), TiO_2 (c), CeO_2 (d), Y_2O_3 (e) and TiN (f) particles. Full arrows indicate the start of cavitation, dashed ones the occurrence of a wide spectrum noise. Depth of logarithmic palette is 60 dB.

As mentioned above and seen in figure 2, the cavitation onset varies depending not only on the pressure oscillation amplitude, but obviously also on the added particles and the trapped gas traces. Moreover, cavitation is only a necessary precondition for the dispersion of nanoparticles, the agglomeration and dispersion certainly depends also on the wetting between particles and melt and further influences. Hence, with the cavitation signals of figure 2 as basis, it was almost clear that a full dispersion of the nanoparticles without any regions or periods of agglomeration was not to be expected.

This is reflected in the micrographs shown in figure 3. The sample analysis revealed that particle admixing varies from dispersed to highly agglomerated phases. The most promising results were achieved with TiN (f) and Al_2O_3 (a), where particles, while still being in relatively small clusters (sub-millimeter in (a) and sub-micron sized in (f)) and not dispersed evenly, could be found in the bulk of the steel matrix. TiB_2 (b), TiO_2 (c) and CeO_2 (d) particles agglomerated near the boundaries in large, submillimeter clusters (dark gray). Y_2O_3 (e) can be found inside the steel matrix in a, most likely, eutectic form covered by MgS (light gray) which could be explained with some form of chemical reaction happening between the liquid steel and the particles. The superconducting magnet results showed that in the Al-Mg alloy SiC particles are fully dispersed (g) while in tin they still are concentrated in micron sized agglomerates (dark gray) (h).

The current findings indicate that one of the main factors of successful admixing of particles is the physical and chemical compatibility between the matrix material and the particles. The compound

should be chemically stable and able to withstand the harsh environment in the melt. In addition the admixing of the reinforcements is strongly dependent on the wettability by the matrix material. Research in this field is scarce due to the fact that the wetting varies with the steel composition – each steel grade could have a different contact angle with a particular substrate in a specific temperature range. Most contact angle measurements have been done with TiN and different types of carbon steel [19,20,21], which can be used to explain the current results and suggest possible improvements. It has been concluded that TiN can be wetted well (contact angle in some cases being as low as 40 degrees) by carbon steel which also can be confirmed with stainless steel in the present experiments.
Results obtained with Al_2O_3, MgO and Ti_2O_3 in [20], where wettability by iron and steel is examined, shows that oxides (103 degrees for Al_2O_3, 90 degrees for MgO) are poorly wetted by the iron melt, but in case of Ti_2O_3 and steel, the contact angle decreased from 140 to 90 degrees. Results from cavitation experiments are similar in tendency – TiO_2 and CeO_2 seem to be badly wetted by the used steel grade. On the other hand, Al_2O_3 shows a reasonably good wetting behavior, and the dispersion could be increased with increased cavitation influence.

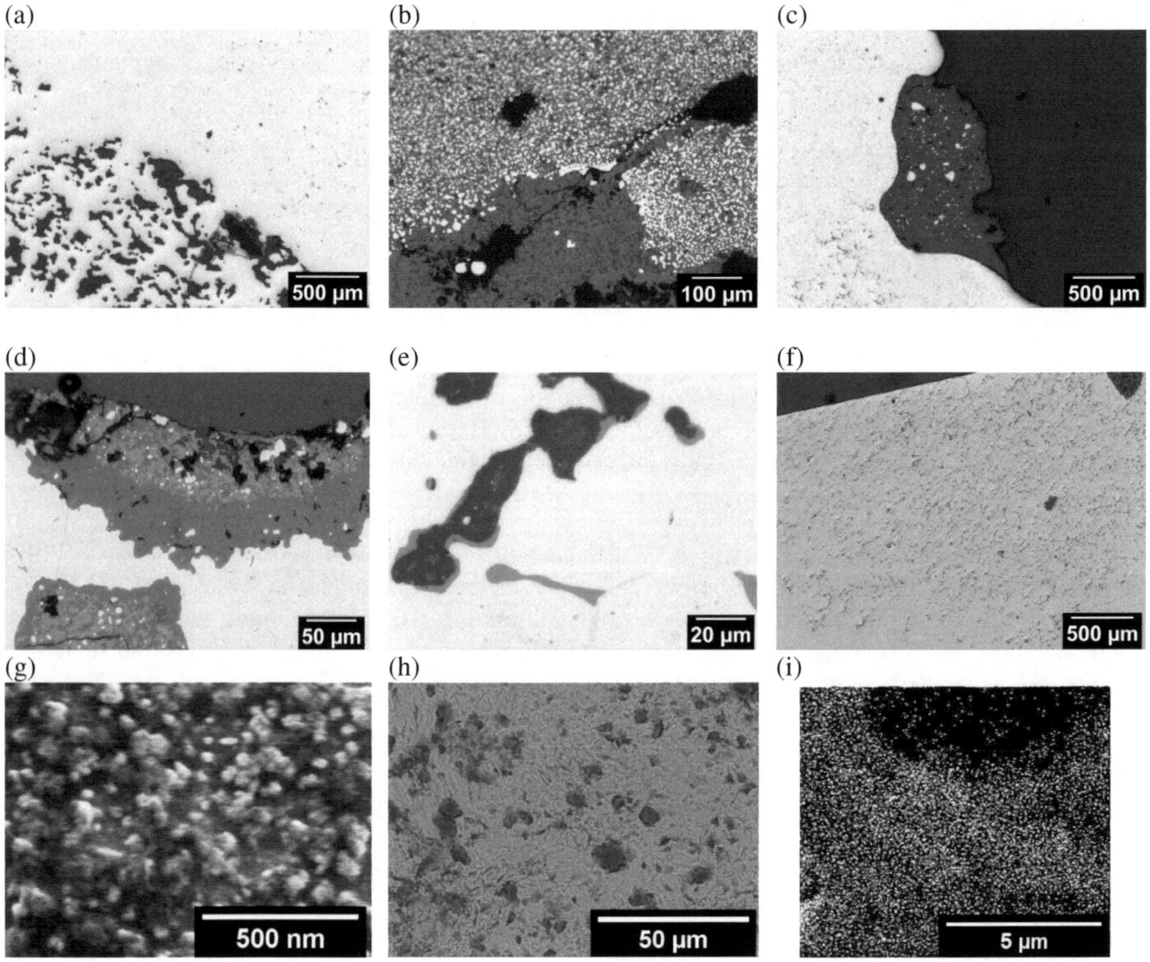

Figure 3. Micrographs with different particle inclusions in a stainless steel matrix – Al_2O_3 (a), TiB_2 (b), TiO_2 (c), CeO_2 (d), Y_2O_3 (e), TiN (f) created using DC magnetic field of 0.5 T. SEM image with samples of Al-6%wt.Mg + 50 nm SiC particles (g), Sn + 50 nm SiC particles (h) and XRF analysis (i) showing Si distribution (white) created in the superconducting magnet. Figures (g,h,i) from [16] with permission.

To improve the dispersion the first idea is to enlarge the period during which the cavitation is applied as the contact angle is expected to decrease with time since the particle surface reacts with the steel and the creation of, e.g., a Ti(N,C,O) precipitation layer on a TiN particle leads to a decrease of the contact angle as TiC is known to have a good wetting behavior by liquid iron [19]. Additionally, the temperature achieved should slightly exceed the melting temperature as in some cases it has been shown that even a 20° C increase in temperature can decrease the contact angle by about 20 degrees [21].
Due to the high level of porosity ($\rho \approx 5000$ kg/m^3 for the samples compared to appr. 7700 kg/m^3 as density of stainless steel) in the used tablets, there is reason to assume that TiB_2, TiO_2 and CeO_2 particles could have agglomerated around the gas bubbles in the melt and then were transported out of the matrix even before the application of cavitation. Taking into account these factors as well as searching for chemically stable reinforcement-steel pairs are the necessary next steps for a successful creation of steel composites by a liquid metal route.

2.3 Pulsed AC field
The continuous application of the AC induction heating may overheat the sample, which may limit the time of processing significantly or require an efficient cooling of the sample. Therefore, the idea of using a pulsed magnetic field which provides extremely high peak values while the time averaged heating is moderate was developed. The current pulses are generated with a high capacitance capacitor battery which is periodically discharged. In these experiments from 1 to 10 pulses per second are applied to the inductor which allows reaching a more than 1T peak magnetic field value. At first, a theoretical model was considered in order to understand the penetration of such pulses into a conducting volume [22] which allowed to derive relations for the pressure oscillation amplitude in the melt resulting from the superposition of an additional DC magnetic field. First experiments with such a pulsed AC field approach are in progress.

3. Summary
The experiments on a contactless electromagnetically induced cavitation and its use for the dispersion of nanoparticles in metal melts, although not yet entirely successful in creating steel composites, provide a valuable insight into the process and the related unresolved questions. By analyzing sound measurements it is possible to conclude that transient cavitation was achieved within the steel melt in all cases. The future aim is to develop a process in which by varying induction heating power, magnetic field, pressure and temperature in the melting chamber a full-scale transient cavitation could be achieved continuously.
It is difficult to disperse ceramic particles in a liquid steel, however not all important factors were yet considered in the present experiments. While TiN and Al_2O_3 revealed a good result and focus should be on improving dispersion of these reinforcements, the other particles should be further investigated as the processing time and temperature plays a key role in dispersing them. Only one type of stainless steel grade has been used in the experiments, however, different composition steels, e.g. various low carbon steels, should certainly be tested, although the mostly missing contact angle data leave the trial-and-error approach as the only option to proceed.
In order to reduce the heating action of the AC magnetic field on the sample, the idea of a pulsed AC current generated by a periodic discharge of a capacitor battery instead of a continuous application of the AC induction heater was developed. Related experiments are in progress.

Acknowledgments
The financial support of the Helmholtz Alliance "Liquid Metal Technologies – LIMTECH" is gratefully acknowledged.

IOP Conf. Series: Materials Science and Engineering **228** (2017) 012020 doi:10.1088/1757-899X/228/1/012020

References

[1] Verhiest K, Almazouzi A, de Wispelaere N, Petrov R and Claessens S 2009 *J. Nucl. Mat.* **385** 308

[2] Dubuisson P, de Carlan Y, Garat V and Blat M 2012 *J. Nucl. Mat.* **428** 6

[3] Franke P, Heintze C, Bergner F and Weißgärber T 2010 *Materials Testing* **52** 133

[4] Bonnet F, Daeschler V and Petitgand G 2014 *Canad. J. Met. Mat. Sci* **53** 243

[5] Aparicio-Fernandez R, Springer H, Szczepaniak A, Zhang H and Raabe D 2016 *Acta Materialia* **107** 38

[6] Abramov O V 1987 *Ultrasonics* **25** 73

[7] Farmer A D, Collings A F and Jameson G J 2000 *Int. J. Min. Proc.* **60** 101

[8] Bittmann B, Haupert F and Schlarb A K 2009 *Ultrasonics Sonochemistry* **16** 622

[9] Suslick K S, Didenko Y, Fang M M, Hyeon T, Kolbeck K J, McNamara W B, Mdleleni M M and Wong M 1999 *Phil. Transact. Royal Soc. A* **357** (1751) 335

[10] Yang Y, Lan J and Li X 2004 *Mat. Sci. Eng. A* **380** 378

[11] Choi H, Jones M, Konishi H and Li X 2012 *Met. Mat. Transact.* A **43** 738

[12] De Cicco M, Konishi H, Cao G, Choi H S, Turng L S, Perepezko J H, Kou S, Lakes R and Li X 2009 *Met. Mat. Transct. A* **40** 3038

[13] Shen M J, Wang X J, Li C D, Zhang M F, Hu X S, Zheng M Y and Wu K 2014 *Mat. Des.* **54** 436

[14] Vives C 1996 *J. Cryst. Growth* **158** 118

[15] Grants I, Gerbeth G and Bojarevics A 2015 *J. Appl. Phys.* **117** 204901

[16] Kaldre I, Bojarevics A, Grants I, Beinerts T, Kalvans M, Milgravis M and Gerbeth G 2016 *Acta Mat.* **118** 253

[17] Neppiras E A 1980 *Phys. Rep.* **61** 159

[18] Cramer E and Lauterborn W 1982 *Appl. Sci. Res.* **38** 209

[19] Xuan C, Shibata H, Zhao Z, Jönsson P G and Nakajima K 2015 *ISIJ Int.* **55** 1642

[20] Xuan C, Shibata H, Sukenaga S, Jönsson P G and Nakajima K 2015 *ISIJ Int.* **55** 1882

[21] Amadeh A, Heshmati-Manesh S, Labbe J C, Laimeche A and Quintard P 2001 *J. Eur. Ceramic Soc.* **21** 277

[22] Grants I, Bojarevics A and Gerbeth G 2016 *AIP Advances* **6** 065014

IOP Conf. Series: Materials Science and Engineering **228** (2017) 012021 doi:10.1088/1757-899X/228/1/012021

Casting technology for ODS steels - the internal oxidation approach

S Miran, P Franke, A Möslang, H J Seifert

Institute for Applied Materials - Applied Materials Physics, Karlsruhe Institute of Technology, 76344 Eggenstein-Leopoldshafen, Germany

Seyedhossein.miran@kit.edu

Abstract. The formation of stainless ODS steel by internal oxidation of as-cast steel has been investigated. An alloy (Fe-16Cr-0.2Al-0.05Y, wt.%) was embedded in a (VO/V_2O_3) powder mixture serving as an oxygen activity buffer and heat treated at 1450 °C for 20 h. After this procedure no oxide scale was present on the surface of the sample but a zone of internal oxidation with a depth of about 2000 μm was formed in its interior. The precipitates within this zone consisted of two types of oxides. Discrete aluminium oxide particles with a size of a few micrometres were formed in outer regions of the specimen. Finer aluminium-yttrium oxides with a size of some hundred nanometres were mainly precipitated in inner regions of the sample. The results can be considered as a promising step towards an alternative production route for ODS steels.

1. Introduction

Reduced activation ferritic-martensitic (RAFM) steels are promising candidate materials for structural applications in fusion reactors. These steels have been developed based on massive industrial experiences of ferritic/martensitic steel [1]. RAFM steels have interesting properties such as low sensitivity to irradiation-induced swelling and helium embrittlement [2].

RAFM steels are very attractive structural materials but their high-temperature strength is limited and one of the effective approaches to solve this issue is oxide dispersion strengthening (ODS) method [3]. The nano-sized oxide particles with high number density can act as pinning points to dislocation movement [4]. Dispersed nano-sized oxides may provide a large number of trap sites for transmutant helium and radiation-induced defects [5]. The improved creep resistance and fatigue strength at high temperatures is another advantage of ODS steels [6].

The typical procedure for manufacturing ODS steels is the powder metallurgy route in which steel powders and Y_2O_3 powders are mechanically alloyed by high energy milling. The input oxides remain unmixed but they are dispersed throughout the microstructure of the alloy [7]. This technique was first developed by J. S. Benjamin and published in 1970 [8].

The next step in this route is hot isostatic pressing (HIP). In HIP process, the simultaneous application of a high pressure at elevated temperatures in a constructed vessel is planned for hot compaction of the powders. The pressure inside HIP chamber is generated usually by applying purified argon gas. In the later steps after HIP, the deformed powders are hot extruded or rolled for the desired shape.

Although this method for ODS steels is well documented, it is considered a complex and costly fabrication technique [9, 10]. In the present study an alternative process for the production of stainless ODS steels has been investigated. Firstly, homogeneous steel was prepared from the melt and in a subsequent step a dispersion of oxide particles was precipitated by internal oxidation in the solid state.

2. Background

Internal oxidation means the fabrication of oxides in the interior of an alloy, and more specific in the context of the present work, that the growth of an external oxide scale at the surface of the sample is avoided. A quantitative description of the internal oxidation has been given by Wagner [11], especially for the case of binary alloys where a less-noble metal (here Al) forms a dilute solution in the more noble matrix metal (here Fe). An English review of this theory has been given by Rapp [12].

For the occurrence of internal oxidation several requirements need to be fulfilled: the oxide of the less-noble element must have a strong negative Gibbs energy of formation, the diffusion of oxygen from the surface of the alloy into the interior must be faster than the diffusion of the less-noble metal into the opposite direction, and the concentration of oxygen in the alloy at its surface should be much smaller than the initial concentration of the less-noble metal [11]. Although the present investigation considers a four-component alloy, these criteria remain basically valid when Al and Y are considered as the less-noble components.

In the current work, the internal oxidation of a selected alloy of Fe-Cr-Al-Y has been investigated. The primary goal for this research is to provide a simple and fairly inexpensive method to disperse oxide particles on the matrix of steel.

The oxidation behavior of the main involved elements (Fe: main component, Cr: alloying element, Al and Y: oxide-forming elements) is a key concept for the successful internal oxidation. In this context, a successful result is the oxidation of oxide forming elements (Al and Y) within the matrix of steel without the formation of Fe or Cr oxide scales.

In order to obtain this goal, the equilibrium oxygen partial pressures, i.e. the oxygen activities (a_{O_2}) at different temperatures for the oxidation of pure elements in the examined system (Fe, Cr, Al and Y) were calculated. The results versus inverse temperature (1000 K/T) have been plotted in figure 1. This plot is known as Ellingham diagram which is a fundamental tool for the understanding of the oxidation behavior of metals. Some selected temperatures in Celsius with vertical line have been shown in figure 1 as well. The thermodynamic databases used for this purpose were four databases referred as [13], [14], [15] and [16].

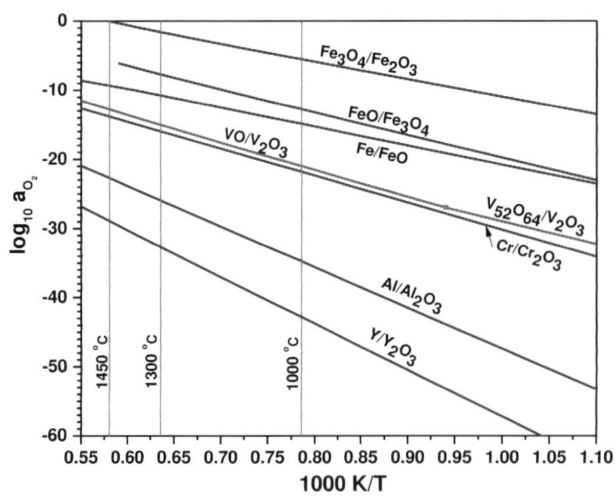

Figure 1. The calculated oxygen activity versus temperature for different metal/oxide equilibria.

IOP Conf. Series: Materials Science and Engineering **228** (2017) 012021 doi:10.1088/1757-899X/228/1/012021

It can be seen (figure 1) that at defined temperature for example at 1000 °C, the equilibrium reactions of (Al/Al_2O_3) and (Y/Y_2O_3) show lower oxygen activity data relative to Fe and Cr equilibrium oxidation reactions. This strengthens the possibility of internal oxidation in a steel alloy.

It should be reminded that in Ellingham type approach, the activities of metallic elements are equal to one, whereas the activity values of Fe, Cr, Al and Y in a real steel alloy are lower than one.

Moreover, this diagram (figure 1) was used to evaluate the oxygen buffer systems to fix the oxygen activities for the experiment. In this work, solid-state powder mixture to adjust the oxygen partial pressures for the purpose of this work was selected. To calculate the candidate buffer system, the dataset for the V-O system [17] in addition to the above stated databases was used.

The most important criterion for the choice of oxygen buffer system is the location of its equilibrium reaction line in the Ellingham diagram. The line has to be above (Al/Al_2O_3) and (Y/Y_2O_3) line and lower than (Fe/FeO) equilibrium line and ideally close to (Cr/Cr_2O_3) line. This is important since Al and/or Y has to be selectively oxidized relative to Fe and Cr as earlier noticed. Based on this concept, the equilibrium mixture of (VO/V_2O_3) seems to be a potential candidate for the purpose of internal oxidation (see figure 1). In this paper, the result of oxidation for a selected Fe-Cr-Al-Y alloy with this buffer has been reported.

3. Experimental procedure

A high purity alloy has been prepared by arc-melting in argon atmosphere (Ar 6.0). The starting materials were the metal pieces of Fe, Cr, Al and Y. The first three elements were 99.99% pure and Y was 99.9% pure. The metallic elements were all obtained from Alfa Aesar GmbH & Co KG.

The produced buttons were flipped and remelted five times in order to improve their homogeneity. Table 1 presents the composition of the alloy after its production which was determined by ICP-OES (Inductively coupled plasma-optical emission spectroscopy) method.

Table 1. The composition of the prepared alloy.

Alloy designation	Nominal composition (wt.%)				Chemical analysis (wt.%)			
	Y	Al	Cr	Fe	Y	Al	Cr	Fe
Fe-16Cr-0.2Al-0.05Y	0.05	0.2	16.0	83.75	0.0537	0.214	15.8	82.6

The chemical analysis (see Table 1) shows a good agreement with nominal compositions. The reported standard deviations (SD) for each element was acceptable as well (SD_Y = 0.0011, SD_{Al} = 0.002, SD_{Cr} = 0.04, SD_{Fe} = 0.1). In as-cast microstructure, the segregation of yttrium or aluminum was not detected.

Afterwards, a cube specimen from as-cast sample with a few millimeters size and a weight of approximately 1 g was produced. In order to standardize the surface conditions, all the sides of the specimens before the oxidation experiment were ground up to grade #4000 (SiC, Struers GmbH).

The oxidation step was performed by embedding the cube specimen into the selected equilibrium oxide powder mixture using (VO/V_2O_3) buffer system. The experiment was carried out for 20 h at 1450 °C.

The heating rate of the furnace from room temperature to the final temperature was fixed to ~ 4 °C/min. The experiment was performed under argon atmosphere (Ar 6.0). The starting constituents of the buffer and its oxygen activity at the defined temperature are given in Table 2. The equilibrium components of oxygen buffer mixtures (VO/V_2O_3) are expected to be formed during annealing (see Table 2).

Table 2. The composition of oxygen buffer system for the oxidation experiment described in this work.

Equilibrium components of oxygen buffer mixture	T(°C)	Provided oxygen activity ($\log_{10} a_{O_2}$)	Equivalent oxygen partial pressure (P_{O_2} in bar)	Starting constituents	Weight ratio of the starting constituents	Total weight
VO/V_2O_3	1450	−12.8	1.61×10^{-13}	V/VO_2	0.35 : 1	2 gr

The reason to carry out the experiments at high temperature and to put the alloy in direct contact to the buffer mixture is to make the kinetics of the transport of oxygen from the buffer to the alloy faster.

After the oxidation experiment, the specimen was cooled down to room temperature inside the furnace. The attached powders were then removed from the surface of specimen by mechanical grinding. The microstructure of the specimen was then imaged with scanning electron microscopy (SEM), FEIXL30S, coupled with EDX detector allowing spot and line scan analyses.

4. Results and Discussions

The result at higher temperature experiment (1450 °C) for Fe-16Cr-0.2Al-0.05Y by a direct contact of the alloy and powder mixture (VO/V_2O_3) for 20 h is shown in figure 2.

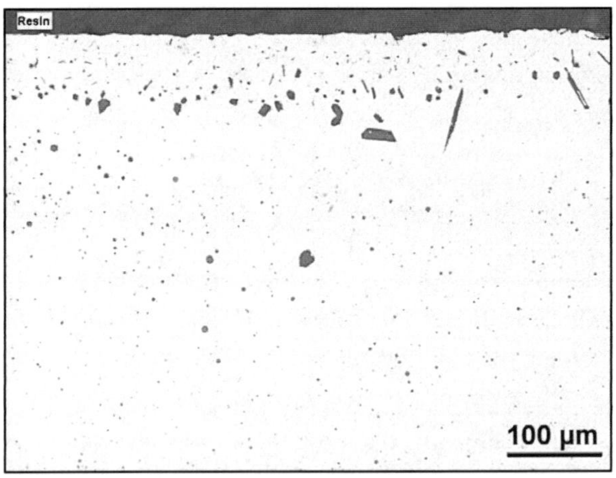

Figure 2. The internally oxidized zone Fe-16Cr-0.2Al-0.05Y with (VO/V_2O_3) buffer for 20 h at 1450 °C on a surface approximately through the middle of sample (after 2 mm grinding from one side). The images are cross-section backscattered-micrographs (magnification of 200x). The coarser oxide particles near the surface and finer oxide particles in the inner regions can be seen.

It can be seen that in outer regions of the sample, the coarser particles are present, whereas the finer particles in the inner regions have been formed. In fact two distinguishing varieties in the size of the particles can be observed. A typical EDX line scan profile for coarser particles is shown in figure 3.

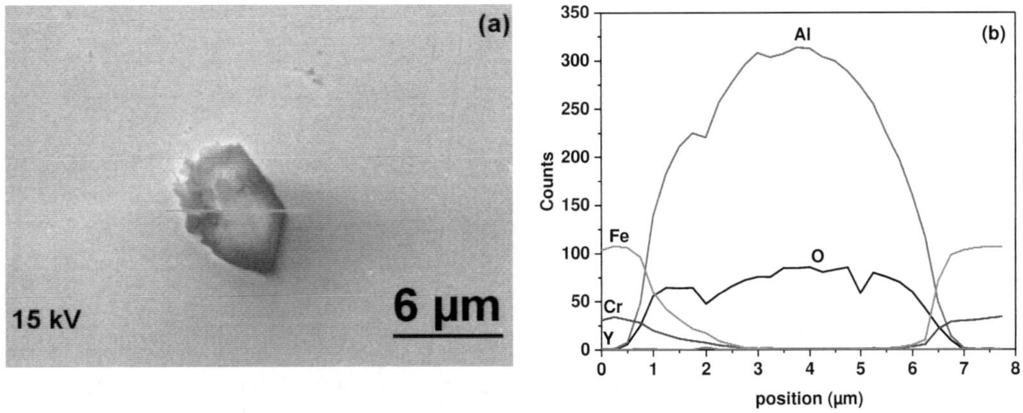

Figure 3. A typical EDX analysis for the coarser formed particles in Fe-16Cr-0.2Al-0.05Y at 1450 °C in outer regions of the sample after 20 h embedded in (VO/V_2O_3) oxygen buffer: (a) secondary electrons-micrograph and indicated path for line scan (magnification of 4800x), (b) the corresponding composition profile using 15 kV acceleration voltage showing the formation of aluminium-oxide particle.

As suggested by the EDX profile in figure 3, the coarser particles (with about a few micrometers size) are aluminum-oxide which are mostly near the surface of the specimen. The finer oxide particles (approximately some hundred nanometers) were located mainly in the interior of the sample consisting of mixed aluminum-yttrium oxides. This has been shown by an EDX line scan profile in figure 4.

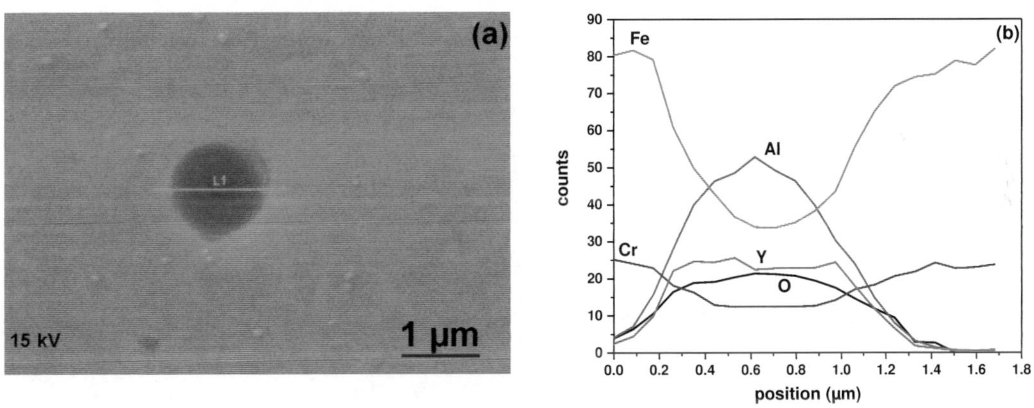

Figure 4 A typical EDX analysis for the finer formed particles in Fe-16Cr-0.2Al-0.05Y at 1450 °C located mainly in inner regions of the sample after 20 h embedded in (VO/V_2O_3) oxygen buffer: (a) secondary electrons-micrograph and indicated path for line scan (magnification of 20500x), (b) the corresponding composition profile using 15 kV acceleration voltage showing mixed aluminum-yttrium oxide particle.

It should be noted that the number of counts for Fe and Cr in the middle of path does not go down to zero (see figure 4b). This effect can be attributed to the small size of the particle which is in the range of the volume activated by the electron beam leading to contributions from the surrounding steel matrix. It is worth mentioning that the internal oxidation depth was approximately 2000 µm. The observed results are the successful internal oxidation examined steels by the formation of desirable oxides particles (Al-O and Al-Y-O precipitates). External scales were not detected for the alloy as well.

5. Conclusions

The oxidation of Fe-16-0.2Al-0.05Y (wt.%) embedded into a (VO/V_2O_3) powder mixture at 1450 °C for 20 h led to an internally oxidized zone with a depth of about 2000 μm. The precipitates within this zone consisted of two types of oxides. Larger oxide particles (a few micrometers) were mainly formed near the surface and they were identified as aluminum oxide particles by EDX method. The finer oxide particles (approximately some hundred nanometers) were located mainly in the interior of the sample. They are identified as mixed aluminum-yttrium oxide particles.

At present, the oxide dispersions are still produced at relatively high temperatures and long oxidation times. As a result, the grain size of the matrix alloy and the particle sizes of the oxides are higher than in ODS steels produced by the powder metallurgical process. The oxide particles in these steels are still smaller by up to two orders of magnitude than in our alloys. Therefore we expect that in our ODS steels, the strengthening effect is clearly below that of the powder metallurgical alloys. Currently experiments are in progress to prepare by internal oxidation samples for tensile tests in order to determine the strengthening effect quantitatively.

Acknowledgement
The financial support of this work by the Helmholtz Alliance LIMTECH is gratefully acknowledged.

References

[1] Shiba K, Tanigawa H, Hirose T, Sakasegawa H and Jitsukawa S 2011 *Fusion Engineering and Design* **86** 2895
[2] Ehrlich K 2001 *Fusion Engineering and Design* **56-57** 71
[3] Kimura A 2005 *Materials Transactions* **46** 394
[4] Mao X, Kim T K, Kim S S, Han Y S, Oh K H and Jang J 2015 *Journal of Nuclear Materials* **461** 329
[5] Lu C, Lu Z and Liu C 2013 *Journal of Nuclear Materials* **442** S148
[6] Bruchhausen M, Turba K, De Haan F, Hähner P, Austin T and De Carlan Y 2014 *Journal of Nuclear Materials* **444** 283
[7] Capdevila C, Bhadeshia H 2001 *Advanced Engineering Materials* **3** 647
[8] Benjamin J S 1970 *Metallurgical Transactions* **1** 2943
[9] Rieken J R Anderson I E, Kramer M J, Odette G R, Stergar E and Haney E 2012 *Journal of Nuclear Materials* **428** 65
[10] Xie R, Lu Z, Lu C and Liu C 2014 *Journal of Nuclear Materials* **455** 554
[11] Wagner C 1959 *Zeitschrift für Elektrochemie* **63** 772
[12] Rapp R A 1965 *Corrosion* **21** 382
[13] TCFE7 TCS Steels/Fe-Alloys Database v7.0
[14] Kjellqvist L, Selleby M and Sundman B 2008 CALPHAD **32** 577
[15] TCBIN TC Binary Solutions Database v1.1
[16] Swamy V, Seifert H J and Aldinger F 1998 *Journal of Alloys and Compounds* **269** 201
[17] Yang Y, Mao H and Selleby M 2015 *CALPHAD* **51** 144

IOP Conf. Series: Materials Science and Engineering **228** (2017) 012022 doi:10.1088/1757-899X/228/1/012022

Two cylinder permanent magnet stirrer for liquid metals

A Bojarevičs, R Baranovskis, I Kaldre, M Milgrāvis, T Beinerts

Institute of Physics University of Latvia, 32, Miera Street, Salaspils, LV-2169, Latvia
imants.kaldre@lu.lv

Abstract. To achieve a uniform liquid metal composition and temperature distribution, stirring is often necessary for industrial processes. Here, a novel permanent magnet system for liquid melt stirring is proposed. It promises very low energy consumption and options for multiple different flow types compared to traditional travelling magnetic field inductors or mechanical stirrers. The proposed system has a simple design: it consists of two rotating permanent magnet cylinders, which are magnetized transversely to the axis of the cylinders. The experimental device was developed and tested under various regimes using GaInSn alloy in a cylindrical crucible. Aluminum stirring by permanent magnets in laboratory scale is tested, and stirring impact on directional solidification of metallic alloys is experimentally investigated.

1. Introduction

In metallurgy, liquid metal flow control - transport and stirring – is an especially challenging task due to the high temperatures and aggressive environments. Molten metal stirring is necessary to achieve uniformity of composition and temperature throughout the volume of liquid [1]. For these tasks, mechanical stirring is limited due to high temperatures and erosion of the walls, leading to melt contamination and short service life. For contactless flow control, traveling or rotating magnetic field AC inductors are commonly used. Due to high power losses in windings caused by the resistance and resulting in useless Joule heating, the efficiency of these systems is low [2]. Permanent magnet systems contrarily do not consume any energy to create a magnetic field. Driving motors only need to overcome forces related to induced secondary magnetic fields in conducting media (both liquid metal and conducting construction elements). Nowadays increased availability of NdFeB permanent magnets allows creating affordable large permanent magnet systems, which can be successfully used to create motion in liquid metals [3-5]. Since magnetic flux decreases rapidly with the distance from the permanent magnet, the size of magnets should be comparable to the distance from the magnets to the liquid metal.

We propose a novel rotating permanent magnet system that provides easy transition between various types of the flow patterns, even switching between mixing in of the particles and stirring, and pumping, dependent what would be required at the production stage. Compared to traditional traveling field AC inductors, the proposed system consumes less energy and, more importantly, provides many options for various configurations. As a substitute for liquid aluminum for some experiments in a laboratory environment, GaInSn eutectic alloy is used. It has similar properties to liquid aluminum and it is convenient to use since it is in a liquid state at room temperature. Ultrasonic velocimetry has been successfully used to measure GaInSn alloy flow velocity.

During solidification of liquid metals, the actual problem is segregation and temperature anisotropy at the solidification interface. Thus, forced convection in the melt can be the beneficiary and help to improve the solidified alloy microstructure [6]. Solidification experiments have been done, and motion influence on the structure has been investigated, demonstrating that flow control is an

IOP Conf. Series: Materials Science and Engineering **228** (2017) 012022 doi:10.1088/1757-899X/228/1/012022

instrument to modify the solidification process [7,8]. In solidification experiments Al-5%wt.Cu and Sn-10%wtPb were used.

2. Experimental setup

The experimental setup was built taking account physical and geometrical similitude so the model results would help to determine potential industrial applications of such system. Principal scheme and picture of the experimental setup are shown in figure 1. The experimental system consists of electric motor, two cylindrical magnets (D=40 mm, H=80 mm or D=50 mm, H=100 mm) type N50 with a remanent flux density of 1.4 T, that are connected by a system of timing pulleys and belt, allowing synchronous rotation either in one direction or opposite directions. Two permanent magnet cylinders, which are magnetized transversely to the axis of the cylinder, are placed on opposite sides of the liquid metal container. A relatively large distance between the magnets ensures minimal magnetic interaction between themselves, therefore providing a smooth and energy efficient system of operation. By changing the magnet's position relative to the container it is possible to create numerous types of flows even with no axial symmetry, which can be very beneficial in certain types of situations. Such versatility and simplicity of the system enables it for a wide range of different applications.

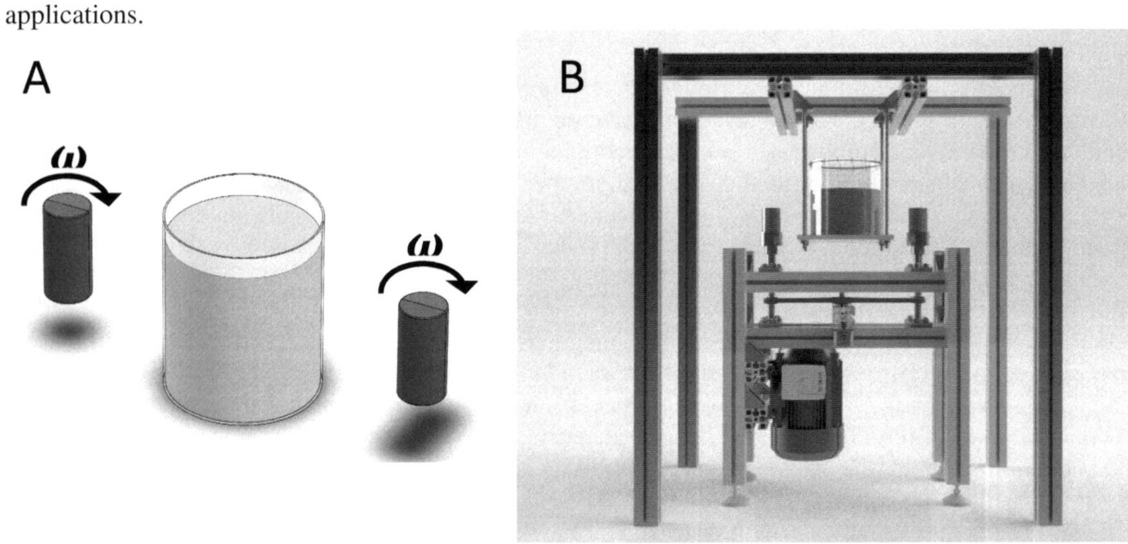

Figure 1. Experimental setup. A) schematic illustration; B) Experimental device.

The system is designed in such way that it is possible to vary magnet positions along all three axes and tilt (θ) relative to the container, thus enabling us to create numerous types of flows. Tilt angle θ is defined as the angle between the main axis of cylinder and z-axis. The magnetic flux density distribution is calculated by Comsol 5.0 software. Numerical results of the magnetic field at a distance between aligned magnet cylinders of 20 cm is given in figure 2. These numerical results were compared to Gaussmeter measurements and agreement is good. The strength of the magnetic field in liquid metal volume ranges from 20 mT at the center and 60 mT near the edge of the container in case of 40 mm magnets which were used in GaInSn stirring experiments.

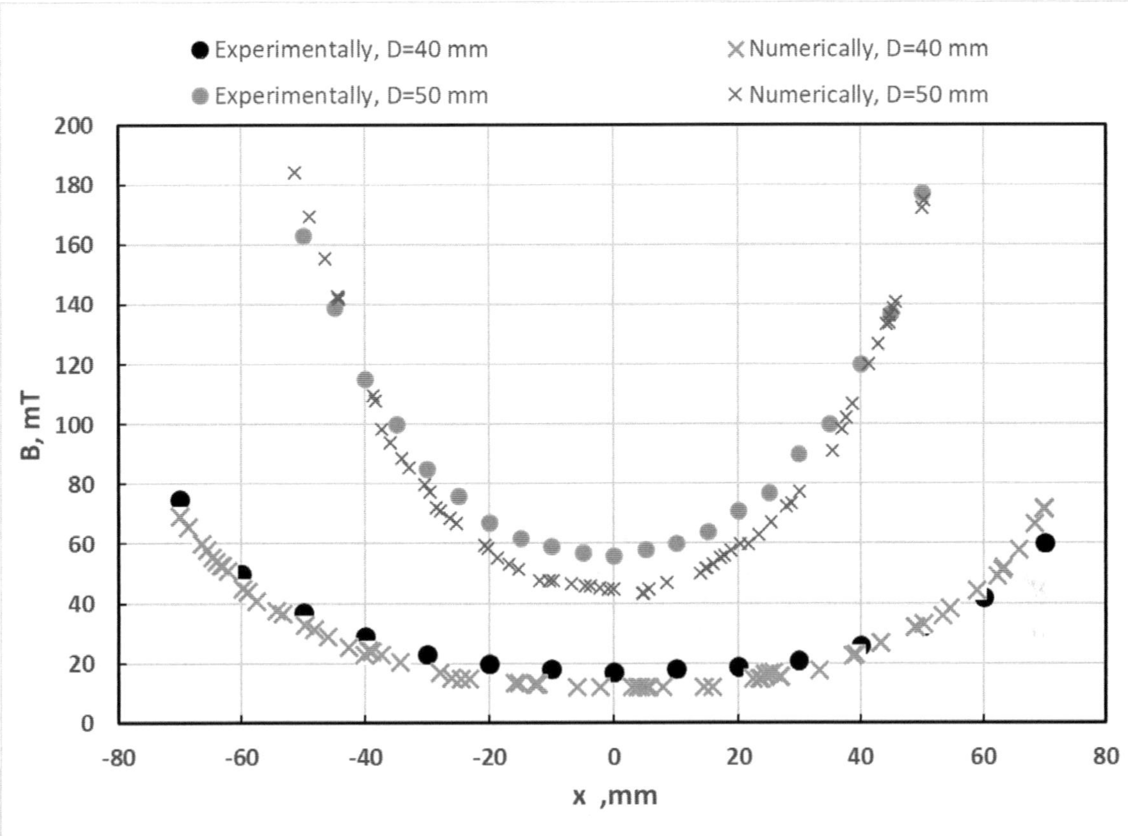

Figure 2. Magnetic field distribution between two permanent magnet cylinders (D=40 mm, H=100 mm and D=50 mm, H=100 mm) at a distance 25 cm and 20 cm between the magnet's axis.

3. Order of magnitude estimates of flow velocity

In this study, we develop an estimation to scale experimental results to different geometries and material properties. Ultrasound velocity measurements were done at the GaInSn alloy experiment at room temperature, and results were used for scaling to different materials and crucible sizes. In this work, we performed several experiments to demonstrate electromagnetic stirring and stirring influence on directional solidification of a metallic alloy. Experimental parameters are summarized in table 1.

Table 1. Summary of experimental conditions

	ρ, kg/m^3	σ, S/m	R, mm	H, mm	B, mT	f, Hz ($\omega, rad\ s^{-1}$)	N	Ω_d
GaInSn	6360	$3.3\cdot10^6$	70	130	20	2.9 (18.2)	0.0014	0.370
Sn-10%wt.Pb	7000	$2\cdot10^6$	30	100	50	10 (62.8)	0.0014	0.014
Aluminum	2400	$4\cdot10^6$	30	100	50	5 (31.4)	0.0211	0.177

Following dimensionless criteria must be met for physical similitude between two cases. Physical similitude requires full similarity of the geometry of the two setups and identity of a set of dimensionless criteria. Order of magnitude estimates of the flow velocity is done neglecting viscosity and assuming that the induced magnetic field is much smaller than the field produced by the magnets.

Firstly, the magnetic interaction parameter (Stuart number) N is the ratio between electromagnetic and inertial forces, secondly, and Ω_d is the dimensionless frequency characterizing the skin effect of an alternating magnetic field. Since there is no given flow velocity scale electromagnetic interaction parameter is defined by applying velocity scale of travelling magnetic field speed $\omega \cdot R$. If these parameters are similar, then similar force distributions and flow regimes can be expected.

$$N = \frac{k}{4} \cdot \frac{\sigma}{\rho} \cdot \frac{B^2}{\omega} \quad and \quad \Omega_d = \sigma \cdot \omega \cdot \mu_0 \cdot R^2 \tag{1}$$

where ρ is the density of the melt, σ is conductivity, ω is magnet angular speed, μ_0 – magnetic permeability, R-characteristic length, in this case radius of metal pool and B is the magnitude of the magnetic field induction in the liquid metal region. Dimensionless coefficient k is introduced to account for geometry of induced current loop's. In following estimate, it is set to 0.5.

$$k = \frac{1}{1 + \left(\frac{\tau}{H}\right)^2} \tag{2}$$

τ is the azimuthal distance over which the phase of the magnetic field changes by 180 degrees. H is the height of the magnet or the melt region (whichever is the smallest). In case similitude criteria are met, the velocity of the liquid metal can be expressed by balancing electromagnetic and inertial forces, which yields (3). This estimate is valid when $\Omega_d \ll 1$ and it does not include viscosity and turbulence.

$$\frac{1}{2} \cdot k \cdot \frac{\sigma}{\rho} \cdot \left(\omega - \frac{u}{R}\right) \cdot R \cdot B^2 = \frac{\rho \cdot u^2}{R} \tag{3}$$

Solving equation respect to u and substituting $\frac{k}{4} \cdot \frac{\sigma}{\rho} \cdot \frac{B^2}{\omega}$ with N yields equation for velocity (4)

$$u = \omega \cdot L \cdot (\sqrt{N^2 + 2 \cdot N} - N) \tag{4}$$

For f=10 Hz in liquid Sn-10%wt.Pb Ω is 0.014 which is a small value. Similar N in GaInSn flow is achieved at f=2.9 Hz, which gives a maximum flow velocity of 7 cm/s.

At field rotation frequency of 10 Hz, we get a maximum velocity in Sn-10%wtPb alloy of 25 cm/s and 40 cm/s for Al-5%wt.Cu alloy. The calculated Peclet number is $4 \cdot 10^2$. Therefore, these velocities are sufficiently large and will have a significant effect on heat transfer in the liquid metal volume and its solidification.

For efficient stirring, not only flow velocity, but also the flow type is important. With proposed system, it is possible to create numerous different flow types even with no axial symmetry. Four characteristic flow types are shown in figure 3. In the case counter rotation (figure 3A), we observe complex flow with fluctuations in velocity and sometimes change in direction. The achieved velocities are comparatively low. The second configuration is co-rotating magnets (figure 3B,) created flow is one fast spinning eddy. This setup has highest tangential velocity, however, there is little z component of flow and in center, velocity approaches 0. If the magnets are tilted horizontally (Figure 3C) same circular motion is obtained, but now the plane of circulation is vertical(xz). By setting magnets at tilted position figure 3D, a combination of above-mentioned flows is obtained. This setup is most versatile of all providing high flow velocity and movement of fluid in all three dimensions.

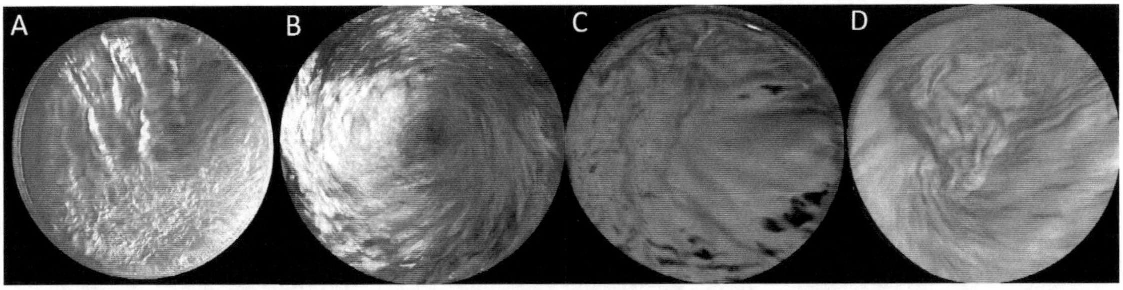

Figure 3. Four characteristic flow types, top view of the free surface A) Counter rotation, $\theta = 0^{0}$; B) co-rotation, $\theta = 0$; C) co-rotation, $\theta = 90^{0}$; D) co-rotation, $\theta = 45^{0}$.

4. Results and discussion

The measured maximal velocity in GaInSn container with horizontal and vertical magnet alignment is shown in figure 4. To calculate velocities in Sn-10%wt.Pb and in aluminum, we use scaling parameters.

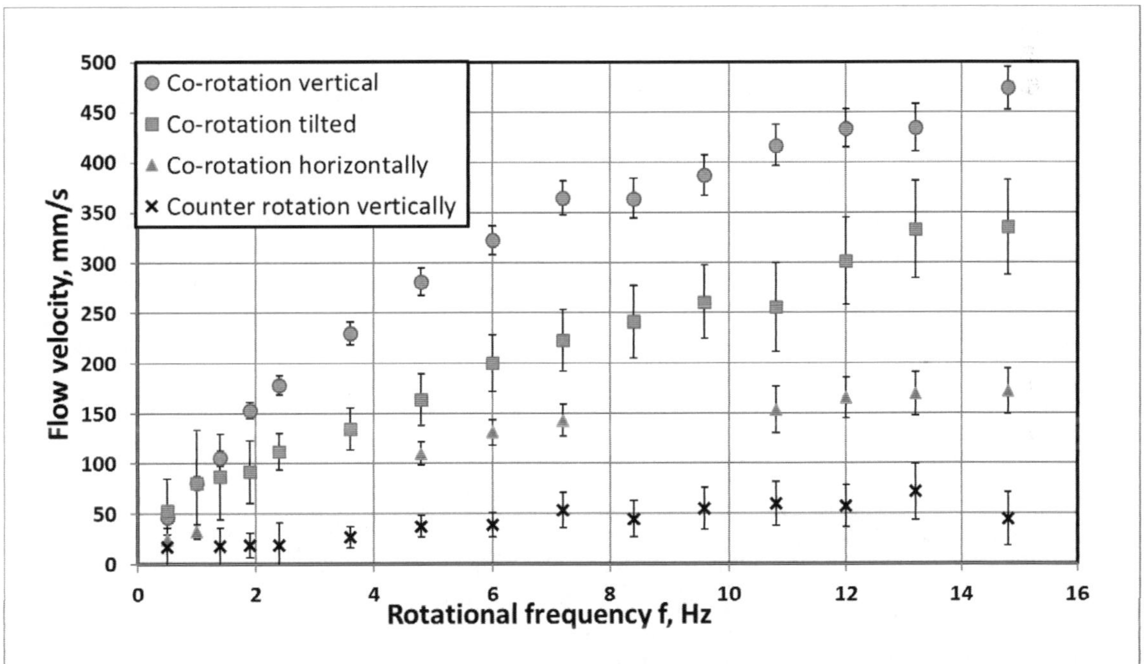

Figure 4. Velocity ultrasound measurements in GaInSn. Characteristic velocity versus magnet rotation frequency.

Velocity measurements in figure 4 are characteristic values of the flow in the most intense direction around 2 cm from crucible wall to avoid any boundary layers. Measurements showed that in all variations higher velocities are in case of co-rotation. Vertical magnet position gives the highest velocities but has the lowest turbulence – so, it can be used in metal continuous directional solidification to refine grain size, but not for temperature and structure homogenization.

Intense turbulence and heavy surface deformations in the case of tilted magnet variation can be perspective to mix in particles or scrap, which is a common problem in industrial manufacturing. Aluminum scrap (~0.1-1 mm in diameter) were added to overheated aluminum, see figure 5.

IOP Conf. Series: Materials Science and Engineering **228** (2017) 012022 doi:10.1088/1757-899X/228/1/012022

Figure 5. Time steps of mixing in of the aluminum scrap in overheated aluminum. Magnets are tilted in a 45-degree angle.

Metal flow pulls in the scrap in liquid metal even when a heavy oxide layer is on the metal surface. In steady metal particles and scrap tends to stay on the metal surface and slowly melts – takes around 1 minute.

Solidification experiments with Sn-10%wt.Pb are carried out to verify the stirring effect on the microstructure of the alloy. The bottom of the crucible was cooled by a copper heatsink which was maintained at a constant temperature with water flow as shown in figure 6. The sample is melted and then directionally solidified under electromagnetic stirring. The temperature gradient is 18 K/cm, growth velocity 0.4 mm/s in reference experiments. Measurements and samples were taken at h=30 mm from the bottom. Also, it was observed that both types of stirring – horizontal and vertical configuration - decreased total solidification time by a factor of 2 when compared to reference with no stirring. With the faster rotation of magnet cylinders severe surface deformation of the liquid aluminum alloy was observed. This confirms the calculated velocity order of magnitude of 40 cm/s.

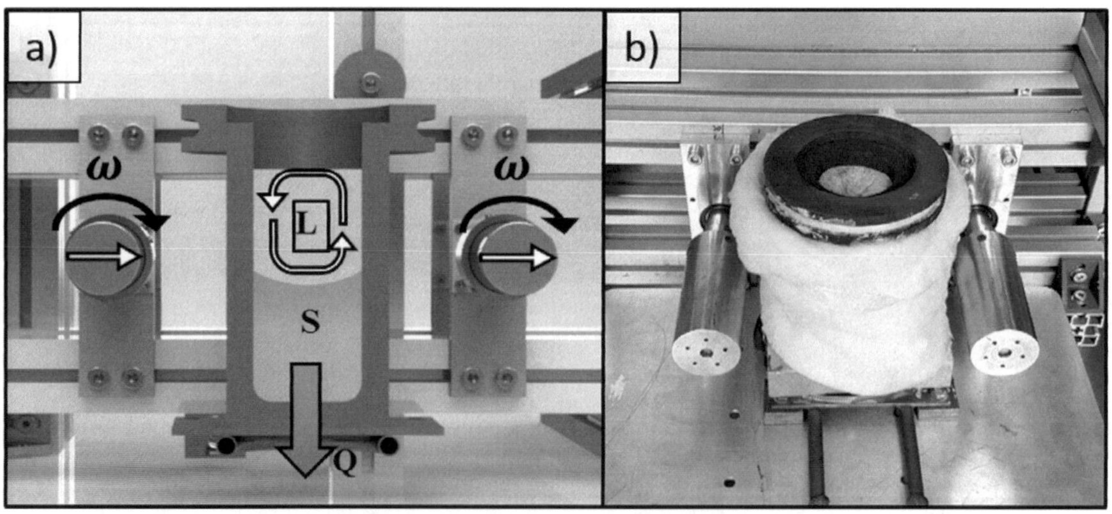

Figure 6. Setup for directional solidification with horizontal magnet configuration: a) concept; b) experimental setup.

Solidification structure is shown in figure 7. Horizontal and vertical cross-sections of the reference sample and the sample solidified under electromagnetic stirring are shown. In figure 5 (b) and (d) oriented electromagnetic stirring is directed from the right to the left side. It can be seen, that due to the stirring, the structure at the same solidification conditions is more uniform. Inducing metal flow solves an actual problem in some metallurgy areas where solidification structure is mainly defined by heat flow direction. Here it is shown that this orientation can be avoided by induced electromagnetic stirring. Although the average grain size is not decreased, the anisotropy of the structure is decreased. Feasibility to stir liquid aluminum alloys by permanent magnet stirrer has also been verified. The microstructure of alloy which is solidified under electromagnetic stirring is compared with the microstructure without stirring. Stirring decreases solidification time significantly due to enhanced convective heat transport.

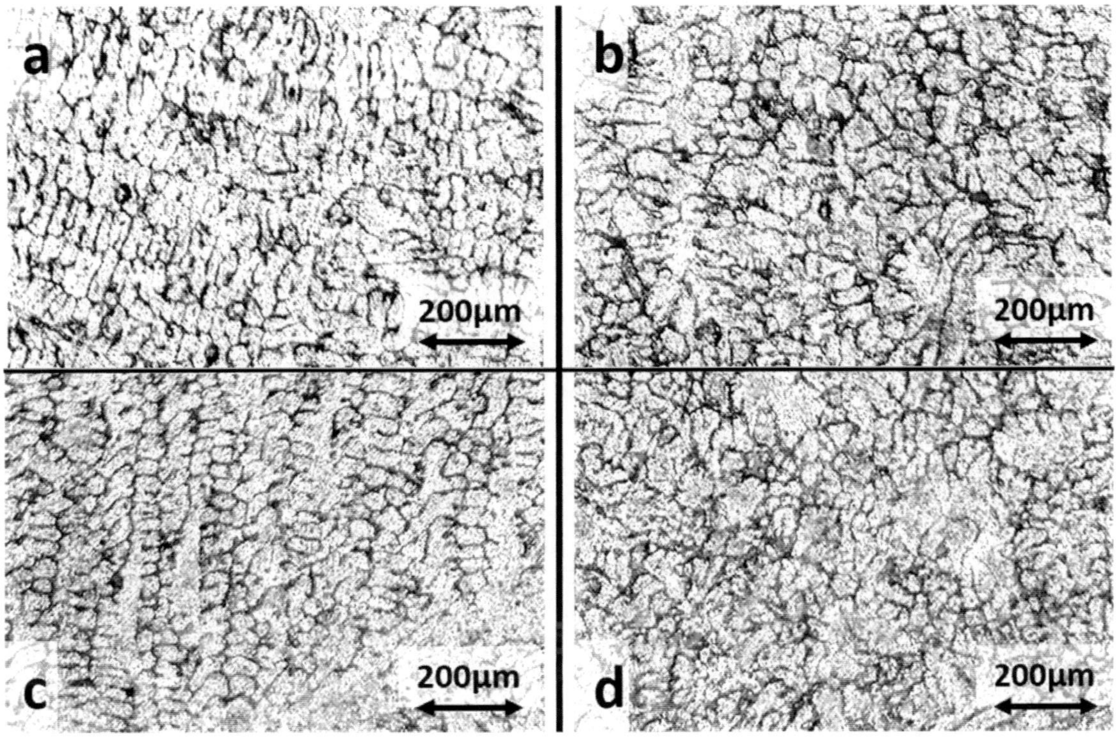

Figure 7. Polished cross-sections of directionally solidified Sn-10%wt.Pb alloy under electromagnetic stirring. Left side: xy plane (a) and xz plane (c) of reference sample (v=0.4 mm/s, grad(T)=18 K/cm); right side: xy plane (b) and xz plane (d) for solidification under stirring.

5. Conclusions

In this work the performance and potential applications of two-cylinder permanent magnet stirrer is investigated. Dimensionless characteristic numbers were introduced allowing scaling GaInSn test model results to different materials and crucible sizes. Two cylinder system performed as expected – multiple types of flow with high intensity can be created and used efficiently by such stirrer on liquid aluminum. Two most prominent flow types were tested in mixing in of the scraps and solidification experiments of Sn-10%wt.Pb to verify forced convection influence on solidification structure. The study showed that scrap mixing in might be a prospective application of such stirrer with the potential to integrate it in industrial processes. As demonstrated in this work this setup might be used for metal homogenization or structure improvement during solidification of various metals and that the solidification rate can be significantly increased with stirring.

Acknowledgment
This research was financially supported by the German Helmholtz Association in the frame of the Helmholtz-Alliance LIMTECH, Seed grant "Test device for liquid aluminum stirring using permanent magnets".

References

[1] Verte L A 1975 *Magnetohydrodynamics in Metallurgy* (Metallurgy Publishing House)
[2] Timofeev V N, Khristinich R M, Boyakov S A and Temerov A A *Linear Induction Machine*, Patent of Russian Federation No. 2069443, Publication 20.11.1996
[3] Beinerts T, Bojarevičs A, Bucenieks I, Gelfgat Y and Kaldre I 2016 *Met. Mat. Transact B* **47** 1626
[4] Bojarevics A, Beinerts T, Gelfgat Y and Kaldre I 2016 *Int. J. Cast Metals Res.* **29** 154
[5] Bucenieks I, Kravalis K and Krishbergs R 2011 *Magnetohydrodynamics* **47** 97
[6] Kaldre I, Fautrelle Y, Etay J, Bojarevics A and Buligins L 2013 *J. Alloys Compounds* **571** 50
[7] Momono T and Ikawa K 1975 *J. Japan Inst. Light Metals* **25** 403
[8] Wang J, Fautrelle Y, Ren Z M, Nguyen-Thi H, Salloum Abou Jaoude G, Reinhart G, Mangelinck-Noël N, Li X and Kaldre I 2014 *Appl. Phys. Lett.* **104** 121916

Patent

LV15144 METAL ALLOY AND SEMICONDUCTOR STIRRING DEVICE, Andris BOJAREVIČS (LV), Jurijs GEĻFGATS (LV), Toms BEINERTS (LV), Matīss KALVĀNS (LV), Reinis BARANOVSKIS (LV)

Measurement techniques for liquid metals

M Ratajczak[1], D Hernández[2], T Richter[3], D Otte[4], D Buchenau[5], N Krauter[1], T Wondrak[1]

[1] Helmholtz-Zentrum Dresden - Rossendorf (HZDR), Institute of Fluid Dynamics, Bautzner Landstraße 400, 01328 Dresden, Germany

[2] Institute for Thermodynamics and Fluid Mechanics, Technische Universität Ilmenau, 98684 Ilmenau, Germany

[3] Institute of Fluid Mechanics, Chair of Magnetofluiddynamics, Measuring and Automation Technology, Technische Universität Dresden, 01069 Dresden, Germany

[4] Institute of Thermal Process Engineering, Karlsruhe Institute of Technology, 76131 Karlsruhe, Germany

[5] HZDR Innovation GmbH, Bautzner Landstraße 400, 01328 Dresden, Germany

E-mail: t.wondrak@hzdr.de

Abstract. The measurement of flow properties of liquid metals, such as flow rate, flow structure and gas distribution, is a challenging task due to the opaqueness, the high temperatures (e.g. 1500 °C for liquid steel or liquid silicon) and the corrosiveness of those fluids. In this paper, a short review about the recent developments of measurement techniques in the framework of the Helmholtz Alliance Liquid Metal Technologies (LIMTECH) is presented. It focuses on the development of contactless inductive measurement techniques exploiting the high electrical conductivity of those melts. These measurement techniques include the contactless inductive flow tomography (CIFT), which is able to reconstruct the mean three-dimensional velocity structure in liquid melts, local Lorentz force velocimetry (local LFV), which enables the local assessment of flows close to the wall, and inductive methods for bubble detection, which are based on mutual inductance tomography (MIT). Additionally, a short overview of contactless inductive flow rate measurement techniques is given. Furthermore, an ultrasound technique called ultrasound transit-time technique (UTTT) will be presented which enables the measurement of position and size of bubbles in large vessels.

1. Introduction

In many industrial applications, flow properties of hot liquid metals play an important role for the processes. Sodium or lead alloys are used as coolants for nuclear power reactors or for solar power stations. Important parameters are the flow rate and the existence of bubbles. In metallurgy, gas is injected into the melt at various stages of melt refining, e.g. for carbon reduction or stirring [1]. Thus, detailed knowledge about the flow structure, the void fraction, the trajectory and the size of bubbles helps to understand and to control the process. A very complex flow regime can be found in the submerged entry nozzle (SEN) and in the mold of a continuous caster of steel, which is operated at about 1500 °C. The flow structure of liquid steel in the mold plays an important role for the quality of the product, because it influences the forming of surface defects and inclusions [2]. Often argon gas is injected into the SEN in order to reduce nozzle clogging and to catch impurities in the melt. Similarly, in the production of mono-crystalline silicon for electronic devices or solar cells, the flow of the liquid silicon in the crucible below the crystallization front is important. For instance, the flow structure has an effect on the number of defects and the concentration of oxygen in the produced crystal. Last but not least, a liquid metal reactor filled with liquid tin is used for cracking of methane and ethane into hydrogen [3]. The operating temperature is in the range of 800 °C to 1200 °C.

Due to the high temperature and the corrosiveness of those melts, contactless methods are preferred. However, well established contactless optical methods like particle image velocimetry or laser Doppler anemometry are precluded due to the opaqueness of the melts. For this type of fluids, ultrasonic measurement techniques can be used. Ultrasound Doppler velocimetry (UDV) [4] measures the flow profile along the ultrasound beam generated by the transducer using scattering particles in the fluid. When the transducer is in direct contact with the fluid, this technique is limited to temperatures of around 200 °C. With the help of wave-guides it is possible to adapt the transducer to temperatures of up to 600 °C due to the increased distance between the transducer and the melt [5]. Ultrasound can also be used to measure the trajectory and the size of bubbles by means of the ultrasound transit-time technique (UTTT) [6]. The advantage of this technique is the applicability to a large fluid volume, while radiation based techniques, such as neutron [7] or X-ray radiography [8], are limited regarding the thickness of the vessel due to the high attenuation of the rays by the liquid metal.

A different type of contactless measurement techniques exploits the high electrical conductivity of metallic melts by using the principle of induction. Typically, a contactless measurement device applies an externally generated magnetic field to the fluid and detects the perturbations of that magnetic field caused either by the change of the electrical conductivity in the volume or by the movement of liquid metal through the magnetic field. The first induction principle can be exploited for level or bubble detection and is the key principle of the mutual inductance tomography (MIT), which is able to detect the conductivity distribution in one cross section of a pipe [9]. The latter one can be used to measure flow rate like the phase shift flow meter [10] or the Lorentz force velocimetry (LFV) [11]. There exist also measurement techniques that are able to resolve the spatial distribution of the flow. One of these is contactless inductive flow tomography (CIFT) [12], which measures the perturbation of the applied magnetic field with magnetic field sensors distributed around the melt. The three-dimensional mean flow structure in the melt is then reconstructed by solving a linear inverse problem.

Additionally, LFV can be extended to detect the flow structure close to the wall by using a permanent magnet whose dimensions are significantly smaller than that of the flow. Only a small volume of the melt is penetrated by the magnetic field and the velocity in this region contributes to the signal. By placing the sensor consecutively at different positions along the wall, the spatial distribution of the force on the magnet is measured which depends on the spatial distribution of the flow velocity close to the wall. This technique is called local Lorentz force velocimetry (local LFV) [13].

A more detailed overview of measurement techniques for liquid metals including invasive probes, such as Vives probes for velocity measurement or resistivity probes for bubble detection, is given in the LBE handbook [14].

This paper will review the research conducted on the development of measurement techniques in the framework of the Helmholtz Alliance Liquid Metal Technologies (LIMTECH). In the beginning, a brief overview of inductive flow rate measurement techniques is given in section 2. This includes not only the new immersed eddy current flow meter (see section 2.2), the transient eddy current flow meter (see section 2.3) and advances in LFV (see section 2.5), but also well established methods such as the phase shift sensor (see section 2.1) or rotary flow meters (see section 2.4) are shortly described for the sake of completeness. The main focus of the work in LIMTECH was the development of the measurement techniques presented in the next sections: local Lorentz force velocimetry (see section 3), contactless inductive flow tomography (CIFT) (see section 4), ultrasound transit-time technique (UTTT) (see section 5) and inductive methods

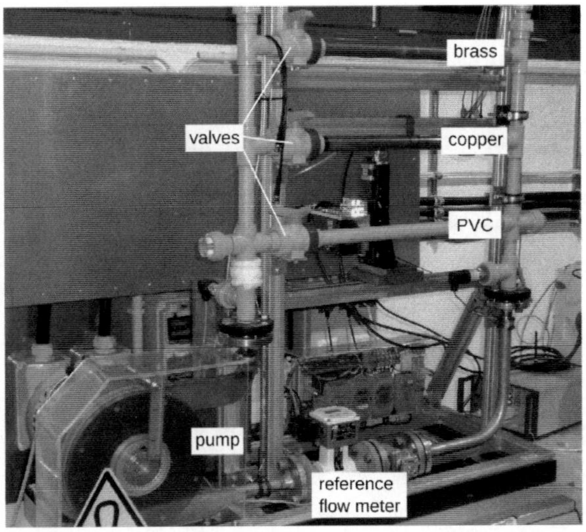

Figure 1. Photo of the GaInSn loop with the pump, the reference flow meter and three different sections made of PVC, brass and copper.

for bubble detection based on the principle of MIT (see section 6). The paper is concluded with a short description of a new level measurement system (see section 7).

2. Inductive flow rate sensors

All flow rate sensors presented in this section were tested at the liquid metal loop seen in Figure 1 which is operated with the eutectic alloy GaInSn at room temperature. The lower part of the loop consists of a circular stainless steel tube with the inner diameter $D = 27\,\mathrm{mm}$. The upper part consists of three independent test sections with a length of 400 mm each. The test sections can be opened and closed independently by valves. They are equipped with three pipes made of brass, copper and PVC which have the same inner diameter as the stainless steel tube in the lower section. The melt is driven by a permanent magnet induction pump with a maximum mean velocity of about $1.5\,\mathrm{m\,s^{-1}}$. As reference, the commercially available flow meter (Copa XL DN25) from ABB is used.

We start with a short description of the phase shift sensor (see section 2.1) and the newly developed eddy current flow meter (ECFM), which is an immersed version of the phase shift sensor (see section 2.2). The next presented sensor is the newly developed transient eddy current flow metering (TEC-FM) sensor, which is available as an external and an immersed sensor (see section 2.3). The advantage of TEC-FM is that it is calibration free. The section is concluded with the description of flow meters which are based on flow induced forces on permanent magnets: the rotary flow meter, the magnetic fly-wheel (see section 2.4) and the Lorentz force velocimetry (see section 2.5).

Before the different flow rate sensors are presented, we shortly delineate the basic principle of inductive flow measurement techniques which is is Ohm's law for moving conductors

$$\vec{j} = \sigma_l(\vec{u} \times \vec{B} - \nabla\phi)\ . \tag{1}$$

If a static magnetic field $\vec{B}$ is applied to the fluid domain with conductivity σ_l, the current $\vec{j}$ in the fluid depends on the flow field $\vec{u}$, the magnetic field $\vec{B}$ and the gradient of the induced electric potential ϕ. The current $\vec{j}$, in turn, generates a flow induced magnetic field $\vec{b}$ which is called the secondary magnetic field. Therefore, the magnetic field $\vec{B}$ is in general the sum of an

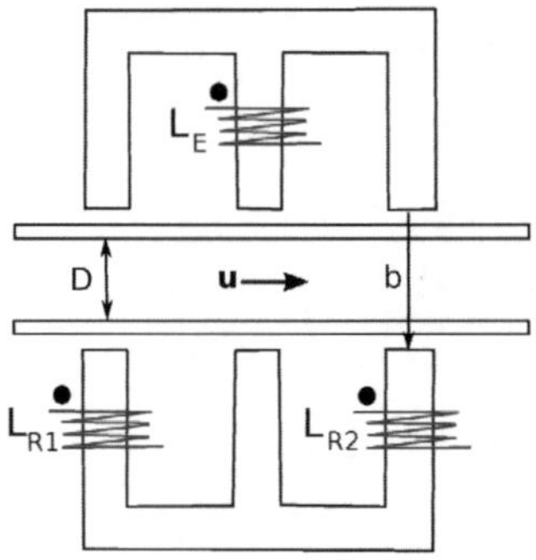

(a)

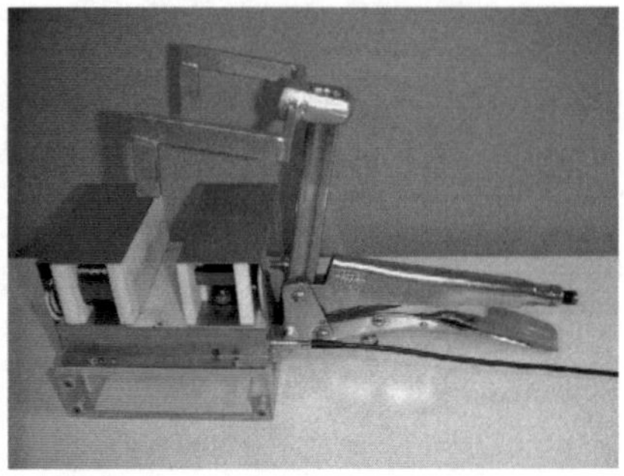

(b)

Figure 2. Schematic diagram (a) and photo (b) of the phase-shift flowmeter with the emitting coil L_E and the receiving coils L_{R1} and L_{R2} around the pipe with the diameter D.

externally applied primary magnetic field $\vec{B}_0$ and the induced secondary magnetic field $\vec{b}$. The ratio between $\vec{b}$ and $\vec{B}_0$ is proportional to the magnetic Reynolds number, defined as

$$\mathrm{Rm} = \mu_0 \sigma_l l u \ , \tag{2}$$

with μ_0 representing the magnetic permeability of vacuum and l and u denoting characteristic length and velocity scales of the flow, respectively. Typically, in most industrial application Rm is smaller than 1 so that the influence of $\vec{b}$ can be neglected.

The electric potential ϕ can be obtained by taking the divergence of Equation (1) and solving the emerging Poisson equation

$$\nabla^2 \phi = \nabla \cdot (\vec{u} \times \vec{B}). \tag{3}$$

The current $\vec{j}$ in the fluid volume generates not only the secondary magnetic field $\vec{b}$ but also under the influence of the magnetic field $\vec{B}$ the Lorentz force density $\vec{f}$ in the fluid

$$\vec{f} = \vec{j} \times \vec{B} \ . \tag{4}$$

The Lorentz force can be used to control the flow in the melt or to measure the flow by means of LFV (see section 2.5) and local LFV (see section 3).

2.1. Phase-shift sensor

The phase-shift sensor is based on a transformer principle operating without any locomotive parts. This contactless AC flow rate sensor can be realized in a lateral and an axial arrangement of the coils. The detector head of the lateral phase-shift sensor (cf. Figure 2) consists of two receiving coils and an emitter coil, placed on opposite sides of the duct containing the liquid metal flow. The emitter coil L_E is fed by a temperature stabilized alternating current I of constant frequency f [10, 15, 16]. In order to minimize the magnetic stray field of the sensor and to enlarge the signal quality, all coils are interspersed with laminated mild iron providing a low magnetic resistance of the sensor arrangement. An alternating magnetic flux is provided by the emitter coil perpendicularly orientated to the flow direction of the electrically conductive melt.

A symmetrical adjustment of the sensor arrangement is achieved for the case that the emitter coil flux is distributed half-and-half over both receiver coils. In order to adjust the symmetry of the sensor, the conductive melt has to be at rest. Considering a symmetrical arrangement of the emitter L_E and receiving coils L_{R1} and L_{R2}, the equal distribution of the alternating magnetic flux is disturbed by the movement of the conducting flow. The information about the averaged

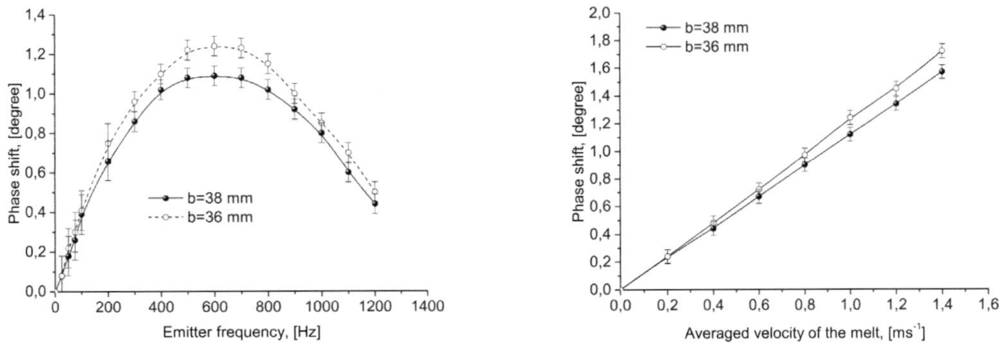

(a) (b)

Figure 3. Frequency response (a) at $u_0 = 1\,\mathrm{m\,s^{-1}}$ and flow induced phase-shift (b) in dependence of the mean velocity measured at the PVC test section for an emitter frequency of $f = 600\,\mathrm{Hz}$ (cf. [15]).

velocity u_0 or flow rate Q is provided by a voltage difference U_D or phase-shift $\Delta\phi$ between the receiver coils. The dependence between the phase-shift and the velocity is given by following function g

$$\Delta\phi = g\,(f, Rm, K) \ . \tag{5}$$

The measured effect is proportional to the magnetic Reynolds number Rm. The flow rate sensor has to be calibrated as indicated by the factor K in Equation (5). Some comparable measurements of the flow-induced phase-shift in dependence of the flow profile at an emitter frequency of $f = 600\,\mathrm{Hz}$ are shown in Figure 3 for two different widths of the coil gap $b = 38\,\mathrm{mm}$ and $b = 36\,\mathrm{mm}$. The results shown in Figure 3(a) indicate that the optimal frequency is about 600 Hz. The phase-shift measurements seen in Figure 3(b) show again a very good linearity with the averaged velocity in the duct.

An alternative realization of an AC electromagnetic flowmeter has been suggested by Schulenberg et al. [16], which is illustrated in Figure 4. The flowmeter consists of two emitter or sending coils L_{E1} and L_{E2} which are supplied with alternating electric currents I_{LE1} and I_{LE2} in opposite direction [10, 15, 16]. The arrangement of emitter coils is comparable to an Anti-Helmholtz coil pair which surrounds a pipe completely filled with liquid metal. Between the emitter coils the resulting induction field is nearly compensated in absence of any fluid motion. As soon as flow occurs, the magnetic field lines are dragged downstream out of the compensated volume where a receiver coil L_R is placed. The flow induced phase-shift $\Delta\phi$ between the induced signal of the receiver coils is proportional to the magnetic Reynolds number. The flow rate sensor has to be calibrated as indicated by the factor K in Equation (5) in the same way as the lateral phase-shift sensor. The flow direction can be detected by the sign of the measured RMS signal from the receiver coil or the phase shift between the emitter and receiver coil.

2.2. Eddy current flow meter

The eddy current flow meter (ECFM) is based on the same principle as the phase shift sensor. The detector head consists of two detection coils L_R and one emitter coil L_E which is fed by an AC current. Contrary to the phase shift sensor, the ECFM can be placed inside a pipe or pool to detect the flow rate in a certain volume around the sensor. This volume is influenced by the frequency of the AC current and the electrical conductivity of the liquid metal. The information about the averaged velocity u_0 or flow rate Q can be obtained by measuring the voltage and/or

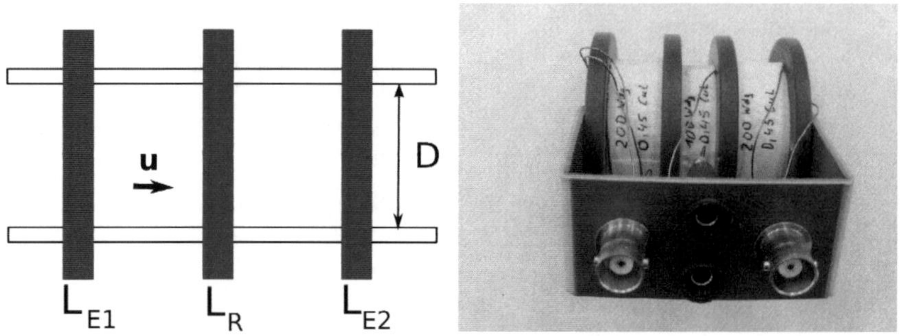

(a) (b)

Figure 4. Schematic diagram (a) and photo (b) of the axial phase-shift flowmeter with the emitting coils L_{E1} and L_{E2} and the receiving coil L_R around the pipe with diameter D.

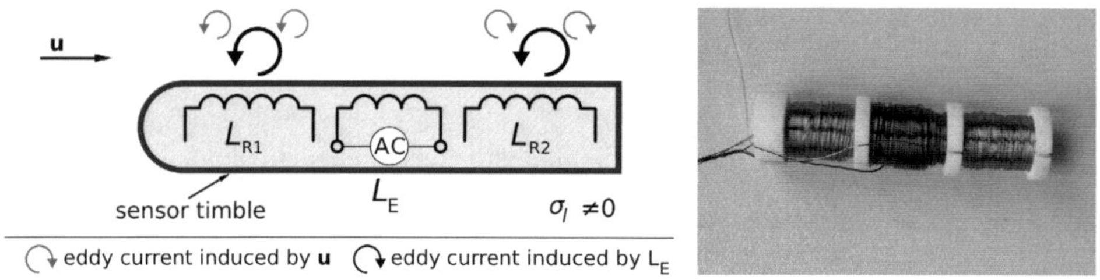

(a) (b)

Figure 5. Schematic diagram (a) and photo (b) of the eddy current flow meter with the emitting coil L_E and the receiving coils L_{R1} and L_{R2}.

phase difference between both detection coils, although the phase difference has proven to be more robust against outside disturbances. A stainless steel thimble protects the sensor coils from direct contact with the liquid metal.

Because of the transformer principle, L_E induces a certain voltage in both detection coils. This voltage is influenced by the presence of eddy currents within the medium around the sensor, which are also caused by L_E. There are two eddy current components, one is caused by the alternating electric field of L_E, the other is caused by the movement of the conductive medium through the magnetic field B of L_E. Because the radial component of B is opposite for L_{R1} and L_{R2}, the motion induced eddy currents also have opposite directions. Eventually this leads to different voltages in L_{R1} and L_{R2}. In Figure 5 the voltage in L_{R1} will decrease and the voltage in L_{R2} will increase with rising u_0. The results are similar to those shown in Figure 3, both voltage and phase difference have a linear dependency of u_0.

2.3. Transient eddy current flow metering

Most inductive flow meters have the disadvantage that the measured voltages depend not only on the flow rate but also on the electrical conductivity of the liquid metal or its temperature. Transient eddy current flow metering (TEC-FM) offers a way to strongly reduce the influence of electrical conductivity on the measurement results. This is achieved by creating a traceable eddy current system within the liquid metal and tracking its movement. In this section, two basic sensor arrangements, the external and immersed TEC-FM are described. Both sensors create the eddy current system by suddenly switching on or off a constant current and thus inducing eddy currents within the liquid metal, which are swept away with flow velocity in flow direction.

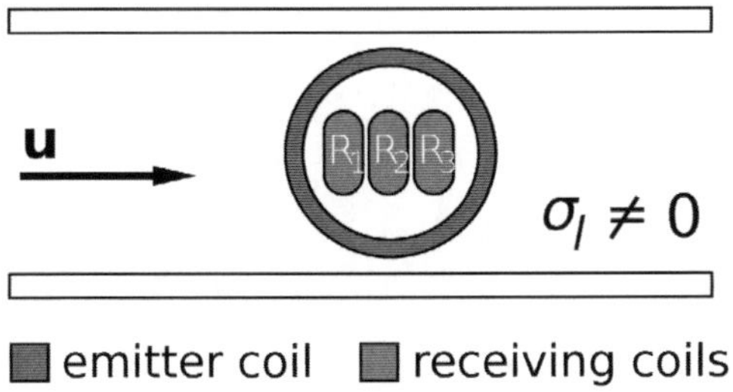

(a)

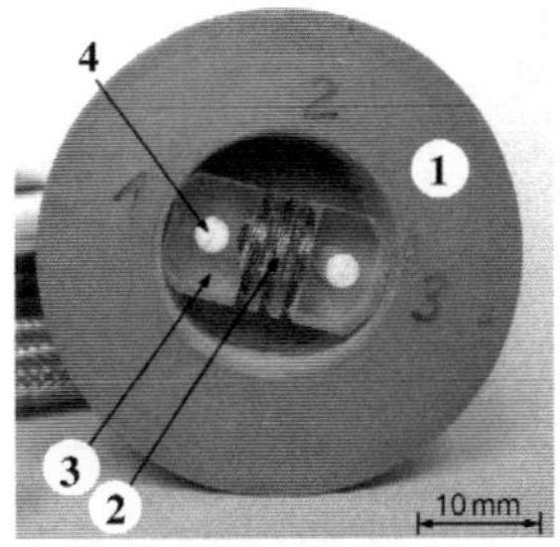

(b)

Figure 6. Schematic diagram (a) and photo (b) of the external transient eddy current flow meter; emitter coil (1), three detection coils R_1, R_2 and R_3 (2), coil holder (3).

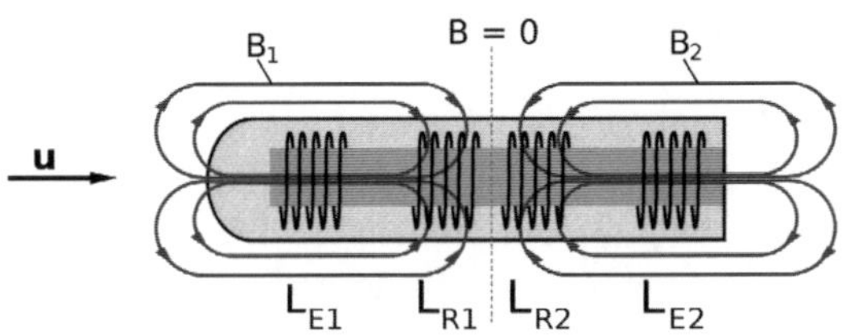

(a)

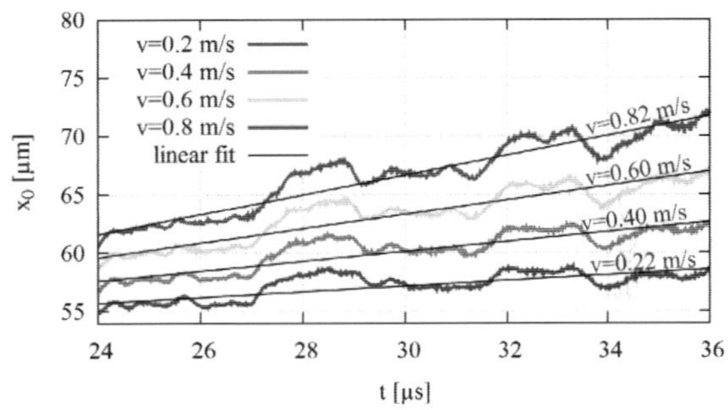

(b)

Figure 7. Schematic diagram (a) with the emitting coils L_{E1} and L_{E2} and the receiving coils L_{R1} and L_{R2} and measurement results (b) for four different flow rates.

The external TEC-FM is placed outside of the boundary of the liquid metal, on a pipe for example. Figure 6 shows a schematic sketch and a photo of the TEC-FM sensor. The emitter coil of the external TEC-FM imprints an eddy current ring into the medium. The position of the magnetic pole x_{pp} of this ring is tracked by measuring and evaluating the voltages of the detection coils. The averaged flow velocity u_0 can be obtained by the time derivative of $x_{\mathrm{pp}}(t)$ [17].

Like the ECFM, the immersed TEC-FM is protected by a stainless steel thimble and can be placed inside a pipe or a pool to measure the local velocity in a certain volume around the sensor. It consists of two emitter and two detection coils. This technique tracks the position x_0 of the zero crossing of the magnetic field B of the imprinted eddy current system. The emitter coils L_{S1} and L_{S2} are fed by DC currents with equal magnitude but opposite directions. When the currents of both coils are switched on or off at exactly the same time, two oppositely directed eddy current rings are induced around the sensor with a magnetic flux density of zero in the middle between them (see Figure 7). This zero crossing is moving with u_0 and can be tracked by the detection coils. The measurement results are almost independent of the electrical conductivity of the liquid metal, so this technique can essentially be called calibration free [18].

When plotting the position of x_0 over time, the slope of the linear regression line of $x_0(t)$ indicates the average flow velocity u_0 around the sensor.

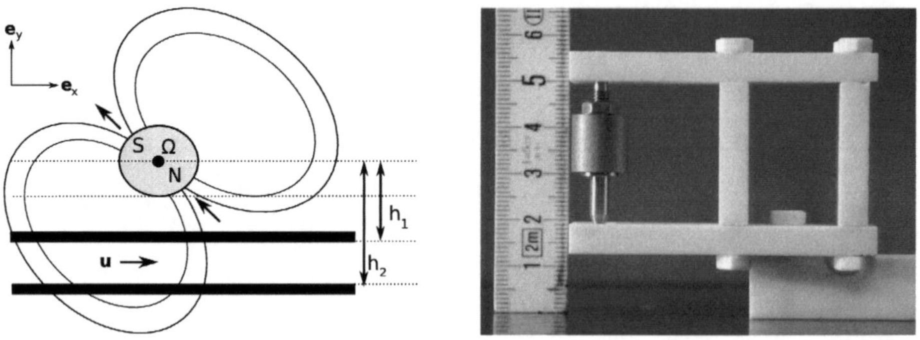

(a) (b)

Figure 8. Schematic diagram (a) and photo (b) of the rotary flow meter with the distances h_1 and h_2.

2.4. Rotary flow meter and magnetic flywheel

Contrary to the sensors presented previously, which used coils, the sensors described in the next sections employ permanent magnets. The rotary flow meter consists of a single cylindrical permanent magnet magnetized perpendicularly to its axis [19]. As shown in Figure 8(a) the sensor is placed close to the pipe and its axis is aligned perpendicular to the flow direction.

The theoretical analysis in [19] for the cases of a long as well as a short length of the cylindrical permanent magnet showed that its resulting equilibrium rotation does not depend on the strength of the magnet. It also does not depend on the conductivity of the liquid metal provided that the related skin-depth due to the rotation exceeds the distance h_2 as defined in Figure 8(a). In this case and, e. g., for a long magnet rotating steadily without a significant friction, the equilibrium rotation rate for a constant velocity u_0 in the channel is given by

$$\Omega = \frac{u_0}{2}\left(\frac{1}{h_1} - \frac{1}{h_2}\right)\ \ln^{-1}\left(\frac{h_2}{h_1}\right)\ . \tag{6}$$

A more complex permanent magnet flowrate sensor is the so-called magnetic flywheel as shown in Figure 9. Already in the sixties Shercliff [20] suggested such an electromagnetic flow meter consisting of a rotatable plate carrying a number of equidistantly spaced permanent magnets with alternating polarity. This solution was further developed [21] and received until today a number of applications at various liquid metal facilities.

2.5. Lorentz Force Velocimetry (LFV)

A typical LFV sensor consists of a pair of permanent magnets which embrace the pipe or channel containing the fluid [11, 22] as shown in Figure 10(a). Under the influence of the static magnetic field the flow generates a flow braking force (see Equation (4)) in the fluid. This force is proportional to the velocity or flow rate of the liquid and, due to Newton's third law, a force of the same magnitude but in opposite direction acts on the source of the applied magnetic field. The force $\vec{F}$ and torque $\vec{T}$ can be calculated by integrating $\vec{f}$ over the fluid domain using following integrals:

$$\vec{F} = \int_{Vol} \vec{j}(\vec{r}\,') \times \vec{B}_0(\vec{r}\,')\ dV'\ , \tag{7}$$

$$\vec{T} = \int_{Vol} (\vec{r} - \vec{r}\,') \times \vec{f}(\vec{r}\,')\ dV', \tag{8}$$

IOP Conf. Series: Materials Science and Engineering **228** (2017) 012023 doi:10.1088/1757-899X/228/1/012023

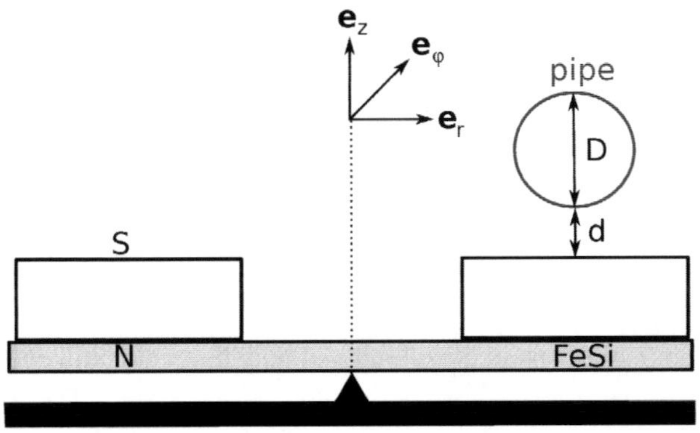

(a)

(b)

Figure 9. Schematic diagram (a) and photo (b) of a flywheel flow meter.

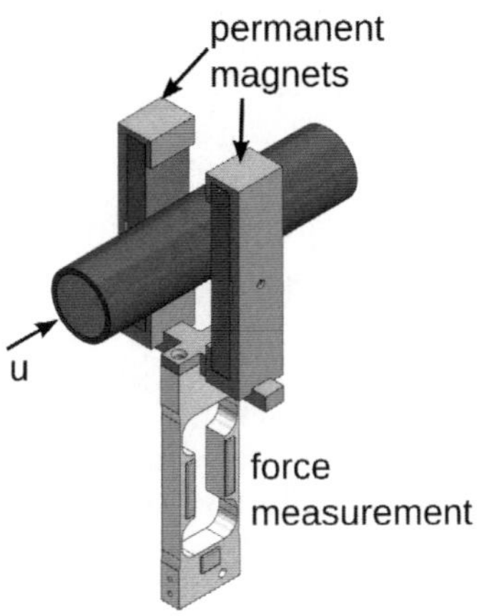

(a)

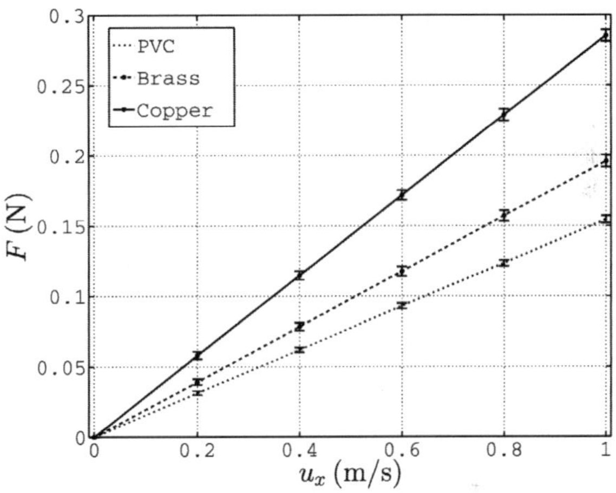

(b)

Figure 10. Schematic sketch (a) of a LFV sensor and the measured force (b) for the three test sections made of PVC, brass and copper of the GaInSn loop shown in Figure 1.

respectively [23]. Here, dV' denotes the volume element, $\vec{r}\,'$ the position in the volume and $\vec{r}$ is the given reference point for the torque which could be either the center of the magnet or the center of the coordinate system which is defined by calibration.

Figure 10(b) shows a clear linear dependence between the measured force and the flow rate for the three different test sections made of PVC, brass and copper of the GaInSn loop shown in Figure 1. It is evident that in the case of conducting walls (brass and copper) the measured force increases depending on the electrical conductivity σ_w of the wall.

This effect was first addressed in the framework of Lorentz force velocimetry using a small-size magnet that was placed in front of the tubes [24]. Here, experiments and simulations showed that when the ratio between the conductivity of the wall σ_w and the liquid σ_l increased, $\vec{F}$ seems to rise asymptotically suggesting a limit of this effect. $\vec{F}$ is in this case the sum between $\vec{F_l}$ and $\vec{F_w}$ which are the Lorentz forces generated in the liquid and the wall, respectively.

In a new study, this effect is analysed in more detail by using an analytic and numerical model of the limiting case where the ratio of conductivities goes to infinity. Our results indicate that the total Lorentz force in the liquid $\vec{F_l}$ approaches asymptotically the Lorentz force term that depends

IOP Conf. Series: Materials Science and Engineering **228** (2017) 012023 doi:10.1088/1757-899X/228/1/012023

on $\vec{u} \times \vec{B}_0$. The force component $F_{l,\phi}$ generated by the induced electric potential ϕ is sensitive to the ratio between conductivity of the wall and the liquid $\varepsilon_w = \sigma_w/\sigma_l$. Perfect conducting and perfect insulating behavior of the walls appear when $\log_{10}(\varepsilon_w) > 4.5$ and $\log_{10}(\varepsilon_w) < -4.5$, respectively. In the first case, $F_{l,\phi}$ is zero and in the second one it reaches its maximum value. In general, the force component generated by the induced potential is always opposite to that one given by $\vec{u} \times \vec{B}_0$. When $F_{l,\phi}$ is diminished by either increasing the electrical conductivity of the walls or by changing the aspect ratio of the flow, we can achieve a considerably increase of the measurable total Lorentz force, and therefore, enhance the sensitivity of the measurements. A publication explaining these effects in detail is in preparation.

3. Local Lorentz Force Velocimetry (local LFV)

By reducing the size of the permanent magnet used for LFV and owing to the rapid decay of magnetic fields, a localized magnetic field distribution in the liquid metal is achieved by using magnets which are significantly smaller compared to the cross-section of the flow. Hence, the measurable force is localized in a small subsection of the fluid giving access to a local velocity assessment. Due to its localized nature, the system composed of the small-sized magnet altogether with the force sensor is called Local Lorentz Force Flowmeter (L2F2). At the beginning of the project this concept of local LFV has already been experimentally tested by successfully identifying obstacles in the flow [13]. The first application of this measurement technique for the reconstruction of the two-dimensional velocity structure in the vicinity of the wall of a confined vessel [25] showed the applicability of this technique to complex flow structures. In these experiments, an one-dimensional Optical Interference Force Sensor (OIFS) was used [26]. This implies that for measuring the second component of the force, the sensor had to be rotated 90 degrees and placed again in the measuring grid introducing another source of error. In order to overcome these problems, the measuring device of the new generation L2F2 is a multicomponent force and torque sensor [27, 28]. This sensor has been specially developed to record simultaneously all three force and three torque components that act on the magnet. By using this sensor connected to a 15 mm cubic permanent magnet (CUBIC), we were able to assess the three-dimensional velocity of GaInSn near the wide face of the mold of a continuous casting model, the Mini-LIMMCAST facility (Liquid Metal Model for continuous casting of steel) [29]. A sketch of the set-up and the measurement grid is depicted in Figure 11(b). The double-roll structure, typical for continuous casting of steel, was clearly identified by both the force and torque signal. It turned out that the torque is less sensitive to ferromagnetic parts in the surrounding. The force F_z perpendicular to the wall showed the area where the liquid metal jet flowing out the SEN impinges onto the narrow faces of the mold. In this area the jet strongly diffuses and deviates in a upper and lower vortices. However, the torque component T_z co-aligned to this force was not accessible at all.

In order to optimize the magnet system to measure this torque, we have developed a numerical model of our experiments at the Mini-LIMMCAST facility in the framework of local LFV. Here, the magnetic Reynolds number Rm and the interaction parameter N are smaller than one. Thus, we are in the kinematic regime of LFV where the influence of the Lorentz force on the fluid is negligible [30]. Additionally, as the flow in the mold is steady, we can obtain the forces and torques directly from Ohm's law for a given velocity and applied magnetic field. This model was first verified with our experimental results with the CUBIC magnet and then used for optimization of the magnet system. Preliminary numerical and experimental results show that a cross-shape permanent magnet (CSM) (see Figure 11(c)) enhances considerably the torque perpendicular to the wall in comparison with the CUBIC magnet system. The torque T_z is expected to be in the order of 20 µN m at the Mini-LIMMCAST facility. Preliminary simulations suggest that this torque component correlates with the curl of the velocity in this direction. A publication giving more details about the simulation and the experiment with the CSM is in preparation.

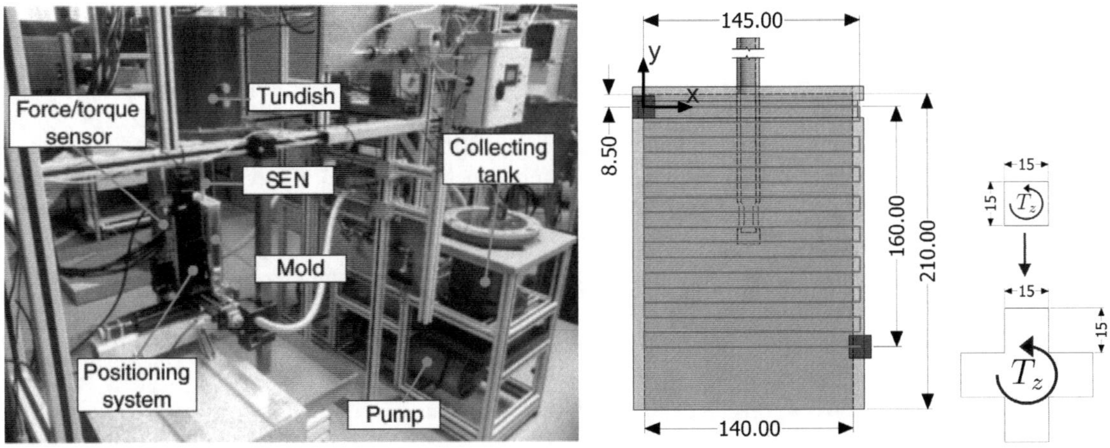

(a) (b) (c)

Figure 11. Picture of the Mini-LIMMCAST facility (a) with the force measurement system (see [29]), schematic sketch (b) of the mold presenting the measuring grid (red) and the location of the 15 mm cubic magnet (magenta) at the beginning ($x = 0$ mm, $y = 0$ mm) and at the end of the experiment ($x = 145$ mm, $y = -160$ mm), (c) schematic sketch of the CUBIC and the CSM magnets.

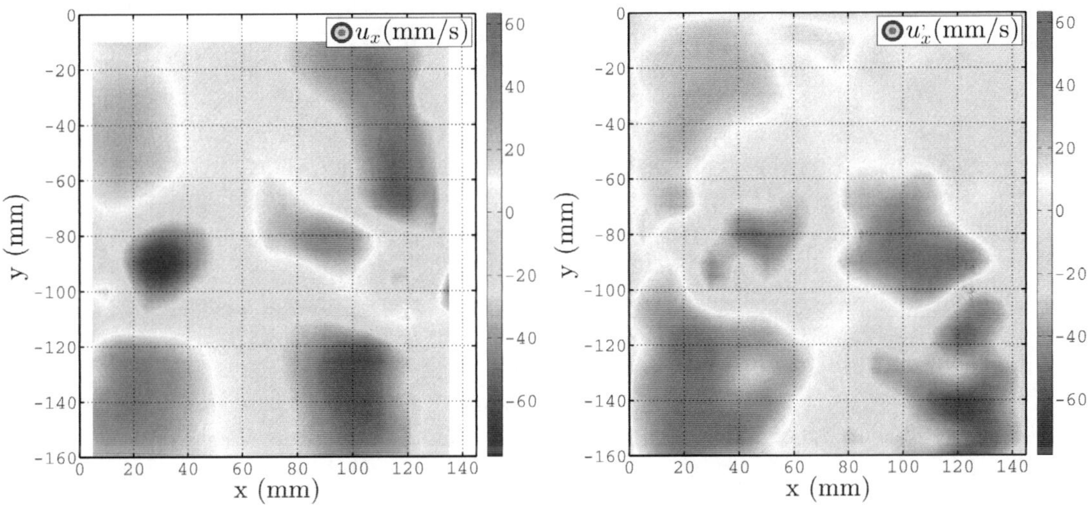

(a) (b)

Figure 12. Comparison between UDV measurement u_x near the wall *(a)* (taken from [29]) and the velocity component u'_x according to (9) using the measured force component acting on the magnet F_x *(b)*.

Before using the CSM in liquid metal, we performed an electromagnetic dry calibration [31] of the L2F2 with a rotating disc made of aluminium. The advantage of this set-up is that we can accurately define the centroid of the magnet system for measuring T_z. This dry calibration technique can also be used to determine the calibration factor for the L2F2. The magnitude of the velocity field u' of the liquid metal is given by

$$u' = K_{liq} \cdot F \ , \tag{9}$$

where F is the magnitude of the measured force component and K_{liq} the calibration factor, respectively. However, one of the biggest challenges in local Lorentz force velocimetry is that K_{liq} does not only strongly depend on the distance between the permanent magnet and the liquid metal, but also on the velocity distribution near the wall spanned by the magnetic field.

In application, the velocity field in a given set-up is unknown and K_{liq} has to be defined *a priori* for a quantitative assessment of the velocity of the liquid metal next to the wall. This calibration factor can be estimated by the proposed dry calibration in the following way

$$K_{liq} = K_{disk} \frac{\sigma_{disk}}{\sigma_l} \ , \tag{10}$$

where K_{disk} and σ_{disk} are the calibration factor and conductivity of the disk, respectively. K_{disk} is the ratio between the velocity field in the disk and the force acting in the center of the magnet at (x_0, y_0). The position of the center of the magnet is sufficiently away from the rim of the disk in which the force is proportional to the radius r from the center of the disk. The CUBIC magnet system used in [29] is located at the same distance a_g away from the surface of the disk as in the experiments at the Mini-LIMMCAST facility. In this case, $K_{disk} = 2.7$ using a disk with 200 mm in diameter. Due to the comparably bigger distance and volume of the magnet system in comparison with the one used in [32], a bigger diameter of the disk has to be considered to avoid edge effects. Finally, the calibration factor in the Mini-LIMMCAST facility according to (10) is $K_{liq} \approx 2.7(25.4/3.3) = 20.8$. A comparison between the velocity field u_x measured by UDV [29] in comparison with the velocity field u'_x according to (9) is shown in Figure 12 having good agreement (rms(u_x)= 27.6 mm/s $\approx$ rms(u'_x)27.2 mm s^{-1}). With the proposed simple calibration procedure, we could obtain a conservative assessment of the magnitude of the velocity field at a depth of 4 mm at Mini-LIMMCAST facility. We believe that, in general, it can be used for a simple and fast approximation for K_{liq} regarding cubic or cross-shaped magnet systems. Further numerical investigation are currently being carried out out in order to define the penetration depth and spatial resolution of local LFV.

4. Contactless inductive flow tomography (CIFT)

In order to reconstruct the flow structure $\vec{u}$ from magnetic field measurements $\vec{b}$ outside the fluid volume V, Biot-Savart's law can be applied to Ohm's law (1) which leads to following integral equation:

$$\vec{b}(\vec{r}) = \frac{\mu_0 \sigma_l}{4\pi} \int\limits_V \frac{(\vec{u}(\vec{r}\,') \times \vec{B}_0(\vec{r}\,')) \times (\vec{r} - \vec{r}\,')}{|\vec{r} - \vec{r}\,'|^3} dV' - \frac{\mu_0 \sigma_l}{4\pi} \oint\limits_S \frac{\phi(\vec{s}\,')\vec{n}(\vec{s}\,') \times (\vec{r} - \vec{s}\,')}{|\vec{r} - \vec{s}\,'|^3} dS' \ . \tag{11}$$

In the volume integral, dV' denotes the volume element and $\vec{r}\,'$ the position vector in the volume. In the surface integral, dS' denotes a surface element and $\vec{n}(\vec{s}\,')$ denotes the normal vector of the surface S at the position $\vec{s}\,'$. The electrical potential ϕ at the boundaries is calculated by applying Green's theorem to the emerging Poisson equation (3) which leads to following integral equation

$$\phi(\vec{s}) = \frac{1}{2\pi} \int\limits_V \frac{(\vec{u}(\vec{r}\,') \times \vec{B}_0(\vec{r}\,')) \cdot (\vec{s} - \vec{r}\,')}{|\vec{s} - \vec{r}\,'|^3} dV' - \frac{1}{2\pi} \oint\limits_S \frac{\phi(\vec{s}\,')\vec{n}(\vec{s}\,') \cdot (\vec{s} - \vec{s}\,')}{|\vec{s} - \vec{s}\,'|^3} dS' \ , \tag{12}$$

if insulating boundaries are assumed. An extension of this integral system for conducting boundaries is given in [33]. In order to reconstruct the velocity field $\vec{u}$ in the volume V, this integral equation system has to be inverted. The intrinsic non-uniqueness problem of the emerging linear inverse problem, which mainly concerns the detailed depth-dependence of the velocity, is circumvented by utilizing the so-called Tikhonov regularization [12, 34].

The main challenge of CIFT is the reliable measurement of the induced magnetic field which is typically 2 to 5 orders smaller than the applied magnetic field. Therefore, a very stable current source for generating the excitation magnetic field and magnetic field sensors with a high dynamic range and a very linear response are needed. In order to make the measurement technique more robust against fluctuations of the environmental magnetic field, an AC excitation field with rather low frequency in the order of 1 Hz is used, such that the skin effect can be neglected. Using AC fields allows the actual induced magnetic field to be separated from the environmental noise suppressing undesired signals from the environment, like moving ferromagnetic materials, or arc welding activities in the vicinity of the experiment, because those are mainly DC signals [35, 33, 36].

At the beginning of the project a new measurement system consisting of 14 induction coils was designed and developed. The induction coils have two major advantages compared to Fluxgate sensors which were used previously: they can be applied in high temperature environments, and they can also work in the presence of strong static external magnetic fields. Induction coil sensors are only sensitive to magnetic fields which are variable in time, thereby suppressing static fields and significantly reducing the dynamic range of the measurement. Two different kind of coils are used: absolute and gradiometric induction coils. Absolute sensors consist of a single winding and detect the absolute value of the magnetic field, whereas gradiometric sensors comprise two windings, wound in opposing directions, which are connected in that way that a homogeneous external field is cancelled out. In order to detect the small flow induced magnetic fields, the absolute sensors have 340 000 turns and the gradiometric sensors have $2 \times 160\,000$ turns. Both type of coils were made of 25 µm wire.

In order to test the robustness of the new sensor system, it was applied to the slab casting mold of the Mini-LIMMCAST facility in the presence of a strong static magnetic field generated by an electromagnetic brake (EMBr). EMBrs are widely used in continuous casting of steel to suppress flow fluctuations, which should lead to a more even solidification shell and less surface inclusions [38]. Figure 13 shows a schematic sketch and a photo of the Mini-LIMMCAST facility with the CIFT measurement equipment (excitation coil and pickup coils) and the EMBr. First experiments regarding CIFT and EMBrs were carried out with a preliminary setup, where just one excitation coil was used directly above the pole shoes of the EMBr, which resembles a previously used setup for the case without an EMBr [39, 40]. These preliminary measurements have delivered promising results, and demonstrated that the induced magnetic field (in the order of 100 nT) is indeed measurable on the background of a 310 mT field from the EMBr. The reconstructed velocity revealed that indeed an asymmetric flow structure was present when the EMBr is switched on while applying insulating boundary conditions [41], which is confirmed by UDV measurements [38].

The data gathered in these campaigns later served as a reference to gain further insight into the role of the significant eddy-currents in this setup by means of numerical simulations [37]. The strong eddy-currents appear even for very low frequencies $f < 10$ Hz which are typical for CIFT.

The experimental setup was numerically modelled and could be accurately validated by the measurement data. Based on this model, forward (calculating $\vec{b}$ from $\vec{B}_0$ and $\vec{u}$) and inverse calculations (solving for $\vec{u}$) were carried out to systematically optimize the experimental setup.

The most relevant finding was that the excitation field from the high-positioned excitation coil is removed from the lower part of the mould by the pole shoes, which leads to âĂIJinvisibleâĂİ

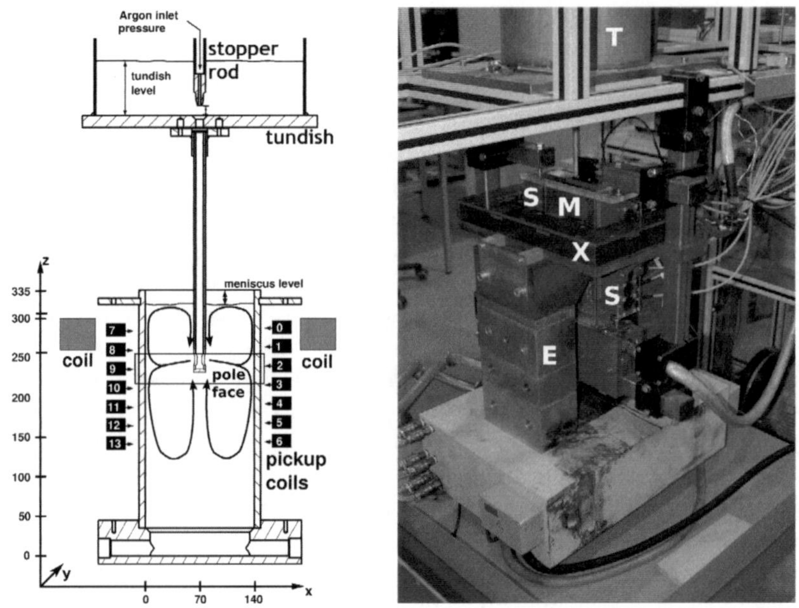

(a) (b)

Figure 13. Sketch and photo of Mini-LIMMCAST with the EMBr and the preliminary CIFT setup from [37]. E: EMBr, M: mould, S: sensors, T: tundish, X: excitation coil.

flow structures at the bottom of the mould. We concluded that a second excitation coil should be installed just below the pole shoes to accurately monitor the flow in this area.

The optimized setup was then installed and new experiments were carried out with low excitation frequency ($f \leq 3\,\mathrm{Hz}$), which matches the parameters of past experiments. Even though the three different cases (EMBr off, EMBr on with insulating or conducting boundaries) should all lead to vastly different magnetic field structures, the two cases with the EMBr turned on revealed the same magnetic field, which consisted completely of chaotic signals. Upon further investigation it could be demonstrated that low-frequency variations of the flow field together with the strong EMBr field generate a new magnetic field that is not related to the main flow structure (as desired by CIFT) but only to the flow variations, which then superimposes the CIFT induced field in a low frequency range [42]. The only possible solution in this case is the use of an excitation field in an higher and unaffected frequency range.

As soon as a higher excitation frequency (at least $8\,\mathrm{Hz}$) is used, all three cases become distinguishable as can be seen in the plots in Figure 14. Unfortunately, the high frequency AC magnetic field generates eddy-currents that are so strong that the slight changes of the liquid metal level and the associated magnetic field in the mould would be visible on top of the flow induced field. This second effect that tends to superimpose our desired signal cannot be avoided and must therefore be compensated by means of calibration. Therefore, reference measurements were carried out where only the liquid metal level was changed without any liquid metal flow. The resulting magnetic field data can then be subtracted from real high-frequency flow measurements. It could be demonstrated that this method leads to accurate results [43].

However, due to the complex structure of the applied magnetic field, which is deformed by the ferromagnetic parts of the EMBr, and the recirculation flow close to the jets, the reconstruction of the flow was difficult to achieve with the present implementation of the forward model for CIFT. Therefore, new version of the forward model using linear and higher order elements is in development which allows the use of CIFT for arbitrary geometries. In order to investigate

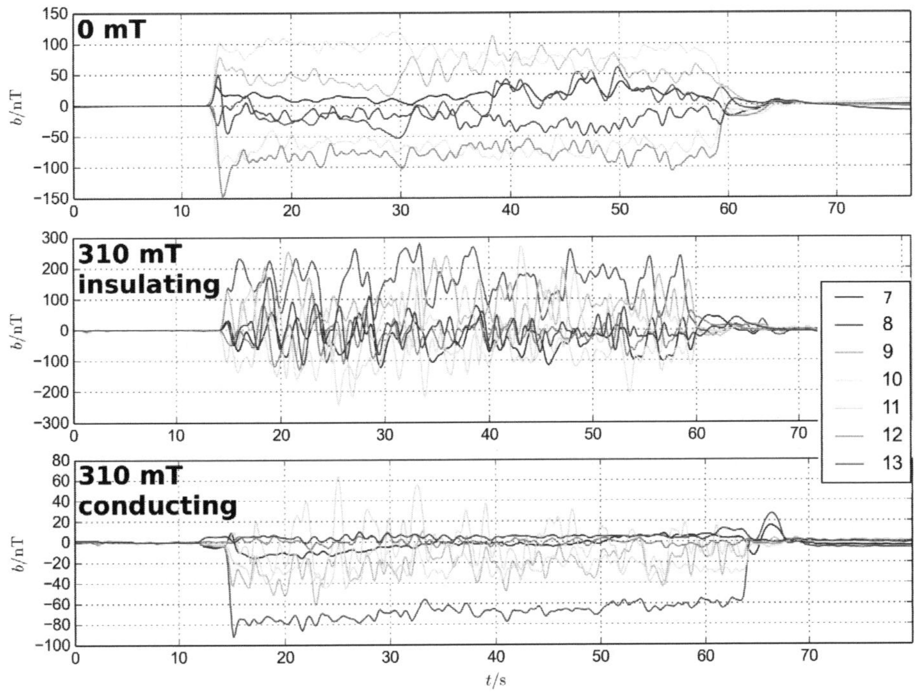

Figure 14. Comparison of the flow induced magnetic field measured by 7 gradiometric induction coils along the left side of the mold (see Figure 13) for the three cases EMBr switched off (a), EMBr switched on with insulating boundaries (b) and EMBr switched on with conducting boundaries (c), with an excitation frequency of 8 Hz.

thoroughly the accuracy of the new model, the calculation of the flow induced magnetic field for a rotating conducting cylinder in a homogeneous axial magnetic field is presently under investigation. In this configuration the flow induced magnetic field as well as the potential at the boundaries can be calculated analytically. Ongoing work is the test of this new model for the existing rectangular mold.

In order to investigate the scalability of CIFT to larger molds, a measurement system with 9 induction coils was developed and applied to the LIMMCAST facility which consists of a mold with the inner diameter of $0.4\,\mathrm{m} \times 0.1\,\mathrm{m}$ [44]. The facility is operated with SnBi at 250 °C which poses new challenges to the instrumentation in terms of stiffness and temperature stability, and thereby resembles industrial facilities more closely.

Because of auxiliary equipment at the right side of the mould, sensors could only be installed at its left side. For the reconstruction, the measured magnetic field was mirrored to the right hand side. Figure 15 shows the magnetic field data and the reconstructed velocity for a typical experimental run with $f = 3\,\mathrm{Hz}$ and $I = 17\,\mathrm{A}$ [42]. Even though the conductivity of SnBi is about 0.4 of that of GaInSn and the distance between the sensors and the melt was increased to 10 cm, the sensors detect approximately the same magnetic field amplitude. It is evident that the flow in the mould is quite stable and just the overall intensity is subject to a small trend. The reconstruction clearly shows the jet leaving the SEN which hits the narrow wall at around $z = 1\,\mathrm{m}$ and then splits up to form the double-roll structure. This successful experiment demonstrates the scalability of CIFT.

A further challenge is the applicability of CIFT to low velocities in the order of $10\,\mathrm{mm\,s^{-1}}$. A first measurement setup for CIFT was developed and installed at a small thermally driven

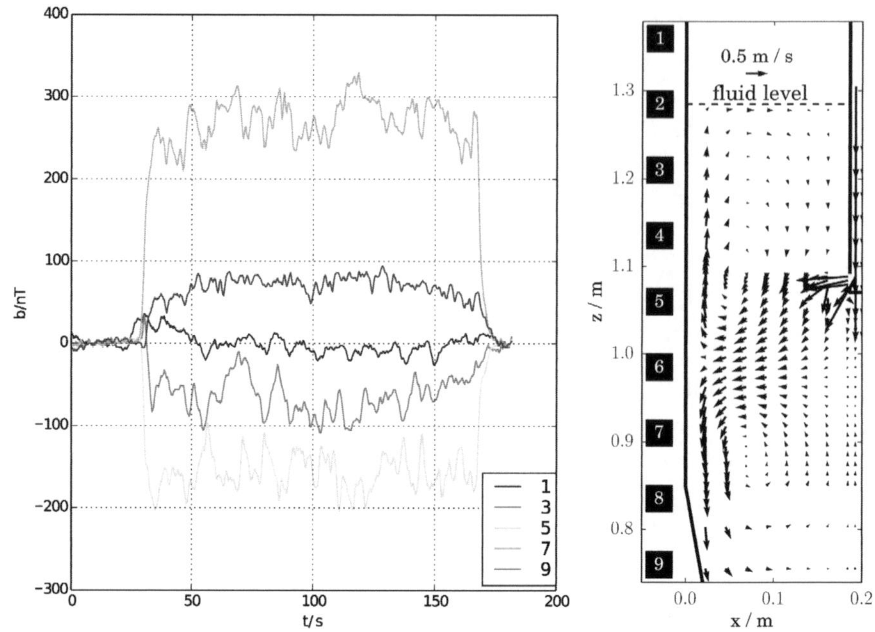

(a) (b)

Figure 15. Induced magnetic field (a) and reconstructed mean velocity (b) for a single run at LIMMCAST.

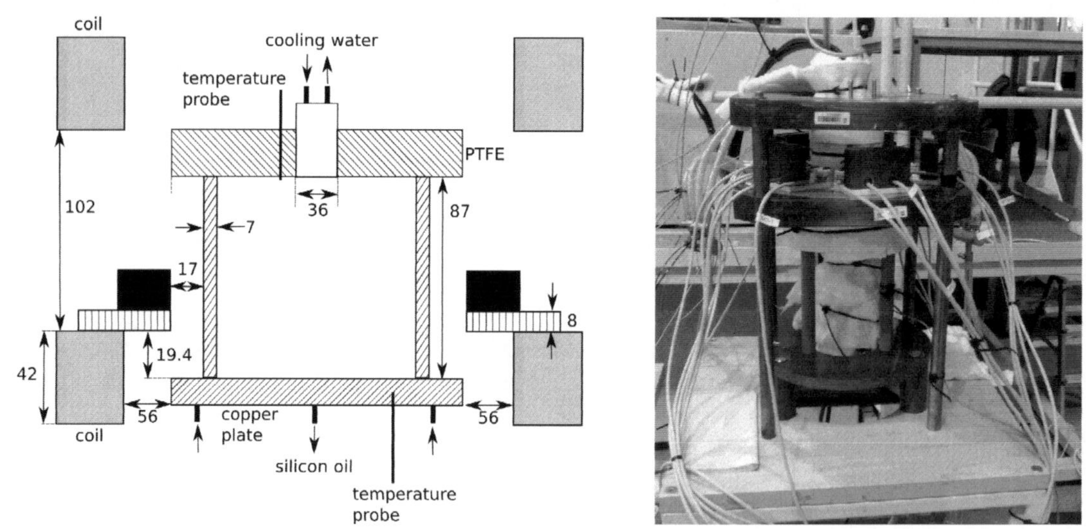

(a) (b)

Figure 16. Schematic sketch (a) and photo (b) of the modified Rayleigh-Bénard configuration.

flow experiment which is motivated by Czochralski crystal growth. In this case, the challenge is the robust detection of the weak flow induced magnetic field which is about 5 orders of magnitude smaller than the applied magnetic field [45]. In addition to the small velocities the long measurement time of more than 4 hours sets high demands on the stability of the measurement system. Due to these difficulties we started to develop the measurement system for one applied magnetic field while a full three-dimensional reconstruction is left for further investigations, which would require an additional applied magnetic field in horizontal direction.

Figure 16 shows a schematic sketch and a photograph of this setup, respectively. It consists of a cylindrical vessel filled with GaInSn, two coils which generate a vertical DC magnetic field of about 4 mT and 20 Fluxgate probes which are arranged over the azimuth at one height. This allows identifying the dynamics of the inclination angle of the large scale circulation (LSC) that is typical for this kind of flows. After removing all ferromagnetic parts and reducing the thermal expansion of the setup to a minimum, we were able to measure the magnetic field over 12 hours with a drift of about 20 nT during night.

Several experiments with temperature differences starting from $\Delta T = 2.3\,\mathrm{K}$ up to $\Delta T = 80.8\,\mathrm{K}$ were performed which showed a variety of flow phenomena. These include a LSC that is pinned at some angle from were it undergoes slight oscillations, and one cessation of the LSC. More details can be found in [45].

5. Ultrasound transit-time technique (UTTT)

Two general types of ultrasound methods have found their way into the research field of velocity measurements and two phase flows. One type of these techniques originates from non-destructive material testing techniques. These are the pulse-echo method and the transit time technique. The pulse-echo method evaluates the height of the echo amplitude and was already used by Matikainen et al. [46] to measure transient gas/liquid interfaces, and by Banerjee et al. [47] and Chang and Morala [48] to analyze single bubbles. The transit time technique focuses on the time of flight of the echo. In connection with advanced flow meters (Moore et al. [49], Mahadeva et al. [50]) the ultrasound transit time technique (UTTT) was developed. This technique is adequate to measure bubble parameters in liquid metal two-phase flows (Andruszkiewicz et al. [6], Richter et al. [51]). The other commonly used type of techniques is ultrasound Doppler velocimetry (UDV), which originates from medical application. UDV is a powerful technique to analyze flow structures (Takeda [52]) and was also successfully used to visualize wake structures of rising bubbles by Eckert et al. [5]. UDV and UTTT were already combined to study flow structures and bubble distribution of a bubble plume by Vogt et al. [53]. A recent technique for two-phase flows is the ultrasound reflector recognition and tracking technique (URRTT) which operates with pulsed ultrasound with the detection of the delay of the pulse reflection from a given bubble surface (Povolni et al. [54]).

The ultrasound transit-time technique (UTTT) operates with ultrasound transducers which emit acoustic pulses with a predefined frequency (in the range of MHz) and a given pulse repetition frequency $\mathrm{f_p}$ (in the range of kHz) through the wall of the vessel into the fluid. Figure 17(a) shows two transducers mounted at opposite sides of the wall of the vessel. The pulses are reflected by an object with a high acoustic impedance, which is either the opposing wall or the surface of the rising bubble, and detected by the emitting transducer. Suppose, the left transducer in Figure 17(a) emits one pulse, then Figure 17(b) shows the recorded amplitude of the echo for this configuration. The echo of both walls as well as of the left surface of the bubble is clearly visible. The time difference between emission and reception of the echos t_W (front wall), t_{BW} (back wall) and t_B (bubble) can be used to determine the distance of the objects, if the vessel diameter D is known.

The position x_B of the bubble surface next to the transducer can be calculated by [6]

$$x_B = D \cdot \frac{t_B - t_W}{t_{BW} - t_W} . \tag{13}$$

If two sensors i and j are mounted on opposite positions at the wall of the vessel, as shown

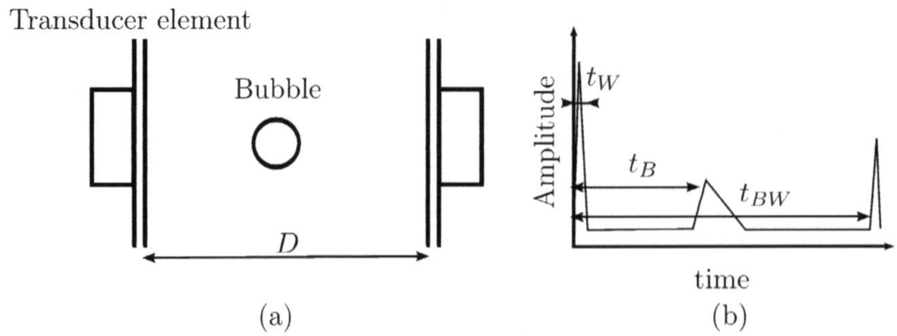

Figure 17. Schematic drawing illustrating all parameters necessary for the data extraction: the vessel cross section illustrating the diameter D (a) and the amplitude signal illustrating the transit times t_W, t_B and t_{BW} (b).

in Figure 17 (a), the diameter of the bubble is derived by

$$d_B = D \cdot \left(1 - \frac{t_{Bi} - t_{Wi}}{t_{BWi} - t_{Wi}} - \frac{t_{Bj} - t_{Wj}}{t_{BWj} - t_{Wj}}\right) . \tag{14}$$

It is obvious that the measured bubble surface shows fluctuations due to the entry and the exit of the bubble into the ultrasound beam. The center of mass x_{Bc} of the bubble is given by

$$x_{Bc} = x_B + \frac{d_B}{2} . \tag{15}$$

By using a test with falling steel balls the error of the method could be calculated to 7 % [55].

In order to measure the motion and trajectory of a bubble in the fluid volume, a combination of several sensors is used. Figure 18(b) shows a typical sensor arrangement. The transducer array is connected to a ultrasonic defectoscope (USIP 40 GE Inspections Technologies GmbH) which has 10 channels. This offers the possibility either to measure with 10 transducer elements, regardless to the number of arrays, to cover bubble trajectories, or to work with 5 pairs of opposing transducer elements to measure the bubble diameters and the trajectory during the rise [6].

While in previous work only the spatial distribution of bubbles was investigated, in the beginning of the project the tracking of single bubbles was implemented. This required a detailed analysis of the accuracy of the measurement. Therefore, the rise of a single argon bubble in water was simultaneously covered by UTTT and a high-speed camera. The intention of these measurements was to provide a deeper understanding of the measured diameter depending on the shape and the motion of the bubble. Figure 19(a) shows the measured trajectory with two pictures of the high-speed camera visualizing the bubble and Figure 19(b) shows the bubble diameters for the same height. The strong variation of the measured bubble diameters is caused by the tilting of the bubble, and thus UTTT measures a smaller apparent bubble diameter [55].

For an exact determination of the bubble diameter, the pair of ultrasound transducers on opposite sides have to be accurately aligned along their axis. As Figure 18(a) shows, there was a slight deviation of the sensor alignment between opposite sensors of the two arrays. Therefore, two new sensor arrays made out of one slide of piezoceramics were build. Each sensor consists of 32 elements with the dimensions 2.5 mm × 5.0 mm. In this case a better alignment of the sensors could be achieved. The high number of transducers allows a much more flexible adjustment of the distance between the single sensors to cover the bubble trajectory. Furthermore two transducer elements can be combined to one excitation element which increases the measurement plane.

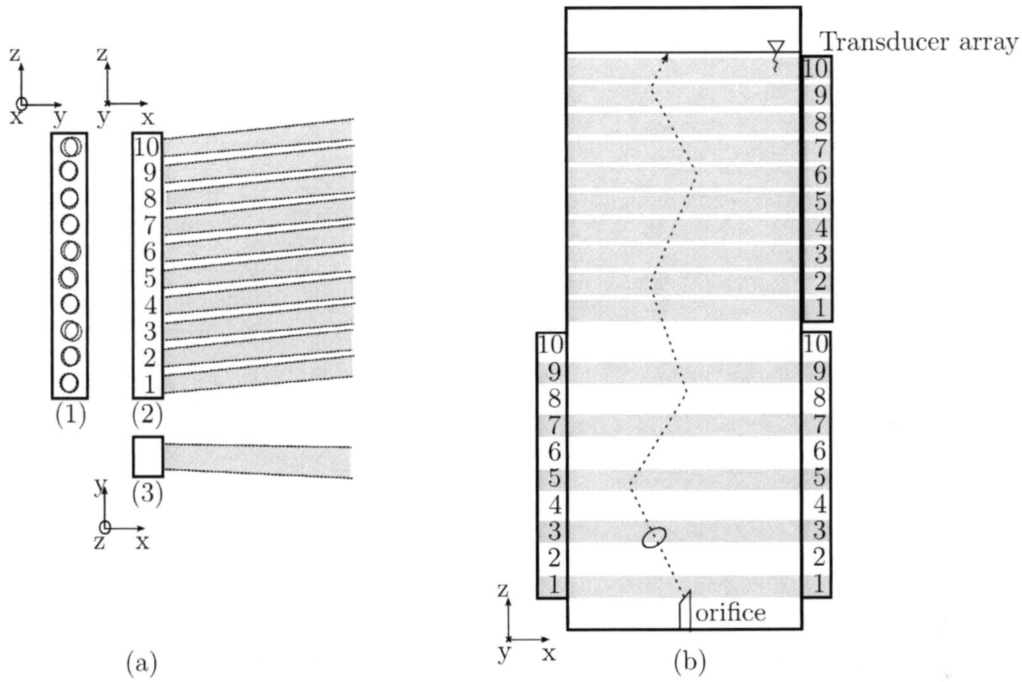

Figure 18. Schematic drawing illustrating a transducer array with 10 elements (a). The dotted lines display the deviation of the actual positions with respect to the nominal ones (Sketch 1). Sketches 2 and 3 visualize the deviation in the alignment of the ultrasonic cones from side and top. Visualization of different combinations of transducer elements (b), above usage of 10 single transducer elements to cover bubble trajectories, below combination of 5 pairs of opposing transducer elements to cover the bubble diameters.

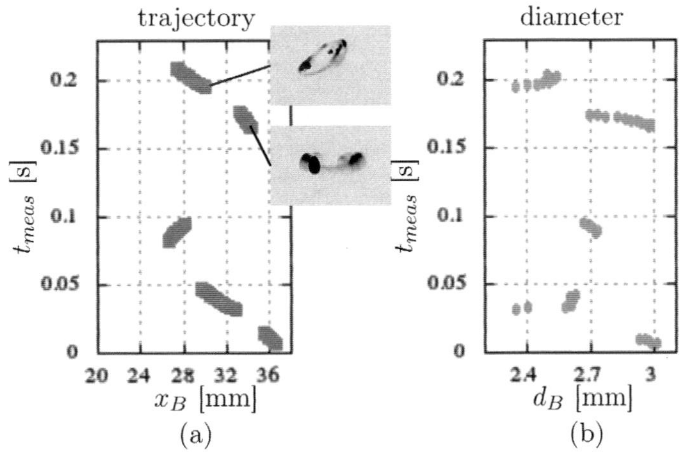

Figure 19. Bubble trajectory with pictures of the high-speed camera (a), bubble diameters of a single bubble rising in water (b).

The measurements of the rise of single argon bubbles in GaInSn with an applied static horizontal magnetic field underlined the suitability of UTTT for the detection of the bubble diameter in liquid metal two-phase flows [51]. Exemplary, Figure 20 presents the trajectory

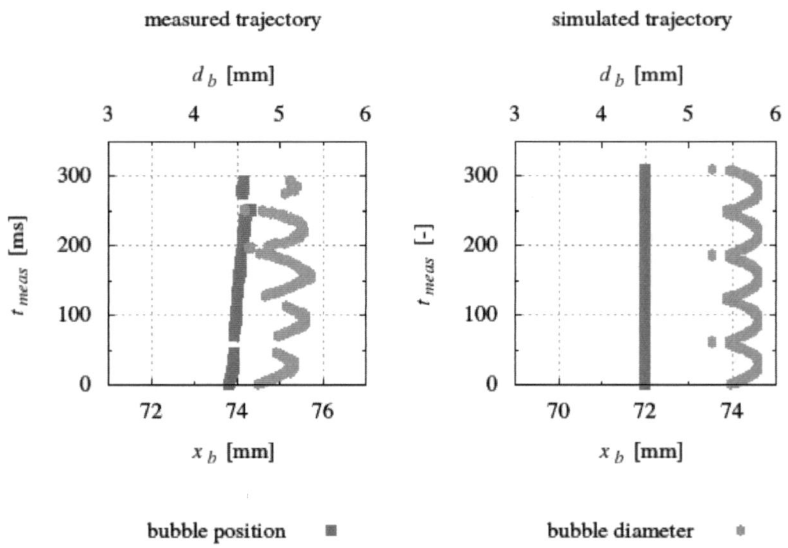

Figure 20. Comparison of measured and simulated bubble trajectory and diameter curves of a bubble rising in GaInSn.

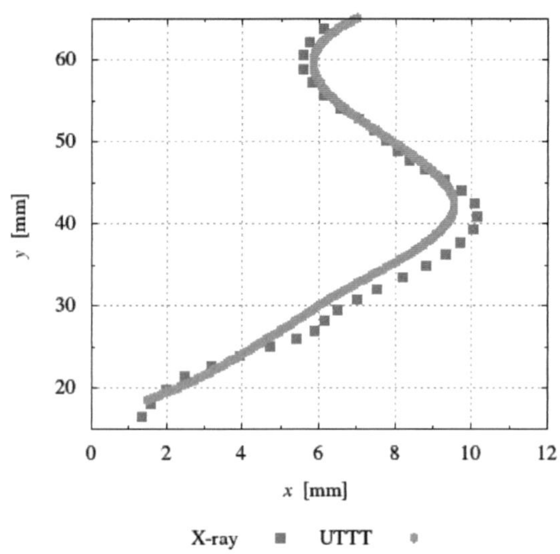

Figure 21. Comparison of bubble trajectories measured by UTTT and X-ray radiography.

and diameter of one argon bubble rising in GaInSn under the influence of a static horizontal magnetic field. As expected, the bubble rises on a nearly straight path. In order to get a better understanding how the shape and the position of the bubble effect the measurement, a ray tracing algorithm was developed which models the ultrasonic beam as a bundle of rays, which are partly reflected by the bubble modelled as an ellipsoid. With this method the trajectory and the diameter of the bubbles were simulated and compared in Figure 20 with the experimental results. A good agreement could be observed.

In order to validate the measurements of UTTT in GaInSn, the trajectory and size of the rising bubbles without an applied magnetic field were recorded by the means of X-ray radioscopy. Figure 21 displays the trajectory of one rising bubble measured by the means of UTTT and X-ray radiography. Again, a good agreement could be achieved.

For the measurement of bubble chains and bubble swarms, the sensor arrays are aligned horizontally. This enables a complete scanning of the cross section of the fluid container for the detection of the position and the diameter of rising bubbles. If two transducer arrays are positioned at two different heights, the rising velocity can also be estimated. Furthermore, the new setup offers the possibility to use UTTT as a void fraction measurement method.

Preliminary measurements at the Mini-LIMMCAST facility suggest that UTTT may be able to get information about the bubble injection through the SEN and the movement of the bubbles in the mold.

6. Inductive bubble detection

Inductive bubble detection methods typically apply a time varying magnetic field to the melt. The induced eddy currents are deformed when a bubble moves through the melt, because the bubbles are electrically insulating. The governing equation is following differential equation

$$\nabla^2 \vec{A} = \mu_0 \sigma \frac{\partial \vec{A}}{\partial t} \tag{16}$$

with the vector potential $\vec{A}$ and the spatially varying electrical conductivity σ in the volume. The eddy currents in the melt generate a secondary magnetic field which perturbs the applied magnetic field and depends on the spatial variation of σ. This secondary magnetic field can be measured outside the melt and depends on the geometry of the conducting volume.

One promising technique is MIT which allows the reconstruction of the conductivity distribution in one cross section of a pipe [9]. In general, N coils are equally spaced around the imaging area which results in $N(N-1)/2$ independent measurements of the mutual inductances between the coils. In order to reconstruct the conductivity distribution, Equation (16) has to be inverted. This leads to a non-linear inverse problem which is difficult to solve.

A sensor using eight coils was already used to visualise the position of the liquid metal strand in the SEN in a real caster [9] and in a cold model [56, 40]. The outer surface of the liquid metal jet in the SEN could be reliably reconstructed. However, bubbles inside the liquid could not be identified. Due to the non-linear inverse problem which has to be solved for MIT it is evidently very difficult to reconstruct the surface of the liquid metal column and the bubbles in the melt at the same time. One solution could be the combination of MIT for visualizing the interior of the liquid metal with the electrical capacitance tomography (ECT) for visualizing the outer shape of the liquid metal strand. In a first feasibility study it could be shown that ECT is able to detect the outer surface of liquid metal [57].

In applications like the stagnant bubble reactor for methane pyrolysis the interesting parameters are the trajectory and the size of the bubbles. The high temperature over 800 °C and the high corrosiveness of liquid tin prevent the use of ultrasonic or invasive measurement techniques. Up to now, no appropriate sensor is available. In order to reduce the complexity of the MIT system for detecting bubbles in a pipe filled with liquid metal, a new sensor consisting only of one excitation coil and one planar gradiometer coil on opposite sides of the pipe was developed [58]. The intention of this sensor was the detection of bubbles which are traversing the sensitive region of the sensor without solving any inverse problem. The sensor was tested in GaInSn and in sodium. Numerical simulation suggested that the recorded signal contains information about the position and the size of the bubble, but it is evidently difficult to extract

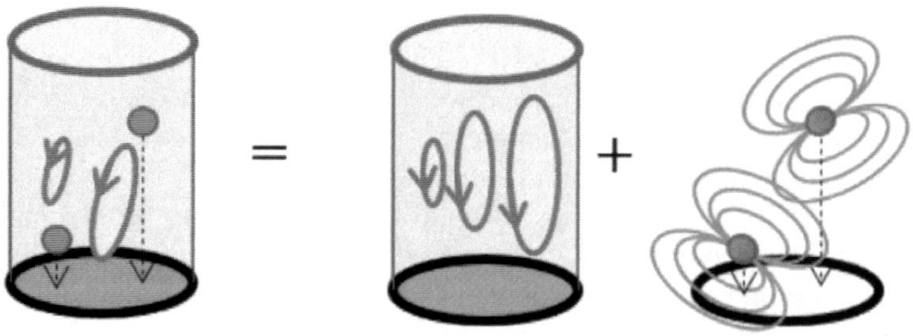

Figure 22. Principle of the dipole: instead of calculating the whole current-distribution for a changed geometry, the field of a reference case is corrected with the dipole fields substituting the bubbles.

this information from one single measured voltage of the detection coil. As a consequence, the inversion is ambiguous. This problem can be overcome by using an array of excitation and detection coils to reconstruct the bubble distribution from different positions.

Instead of recalculating the current distribution in the observed volume for every bubble distribution, only the reference case without any bubble is calculated. If a comparably small bubble is placed in the observed volume, the main difference in the current distribution is the absence of electric currents in the bubble volume. As simplification, the change of the secondary magnetic field is modelled by an appropriate magnetic dipole which compensates the magnetic field from the reference case as seen in Figure 22. In the case of a spherical bubble in a sinusoidal excitation field, the equation for the momentum is given by

$$m_{dipole}(t) = -\frac{8\mu_0\pi\sigma_l\omega B_0}{15} \cdot R_{bubble}^5 \cdot cos(\omega t) \ , \tag{17}$$

where R_{bubble} is the radius of the sphere and B_0 is the amplitude of the excitation field at the center of the bubble.

This model has some significant advantages as the dipole has a directed field and scales linearly with the number of bubbles. For the reconstruction of the bubble distribution a distribution of dipoles has to be found which corresponds to the detected signal. The measured magnetic field H_{total} at the detection coils can be written as following sum:

$$H_{total} = H_{excitation} + H_{reactor} - \sum H_{dipoles} \ . \tag{18}$$

$H_{excitation}$ is the magnetic field of the excitation coil, $H_{reactor}$ is the field generated by the eddy currents and H_{dipol} are the fields of the magnetic dipoles substituting the bubbles.

For a reliable measurement of the magnetic field of the dipole, gradiometric detection coils are used and the excitation and detection coils are placed in a special arrangement so that the magnetic field from the excitation coil is nearly eliminated at the detecting coils. In this case, only the magnetic field of the dipoles is measured.

Figure 23 shows a schematic sketch and a photo of the first experimental setup. The entire sensor consists of two planes with three coils each. Each coil consists of an excitation (red) and receiving (green) coil. For each plane three differential measurements from three perspectives can be obtained. This allows the detection of single bubbles in the cross section of the volume. The two planes are used to detect the movement of the bubble in vertical direction. The first experiment at room temperature is set up in order to test the new sensor. A non-conductive phantom of a bubble at the tip of a metallic rod can be inserted from above at any place inside the vessel. Objectives are the optimal positioning of the coils and the analysis of bubble swarms.

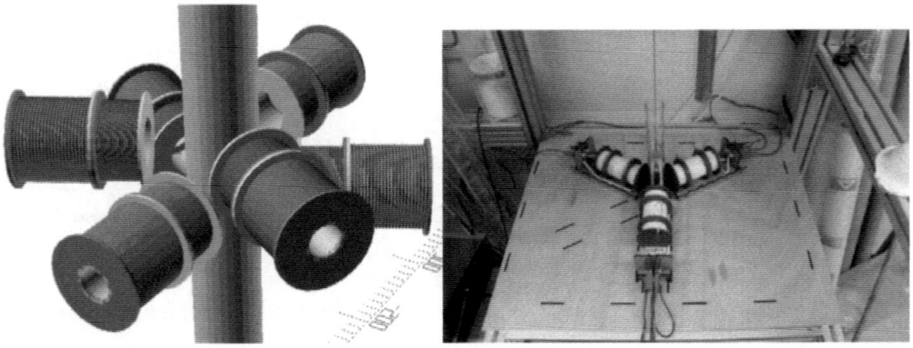

Figure 23. Schematic sketch of the sensor and a picture of the experimental setup using only three coils.

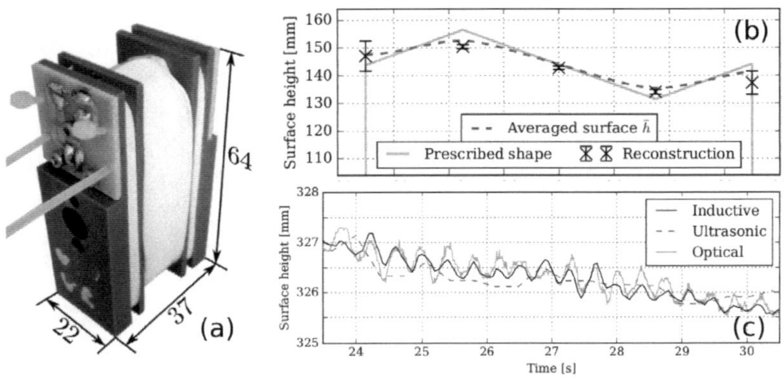

Figure 24. Inductive sensor (a) with dimensions in mm; test measurement (b) at a container with a zig-zag surface; experimental validation (c) in comparison to other measurement techniques [59].

The challenge to adapt the new sensing system to the high temperature are left for further investigations. Among them are the construction of high temperature coils, thermal shielding as well as filtering out the electromagnetic field generated by the heaters.

7. Inductive level measurement sensor

One industrial problem is the accurate and fast determination of the melt level in various processes incorporating molten metals or semiconductors. For this we developed a sensor which consists of an oval shaped excitation coil at the center of the body and two oval shaped receiver coils connected to form a gradiometric coil, positioned symmetrically from the center coil (Figure 24(a)). The sensor permeates the melt with an AC magnetic field, which generates eddy-currents, that in turn give rise to a secondary magnetic field. The primary field is suppressed by the gradiometric receiving coils, which results in a good signal-to-background ratio for the secondary field. The conversion from magnetic field amplitude to liquid level is carried out by calibration and subsequently fitting a sigmoid function to the data [59].

In order to evaluate the sensor we performed static and dynamic tests. The static test consisted of a liquid metal container with the dimensions of the Mini-LIMMCAST slab mould, filled with GaInSn and a zig-zagged PVC block on top of the liquid, which creates a non-flat surface shape which can be measured by the sensor. We positioned the sensor successively at five evenly distributed horizontal positions along the wide face of the container and reconstructed the level at this position (Figure 24(b)). While the extrema are underestimated because of the nature of the eddy-currents which occupy the whole available volume, the measured levels are very close to the calculated averaged surface.

The dynamic test was conducted at Mini-LIMMCAST itself and evaluates the sensor performance in comparison to an ultrasonic sensor (sampling frequency 2.8 Hz) and a laser scanner (93 Hz), as seen in Figure 24(c). During Mini-LIMMCAST operation surface waves are propagating between the two narrow faces, resulting in a seesaw motion of the surface. While the ultrasonic sensor has a too low sampling frequency to detect the surface waves at Mini-LIMMCAST, the optical and our inductive sensor show a striking similarity, thereby suggesting a sub-millimeter resolution of our sensor at a time-resolution of 20 Hz until now. A patent application has been filed and is under review.

8. Conclusion

In this short review, the recent developments of flow measurement techniques in the framework of the Helmholtz Alliance LIMTECH were presented. At the beginning, an overview of different inductive flow rate sensors for ducts and pipes were given. These techniques include TEC-FM, which is calibration free, and LFV, which was tested with different wall conductivities. The extension of LFV, which is called local LFV, allows the investigation of flow structures close to the wall. It was applied for the first time to a complex flow in a cylindrical vessel and to the flow structure at the wide face of a slab casting mold. A new three dimensional sensor is capable of measuring all three force and torque components at the same time. By applying a dry calibration of the new sensor, the velocity distribution along the narrow face could be calculated and compared to UDV measurements.

A different flow measurement technique is CIFT which allows the reconstruction of the mean three-dimensional velocity structure in liquid metals. By using gradiometric induction coils for the magnetic field measurement in combination with an AC excitation magnetic field with low frequency, the magnetic field measurements are less sensitive to fluctuations of the environmental magnetic field. This was demonstrated by the reliable measurement of the small induced magnetic field along the narrow faces of the wall of a slab casting mold even in the presence of a strong static magnetic field generated by an electromagnetic brake. Additionally it could be shown that the dynamics of a typical large scale circulation in a temperature driven experimental setup can be reconstructed. The challenge in this case is the magnitude of the velocity which is one order of magnitude smaller than in the case of continuous casting.

For the measurement of two phase flows in large vessels, UTTT can be used which allows the measurement of trajectories and diameters of bubbles. In order to understand the dependence between the shape of the bubble and the diameters measured by UTTT, experiments in water as well as in GaInSn were conducted. These measurements were compared with optical and X-ray measurements. Additionally, a simulation of the ultrasound beam based on ray tracing was developed. These developments provide a sound basis for the investigations of complex two-phase flow regimes like in a continuous casting mold.

In order to investigate bubble behavior in high temperature melts, a quite new inductive method based on MIT is under development. While for MIT a non-linear inverse problem has to be solved in order to reconstruct bubbles, the new sensor system will use a different strategy, by modelling the bubbles as a single magnetic dipole which perturb the applied magnetic field.

While the LIMTECH Alliance is about to end, the developed techniques are taken to the next step of industrial application with experienced industrial partners. This includes a collaboration with SolarWorld Innovation GmbH in Freiberg regarding Czochralski crystal growth and with an manufacturer of continuous steel casting plants. Additionally, an innovative training network called TOMOCON will start in summer 2017. The aim of this project is the integration of tomographic sensors into process control. One of the four planned demonstrators will be the control of continuous casting using CIFT and MIT.

List of acronyms

CIFT	contactless inductive flow tomography
CSM	cross-shape magnet
ECFM	eddy current flow meter
ECT	electrical capacitance tomography
EMBr	electromagnetic brake
GaInSn	gallium-indium-tin
L2F2	local Lorentz force flowmeter
LFV	Lorentz force velocimetry
LIMMCAST	liquid metal model of continuous casting
LIMTECH	liquid metal technologies
LSC	large scale circulation
MIT	mutual inductance tomography
SEN	submerged entry nozzle
SnBi	tin-bismuth
TEC-FM	transient eddy current flow meter
UDV	ultrasound Doppler velocimetry
UTTT	ultrasound transit-time technique

Acknowledgments

Financial support of this research by the German Helmholtz Association in the frame of the Helmholtz-Alliance LIMTECH is gratefully acknowledged. The authors are also grateful to Deutsche Forschungsgemeinschaft (DFG) for the financial support in the framework of Research Training Group Lorentz Force Velocimetry and Lorentz Force Eddy Current Testing (GRK 1567) at Technische Universität Ilmenau. We thank Shenyi Qin, Till Zürner and Robert Martin for their valuable contributions to CIFT, as well as Frank Stefani and Sven Eckert for fruitful discussions, and Klaus Timmel for technical support.

References

[1] Manning C P and Fruehan R J 2001 *JOM* **53** 34–43
[2] Zhang L and Thomas B 2003 *ISIJ International* **43** 271–291
[3] Seerban M, Lewis M, Marshall C and Doctor R 2003 *Energy Fuels* **17** 705–713
[4] Yasushi Takeda 1987 *Nuclear Technology* **79** 120–124
[5] Eckert S, Gerbeth G and Melnikov V I 2003 *Experiments in Fluids* **35** 381–388 ISSN 0723-4864
[6] Andruszkiewicz A, Eckert K, Eckert S and Odenbach S 2013 *The European Physical Journal Special Topics* **220** 53–62
[7] Saito Y, Shen X, Mishima K and Matsubayashi M 2009 *Nuclear Instruments and Methods in Physics Research Section A: Accelerators, Spectrometers, Detectors and Associated Equipment* **605** 192 – 196
[8] Timmel K, Shevchenko N, Röder M, Anderhuber M, Gardin P, Eckert S and Gerbeth G 2015 *Metallurgical and Materials Transactions B* **46** 700–710
[9] Ma X, Peyton A, Binns R and Higson S 2005 *IEEE Sensors Journal* **5** 224–232
[10] Priede J, Buchenau D and Gerbeth G 2011 *Measurement Science and Technology* **22** 055402
[11] Thess A, Votyakov E V and Kolesnikov Y 2006 *Physical Review Letters* **96** 164501 ISSN 0031-9007
[12] Stefani F, Gundrum T and Gerbeth G 2004 *Physical Review E* **70** 056306 ISSN 1539-3755
[13] Heinicke C 2013 *Experiments in Fluids* **54** ISSN 0723-4864
[14] OECD Nuclear Energy Agency 2015 *Handbook on Lead-bismuth Eutectic Alloy and Lead Properties, Materials Compatibility, Thermal-hydraulics and Technologies* 2nd ed (OECD)
[15] Buchenau D, Eckert S, Gerbeth G, Stieglitz R and Dierckx M 2011 *Journal of Nuclear Materials* **415** 396–403
[16] Schulenberg T and Stieglitz R 2010 *Nuclear Engineering and Design* **240** 2077–2087
[17] Forbriger J and Stefani F 2015 *Measurement Science and Technology* **26** 105303

[18] Krauter N and Stefani F 2017 *Measurement Science and Technology* ArXiv: 1703.09116
[19] Priede J, Buchenau D and Gerbeth G 2011 *Journal of Applied Physics* **110** 034512
[20] Shercliff J A *The Theory of Electromagnetic Flow-Measurement* (Cambridge Science Classics) ISBN 978-0-521-33554-6
[21] Bucenieks I 2002 *Proceedings of the 5th PAMIR Int. Conference, Sept. 16-0, Ramatuelle, France* pp II–103
[22] Thess A, Votyakov E, Knaepen B and Zikanov O 2007 *New Journal of Physics* **9** 299 ISSN 1367-2630
[23] Thess A and Böck T 2013 *IEEE Transactions on Magnetics* **49** 2847–2857 ISSN 0018-9464
[24] Kazak O, Heinicke C, Wondrak T and Böck T 2015 *Magnetohydrodynamics* **51** 579–588
[25] Heinicke C and Wondrak T 2014 *Measurement Science and Technology* **25** 055302 ISSN 0957-0233
[26] Füßl R and Jäger G 2009 *Fundamental and applied metrology: IMEKO XIX world congress, September 6 - 11, 2009, Lisbon, Portugal* (Budapest: IMEKO) pp 235–238 ISBN 978-963-88410-0-1
[27] Schleichert J, Rahneberg I and Fröhlich T 2013 *International Journal of Modern Physics: Conference Series* **24** 1360017
[28] Schleichert J, Rahneberg I, Marangoni R R and Fröhlich T 2016 *tm - Technisches Messen* **83** ISSN 0171-8096
[29] Hernández D, Schleichert J, Karcher C, Fröhlich T, Wondrak T and Timmel K 2016 *Measurement Science and Technology* **27** 065302
[30] Heinicke C, Tympel S, Pulugundla G, Rahneberg I, Böck T and Thess A 2012 *Journal of Applied Physics* **112** 124914
[31] Hernández D, Karcher C and Wondrak T 2016 *Proceedings of the 10th PAMIR International Conference, June 20-24, 2016, Cagliari, Italy* (Cagliari: PAMIR Centre) pp 198–202 ISBN 978-88-905519-3-2
[32] Hernández D, Böck T, Karcher C and Wondrak T 2017 *Magnetohydrodynamics* **53** 3–13
[33] Ratajczak M, Wondrak T, Zürner T and Stefani F 2015 *2015 IEEE Sensors* 7370342
[34] Stefani F and Gerbeth G 2000 *Inverse Problems* **16** 1–9
[35] Wondrak T, Eckert S, Gerbeth G, Stefani F, Timmel K, Peyton A, Terzija N and Yin W 2014 *Steel Research International* **85** 1266–1273
[36] Wondrak T, Ratajczak M, Gundrum T, Stefani F, Krauthäuser H G and Jacobs R T 2015 *2015 IEEE International Symposium on Electromagnetic Compatibility* 306–311
[37] Ratajczak M, Wondrak T, Stefani F and Eckert S 2015 *Magnetohydrodynamics* **51** 461–471
[38] Timmel K, Eckert S and Gerbeth G 2011 *Metallurgical and Materials Transactions B* **42** 68–80
[39] Wondrak T, Galindo V, Gerbeth G, Gundrum T, Stefani F and Timmel K 2010 *Measurement Science and Technology* **21** 045402 ISSN 0957-0233
[40] Wondrak T, Eckert S, Gerbeth G, Klotsche K, Stefani F, Timmel K, Peyton A, Terzija N and Yin W 2011 *Metallurgical and Materials Transactions B* **42** 1201–1210
[41] Ratajczak M, Gundrum T, Stefani F and Wondrak T 2014 *Journal of Sensors* **2014** 739161
[42] Ratajczak M, Gundrum T, Stefani F and Wondrak T 2015 *J. Manuf. Sci. Prod.* 41–48
[43] Ratajczak M, Wondrak T and Stefani F 2016 *Philosophical Transactions of the Royal Society A* **374**
[44] Timmel K, Eckert S, Gerbeth G, Stefani F and Wondrak T 2010 *ISIJ International* **50** 1134–1141
[45] Wondrak T an Pal J, Stefani F, Galindo V and Eckert S 2017 *Flow Measurement and Instrumentation* ArXiv: 1702.04880
[46] Matikainen L, Irons G, Morala E and Chang J S 1986 *Review of Scientific Instruments* **57** 1661–1666
[47] Banerjee S and Lahey Jr R 1981 *Advances in Nuclear Science and Technology* (Springer) pp 227–414
[48] Chang J S and Morala E 1990 *Nuclear Engineering and Design* **122** 143–156
[49] Moore P I, Brown G J and Stimpson B P 2000 *Measurement Science and Technology* **11** 1802
[50] Mahadeva D V, Baker R C and Woodhouse J 2009 *Instrumentation and Measurement, IEEE Transactions on* **58** 1602–1609
[51] Richter T, Keplinger O, Strumpf E, Wondrak T, Eckert K, Eckert S and Odenbach S 2017 *Magnetohydrodynamics* **53** 383–392
[52] Takeda Y 2012 *Ultrasonic Doppler velocity profiler for fluid flow* vol 101 (Springer)
[53] Vogt T, Andruszkiewicz A, Eckert S, Eckert K, Odenbach S and Gerbeth G 2012 *Metallurgical and Materials Transactions B* **43** 1454–1464
[54] Povolny A and Kikura H 2016 Ultrasound reflector recognition and tracking technique for two-phase flow *2016 24th International Conference on Nuclear Engineering* p V005T15A041
[55] Richter T, Eckert K, Yang X and Odenbach S 2015 *Nuclear Engineering and Design* **291** 64–70
[56] Terzija N, Yin W, Gerbeth G, Stefani F, Timmel K, Wondrak T and Peyton A J 2011 *Measurement Science & Technology* **22** 015501
[57] Wondrak T and Soleimani M 2017 *Measurement Science and Technology* **28** 064001
[58] Gundrum T, Büttner P, Dekdouk B, Peyton A, Wondrak T, Galindo V and Eckert S 2016 *Sensors* **16** 63
[59] Zürner T, Ratajczak M, Wondrak T and Eckert S 2017 *Measurement Science and Technology* Accepted

IOP Conf. Series: Materials Science and Engineering **228** (2017) 012024 doi:10.1088/1757-899X/228/1/012024

On an Analog Controlled Precision Heat Power Source

M Seilmayer, V K Katepally

Institute of Fluid Dynamics, Helmholtz-Zentrum Dresden-Rossendorf, Bautzner Landstraße 400, 01328 Dresden, Germany

E-mail: m.seilmayer@hzdr.de

Abstract. The analog controlled precision heat power source is the main part of a sensor development to estimate real time properties of the primary coolant in a 300 kW process cooling system. The measurement of physical properties like thermal conductivity or thermal diffusion requires a precise and accurate heater. The constant heat flux is then applied to the liquid under test to acquire its different properties. Here, commonly established control methods of heating with constant current or constant voltage may fail because the heating resistor changes its resistance with temperature. The idea is to utilize a power monitor circuit like the LT2940, which contains an analog multiplier with a control loop around it. The initial design and its assumed uncertainties will be discussed. The first version of the power controller shows an outstanding performance in terms of precision in a steady state. Compared to conventional switching mode power sources the approach with an analog controlled heater avoids EMI issues as well. The main goal of the present design is a precise source of heat power with *less* than 0.5% of error.

1. Introduction

Liquid metal experiments concerning effects like the “Tayler instability” [1], “Azimuthal Magnetorotational Instability” (AMRI) [2] or Rayleigh-Bérnard convection in conjunction with magnetic fields [3] require a stable and reliable thermal boundary condition. It is quite obvious that a stable temperature at the wall must be guaranteed to achieve good results. In addition to that liquid metal experiments also require an accurate magnetic field or precise electrical current, which is generated by high performance current sources. An illustrative example is the AMRI experiment, in which a central water cooled copper rod carries a DC current of up to 20 kA. The resulting constant magnetic field B_φ then interacts with the flow. Here, spectral analysis of the very first velocity measurements indicated that slow temperature oscillations of the water cooling system inflect the flow significantly. The worst case would be, if these characteristic frequencies overlap the AMRI itself, which is also indicated by similar low frequencies. Above mentioned thermal interaction takes place because of the heat transfer from the inner conductor to the inner cylindric wall of the set-up. A proper vacuum insulation along the axis stops this effect but the global influence on the resistive load and the current measurement inside of the power supply still remains. The immense ohmic heating from the copper supply rods and heat power from the current sources entails a large scale process cooling system. For the mentioned experiments in the liquid metal laboratory a typical 300 kW cooling system is installed to dissipate the waste heat from the lab to the air outside the building.

The industry standard PID control approach lead to very poor dynamics with $\approx$ 10 K overshoot and about 3 K long term fluctuations of the outflow temperature with time constants in the order of five to twenty minutes. Consequently, slowly changing outflow modulates the thermal boundary which affects the flow and final quality of the experiment. An other consequence of the introduced frequencies is a modulated magnetic field as a response to the resistive current measurement in the power supplies. To improve the stability, precision and dynamics of the process cooling system the control strategy is changed to a predictive approach in which the power demand is calculated out of the coolants flow rate $\dot{V}$ and temperature differences $\Delta\vartheta$. The concept is based on the idea of a calorimetric balance

$$\underbrace{\dot{V}_{\mathrm{P}} \cdot \rho_{\mathrm{Gly}}(\vartheta_{\mathrm{P}}) \cdot c_{\mathrm{p,Gly}}(\vartheta_{\mathrm{P}}) \cdot \Delta\vartheta_{\mathrm{P}}}_{\text{primary power transported to the air}} = \underbrace{\dot{V}_{\mathrm{S}} \cdot \rho_{\mathrm{H_2O}}(\vartheta_{\mathrm{S}}) \cdot c_{\mathrm{p,H_2O}}(\vartheta_{\mathrm{S}}) \cdot \Delta\vartheta_{\mathrm{S}}}_{\text{secondary power taken from the lab}} \tag{1}$$

between absorbed and exhausted heat. The primary coolant is a water glycol mixture to prevent freezing in winter and the secondary coolant is pure water. The power is balanced by calculating equation (1) to set the primary flow rate $\dot{V}_{\mathrm{P}}$.

Up to now the density ρ and the specific heat capacity c_{p} is assumed to be constant, which gives defective accuracy especially for the primary side. However, these values do vary since the primary coolants temperature ϑ_{P} ranges from $-20°\mathrm{C}$ in winter to $25°\mathrm{C}$ in summer, which lead to different values for the thermal carrying capacity $\rho \cdot c_{\mathrm{p}}$ according to the weather or season. The characteristic curves from the datasheets of the liquids do apply to laboratory conditions. But, in a large scale cooling system there are impurities like rust, dissolved gas or other undefined solutes, which drastically change the thermophysical properties like ρ, c_{p} or the heat conductance λ of the fluids. To minimize the effect of thermal and chemical dependencies on the outflow temperature it becomes necessary to measure the parameters of interest in real-time.

In Aller et al. [4, Ch. 66] the "transient hot wire" method is recommended to determine the required values. He also states that "*(d)espite* [...] (the) *wide* (scientific) *application to measurements in gases, solids, and liquids, there has been* ***no significant commercial*** *development of an instrument of this kind* (until the year 2014) [...]" [4, p. 66-8], which continues until now. This is because chemical or physical instruments (e. g. for determining density or viscosity) are designed as table top apparatus which typically do not support on-line or real-time measurements, in general. In comparison, the present sensor design might be a little less precise but it will fit into the automated industrial environment of the process cooling system.

The general idea behind such sensor principles is to introduce a constant well known heat flux P into a liquid, which leads to a characteristic temperature field around the heater. Next, the thermophysical parameters of the fluid under test can be estimated by properties of the transient temperature field. The transient hot wire method according to J. J. Healy et al. [5] or M. J. Assael [6] is the method of choice. It assumes a constant heat flux from the cylindrical surface of a wire into the liquid and at the same time the wire acts as the temperature sensor. The idealized model equation

$$\vartheta(r,t) = \frac{q}{4\pi \cdot \lambda} \ln\left(\frac{4\alpha \cdot t}{r_0^2 \cdot C}\right) + \vartheta_0 \tag{2}$$

of this method indicates a linear relation between logarithmic temperature and time. Here r_0 denotes the wire radius and C is an integration constant. The wire is assumed as an infinite line sourcing a constant flux q per unit length. Finally, the thermal conductivity λ and thermal diffusivity $\alpha = \lambda/(\rho \cdot c_{\mathrm{p}})$ can be derived as parameters from the linear model fit of $\log t \propto \vartheta$.

In summary, it can be stated that the introduced transient hot wire method and many anemometric flow meters incorporate constant power like the "Two-dimensional Flow Meter" by Makinwa et al. [7] or some integrated types of gas detectors like Fung et al. [8] or Rueda-Guerrero et al. [9]. But the key component of such a sensor system is the precision heater, which also determines the overall accuracy of the measurement. The present paper focuses on this part and sketches out the development and first results of the prototype in discrete standard electronics.

A controlled electrical heater require a suitable power supply which can keep the load power constant during the temperature and resistivity fluctuations of the heater. Since voltage and current are connected by the well known Ohm's relation the quotient $R_\mathrm{L} = U_\mathrm{L}/I_\mathrm{L}$ represents the load resistance. In a more general meaning R_L describes a linear system, in which I could be the input (measured value) and the voltage U would be the reaction on this input. Most of the power supplies support constant current (CC) and constant voltage (CV) operation modes. Additionally, the linear behaviour of the load R_L supports the stable performance of these output modes.

In contrast to that, keeping the product of potential and flux constant is very different. First of all, the well known relation of load power

$$P_\mathrm{L} = U_\mathrm{L} \cdot I_\mathrm{L} \rightarrow \text{const.} \tag{3}$$

$$\text{with } P_\mathrm{L} = R_\mathrm{L}(\vartheta) \cdot I_\mathrm{L}^2, \tag{4}$$

is the state of the electrical system. Additionally, equation (4) indicates that the load itself can vary with temperature ϑ or even with other parameters. It follows, that a suitable power regulation should keep the output power constant even if the load $R_\mathrm{L}(\vartheta)$ changes.

A comprehensive study of different integrated analog approaches can be found in [9]. In the following, the common variants are discussed very briefly. First, a fully integrated CMOS approach on a precision constant power (CP) source was developed by S. Chan et al. [10] with an accuracy of about 1% over a wide dynamic load range ($\approx 50\%$) and approximately $P_\mathrm{max} \approx 10\,\mathrm{mW}$. Next to that, the discrete counterpart with similar accuracy was build by N. Tadić et al. [11]. Here, the working principle is based on a resistive mirror [12], with which the output power

$$P = \frac{1}{k} U_\mathrm{ref} \cdot I_\mathrm{ref} \tag{5}$$

can be set according to a reference voltage and reference current. Keeping one of them constant will result in a voltage (VCP) or current controlled (CCP) power source with a characteristic system constant k. The main drawback on the global error is the precision (drift and accuracy) of two different references for voltage and current which would individually influence the output power.

Nevertheless, the control of electrical power requires the measurement of both voltage and current at a moment along with a separate control or calculation of its product. On the one hand side, digital signal processing makes it easy to calculate $U \cdot I$ but requires a precise sampling of current and voltage. In the present case a minimum of 10 bit quantisation is required to measure U_L and I_L to estimate the product $U \cdot I \leq 100\,\mathrm{mW}$ with adequate precision of better than $\pm 0.5\,\mathrm{mW}$. Then, setting the output would require an additional digital to analog conversion stage which brings a third quantisation error into the system. With respect to the subsequent measurement of thermal properties this value of accuracy would lead to an undesired large uncertainty in parameter estimation of any fluid property. Next to that, the digital signal processing (acquiring signals and calculating) would take a certain time so difficulties with the control loop stability must be considered within the design. One exception represents the sensor development by Makinawa et al. [7]. Here, a control loop with Σ–Δ-modulation of two opposing heaters enables CP mode operation in conjunction with orthogonal arranged thermopiles. The latter's translate the transduced heat flux into a differential voltage. According to its sign a comparator switches the related heater. With time, this procedure balances the power distribution across the individual heaters. The significant output here is a bit stream which represents the power balance within the sensor.

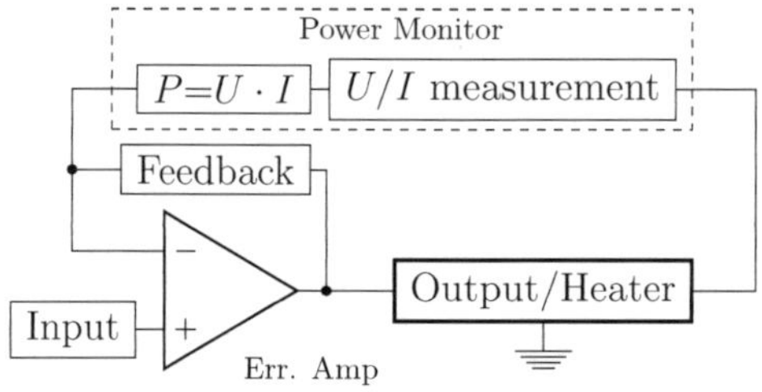

Figure 1. Simple Control loop

However, except Makinawa's approach the analog access is indispensable to gain sufficient stability, bandwidth, accuracy and precision of such a system. The present design with discrete components utilizes an analog power monitor to process the load voltage and load current. The product of the two measured quantities is then determined by the integrated analog multiplier unit. In contrast to Maxims tech report [13] and all other discussed approaches the used heater is a standard n-MOS transistor, because the transistor case (TO220) offers the easiest way to mount the heater to the sensor cell itself. The first development stage of the thermophysical properties sensor incorporates the direct measurement of $\lambda \approx \mathcal{O}\left(0.1\,\mathrm{W}/(\mathrm{K}\cdot\mathrm{m})\right)$ by heating a column of heat transfer oil from the top. This leads to a stratified stable liquid column which conducts the heat flux without fluid motion. This strategy prevents convection which would distort the measurement otherwise.

2. Design Idea

Figure 1 shows the typical structure of a constant power source. A scaled input is compared with the measured values by the error amplifier, which sets the actuator in the system (heater). As long as the measured power is lower than the input, the error amplifier will increase the output power on the heater. The key component is the integrated power monitor circuit (PowMon), which determines the electrical power in terms of an output voltage or output current. Such a device typically contains an analog multiplier to calculate the product of the load voltage and load current. Next to that, the PowMon also includes an input stage with a high-side current measurement and in the best case it provides a differential load voltage input as well. Since the main part in the PowMon – the analog computational unit (ACU) – has a limited band width, see [14] for details, an additional integrating RC feedback network supports the error amplifier and limits the frequency response.

To meet the industrial signal standards an input voltage range of $U_{\mathrm{in}} = 0 \ldots 10\,\mathrm{V}$ is defined and with respect to the measurement of thermal conductivity a total output power of $P_{\mathrm{out}} \leq 100\,\mathrm{mW}$ should be sufficient. The final transfer function

$$P_{\mathrm{out}} = 10\,{}^{\mathrm{mW}}/_{\mathrm{V}} \cdot U_{\mathrm{in}}, \tag{6}$$

is assumed to be linear.

3. Circuit Design

There are different analog multiplier IC's on the market, which are compared in Table 1. The most flexible device is the AD538 by Analog Devices [15]. It is one of the classic ACUs, which has the parametric transfer function $U_{\mathrm{o}} = U_{\mathrm{Y}} \cdot (U_{\mathrm{Z}}/U_{\mathrm{X}})^{m}$. This device gives the maximum flexibility but needs signal conditioning around to suit the task. These additional components (like the high side current measurement) would introduce additional uncertainties and errors. Next to that the MAX4210E [16] from Maxim Integrated must be mentioned. The fully integrated device

Table 1. Comparison of different analog multipliers. The numbers are valid for $T = 25°C$ and the desired conditions in this design

		AD538BD [15]	MAX4210E [16]	LT2940 [17]
Input	Range	±10 V	150 mV	8 V/200 mV
	Offset/Bias		3 μA	100 nA/100 μA
Output	Reading err.	0.1%	< 0.2%	< 0.5%
	FS error	< 1.5%	< 2.5%	< 2%
	Offset	0.25 mV	2 mV	6 μA
Type		ACU	PowMon	PowMon
Comment		Inputs referenced to ground. Errors in the 100 : 1-range	High side sensing.	Voltage/Current values. Error is valid for $\|U_V \cdot U_I\| < 0.4^2$

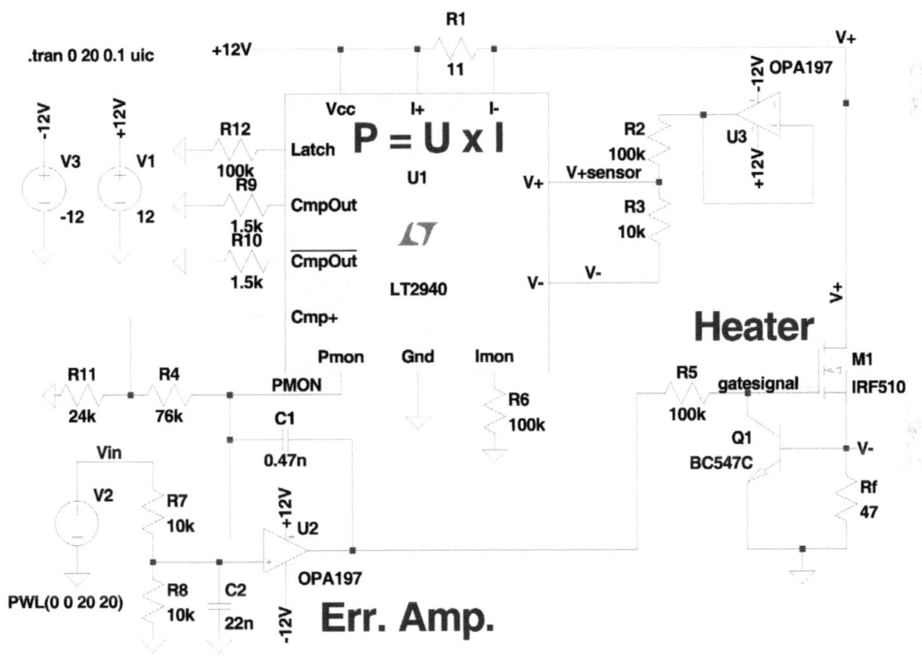

Figure 2. Design Idea Simulation. This is the LT-Spice simulation scheme. The different sections "Input Stage", "Error Amp.", "Heater", "Signal conditioning" and "Power Calculation" can be distinguished.

measures the high side input current and multiplies it with an input voltage. The main drawback is that differential voltage inputs for a floating load are missing. So additional signal conditioning becomes necessary which also introduces new sources of errors. Finally, the decision goes for the LT2940 from Linear Devices, this IC is also fully integrated and is also able to measure a high side current. But compared to the MAX4210E it provides differential voltage inputs with which the voltage drop over the heater can be measured directly.

In contrast to the typical approach the heater in the present design is a standard n-MOS

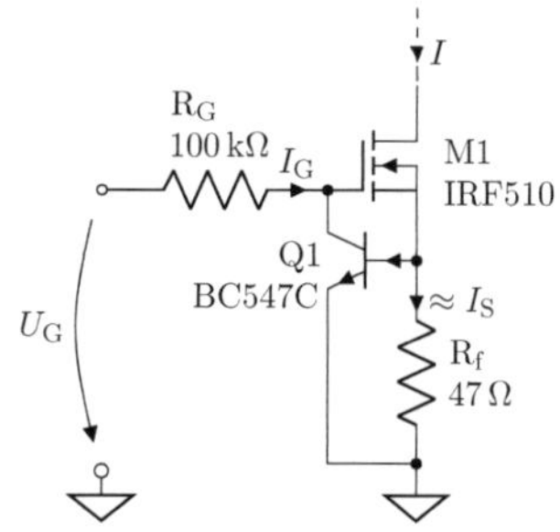

Figure 3. Power limitation with feedback transistor Q_1.

transistor. The specific type of M_1 (see Figure 2) is almost free of choice but with respect to global stability small gate-source and gate-drain capacities must be guaranteed. However, the IRF510 was chosen as heater, because of the TO220 case, which can be mounted very easily to a material to be heated.

Figure 2 shows the scheme of the LT-Spice simulation. In a stationary operation scheme the gate current I_G of M_1 can ideally be assumed as zero. So all the significant current is measured by R_1, which is rather large to gain a sufficient voltage drop. The npn-transistor Q_1 in conjunction with $R_f = 47\,\Omega$ limits the total power of the system to

$$\begin{aligned} P_{max} &= U_{BE}/R_f \cdot U_{CC} \\ &\approx 130\,\text{mW}, \end{aligned} \tag{7}$$

which gives enough reserve to the design goal of $P_{out} \leq 100\,\text{mW}$. Notice the collector-emitter current path at Q_1 as seen in Figure 3. Here $R_G = 100\,\text{k}\Omega$ sets the maximal collector current to $I_{CE} \leq 114\,\mu\text{A}$ so a minimal basis-emitter voltage $U_{BE,on} = 0.55\,\text{V}$ [18] of the BC547 can be assumed. If Q_1 becomes conductive the gate voltage of M_1 tends to ground and so it acts as a protection circuit that limits the total power of the device. Additionally, R_f represents a current feedback on the gate-source voltage and makes the output stage transfer (see Figure 4) function linear.

The input stage V_2 is made in a way that it fits the industrial requirements with a control voltage range of $0 \leq U_{in} \leq 10\,\text{V}$. The input frequency response is limited by C_2 and R_8 respectively and the operational amplifier (OPV) U_3 operates as an impedance converter to reduce the measurement error. However, the scaling of the voltage measurement is done with the voltage divider R_2 and R_3 to fit the minimum error conditions of the LT2940 input stage. Finally, the measured power is represented by a current I_{Pmon}, which has an operating limitation of $|I_{Pmon}| \leq 200\,\mu\text{A}$ [17]. The power monitor can source and sink the output current. In the present design the measured power is always positive, so a single resistor $R_{Pmon} = R_4 + R_{11} = 100\,\text{k}\Omega$ against ground translates the power proportional current into a corresponding voltage.

3.1. Proof of Concept

The power monitor measures the scaled input signals and processes its values. With respect to the desired overall transfer function (6) the scaling

$$K_V = \frac{R_3}{R_2 + R_3} = \frac{1}{11} \tag{8}$$

of the measured voltage $U_V = U_+ - U_-$, the scaling of the current

$$K_I = R_1 = 11\,\Omega \tag{9}$$

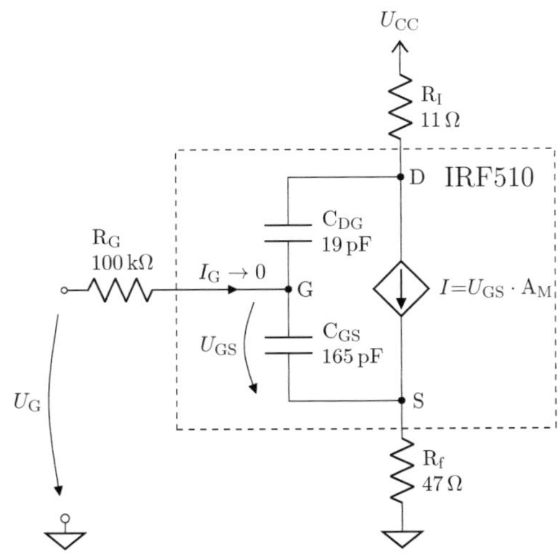

Figure 4. Output stage with a simple n-MOS transistor model. The gate-source capacity $C_{GS} = 165\,\mathrm{pF}$ and gate-drain capacity $C_{DG} = 15\,\mathrm{pF}$ are defined according to the datasheet of the transistor IRF510 [19]. The transconductance $A_M \approx 0.022\,\mathrm{A/V}$ was determined with the simulation using the correct SPICE model of the transistor.

and the linear transfer coefficient of the LT2940

$$K_{Pmon} = 500\,\mathrm{\mu A/V^2} \tag{10}$$

must be defined. Additionally, the control voltage $U_{con} = U_{in}$ is scaled down by the factor of two to meet the required input range. Performing classical network analysis according to Section 3.2 we find the stationary transfer function

$$\begin{aligned} P_{out} &= \frac{1}{2 \cdot K_{Pmon} \cdot K_V \cdot K_I \cdot R_{Pmon}} \cdot U_{in} \\ &= 10\,\mathrm{mW/V} \cdot U_{in}, \end{aligned} \tag{11}$$

which now equals the design goal equation (6).

3.2. Stability Analysis

During simulation it turned out that the stability of the circuit is strongly effected by the intrinsic gate-source and gate-drain capacities of the transistor M_1. Performing classical network analysis to the circuit in Figure 2 and with respect to a simplified transistor model (see Figure 4) we find the frequency depending transfer function

$$\frac{P_{out}}{U_{in}} = \frac{\frac{1}{2+sC_2R_8} \cdot \left(\frac{1+sC_1R_{Pmon}}{1+sC_1R_{Pmon}\cdot(A_{OP}+1)} \right)}{\frac{s\cdot(C_{GS}+C_{GD})\cdot R_G + R_f\cdot(A_M+sC_{GS})+1}{U_{CC}A_{OP}\cdot(A_M-sC_{GD})} + \frac{\frac{K_VK_I}{(s/\omega_{LT}+1)^2}K_{Pmon}R_{Pmon}}{1+sC_1R_{Pmon}\cdot(A_{OP}+1)}} \tag{12}$$

in which $s = i\omega$ denotes the complex frequency variable and with respect to the input stage the value of $R_8 = R_7$. The relevant physical parameters for the transistor model were taken from the corresponding datasheet [19]. The internal R_{DS} resistance of the transistor M_1 is ignored since the current feedback resistor R_f is governing the circuit. The current monitor LT2940 shows a

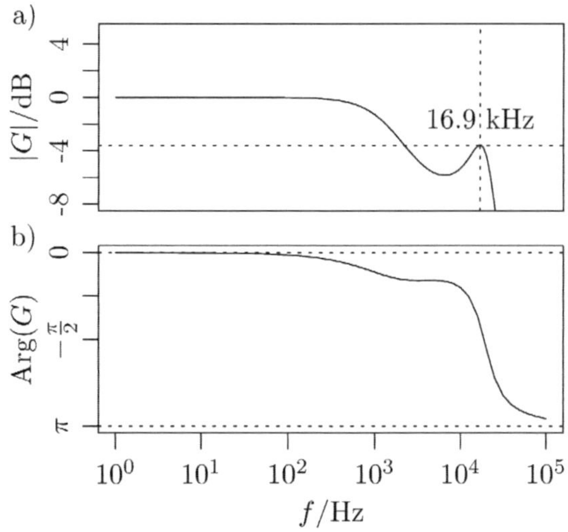

Figure 5. Bode plot of the transfer function. Panel a) gives the amplitude function and b) the corresponding phase diagram. The displayed transfer function $G(s) = 20 \cdot \log_{10}\left(P_{\text{out}} \cdot U_{\text{in}}^{-1} \cdot 0.1\,\text{mW}^{-1} \cdot \text{V}\right)$ represents the gain of equation (12) scaled by the theoretical value of $10\,\text{mW/V}$. The resonance peak at $f = 16.9\,\text{kHz}$ has $-3.6\,\text{dB}$ gain at $-73.5°$ phase. It follows, that the idealized circuit remains stable over all frequencies.

first order low pass behavior for K_I and K_V each. For these parameters the corner frequency of about $\omega_{\text{LT}} \approx 2\pi \cdot 200\,\text{kHz}$ is estimated from the datasheet [17, p. 7].

In the limit of a stationary operation with $s \downarrow 0$ the product of the transfer functions of the error amplifier U_2 and transistor M_1 can be taken to infinity, with $A_\text{M} \cdot A_\text{OP} \gg 1$. Equation (12) finally reduces to equation (11) which is only governed by the parameters of the analog PowMon.

Finally, the Bode plot of equation (12) is displayed in Figure 5. The graph shows the ideal behavior of the system, which only depends on idealized components. From this perspective, stability is reached over all frequencies up to $f \approx \mathcal{O}(100\,\text{kHz})$. Note, the LT2940 has a global low pass behavior with a corner frequency of about 200 kHz which influences the resonance frequency. From simulation and equation (12) it turns out, that

$$\begin{gathered} \frac{1}{2\pi \cdot \text{R}_8\text{C}_2} < \frac{1}{2\pi \cdot \text{R}_{\text{Pmon}}\text{C}_1} \\ 0.7\,\text{kHz} < 3.4\,\text{kHz} \end{gathered} \tag{13}$$

with $\text{C}_1 = 0.47\,\text{nF}$, $\text{C}_2 = 22\,\text{nF}$, $\text{R}_8 = 10\,\text{k}\Omega$ and $\text{R}_{\text{Pmon}} = 100\,\text{k}\Omega$ gives a stable performance. Lower values for C_1 cause stronger ringing effects in the circuit in case of a step response to U_{in}. Rather small values for C_1 below 90 pF mark the boundary of validity for the analytical model. For such values a significant oscillation occurs which can be proved with simulation and approximately with stability analysis of (11).

However, with respect to above suggested values the control loop is assumed to be stable over a wide frequency range up to $f \approx \mathcal{O}(100\,\text{kHz})$. In the limit of a stationary operation $s \downarrow 0$ the transfer function can be described by a simple linear equation

3.3. Component Uncertainties

With regard to the implementation the question about precision and accuracy comes up. To get an idea about the confidence or uncertainty of statistical measures the parameters

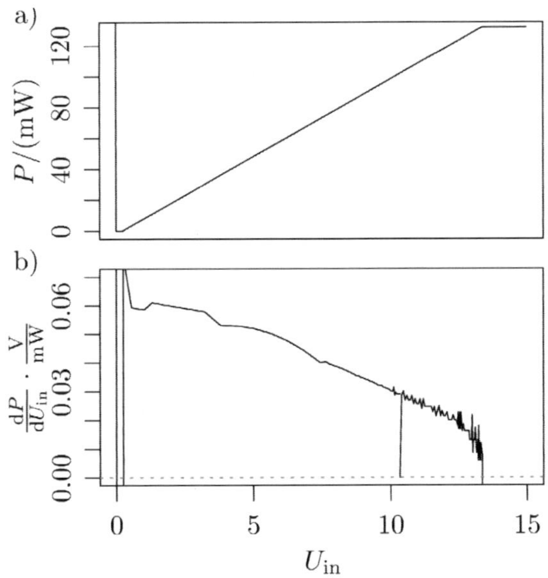

Figure 6. Simulation results. Panel a) shows the overall transfer function in dependence of the input voltage. The bottom graph illustrates the difference $\mathrm{d}P/\mathrm{d}U_{\mathrm{in}} - 10\,\mathrm{mW/V}$ to the ideal slope.

$$\frac{\Delta \mathrm{K_V}}{\mathrm{K_V}} = 0.0182\%, \tag{14}$$

$$\frac{\Delta \mathrm{K_I}}{\mathrm{K_I}} = 0.1\% \tag{15}$$

of the final transfer function (11) must be derived. The tolerance values of the corresponding components are listed in Table 2. The relative deviation of the output power then sums up to

$$\frac{\Delta P_{\mathrm{out}}}{P_{\mathrm{out}}} = \frac{\Delta \mathrm{K_V}}{\mathrm{K_V}} + \frac{\Delta \mathrm{K_I}}{\mathrm{K_I}} + \frac{\Delta \mathrm{R_{Pmon}}}{\mathrm{R_{Pmon}}} + \frac{\Delta \mathrm{K_{Pmon}}}{\mathrm{K_{Pmon}}} \approx 0.25\%. \tag{16}$$

Here the main contribution comes from the current scaling $\mathrm{K_I}$ and $\mathrm{R_{Pmon}}$ itself. A higher precision can be achieved, if $\mathrm{R_{Pmon}}$ is a variable resistor. Since the slope of the output power only depends on a product of coefficients it should be sufficient to trim the power source by $\mathrm{R_{Pmon}}$ to get a minimum deviation from the predefined slope of $10\,\mathrm{mW/V}$.

3.4. Simulation of the Stationary Transfer Function

With respect to uncertainties of the resistors and the non-linearities of real component models a LT-Spice simulation is carried out to prove the functioning of the circuit. The simulation takes into account the provided models for the LT2940 and the transistors. For the operational amplifier the OPA197 was chosen, because of its low bias JFET characteristics. Figure 6 displays the simulated transfer function with respect the input voltage. There is a good agreement to a linear correlation between input and output. A detailed view to the slope in Figure 6b shows the derivative

$$\frac{\mathrm{d}P}{\mathrm{d}U_{\mathrm{in}}} \propto U_{\mathrm{in}} \tag{17}$$

Table 2. Parameter tolerance

No.	Value	Tolerance [%]
R_1	$11\,\Omega$	0.1
R_2	$100\,k\Omega$	0.01
R_3	$10\,k\Omega$	0.01
R_{Pmon}	$100\,k\Omega$	0.1
K_{Pmon}	$500\,\mu A/V^2$	0.03

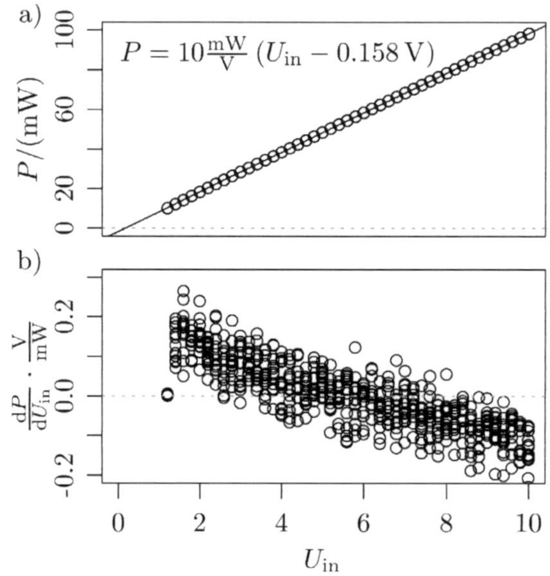

Figure 7. Output power P vs. input voltage U_{in}. A mainly linear correlation between output power and input voltage can be observed. The linear function in panel a) indicates a clear offset, which is a result of bias currents and voltages.

of the transfer function, which is also proportional to the input voltage. This means that the true transfer function would be a polynom in the second order. Next to that, the slope is slightly too large over the range of U_{in} so that the tuning resistance $R_{Pmon} = 100\,k\Omega$ is a bit too big. Finally, there is an offset of 0.15 V for U_{in} shifting the resulting transfer function systematically.

4. Results

The first prototype of the present design idea utilizes the LT2940 for power measurement. A surrounding control loop compares the measured heat power with an input voltage to keep the former constant. The input power stage is made up of a standard linear regulator of the 78xx series to stabilize the supply voltage and a protection circuit to prevent the user from flipping the supply voltage. The control loop corresponds to the simulation design from Figure 2.

The measurement of the transfer function in terms of input voltage U_{in}, voltage drop U_I over R_1 and the voltage over the heater U_H was carried out with a Keithley 2701 and the corresponding multiplexer card. The input voltage was set as a step function with a Keithley 2400 source meter. To prevent non-linear behavior the power source was measured within an input voltage range of $U_{in} = 1.2\,V \ldots 10\,V$.

IOP Conf. Series: Materials Science and Engineering **228** (2017) 012024 doi:10.1088/1757-899X/228/1/012024

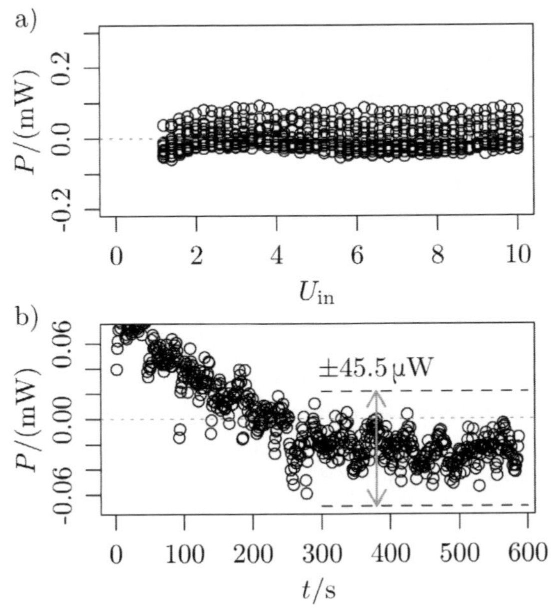

Figure 8. Second order model. The input voltage U_{in} is a periodic ramp function with $T = 46\,\text{s}$ so the displayed graphs cover several cycles of the transfer function. Panel a) shows the residuals vs. the input voltage of the second order model function (19) in the same scaling as Figure 7b). Sub-figure b) displays the drift of its residuals with the time, which relax from 300 s. The arrow covers a precision range of $\pm 45.5\,\mu\text{W}$ with an uncertainty level of $\alpha = 0.01\%$.

The linear model

$$P = 10\,\text{mW/V} \cdot (U_{\text{in}} - 0.158\,\text{V}) \pm 0.36\,\text{mW} \tag{18}$$

fits to the data and indicates a clear offset of about 0.158 V. Hence, such an error is systematic which should be taken into the final transfer function. The main contribution to the offset is due to the bias output current of the LT2940 (see Table 1). In contrast to that the offset and bias currents or voltages of the operational amplifier are rather small and therefore neglectable. Ignoring the obvious trend in the remaining residuals, the final statistical accuracy is $\pm 0.36\,\text{mW}$. Hereafter, such an estimate is taken of the $\alpha = 0.01\%$-quantil of the data.

The second order transfer function

$$\begin{aligned} P = & -0.0143\,\text{mW/V}^2 \cdot U_{\text{in}}^2 \\ & + 10.16\,\text{mW/V} \cdot U_{\text{in}} - 1.925\,\text{mW} \pm 0.135\,\text{mW}, \end{aligned} \tag{19}$$

gains a much higher accuracy of about $\pm 0.135\,\text{mW}$. From Figure 8 it follows, that there is a clear drift after the turn on, which stabilizes for the last 300 s of the measurement. For this 300 s tail a precision of about $\approx 45.5\,\mu\text{W}$ can be concluded, if the system is stable.

5. Conclusion

With a quite simple analog circuit it is possible to control the output power of a heater with high precision. In comparison to the approaches of S. Chan et al. [10] and N. Tadić [11] the present design can get a precision of better than 0.36 mW which represents approximately one order of magnitude less error.

The power monitor IC LT2940 helps to estimate the power in terms of a product of two measured voltages. The calibrated circuit then achieves a maximum precision of about 45.5 µW under the assumption of a quadratic transfer function in a saturated state. Even for the linear transfer function (18) the deviation from the ideal state is better than 0.5 mW, which was assumed in the beginning. All this holds if the constant offset is a systematic measure which can be easily corrected through the input $U_{\text{in}} - u_{\text{offset}}$. So the statements above apply for statistical errors and uncertainties like noise or bias effects of the sub-circuits.

In the next design the non-linear transfer function will be compensated by a digital input stage, which calculates the correct control voltage. So it becomes possible to compensate the second order term along with the input offset as well. A digital signal processor at the input stage also makes the calibration with $\mathrm{R}_{\mathrm{Pmon}}$ obsolete and gives the possibility to handle the long term drift.

Acknowledgements

The present work belongs to an ongoing improvement of the lab facilities which was in part supported by the Helmholtz Alliance "Liquid Metal Technologies – LIMTECH" and is gratefully acknowledged.

References

[1] Seilmayer M, Stefani F, Gundrum T, Weier T, Gerbeth G, Gellert M and Rüdiger G 2012 *Physical Review Letters* **108** 244501 ISSN 0031-9007, 1079-7114
[2] Seilmayer M, Galindo V, Gerbeth G, Gundrum T, Stefani F, Gellert M, Rüdiger G, Schultz M and Hollerbach R 2014 *Physical Review Letters* **113** 024505
[3] Tasaka Y, Igaki K, Yanagisawa T, Vogt T, Zuerner T and Eckert S 2016 *Physical Review E* **93** 043109
[4] Webster J G and Eren H (eds) 2014 *The Measurement, Instrumentation and Sensors Handbook* 2nd ed (Boca Ratón (Florida): CRC Press) ISBN 978-1-4398-4889-0
[5] Healy J J, de Groot J J and Kestin J 1976 *Physica B+C* **82** 392–408 ISSN 0378-4363
[6] Assael M J, Nieto de Castro C A, Roder H M and Wakeham W A 1991 *Measurement of the Transport Properties of Fluids* (*Experimental thermodynamics* vol 3) ed International Union of Pure and Applied Chemistry (Oxford, England: Blackwell Scientific Publications) ISBN 978-0-632-02997-6
[7] Makinwa K A A and Huijsing J H 2002 *IEEE Transactions on Instrumentation and Measurement* **51** 840–844 ISSN 0018-9456
[8] Fung S K H, Tang Z, Chan P C H, Sin J K O and Cheung P W 1996 *Sensors and Actuators A: Physical* **54** 482–487 ISSN 0924-4247
[9] Guerrero L E R 2014 *An Algorithmic Readout Approach for Thermal Conductivity Based CO2 Sensors* (Delft)
[10] Chan S S W and Chan P C H 1999 *IEEE Journal of Solid-State Circuits* **34** 432–439 ISSN 0018-9200
[11] Tadić N, Zogović M and Gobović D 2014 *IEEE Sensors Journal* **14** 1988–1996 ISSN 1530-437X
[12] Filanovsky I M and Baltes H 1991 *IEEE Circuits and Devices Magazine* **7** 36– ISSN 8755-3996
[13] Sackett D 2009 Constant-Power Source - Application Note - Maxim Application Note AN4470 Maxim Integrated
[14] Sheingold D H 1978 *Multiplier Applications Guide* (Norwood, USA: CRC Press)
[15] Analog Devices 2011 AD538 - Real-Time Analog Computational Unit (ACU) Datasheet Analog Devices Inc. Norwood
[16] Springer A 2005 MAX4210-MAX4211 Datasheet Maxim Integrated San Jose, USA
[17] Linear Technology Corporation 2009 LT2940 - Power and Current Monitor Datasheet Linear Technology Corporation Milpitas, USA
[18] ON Semiconductor 2007 BC546B, BC547A, B, C; BC548B,C – Amplifier Transistors, NPN Silicon Datasheet Semiconductor Components Industries, LLC Phoenix, Arizona, USA
[19] Vishay Electronic GmbH 2015 IRF510, SiHF510 – Power Transistor Datasheet 91015 Selb, Germany

IOP Conf. Series: Materials Science and Engineering **228** (2017) 012025 doi:10.1088/1757-899X/228/1/012025

Electromagnetic interaction between a rising spherical particle in a conducting liquid and a localized magnetic field

Z Lyu, N Tran, T Boeck, C Karcher

Institute of Thermodynamics and Fluid Mechanics,
Technische Universität Ilmenau, 98693 Ilmenau, Germany

E-mail: ze.lyu@tu-ilmenau.de

Abstract. Lorentz force velocimetry (LFV) is a non-contact electromagnetic flow measurement technique for electrically conductive liquids. It is based on measuring the flow-induced force acting on an external permanent magnet. Motivated by extending LFV to liquid metal two-phase flow measurement, in a first test we consider the free rising of a non-conductive spherical particle in a thin tube of liquid metal (GaInSn) initially at rest. Here the measured force is due to the displacement flow induced by the rising particle. In this paper, numerical results are presented for three different analytical solutions of flows around a moving sphere under a localized magnetic field. This simplification is made since the hydrodynamic flow is difficult to measure or to compute. The Lorentz forces are compared to experiments. The aim of the present work is to check if our simple numerical model can provide Lorentz forces comparable to the experiments. The results show that the peak values of the Lorentz force from the analytical velocity fields provide us an upper limit to the measurement results. In the case of viscous flow around a moving sphere we recover the typical time-scale of Lorentz force signals.

1. Introduction

Two-phase flows in an electrically conductive liquid occur in a number of metallurgical processes. For example in continuous casting of steel, argon bubbles are injected in order to prevent clogging of the submerged entry nozzle, to mix the melt in the mold, and to remove slag particles via the free surface. Hence, liquid metal two-phase flows are not only of fundamental interest but also of practical importance. Among the various methods to measure them [1], one promising candidate is Lorentz force velocimetry (LFV) [2, 3].

LFV is based on the electromagnetic induction when a static localized magnetic field is applied to a conductive liquid. The current density $\mathbf{j}$ is induced according to Ohm's law for moving conductors, i.e.

$$\mathbf{j} = \sigma(\mathbf{E} + \mathbf{u} \times \mathbf{B}), \tag{1}$$

where σ is the electrical conductivity of the liquid, $\mathbf{E}$ is the electrical field, and $\mathbf{u}$ is the liquid velocity, respectively. In turn, according to Ampère's law, this current density $\mathbf{j}$ is connected with the so-called secondary magnetic field $\mathbf{b}$, i.e.

$$\mathbf{j} = \frac{1}{\mu_0} \nabla \times \mathbf{b}, \tag{2}$$

where μ_0 is vacuum magnetic permeability. In general, the magnetic field $\mathbf{B}$ is the sum of the applied magnetic field $\mathbf{B_0}$ and the secondary magnetic field $\mathbf{b}$. The ratio between $|\mathbf{b}|$ and $|\mathbf{B_0}|$ is roughly proportional to the magnetic Reynolds number Re_m, defined as

$$Re_m = \mu_0 \sigma u L, \tag{3}$$

where u is the liquid velocity, and L is the characteristic length of the flow. In most industrial applications (including our experiment), $Re_m \ll 1$, so that $\mathbf{B} \approx \mathbf{B_0}$ [4]. Therefore, the Lorentz force density can be defined by the simplified expression

$$\mathbf{f_L} = \mathbf{j} \times \mathbf{B_0}. \tag{4}$$

Upon integrating this Lorentz force density over the volume, we obtain the braking force induced in the moving conductor according to the relation,

$$\mathbf{F_L} = \int_V \mathbf{j} \times \mathbf{B_0} \mathrm{d}V. \tag{5}$$

The magnitude of this force can be determined by simple scaling arguments, whereby the force is proportional to the flow velocity u,

$$F_L \propto \sigma u B_0^2 L^3. \tag{6}$$

In LFV we measure the counter force to this flow-braking Lorentz force, which acts on the source of the magnetic field, i.e. the external magnet system. When the small permanent magnet is used, the measured force can be used to evaluate the local velocity in the vicinity of the magnet. Earlier work has demonstrated the capability of LFV to detect a particle rising in liquid metal at rest [5]. The aim of the present work is to check if our simple numerical model can provide Lorentz forces comparable to the experimental results for a particle rising.

This paper is structured as follows: in Section 2 the experimental setup is explained. In Section 3 we present our numerical model. In Section 4, we present the experimental and numerical results and discussions. The concluding remarks are given in Section 5.

2. Experimental setup

The diagram of the experimental setup for a sphere rising in liquid GaInSn is shown in figure 1. As described in [5], a cylindrical glass tube with an inner diameter of $D_{in} = 20$ mm is used. The tube is filled with alloy GaInSn as working fluid. A fishing line is pulled straight through the tube by springs. The line goes through the central hole of a sphere of 6mm diameter and thus, the particle rising path is restricted by the fixed fishing wire. The particle is made of plastic and not conducting, and the effects of magnetic permeability is neglected. The particle rises freely due to buoyancy. A cubic permanent magnet (NdFeB 48 material) of 12 mm side-length is attached to the LFV sensor. The magnetization points directly into the tube. The sensor detects the vertical component of the flow-induced Lorentz forces. Its resolution is 1 μN with cut-off frequency 6.3 Hz. Upward forces are set to be positive in this study. Without the sphere motion, LFV detects only the gravity force of the permanent magnet. When the sphere rises through the near-magnet region, some local liquid flow and eddy currents occur due to the displacement effect of the sphere, and Lorentz forces are generated. Hence, using this setup, a difference measurement of the Lorentz forces can be made.

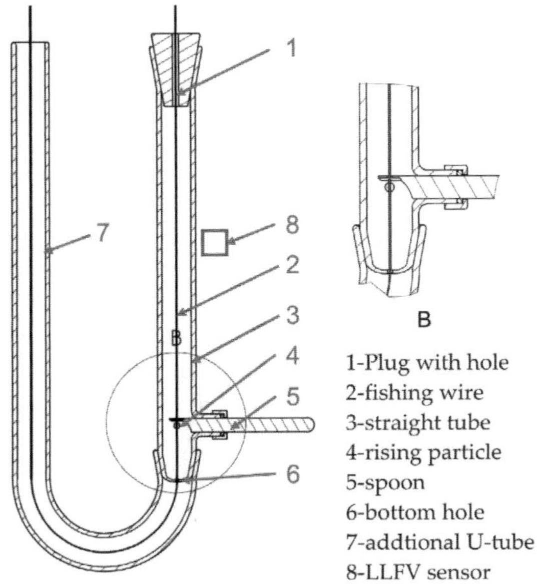

Figure 1. Schematic of the experimental setup of a sphere rising in liquid GaInSn [5].

In the experiment, shortly after we start the LFV measurement, the spoon (see figure 1) is removed and the particle starts to rise, following the straight fishing line. When the displacement flow by the particle rising affects the region permeated by the magnetic field, the Lorentz force is generated. In LFV we measure the counter force to this flow-braking Lorentz force. After seeing that the particle reaches the top free surface, we stop LFV measurement. And then the particle is pulled down to the bottom of the tube and we wait till the liquid comes to rest again. Then we repeat this process. It is difficult to observe the particle itself or to measure the rising velocity u_0 of the particle. We assume that the averaged particle velocity is similar to that in [6, 7], which lead us to $u_0 = 0.22$m/s. This velocity is used in the following experimental and numerical results. We assume the liquid velocity is in the similar range as particle velocity. Therefore, the particle Reynolds number Re_d can be estimated as follow

$$Re_d = \frac{u_0 d_0 \rho}{\mu} = 3570.6, \tag{7}$$

where $u_0 = 0.22$ m/s is the mean velocity of the particle, $d_0 = 6$ mm is the diameter of the particle, $\rho = 6.492$ g/cm^3 is the density of GaInSn, and $\mu = 0.0024$ Pa·s is the dynamic viscosity of GaInSn, respectively. For the Hartmann number Ha we obtain

$$Ha = B_0 d_0 \sqrt{\frac{\sigma}{\mu}} = 10.7, \tag{8}$$

where $B_0 = 47$ mT is the magnetic flux density at the center of tube, and $\sigma = 3.46 \times 10^6$ S/m is the conductivity of GaInSn at 20 ^{o}C. This value demonstrates that in the present experiment the Lorentz force dominates the friction effects. As expected, the magnetic Reynolds number Re_m is small according to equation (3),

$$Re_m = \mu_0 \sigma u_0 d_0 = 0.0057, \tag{9}$$

which suggests that the secondary magnetic field can be neglected in the numerical simulation due to $Re_m \ll 1$. Finally, to describe the ratio of the Lorentz forces to the inertial forces, we have the interaction parameter (Stuart number) N,

$$N = \frac{\sigma B_0^2 d_0}{\rho u_0} = 0.0321, \tag{10}$$

which is small in our experiments. We use the so-called "kinematic approach" [8] for our numerical simulation, although Ha is not small, which may modify the flow near the boundaries. Here the velocity is prescribed by the analytical solutions of the flow around a moving sphere.

3. Numerical model

The geometry of numerical model is shown in figure 2, which is equivalent to the experiment. Here $L_1 = 180$mm and $d_1 = D_i = 20$mm are the length and diameter of the cylindrical domain of the conducting liquid, respectively. The diameter of the rigid sphere is $d_2 = d_0 = 6$mm. The side-length of the cubic permanent magnet is $w = 12$mm. The axial distance between the magnet and the sphere is $L_2 \in [-90, 90]$mm, and $L_3 = 4$mm represents the distance between the magnet surface and the liquid boundary.

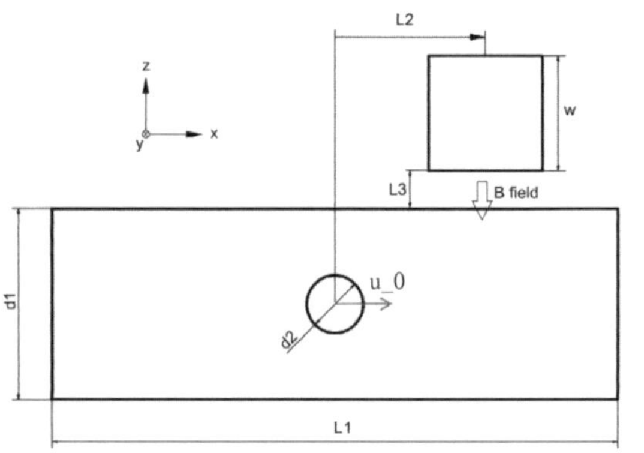

Figure 2. Schematic of the geometry in simulation (view on the $x - z$ plane).

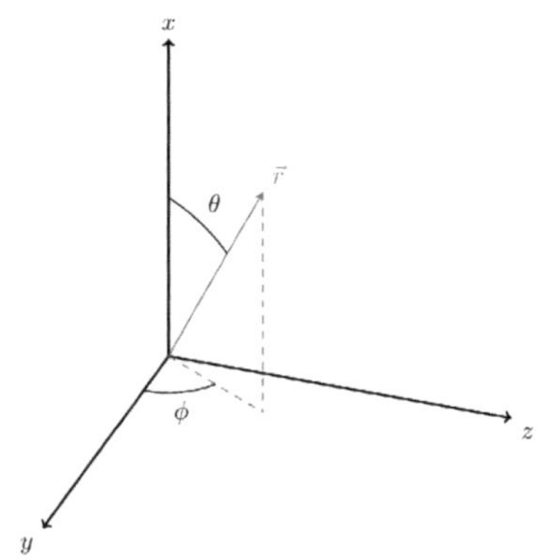

Figure 3. Spherical coordinate.

In the present case, we want to compute the electromagnetic induction of a pre-defined velocity field and a localized magnetic field. Thus the governing equations read

$$\nabla^2 \Phi = \nabla \cdot (\mathbf{u} \times \mathbf{B_0}), \tag{11}$$

$$\mathbf{j} = \sigma[-\nabla\Phi + (\mathbf{u} \times \mathbf{B_0})], \tag{12}$$

$$\mathbf{f_L} = \mathbf{j} \times \mathbf{B_0}, \tag{13}$$

where $\mathbf{u} = (u_x, u_y, u_z)$ is the velocity field in the reference system, and Φ is the electrical potential. The imposed magnetic field of the permanent magnet is calculated from the analytical solution presented by Furlani [9], which is determined upon integration of

$$\mathbf{B_0}(x,y,z) = \frac{\mu_0 M_s}{4\pi} \sum_{k=1}^{2} (-1)^k \int_{y_1}^{y_2} \int_{x_1}^{x_2} \frac{[(x-x')\mathbf{x} + (y-y')\mathbf{y} + (z-z_k)\mathbf{z}]}{[(x-x')^2 + (y-y')^2 + (z-z_k)^2]^{3/2}} dx'dy'. \tag{14}$$

In equation (14), M_s is the surface magnetization of the magnet, (x_1, x_2), (y_1, y_2), (z_1, z_2) are the coordinates of the magnet's corners, **x**, **y**, **z** are the unit vector in the x,y,z directions, respectively. To describe the velocity field using the analytical solutions, a spherical coordinate system is introduced and fixed on the center of the sphere (see figure 3):

$$x = r\cos\theta, \quad y = r\sin\theta\cos\phi, \quad z = r\sin\theta\sin\phi. \tag{15}$$

We use the analytical solutions of the flow generated by a sphere moving in the $x-$direction. One option to compute the velocity field is to use the analytical solution of the potential flow around a moving sphere [10]. The velocity potential ϕ' is

$$\phi' = -\frac{u_0 R^3 \cos\theta}{2r^2}, \tag{16}$$

where u_0 is the velocity of the sphere, R is the sphere radius, r is the radial coordinate, and θ is the polar angle between radial direction and flow direction, respectively. Then the velocity field reads

$$u_x = \frac{\partial\phi'}{\partial x}, \quad u_y = \frac{\partial\phi'}{\partial y}, \quad u_z = \frac{\partial\phi'}{\partial z}. \tag{17}$$

The second and third options are to use the Stokes solution of viscous flow due to a moving sphere at low Reynolds-number [11]. In "free-slip" condition the radial and angular components of velocity are written as

$$u_r = \frac{u_0 R\cos\theta}{r}, \quad u_\theta = -\frac{u_0 R\sin\theta}{2r}, \quad u_\phi = 0. \tag{18}$$

Under "no-slip" condition the radial and angular components of velocity are

$$u_r = u_0\cos\theta\left(\frac{3R^3}{2r} - \frac{R^3}{2r^3}\right), \quad u_\theta = u_0\sin\theta\left(-\frac{3R}{4r} - \frac{R^3}{4r^3}\right), \quad u_\phi = 0. \tag{19}$$

Then we implement equation (18) or (19) into our geometry as

$$\begin{aligned} u_x &= u_r\cos\theta - u_\theta\sin\theta, \\ u_y &= u_r\sin\theta\cos\phi + u_\theta\cos\theta\cos\phi, \\ u_z &= u_r\sin\theta\sin\phi + u_\theta\cos\theta\sin\phi. \end{aligned} \tag{20}$$

Finally, we may solve equations (11-13) by implementing equations (14-20), respectively, and receive the re-action force $\mathbf{F_L}'$ on the permanent magnet as

$$\mathbf{F_L}' = -\int_V \mathbf{f_L} \mathrm{d}V. \tag{21}$$

The simulations in this paper were done by the commercial code COMSOL Multiphysics. The finite Element Method (FEM) is used in order to determine the forces acting in the fluid. The solution domain is a cylinder with a sphere cut out. For this model we use the hybrid tetrahedral mesh. In the computational domain, the elements are uniformly distributed over the x-axis. Smaller elements are applied in the region where the gradient of velocity field is large. The influence of discretization was examined by using meshes with different numbers of elements. The mesh used in subsequent computations in section 4 is shown in figure 4, where the surface mesh of the sphere can be seen as well. The effect of mesh quality on the Lorentz forces is checked and shown in figure 5. Here we define the deviation of force amplitude De by

$$De = \frac{F_x - F_{x0}}{F_{x0}}, \tag{22}$$

where F_{x0} is the peak value of the force in the case of maximum number of mesh elements. The influence of mesh quality on Lorentz forces is below 10^{-4}.

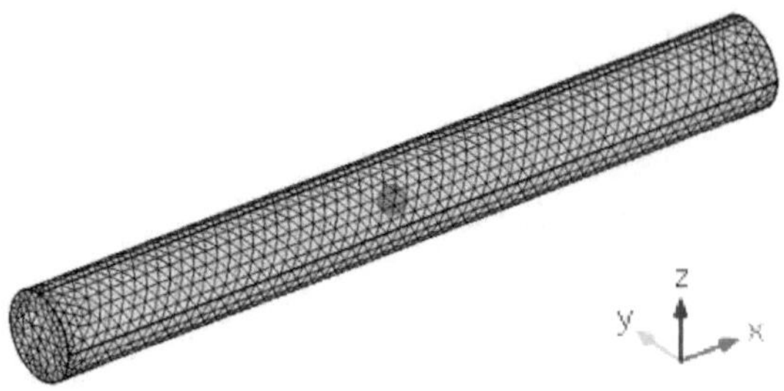

Figure 4. Mesh used in simulations reported in section 4 with 5.9×10^5 elements.

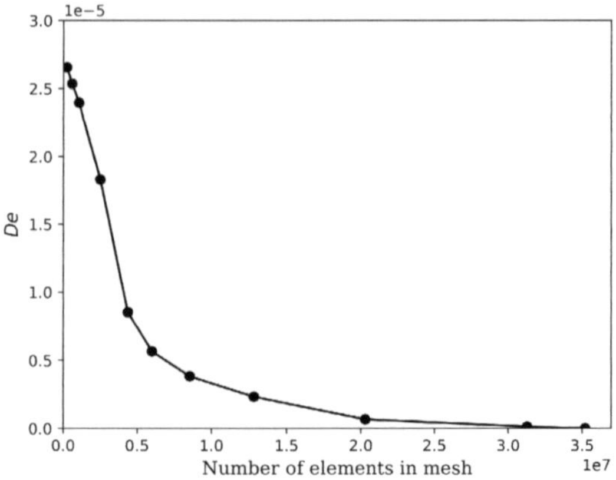

Figure 5. Mesh study with velocity field in equations (18,20) and magnet at $L_2 = 0$ mm.

4. Results and discussion

The LFV measurement results of three cases are shown in figure 6 by setting the force peaks to the same time (1.667s). The Lorentz force signals vary significantly when the particle moves through the test region. The LFV sensor detects the displacement flow in the near-magnet region. The time-scales of these signals agree with each other. However the peak values of them are different, varying from 54 μN to 85 μN. The reproducibility of the force measurement is not good. This may be due to the lateral displacement of the particle during rising, since the hole through the particle is of 1mm diameter and the fishing line is of 0.1mm diameter. Another reason can be unsteady turbulent wake structures downstream of the particle [7, 12]. Additionally, we observe an asymmetry of the experimental signals, which is likely to be caused by the asymmetry between the flow upstream and downstream of the sphere. The long tail of signals represents the flow in the wake.

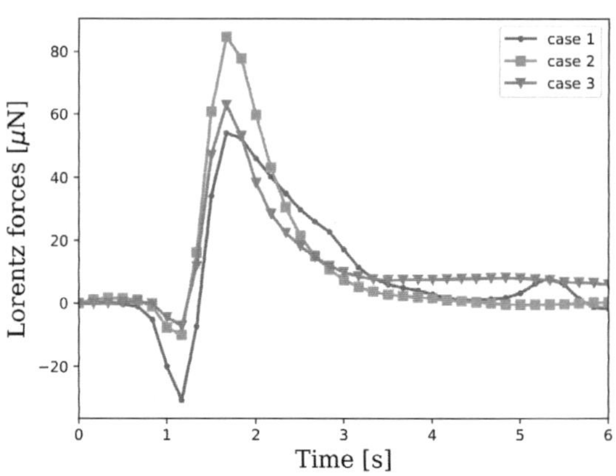

Figure 6. Lorentz force measurements of a particle rising in liquid GaInSn at rest.

As for the numerical simulation, the analytical solution of the magnetic flux density is validated by comparing it to the measurement results. We observe a good agreement between them, which are both shown in figure 7. At the point of $z = 0$mm, the magnetic flux $B_z = 47$mT, which leads to the interaction parameter $N = 0.0321$ in equation (10). However, $B_z = 284$mT at the boundary of the liquid, which leads to a much higher interaction parameter near the wall. However, the velocity is smaller there.

The three different analytical solutions of the flow around a sphere, i.e. potential flow around sphere (equation 16,17), free-slip viscous flow around sphere (equation 18,20), no-slip viscous flow around sphere (equation 19,20), are implemented and shown in figure 8 on the line ($x = 0,\ y = 0$). It should be noticed that the velocity is non-zero on walls, since the distance to the wall is not very large. The axial velocities are in opposite direction in these cases. The viscous forces exert shear to the surrounding liquid and thereby cause the upward velocities on the central plane $x = 0$, which decays with r^{-1}. For potential flow there is a downward motion on the symmetry plane $x = 0$ since forces are transmitted by pressure only. The decay r^{-3} of the velocity with the distance is also far more rapid than that of a viscous flow. We do not expect the different analytical flows to correspond closely to the actual flows in the experiment. Nevertheless, they may approximate the behavior of the real flows in some respects. E.g., the potential flow should be a reasonable solution upstream of the particle when the Reynolds number is large, and the viscous no-slip flow may provide a reasonable bound on the velocity magnitude upstream of the sphere at smaller Reynolds numbers. Downstream of the sphere the actual flow should certainly differ significantly from all three analytical solutions due to flow separation and wake formation. It may be presumed that the slowly decaying viscous velocity distributions provide an indication of the maximal force signals.

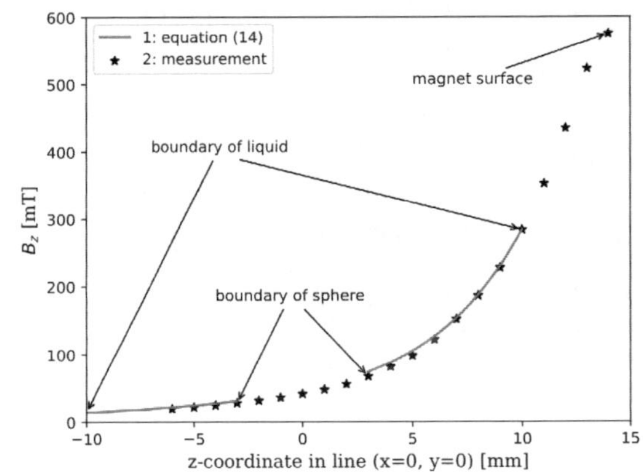

Figure 7. The magnetic flux density B_z.

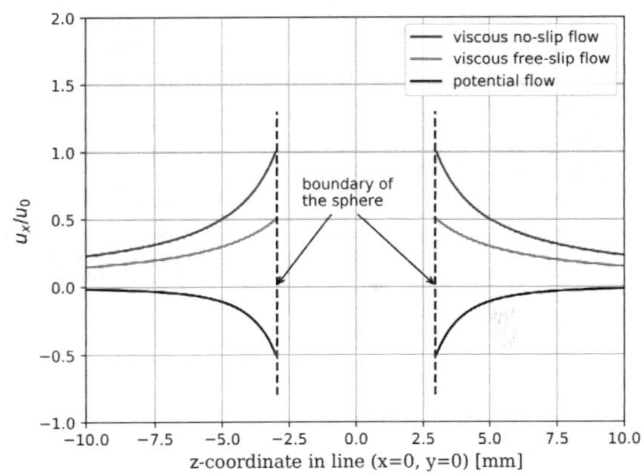

Figure 8. The velocity profile u_x.

We set the permanent magnet to different locations, and we observe the development of the eddy currents distributions and the Lorentz force distributions, which are shown on the $x - z$ symmetry plane in figure 9 and 10, respectively. The intensities of the eddy current in figure 9 increase when the magnet approaches the sphere, and they penetrate much deeper into the liquid. The intensities of the eddy current reach the maximum value, when the magnet is at the same position as the sphere. Similarly in figure 10, the Lorentz force densities increase, and penetrate deeper into the liquid as the magnet approaches the sphere.

The peak values of the Lorentz forces of the different velocity fields are shown in table 1. The Lorentz force of viscous no-slip flow is the highest one, because in this case the velocity spreads further to the wall, and the flow in the near-magnet region contributes more to the total force. The peak value of the Lorentz force of the potential flow has opposite sign to that of the

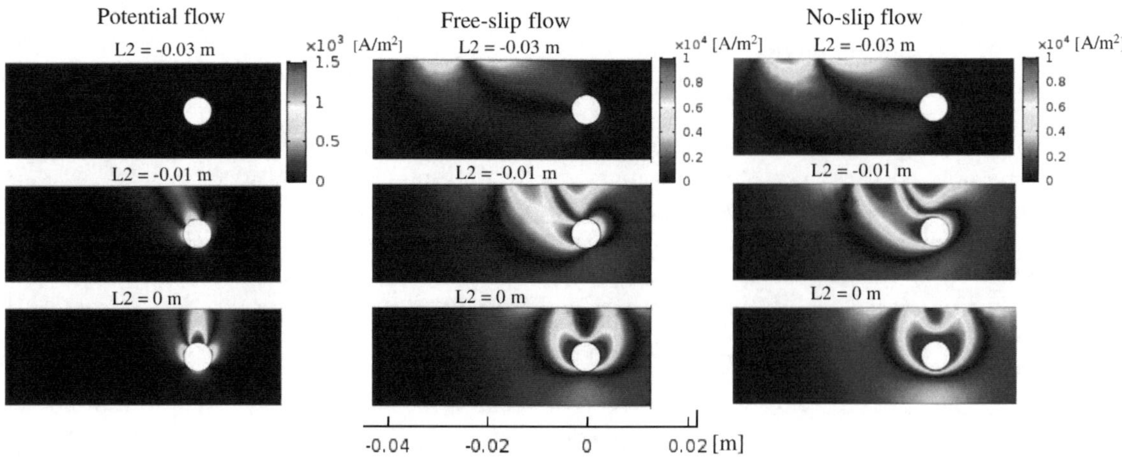

Figure 9. The simulation results of the eddy current distributions on the $x - z$ plane (magnitude of $\mathbf{j}$ is shown).

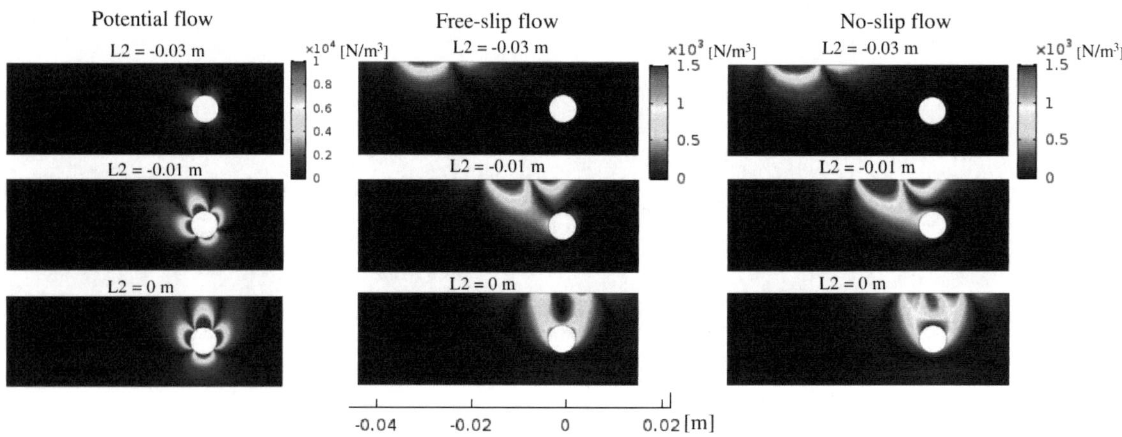

Figure 10. The simulation results of the Lorentz force density distributions on the $x - z$ plane (magnitude of $\mathbf{f_L}$ is shown).

Table 1. The simulation result of the Lorentz forces.

The velocity field	The peak values of the forces on magnet (μN)
potential flow	-49.1
free-slip viscous flow	471.0
no-slip viscous flow	725.0

viscous flow, because of the different velocity fields in figure 8. The disagreements between the peak values of the Lorentz forces in the experiments and in the simulation may be due to the presence of the wake and insufficient decay of the viscous velocity fields towards the walls. In addition, the velocity field could be affected by the Lorentz forces, although the effects should not be very significant since N is small. Nevertheless, in the upstream region the computational results provide a reasonable range for the measured Lorentz forces.

From simulation we receive total Lorentz forces that depend on the axial distance L_2 between the magnet and the sphere as well. In order to compare the experimental and numerical results, the time t in measurement is transformed to a distance L' by

$$L' = u_0(t - t_0) \tag{23}$$

where $t_0 = 1.667$s is the reference time point of force peaks in figure 6. The Lorentz forces of measurements and simulations are normalized by their peak values,

$$F_{norm} = \frac{F_L}{F_{L,max}}, \tag{24}$$

where F_{norm} is the normalized Lorentz force, and $F_{L,max}$ is the peak values of the re-action force on permanent magnet in experiment or simulation. Thus they can be compared in figure 11. We observe an agreement of the time-scale between the experiment and the Stokes solution of flow around a moving sphere, both about 1s. All the forces of simulation are symmetrical because of the symmetries of **B** field and the velocity field of the analytical solutions. However, the measurements are asymmetrical due to the fact that the flows upstream and downstream of the sphere are not the same. The large force variations downstream of the sphere may be related to the unsteady wake structures.

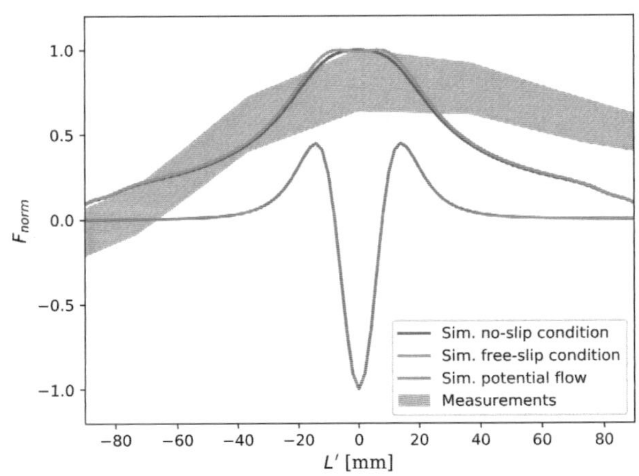

Figure 11. Comparison of Lorentz force between the experiments and simulations.

5. Conclusion

In this paper, we present Lorentz force measurements of a spherical particle rising in liquid metal initially at rest. Three different types of analytical solutions of flow around a moving sphere are applied, respectively, and the Lorentz forces are compared to experiments. We observe an agreement of the typical time-scale of Lorentz forces using Stokes solution of viscous flow around sphere. In view of many simplifications, the peak values of Lorentz force from analytical velocity fields provide us a maximum range for the measurement results. Accurate numerical simulations of the flow will be required to reproduce the force signals quantitatively.

Acknowledgements
The authors acknowledge financial support from Deutsche Forschungsgemeinschaft (DFG) in the framework of the Research Training Group GRK 1567 "Lorentz Force Velocimetry and Lorentz Force Eddy Current Testing" and Helmholtz Alliance LIMTECH within the seed grant program of Young Investigators Group.

References
[1] Shercliff J A 1987 *The theory of electromagnetic flow-measurement* (CUP Archive)
[2] Thess A, Votyakov E and Kolesnikov Y 2006 *Physical Review Letters* **96** 164501
[3] Heinicke C and Wondrak T 2014 *Measurement Science and Technology* **25** 055302
[4] Davidson P A 2001 *An introduction to magnetohydrodynamics* vol 25 (Cambridge university press)
[5] Lyu Z and Karcher C 2017 *Magnetohydrodynamics* **53** 67–77
[6] Zhang C, Eckert S and Gerbeth G 2005 *International journal of multiphase flow* **31** 824–842
[7] Fröhlich J, Schwarz S, Heitkam S, Santarelli C, Zhang C, Vogt T, Boden S, Andruszkiewicz A, Eckert K, Odenbach S *et al.* 2013 *The European Physical Journal Special Topics* **220** 167–183
[8] Thess A, Votyakov E, Knaepen B and Zikanov O 2007 *New Journal of Physics* **9** 299
[9] Furlani E P 2001 *Permanent magnet and electromechanical devices: materials, analysis, and applications* (Academic Press, London, U.K.)
[10] Batchelor G K 2000 *An introduction to fluid dynamics* (Cambridge university press)
[11] Brennen C E 2005 *Fundamentals of multiphase flow* (Cambridge university press)
[12] Zhang J, Ni M J and Moreau R 2016 *Physics of Fluids (1994-present)* **28** 032101

IOP Conf. Series: Materials Science and Engineering **228** (2017) 012026 doi:10.1088/1757-899X/228/1/012026

Neutron radiography for visualization of liquid metal processes: bubbly flow for CO_2 free production of Hydrogen and solidification processes in EM field

E Baake[1,*], T Fehling[1], D Musaeva[1,2] and T Steinberg[1]

[1]Institute of Electrotechnology, Leibniz University of Hannover, Wilhelm-Busch Str. 4, DE 30167 Hannover, Germany
[2]Kazan State Power Engineering University, Krasnoselskaya Str. 51, RU 420066 Kazan, Russian Federation
[*]baake@etp.uni-hannover.de

Abstract. The paper describes the results of two experimental investigations aimed to extend the abilities of a neutron radiography to visualize two-phase processes in the electromagnetically (EM) driven melt flow. In the first experiment the Argon bubbly flow in the molten Gallium – a simulation of the CO_2 free production of Hydrogen process – was investigated and visualized. Abilities of EM stirring for control on the bubbles residence time in the melt were tested. The second experiment was directed to visualization of a solidification front formation under the influence of EM field. On the basis of the neutron shadow pictures the form of growing ingot, influenced by turbulent flows, was considered. In the both cases rotating permanent magnets were agitating the melt flow. The experimental results have shown that the neutron radiography can be successfully employed for obtaining the visual information about the described processes.

1. Introduction

Liquid metals are used in various industrial applications and processes. One new approach to produce hydrogen without CO_2 emission is the thermal cracking of methane. To deliver the high temperature ambience for methane cracking, liquid tin is used at temperatures above 660 °C. In order to study multi-phase phenomena in the reactor, a method is demanded, that is able to penetrate even heavy metal atoms to visualize gas bubbles in opaque media.

Another process, where visualization of physical phenomena in opaque liquid metal is demanded, is an influence of electromagnetic stirring on a solidification processes. Electromagnetic (EM) stirring of a solidifying melt is a well-known approach in industry to enhance material quality. However, physical effects like development of vortices in the liquid part of the melt and the effect on the development of the solidification front can be studied very carefully using a technology, that is able to visualize these effects.

In the history of liquid metal research many methods have been investigated to characterize effects in the melt. Outstanding investigations have been performed by Tarapore and Evans in 1977 [1], who investigated non-invasively the velocity distribution of stirring effects by using stroboscobic photography of the free surface of liquid metal. Later Moore and Hunt (1983) [2] used invasive methods like drag probes and hot film anemometer to estimate velocities inside the liquid metal. The approach from non-invasive to invasive methods shows the big problem to visualize physical effects in opaque liquid melt. Later Umbrashko (2006) [3] and Kirpo (2007) [4] investigated turbulence fluctuations in induction crucible furnaces (ICF) using potential probes and numerical simulation tools. Till now the most common method, which could be implemented for obtaining visual information about the two-phase flows was X-ray attenuation [5-7]. Applicability of this method is limited by the thickness of the melt volume under investigation. In this kind of systems, a development of turbulence is limited as well.

One of the solutions in this area was an application of the well-developed methods of the neutron radiography. Here the samples thickness limited as well by the neutron flux and attenuation of the material, but not in such a critical way, so the turbulent phenomenon can find a place. The first experiments on this field were performed three decades ago by American scientific group from Minneapolis. Such a processes as diffusion and dispersion in porous media [8] were investigated by means of neutron radiography and have showed its potentialities. The solidification process of the metal and segregation processes were visualized as well with beam of thermal neutrons [9]. Newly, Wondrak et. al (2014) [10] developed the non-invasive Contactless Inductive Flow Tomography (CIFT) using tomography and numerical simulation to visualize velocity distribution inside liquid metal. In this time Saito et al. [11] started to investigate liquid metal flow and especially the formation of gas bubbles in liquid metal using neutron radiography. In this experiment gas bubbles from vaporized water in Pb-Bi eutectic has been detected. This experiments and first successful experiments of Sarma et al. in 2015 [12], aimed to visualize particle admixture in liquid metal gave the idea to try to use neutron radiography for studying gas bubble dynamics in liquid metal and EM influenced solidification processes more carefully.

Whereas the neutron radiography technology improves during all this time, there are no systematic investigations of liquid metal bubbly flows and solidification processes by Neutron radiography so far. In general, many previous experimental studies exist carried out using other diagnostic techniques. The goal of this research is to test the abilities of the neutron radiography for visualization the two-phase process. This paper contains on the one hand experiments on visualizing gas bubble dynamics in liquid metal and on the other hand solidification of metal under the influence of externally applied EM fields.

2. Motivation

2.1. Visualization of a bubbly flow in a melt for CO_2 free production of Hydrogen

For production of ecologically friendly hydrogen compared to common steam reforming process, the thermal cracking of methane can be considered as an alternative route. The thermal cracking takes place in a high temperature environment. At temperatures above 600°C methane (CH_4) decomposes to $C + 2H_2$ (Δh_0=74,85 kJ/mol). To explore the potential of carbon dioxid free production via thermal decomposition of methane, experiments were realized by a small scale experimental bubble column reactor. Here methane bubbles are inserted via an orifice and rise up in a column filled with liquid tin at temperatures higher than 600°C. One problem in this setup is the missing possibility of viewing the dynamics inside the reactor. Especially the optimization of the bubbles residence offers the chance to increase the hydrogen yield. Based on that fact, numerical simulations (figure 1, a-d) have been done to show the dynamics and to calculate possible improvements. The numerical modelling includes studies for different initial bubble sizes, volume flows, inlet types and geometry as well as feasibilities of electromagnetic stirring. A detailed description of the studies can be found in [13]. To get a numerical result of the bubble dynamics models for the physics of the two-phase flow have to be used. In the discovered cases an Euler-Euler-description is used based on an initial observer formulation of the two phases. Based on this formulation an Eulerian method [14] is used with regards to a more exact calculation for greater bubble volume fractions and greater Stokes numbers, especially at the surface and for different velocities of the phases. In this method a separated set of equations is solved for every phase. Parameters for phase exchange to describe the gas liquid interface (Schiller-Naumann) as well as heat transfer coefficients (Ranz Marshall) can be specified by correlations.

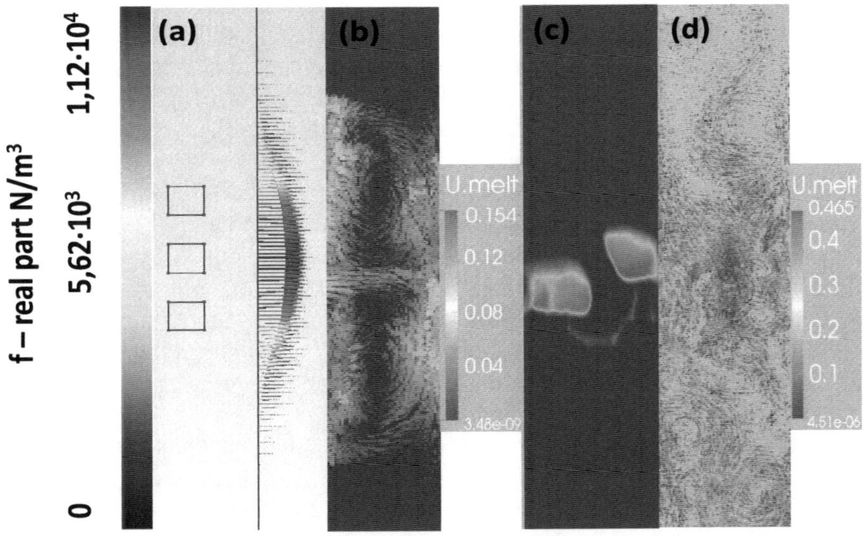

Figure 1. Numerical simulation results: Lorentz force for an AC EM-field in left half (a) and resulting velocity [m/s] in right half (b) of the column, split for a 20 mm bubble (c) and velocity field around the split bubble (d) for whole column diameter.

To get an idea about the residence times of different single bubble sizes in the reactor different residence times for initial bubble diameters have been discovered in the numerical simulation. It can be seen that bubble residence times decrease with bubble diameter down to a minimum until the occurrence of oscillations that decrease the velocities. Further increasing the diameter leads to bubble shape changes which decreases the residence time again. These effects can occur for different gas-liquid flows. In actual studies also the effect of an electromagnetic field on the bubbles dynamics has been discovered [23]. It has been shown that a bubbles breakup and as a result an increase of residence time can be achieved. This has been done for different initial bubble sizes to discover possibilities for breakup as a function on bubble diameter. Figure 1 shows the dynamics under influence of an AC electromagnetic field generated by a coil in the middle of the reactor. The resulting Lorentz forces generate whirl structures in the melt. Afterwards a single bubble is introduced into the reactor. Depending on frequency and current a split into blebs can be achieved due to turbulent shear forces. Numerical simulations investigated breakup possibilities and influence on residence time for different bubble diameters. Another potential is given the usage of packed and multi-inlets, that has also been simulated. It has to be mentioned, that also wall effects also reduced the averaged magnitude of bubble velocities. Due to this and the effect of carbon sediments at the reactor surface, also greater bubbles near the reactor diameter have been discovered. This also leads to the question which bubble sizes are generated in continuous volume flows and how they behave in bubble collectives. Simulations show the effect of wobbling appearance when increasing the volume flow. To get a view on the effects of volume flows experiments with neutron radiography have been done.

2.2. Visualization of a formation of a solidification front

The structures which form during the solidification of a material play an enormous role in the properties of the solid material. It is well known, that application of EM stirring to the solidifying material can provide more homogeneous and fine structure, which is more desirable in the metallurgy. Due to intensive turbulent flows an efficient EM stirring leads to an absence of high temperature gradients in the melt, pinning of growing grains due to the melt motion, partial re-melting of the structure under formation [15, 16]. At the same time, a development of such undesired effects as a macro-segregation as well can be a consequence of the intensive melt motion [17] Consequently, an optimization of application of EM stirring to the crystallizing metal is required. Different magnetic fields, such as rotating, travelling, pulsating, and their different combinations, are available for the optimization of stirring process and for creating the flow patterns in the melt, adjusted according to the requirements. One of optimized stirring methods is an application of EM field with pulses. In such a conditions intensive stirring periods are interleaved with pause periods, when the segregation effects could be reduced. Bright illustration of an effect of pulsed EM stirring was obtained in the numerical and experimental investigations and described by S. Eckert et. al (2007) [18], B. Willers et. al (2008) [19] and X. Wang et. al (2009) [20]. These scientific groups observed an influence of an application of travelling and rotating magnetic fields modulated at low-frequency pulses to solidifying metal. It was shown, that an optimized forcing of the melt can lead to more homogeneous and fine metal structure. Nevertheless, up to now it is an open question – how the turbulent electromagnetically driven flows interact with a solid phase? The knowledge about a process, happening in the solidifying material is a key for improving existing technologies of high quality materials production and should have to be extended by means of new investigation techniques.

Due to the liquid metals opaqueness an obtaining of spatial information about a solidification process is one of the challenging problems. Up to now the most common method, which could be implemented for obtaining visual information about the interaction of solid and liquid phases during an EM stirring was X-ray attenuation. But a limitation of a thickness of the systems for X-ray analysis does not give ability to analyse its influence of 3 dimensional turbulent flows on an ingot formation, although particularly intensive turbulent flows and flow structures with high wave number have the strongest influence on the growing crystals [16]. So the search and improvement of visualization techniques for metallurgy is one of the current scientific tasks. A principle of neutron radiography seems to be feasible for obtaining an information about a solidification process development through time and observing the influence of EM stirring. So the technique was applied to visualize and analyse a change of a solidification front form, as a consequence of an intensive EM stirring.

3. Neutron radiography

Nowadays the method of neutron radiography develops rapidly and extends the ranges of physical phenomena which can be analysed with its application. The principal on which it is based can be described as following: in the real-time image system for visualisation of dynamic processes the 2D distribution of the transmitted neutron beam intensity through a sample is measured. The illuminating neutron beam, controlled to be almost parallel, meets on its way the sample and the neutron flux, passing through the material decreases. Then on the back side of the sample the neutron beam with decreased flux passes through the scintillator which converts neutrons to the light emission. The light output could be processed with the imaging detectors covering the illuminated sample area. In most cases this is a digital camera system looking onto a thin neutron scintillator screen. The recorded stack of digital pictures can be converted into a video sequence. The required exposure time determines the frame rate.

The most important requirement here is an attenuation coefficient of analysed sample which should have to be adjusted so to provide a high contrast in the transmission image. For that can be suitable very few materials, such as Gadolinium, Cadmium, and Boron. Materials with low attenuation coefficient are virtually transparent to the thermal neutrons. Although the combination of materials with different attenuation coefficients could be used as well for imaging of some processes, such as flow tracers in porous media [8]. When applying the neutron radiography to the pure materials one can use the dependence of the thermal neutron flux on the density of the material, so such processes as solidification or melting of the materials with initially high enough attenuation coefficient can be visualized. According to the equation (1) the neutrons transmission through the melt volume depend on the density ρ, the thickness x of the sample, neutron flux I and could be expressed by exponential relation between these parameters [21]:

$$\frac{I}{I_0} = \exp(-\mu\rho x) \qquad (1)$$

where μ [cm^2/g]– is the thermal neutron cross-section of the object, I_0 [neutrons·cm^{-2}·sec^{-1}·mA^{-1}] – incident thermal neutron flux. In case of use of different materials, as well as solid and liquid phases of the solidifying material, a difference in a density and, consequently, difference in the attenuation, provides the neutron shadow images.

In comparison to X-ray, neutron radiography has contrary characteristics in relation to attenuation coefficient. Whereas, X-rays interact mainly with the electrons of an atom and thus have low penetration depths for heavy materials like metals, neutrons interact with atomic core directly and shows high attenuation for atoms like Hydrogen, Boron, Cadmium, etc.

Saito et. al [11] used neutron radiography and Particle Image Velocimetry (PIV) techniques to study the effect of recirculation flow on the bubble behaviour. The dynamics of molten lead bismuth as liquid metal and nitrogen as gas bubbles in a rectangular tank were studied. Additionally, particles made of gold-cadmium ($AuCd_3$) inter-metallic alloy were used as tracers for calculation of flow field using PIV. From these results basic characteristics of liquid-metal and gas dynamics have been clarified. Sarma et al. [12] used neutron radiography and PIV method to study the motion of particles in the liquid metal, which was stirred by a magnetic field applied from four counter rotating permanent magnets. Carried out results show that neutron radiography is a suitable method for investigations of multi-phase flow. Animated from these results the application of neutron radiography to visualize liquid-gas multi-phase flows and solidification processes seems very promising.

In this paper two experiments directed on visualization of two-phase process (gas bubbles flow in the melt and a solidification) by means of pure Gallium, forced by an alternating electromagnetic field, is described and discussed. Abilities of a thermal neutron beam were tested to provide information about the influence of the melt flows on the processes.

4. Experimental setups and procedures

Experimental setup (figure 2) was designed to provide 2D snapshots of the processes, happening in the melt bath. For that the vessel, containing the liquid Ga, was made in a rectangular form to be placed before the beam of neutrons. The construction of the experimental unit was inspired by those, used by the scientific group and described in [12]. Liquid metal flows were agitated by the stirrer on rotating permanent magnets. Material properties of Gallium and magnets are presented in the table 1. According to the aim of the experiment a construction of the setup differed and described below for the both cases.

Table 1. Properties of materials, used in the experiment.

Variable	Symbol	Value
	Gallium	
Melting temperature, ℃	t_{melt}	29,8
Weight, kg	m	1,4
Electric conductivity, S/m`	σ	$3{,}7\cdot 10^6$
Density, kg/m^3	ρ,	
	at $t > t_{melt}$	6080
	at $t < t_{melt}$	5910
Attenuation coefficient for thermal neutrons, cm^{-3}	K	0,49
	NdFeB Magnets	
Remanence on the surface, T	Br	1,17
Coercivety, A/m	Hc	$13{,}5\cdot 10^5$

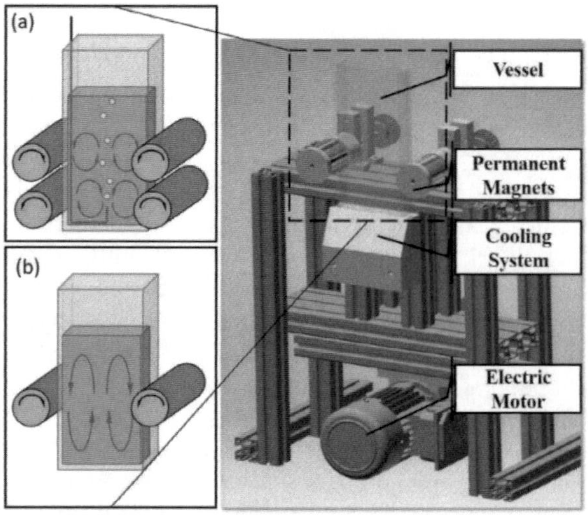

Figure 2. Sketch of the experimental setup and the direction of magnets rotation and flows in the liquid Ga: (a) – gas bubbly flow experiment; (b) – solidification experiment.

4.1. Bubbly motion under the influence of EM driven melt flow

The vessel with the size of 300 mm x 140 mm x 25 mm (figure 2, a) was made of PEEK material to avoid chemical reactions with the frame of the vessel. Gallium has been chosen for liquid metal, because of its low melting point and working safety. Specially designed particles with a lead kernel and glued boron carbide, fitting to attenuation coefficient and density to fulfil admixture ability, were used as particles and tracers for PIV simultaneously. Instead of methane Argon gas has been used, due to its easy availability in industrial processes and in laboratories like Swiss Spallation Neutron Source (SINQ) at Paul Scherrer Institute, Villigen, Switzerland. The attenuation coefficient for Ga is with 0,49 appr. 16 times higher compared to coefficient of Ar (0,03). Since the neutron beam is attenuated much more by Ga and less by Ar, Ar bubbles appear as bride spots on the camera (figure 3, a). For gas inlet capillary tubes made of stainless steel with an outer diameter of 1.5 mm and inner diameter between 0.5 – 1.3 mm connected to an automated gas controller have been used. The formation of bubbles has been studied by differently sized inlet diameters of capillary tubes, as well as their velocity and trajectory in the reactor. Furthermore, impact on bubbly flow by applied electromagnetic field by two and four counter-rotating magnets has been studied.

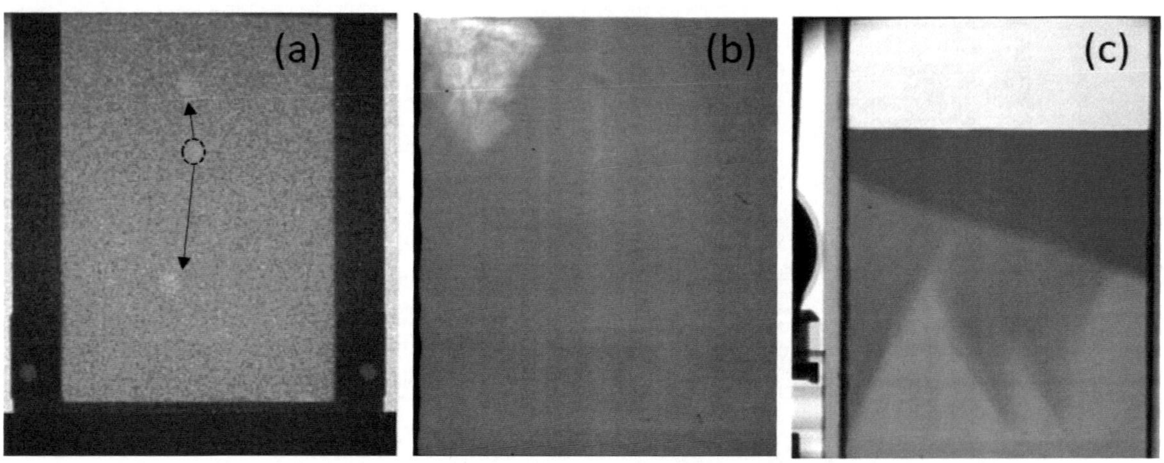

Figure 3. Raw neutron images: (a) – gas bubbly flow experiment; (b) – crystals of Ga on the free surface of the melt, appeared due to the cooling of the camera; (c) – Ga solidification experiment.

In the experiment, four counter rotating magnets have been used to create a mixing inside of the melt. The rotation of permanent magnets creates four vortices in a plane. The structure has been chosen that the velocity in the middle of the vessel is opposed to the ascending motion of the bubbles. This experimental setup has been used from Sarma, Scepanskis et. Al. [12] for investigation of the distribution of particles inside of liquid metal flow. Due to a joint experimental stay at PSI this setup has been used and further developed for bubbly flow and solidification experiments. Two effects should be investigated: 1) Breakup of bubbles due to shear stress induced by the rotation of the magnets, 2) impact on the size of the bubbles, their velocity and thus their residence time.

The camera, used to obtain the neutron shadow images, was intensively cooled during the experiment. The cooled air, reaching the free surface of the liquid Ga provided the conditions of its solidification. Although it was an undesired effect according to the aims of the gas bubbly flow investigation, the both solid and liquid phases of Ga were recognized on the obtained pictures (figure 3, b). This became a seed for the idea and a preliminary experiment for the investigation of solidification front formation described below.

4.2. Gallium solidification experiment

The experimental setup (figure 2, b) was used to obtain 2D snapshots of the Ga solidification front formation influenced by an electromagnetic field. For the material of the vessel a window glass ($SiO_2^{72\%}$ $Na_2O^{14,2\%}$ $CaO^{10\%}$ $MgO^{2,5\%}$ $Al_2O_3^{0,6\%}$) was chosen, due to its transparency for the beam of thermal neutrons [21]. Inner dimensions of the vessel were H x L x W: 20 x 10 x 2 cm. A weight of Ga, placed in the vessel, was 1,4 kg (corresponds to 12 cm height of the melt). The vessel with Ga was preliminary pre-heated and held with the temperature of 40℃. The vessel was fixed on the cooling system, providing a temperature of 12℃ during the experiment to realize a temperature gradient for directional solidification of the metal (2,3 K/cm).

The formation of solidification front and its movement was observed for three different types of conditions: solidification of the melt in constant magnetic field (DC), a convection, forced by AMF and a melt flow, driven by AMF applied with pulses.

As in the prototype setup magnetic fields were applied with use of NdFeB permanent magnets, placed beside the vessel as it shown on the figure 2, b. The magnets sets had a cylindrical form (diameter / length – 60 / 90 mm) and consisted of two pairs of magnets of different polarization. An electric motor was used to bring the magnet sets to rotation. Movement of the magnets sets was providing an appearance of AMF and Lorentz force in the metal, what was bringing the melt to the motion. The alternating frequency of magnetic field was equalled to the frequency of changing of the magnets poles with the angular frequency of rotation f_{ac} = 23,3 Hz. The disposition of magnets made possible to realize the melt flow during solidification without placing anything except the metal on the way of the neutron beam. Distance between the side of the magnet and the melt was 7 mm.

An application of PMF was realized by switching the electric motor on and off with predetermined frequency $f_p = 1 / (T_{act} + T_0)$, where T_{act} – is the period of stirring and T_0 – is the period, when the magnets do not rotate. Influence of a PMF with four values of f_p = 0,025; 0,05; 0,1 and 0,25 was analysed. For all values of f_p the duration of both periods was equal $T_{act} = T_0$. A development of a form of solidification fronts through the experiment time and a density of the grown ingots was compared for all cases.

The figure 2, b demonstrates the stream-lines of the melt flow, formed under the influence of the magnets rotation. Such a melt flow is a reproduction of the liquid metal circulation forced by application of a travelling magnetic field in an ingot-forming equipment [22].

Material's solid and liquid phases having a difference in a density and, consequently, difference in ability for neutrons attenuation, provided the neutron shadow images. Figure 3, c demonstrates the initial raw image, where with darker grey the melt and with lighter colour – the solid part of the metal can be recognized. As well partially can be seen the magnets, which are demonstrating a high contrast due to the high neutrons attenuation.

IOP Conf. Series: Materials Science and Engineering **228** (2017) 012026 doi:10.1088/1757-899X/228/1/012026

5. Experimental results and discussions

5.1. Bubbly flow in the melt

The neutron beam carries information about the material inside of the vessel. Since the attenuation coefficient of Argon is less than the attenuation coefficient of Gallium, the beam is more intense in areas of the gas bubbles. The pictures with the information are available as raw images. For post-processing the free software ImageJ has been used. Standard filters for image processing are available as part of the library of the software. The part of the picture containing the melt and the gas bubbles has been cropped. In order to highlight the gas bubbles a bandpass filter has been applied, following by an edge detection algorithm. The edge detection has been applied two times to highlight the shape of the bubbles. This procedure has been done for the whole video sequence. Thus, the development of bubbles ascending from capillary tubes, as well as their velocity can be detected (figure 3, a). Figure 3 a) gives a scale of the bubble size according to the vessel, which is 140 mm in width. Bubbles shown in the picture have in this case a size of around 6 mm.

Figure 4 shows the bubble rise velocity in dependency of Ar volume flow rate. The capillary tube has an outer diameter of 1,5 mm and an inner diameter of 0,1 mm. Velocities are calculated from bubble formation at the capillary tube to frame of the picture. In order to follow the trajectory, the middle point of each bubble in the image has been calculated and followed from previous to next picture linearly. Velocities presented are the average value of a bubble ascending from bottom to top. In the experiments bubbles ascended in a bubble chain of 3 – 5 for gas volume flow rates below 100 sccm. At flow rates of above 100 sccm continuous chains started to appear. All in all, bubbles ascend in a zig-zag movement. The CCD camera used for the experiments was used with the following acquisition parameters: frame rate of 35.7 fps, an exposure time of 0.028 s and a pixel binning of 4x4. Since bubble formation is a very sensitive process depending on gas flow, pressure from liquid metal on the inlet, which is related to flow situation in the vessel and depends on the previous bubble flow, angle of capillary tube etc. velocity of ten bubble formation has been evaluated and averaged. In figure 4, the velocity including error bars, which represents the maximum and minimum bubble velocity of each flow rate are shown. Although, the bubble velocities are highly fluctuating, at the average of appr. 23%, a tendency of increasing velocity by increasing volume flow can be detected.

The bubble size formation for increasing volume flow rate is likewise highly fluctuating. The detected tendency by increasing volume flow rate is a decrease in bubble diameter. A smaller bubble size is proportional to reduced Archimedes force, which is the driving force for the velocity. Thus, with a smaller bubble size, the tendency of bubble velocity should also be decreasing. Bubble velocity is also influenced by previous bubbles. The conditions in the wake of a bubble and the flow situation in the vessel may lead also to a faster bubble rise and an increase in bubble velocity. Applying induced forces by rotation of permanent magnets, shows a stabilization of rising velocity independent of bubble size distribution at appr. 20,5 cm/s (figures 6 and 7). The bubble formation is still highly oscillating, although the bubble size distribution compared to bubble formation at the same flow rate of 100 sccm (Standard Cubic Centimeters per Minute) without applied magnetic field appears in average much smaller. In this case the liquid metal flow is directed to the inlet and changes situation for bubble formation. In the post-processing of the pictures it can be seen that the shape of bubbles is squeezed and stretched at high liquid metal velocities. Nevertheless, a complete breakup of bubbles at the beginning could not be detected until the maximum rotation frequency was reached. On the one hand, the bubble size seems to be slightly reduced by small breakups, which are almost not visible with the resolution in neutron radiography for Ar and Ga. On the other hand, the different conditions of pressure and velocity at the orifice can cause a more compact formation of bubbles which are evaluated as smaller bubbles.

IOP Conf. Series: Materials Science and Engineering **228** (2017) 012026 doi:10.1088/1757-899X/228/1/012026

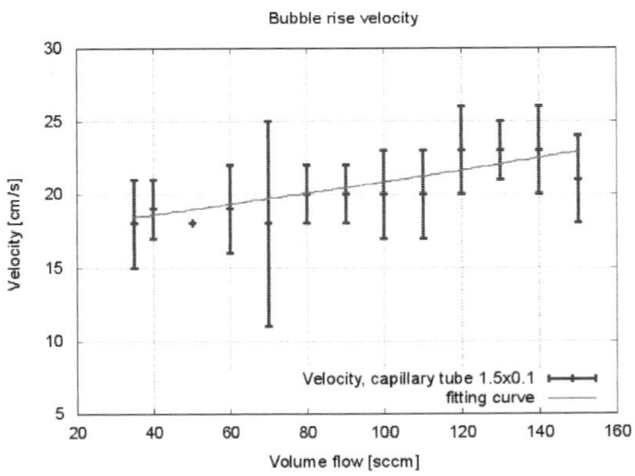

Figure 4. Bubble rise velocity in dependence of volume flow rate. Inlet diameter 0.1 mm. Averaged velocity at each volume flow of ten bubbles.

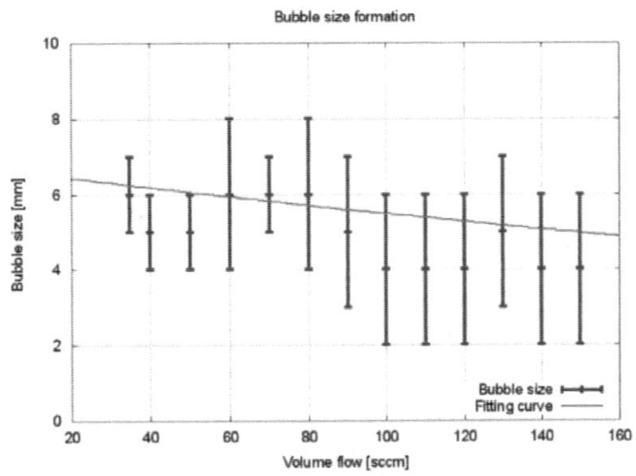

Figure 5. Bubble size in dependence of volume flow rate. Inlet diameter 0.1 mm. Averaged size at each volume flow of ten bubbles.

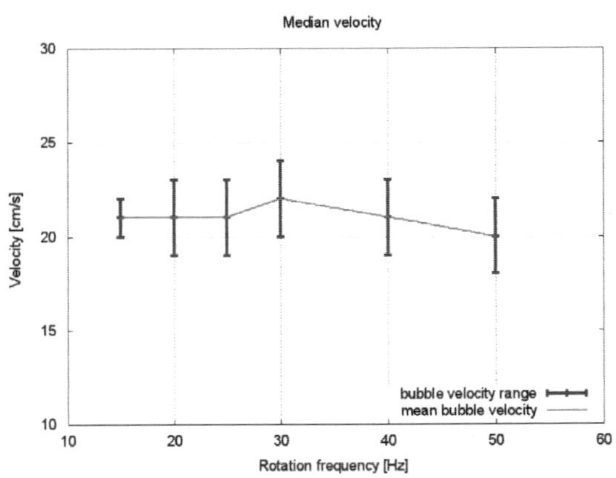

Figure 6. Bubble rise velocity in dependency of rotation frequency of permanent magnets. Inlet diameter 0.1 mm. Flow rate 100 sccm. Averaged velocity at each volume flow of ten bubbles.

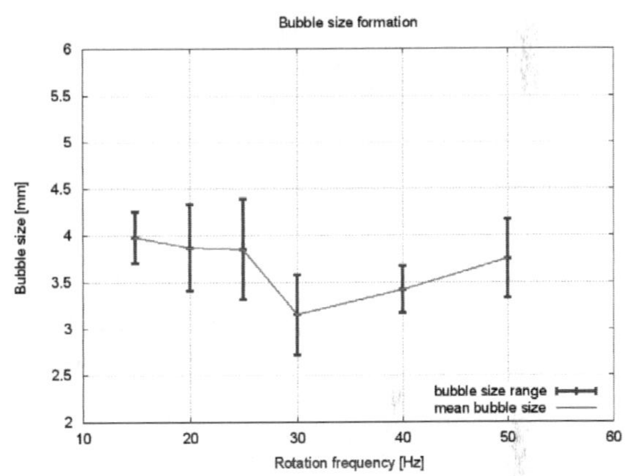

Figure 7. Bubble size formation in dependency of rotation frequency of permanent magnets with error bars representing maximum and minimum size at each flow rate.

5.2. Gallium solidification

After the image post-processing of the raw images the pictures of the ingot growth formation at different time-steps as well as the videos of the crystallization front movement were obtained. Figures 8 shows the evolution of the Ga solidification in constant field and convection, forced by AMF and PMF in time at 20th, 60th and 140th minutes of the process. Here the red colour (darker grey for black/white figure) and its shades correspond to the melt with higher temperature, while the blue colour (light grey) demonstrates the cooled solid phase. According to the equation (1) the colour saturation of different parts of solidified sample corresponds to the different dense of the material.

As it can be seen from the figure 8 (1, a), in the case of constant field two crystallization centres can be recognized – approximately in the middle on the vessel's bottom and on the side of the tank. A development of the first centre was preconditioned by the cooling. Although, an appearance of the second solidification centre is not so clear and can be a consequence of impurity of the glass or a proximity to the melt's free surface and intensive heat transfer from it.

IOP Conf. Series: Materials Science and Engineering **228** (2017) 012026 doi:10.1088/1757-899X/228/1/012026

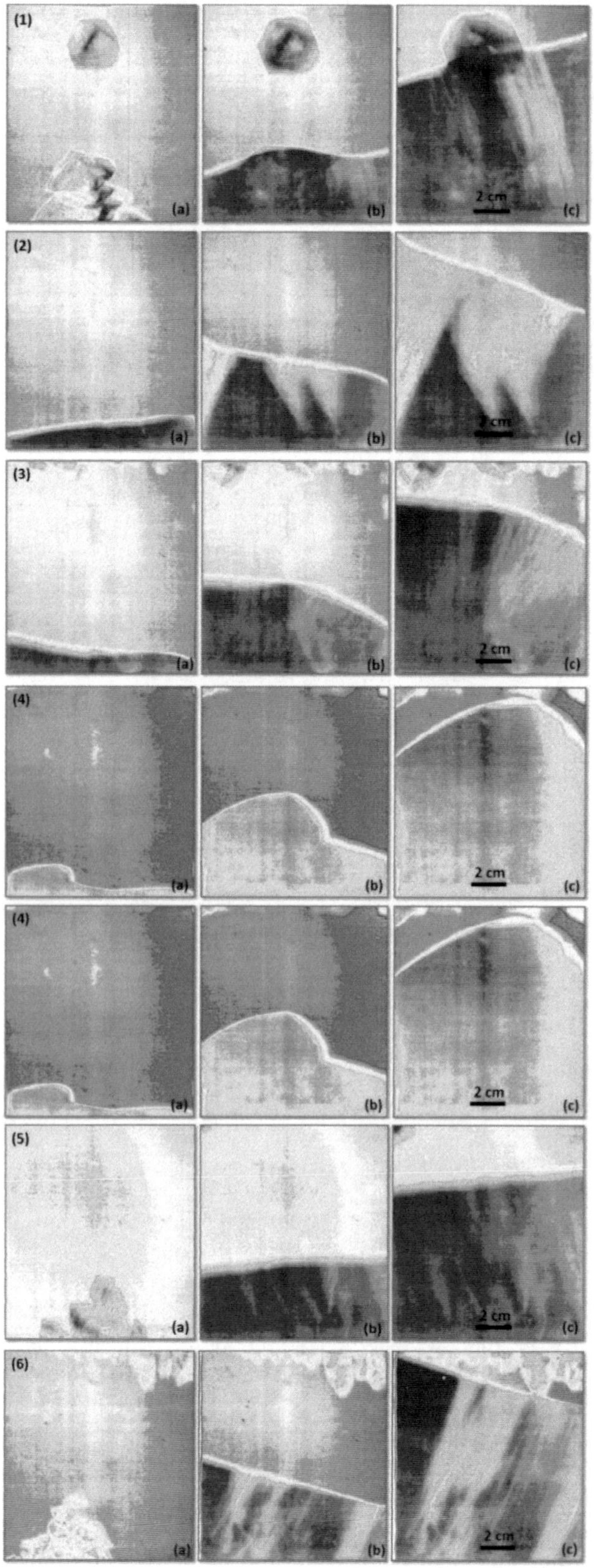

Figure 8. Gallium solidification in conditions of (1) buoyancy driven convection (no rotation of the magnets), (2) forced convection in AMF field (f_{ac} = 23,3 Hz) and in PMF: (3) f_p = 0,025 Hz; (4) f_p = 0,05 Hz; (5) f_p = 0,1 Hz; (6) f_p = 0,25 Hz: (a) 20th minute from the process begin; (b) 60th min; (c) 140th min

On the bottom crystal can be recognized the orthorhombic crystal structure, typical for the α-Gallium as well as several layers of the crystals and their direction of the growth. After 40 minutes from the start of the cooling process the solidification front develops in another way (figure 8 (1-b)). In fact, a growth of initial crystal on the bottom (at 20th min high is 4 cm) took less than 5 min, while for the formation of the main ingot with high 4,2 cm it took one hour. The reason for this phenomenon is an unstable nature of solidifying Ga and corresponding to it an overcooling of the melt.

Due to the intensive melt motion (up to 25 mm/sec, measured with ultrasound Doppler system) the formation of the initial crystal on the bottom for the case of forced convection was prevented. As it shown on the figure 8 (2-6, a-c) a flatter form of the solidification front has appeared at the beginning. For the cases of AMF and PMF with frequencies (figure 8 (2, 4, 6)) the tilt has developed. A reason for the formation of the tilt could be consequence of not homogeneous cooling or the nature of the solidification process, when the already formed solidification front catches the crystals from the flowing liquid. An evidence for the second statement to be true is the behaviour of the tilt growth in the case of mixed melt: regardless to the absence of the influence of the field from the left magnet (see figure 8 (2, 4, 6 - c)) the tilt did not become flat. The flow, produced by the field from the right magnet washes new crystals to the solid front, which grows keeping the tilt angle and the tilt, which once appeared develops with time with increasing of its angle. Triangle forms on the figure 8 (2, a-c) could be the wake from the bulky orthorhombic crystal. Although some of the cases demonstrate denser and more homogeneous structure of the ingot (figure 8 (3-5)), they cannot be compared, due to the individual calibration of the picture contrast for each sample.

As it can be seen for all ingots, formed under the influence of PMF (figure 8 (3-6)) the solidification was starting from the free surface as well as from the cooled bottom. There is a certain reason for such a behaviour: due to the density difference (for solid Ga is lower as for liquid metal) the small crystals, solidified during the stirring were floating to the free surface during the pause between pulses of forcing. The flattest form of a solidification front, corresponding to the most stable process, was obtained for the case of f_p = 0,1 Hz (figure 8 (5)). Such a frequency seems to be optimal for PMF application, becoming a compromise between ST and permanently applied AMF.

6. Conclusions

Carried out results show, that neutron radiography can be used to investigate bubble dynamics in liquid metal. Bubble formation, as well as bubble velocity development is highly fluctuating process. For detection of terminal velocity for bubble columns with a height of appr. 1 m, the size of imaging frame is too small. Thus, for reliable data for bubble column reactors a realistic column with a depth of maximum 25 mm has to be manufactured. Together with a sensor at the inlet, which detects bubble formation the camera can be adjusted to the surface and terminal velocity can be calculated more reliable. In order to study bubble formation and velocity distribution of the liquid melt, particle tracers and PIV can be utilized. Regarding separation of bubbles, more powerful magnets and adjusted gears for higher rotation speed will lead to higher induced shear stresses and mixing of bubbles.

Thermal neutron radiography is suited for visualization of solidification process in different conditions. Nonetheless the range of materials, which can be employed to the experiment is limited, this technique provides essential information about the complex process of formation of solidifying material. Such low-temperature metals as Gallium, Caesium, and Tin provide high-contrast neutron images. At the same time melts of these metals could be stirred by application of EM field. Information about the mechanisms of interaction between the forced melt flows and a solidification front and the form of solid/liquid interface, demonstrating thermal conditions, can be obtained by means of neutron radiography. Another advantage of this investigation method in comparison to the analogies is possibility to work with relatively thick samples, so an influence of a three – dimensional turbulence phenomena could be analysed. The results of Ga solidification in conditions of buoyancy driven and forced by alternating magnetic field convection have shown that the process of solid front formation can be influenced by the intensive melt flow. More flat form and absence of the initial bulky crystals can be stated for the case of forced convection.

Acknowledgments
Authors wishing to acknowledge the Helmholtz Association in the framework of the Helmholtz-Alliance LIMTECH (Liquid Metal Technology) for financial support of the investigation and colleagues from the Paul Scherrer Institute for their help during the experiments.

References
[1] Tarapore E D and Evans J W 1977 *Met. Mater. Trans.* B **7B** 345–351
[2] Moore D J and Hunt J C R 1983 *Progress in Astronautics & Aeronautics* **84** 359–373
[3] Umbrashko A 2010 *Heat and Mass Transfer in Electromagnetically Driven Recirculated Turbulent Flows* PhD thesis (Physics) (Riga: Latvian University) p 108
[4] Kirpo M 2008 *Modeling of Turbulence Properties and Particle Transport in Recirculated Flows* PhD thesis (Physics) (Riga: Latvian University) p 188
[5] Koster J N 1997 *JOM* **49** 31–35
[6] Boden S, Eckert S, Willers B and Gerbeth G 2008 *Metall. Mater. Trans.* A **39** 613–62
[7] Shevchenko N, Boden S, Eckert S, Borin D, Heinze M and Odenbach S 2013 *Eur. Phys. J. Special Topics* **220** 63–77
[8] Jasti J K and Fogler H S 1992 *AIChE Lournal* **38** 481–488
[9] Kim B 2002 *Development of Macrosegregation During Solidification of Binary Metal Alloys* PhD thesis (Mech. Eng.) (USA: Pennsylvania State Univ.) p 255
[10] Wondrak T, Eckert S, Gerbeth G, Stefani F, Timmel K, Peyton AJ, Terzija N and Yin W 2014 *Steel Research Int.* **85** 1266–73
[11] Saito Y, Mishima K, Tibita Y, Suzuki T and Matsubayashi M 2004 *Appl Radiation and Isotopes* **61** 983–691
[12] Sarma M, Scepanskis M, Jakovics A, Thomsen K, Nikoluskins R, Vontobel P, Beinerts T, Bojarevics A and Platacis E 2015 *Physics Procedia* **69** 457–463
[13] Fehling T, Steinberg T and Baake E 2017 *Int. J. of Appl. Electromagnetics and Mechanics* **53** S1, S111–S120
[14] Drew D A 1983 *Ann. Review Fluid Mech.* **15** 261–291
[15] Yang Z, Seo P K and Kang C G 2005 *J. Mater. Sci. Technol* **21** 219–225
[16] Vives C 1986 *J. Cryst. Growth* **76** 170–184
[17] Buschow K H J 2001 Encyclopedia of Materials: Science and Technology (Amsterdam: Elsevier Science Ltd.) 4733–39
[18] Eckert S, Nikrityuk P A, Räbiger D, Eckert K and Gerbeth G 2007 *Metall. Mater. Trans.* B **38** 977–988
[19] Willers B, Eckert S, Nikrityk P A, Räbiger D, Dong J, Eckert K and Gerbeth G 2008 *Metall. Mater. Trans.* B **39B** 304–316
[20] Wang X, Fautrelle Y, Etay J and Moreau R 2009 *Metall. and Mater. Trans.* **40B** 82–90
[21] Sears V F 1992 *Neutron News* **3** 26–37
[22] Kolesnichenko A F, Podoltsev A D and Kucheryavaya I N 1994 *ISIJ Int* **34** 715–721
[23] Fehling T, Steinberg T and Baake E 2017 *Proceedings of XVIII International UIE Congress: Electrotechnologies for Material Processing* 416-421

Institute of Physics
Dirac House, Temple Back
Bristol BS1 6BE UK

ISSN: 1757-8981
ISBN 978-1-5108-4648-7

CURRAN ASSOCIATES INC.
proceedings.com

9th International Conference on Fibre Reinforced Concretes, Textile Reinforced Concretes and Ultra-High Performance Concretes (FIBRE CONCRETE 2017)

IOP Conference Series: Materials Science and Engineering
Volume 246

Prague, Czech Republic
13 - 16 September 2017

IOP
Institute of Physics